PALYNOLOGY
OF ARID LANDS

PALYNOLOGY OF ARID LANDS

AHARON HOROWITZ
Palynological Laboratory
Institute of Archaeology, Tel Aviv University
Ramat Aviv, Tel Aviv 69978, Israel

1992

ELSEVIER
AMSTERDAM • LONDON • NEW YORK • TOKYO

ELSEVIER SCIENCE PUBLISHERS B.V.
Sara Burgerhartstraat 25
P.O. Box 211, 1000 AE Amsterdam, The Netherlands

Library of Congress Cataloging-in-Publication Data

Horowitz, Aharon.
Palynology of arid lands / Aharon Horowitz.
p. cm.
Includes bibliographical references (p.) and indexes.
ISBN 0-444-88277-4
1. Palynology. 2. Arid regions plants--Pollen. 3. Palynology--Israel. I. Title.
QE993.H67 1992
561'.13'0913--dc20 92-8579
CIP

ISBN: 0 444 88277 4

This book is printed on acid-free paper.

Printed in The Netherlands

To Tzutz, who loves arid lands

Contents

Preface

Quaternary palynology had commenced and developed to a high degree of specialization by scientists living almost exclusively in countries enjoying a temperate climate, where vegetation is lush and natural forests cover the land. Pollen production is very high in these regions, rates of accumulation of continental deposits are quite low and most sediments contain high concentrations, in the order of thousands of palynomorphs per gram sediment. Most depositional environments have rather low pH due to flushing by abundant rains and decomposition of the rich flora, resulting in excellent preservation environment for palynomorphs and easy extraction for study. These virtues had considerably enhanced development of palynology in Europe and North America, resulting in a great wealth of studies.

Later on, palynologists have moved to study tropical regions, for essentially similar reasons: rich vegetation, high pollen production, acid environments of deposition and rather low rates of sediment accumulation. At the same time, arid lands had to a high degree been avoided due to their sparse vegetation and poor pollen production, high rates of deposition in most continental basins, predominantly clastic sediments and abundance of carbonate and sulphate cements. All these factors made extraction very difficult and pollen spectra poor, both in number of grains and state of preservation. Another factor is the poverty in arboreal pollen in arid lands. This is in contratst to forested regions, where AP are the primary paleoenvironmental and stratigraphic markers, while the non-arboreal pollen are used for defining local conditions. Indeed, one of the basic rules a student of palynology is faced with is: "It doesn't matter how many non-arboreal pollen are in the samples, as long as you identify and count at least 150 arboreal ones". This rationale is quite understandable in forested regions, but cannot hold for arid lands, where trees are subordinate.

It is usually possible to identify arboreal pollen to the genus or species level, as opposed to non-arboreal, which can only be identified to family or subfamily level, rarely genus and hardly ever species. The state of preservation in lime rich sediments may hinder exact identifications, thus further limiting the significance of pollen spectra obtained from arid lands. For all these reasons and another important one, the much higher concentration of research institutions in temperate countries, palynology of arid lands remained far backward as compared with temperate and even tropical regions.

This situation is now changing, thanks to scientists who refuse to accept the natural limitations on arid lands' palynology, while realizing the great potential of these regions. Arid and semi-arid regions are more sensitive to environmental changes as compared with temperate or tropical, thus enabling finer reconstructions of global trends. The first steps were in extracting pollen by application of new methods to sediments usually discarded by the "classical" palynology, such as clastics, residual and other rocks. When those yielded sufficient pollen for statistically significant analysis, students were faced with conceptual problems concerned with different ways of interpreting pollen arriving by long distance transportation, pollen from multiple sources and occasionally different ages and so on. These methods, both technical and conceptual, are still being developed.

I hope that this book will bring forth the state of art in this slightly different field — palynology of arid lands — so as to enable students to become familiar with different techniques and concepts involved and to draw attention to the potential lying in palynological studies of regions hitherto only superficially touched upon.

The first part deals with natural agents relevant to palynology in arid lands, from pollen release by parent plants through transportation, deposition and preservation in various types of rocks; the second describes sampling strategies, various treatments for extracting microfossils from host rocks, identifying, counting and presenting the results; the third discusses concepts and ways for interpretations of results; the fourth comprises several case studies, to demonstrate prospects and possibilities.

Since arid lands palynology is not necessarily confined to "Quaternary Palynology" (I prefer the term "late Cenozoic Palynology", since in the last years many researchers are using parent plants names also for pollen recovered from Neogene rocks), a discussion of pre-late Cenozoic arid lands palynology is also included, with examples from the late Paleozoic and early Mesozoic deserts.

Arid lands palynology is extremely important for understanding the development of man and his cultures, since through most ages people had inhabited semi-arid and arid lands. A chapter is therefore devoted to palynological analyses of archeological sites, accompanied by a case study.

It is evident that, as for any rather new ramification in science, precautions are many while examples are quite few. It is only hoped that future research will change this situation and newly developed methods, accompanied by an extensive data bank, would overcome present difficulties and make for better and sounder foundations.

The book is intended for students and scientists who wish to broaden their horizons beyond "classical" palynology; geologists who want to understand the basis for arid lands palynostratigraphy, both "Quaternary" and "pre-Quaternary"; climatologists who need knowledge of paleoclimate evolution in these sensitive regions and the validity of its data base; biologists and geomorphologists, who are interested in evolution of paleoenvironments and landscapes of arid lands; botanists, who deal with evolution of plant communities in dry regions; and anthropologists, who need information concerning past habitats of man, his activities and the mutual influence and relationships with environments.

AHARON HOROWITZ

Acknowledgements

Some 30 years ago, when I arrived at the State University of Leiden, The Netherlands, to study palynology with Thomas van der Hammen, this branch of science was hardly practiced away from forested lands. It is however my teacher who had inspired me not to give up and, in much the same way as he was pioneering tropical palynology, encouraged me to attempt pollen analyses in arid lands. I am deeply and sincerely grateful for his help and encouragement.

Collection of published material was tedious, since many publications appeared in journals which are hard to get and many studies only occur as unpublished reports. The library at the Laboratoire de Palynologie, Université des Sciences et Techniques du Languedoc, Montpellier and the kind and efficient librarian, Claire Millerand, had been of great help in this respect, as were Jean-Pierre Suc and Niki Drivaliari, who made my stay at Montpellier pleasant. Thanks are also due to the librarian of the Geological Survey of Israel, Varda Arad.

Many colleagues have considerably contributed by sending copies of papers, photographs and exchanging information. It is impossible to name them all, but some who made special efforts are Gideon Almagor, Vaughn Bryant, Claude Caratini, Hector D'Antoni, Francine Darmon, Ann El-Moslimany, Pat Fall, Suzanne Fish, Eli Ganor, Henry Hooghiemstra, Vera Markgraf, Paul Martin, Mary O'Rourke, Jim Ritchie, Erhard Schulz, Louis Scott, Gurdip Singh, Amihai Sneh, Geof Spaulding and Eitan Tchernov, among others. Michael Dvorachek, of the Geological Survey of Israel, Jerusalem, had done the SEM microphotographs. All the line art was skillfully and devotedly done by Ora Paran, at the Drawing Department, Institute of Archaeology, Tel Aviv University. The Institute of Archaeology at Tel Aviv University, its Head, Jacque Yakar and the secretary Pnina Rodan had provided all possible assistance. Financial help was extended by the Faculty of Humanities, Tel Aviv University and the Dean, Moshe Gill.

I am indebted to Yoram Eshet, Geological Survey of Israel, for writing Chapter 11.3 and discussing problems of paleoaridity. Mina Weinstein-Evron of the Palynological Laboratory at Haifa University, a colleague and friend, has critically read the manuscript and extended considerable help with Chapter 10, based on her substantial knowledge in the archeopalynology of Israel. Apologies are extended to those who had been accidentally forgotten and to people who may have missed me some time during the last three years, when I was too busy with this book for prompt attention.

Thank you all.

A. HOROWITZ

Acknowledgements for reproductions of previously published material

Thanks are extended to the following publishers, editors, authors and colleagues who permitted reproduction of previously published figures and tables (numbers in parentheses are as appearing in the original publications):

Academic Press, Inc.:

Figs. 2.2.1 (6.2); 2.2.2 (6.2); 7.4.2 (6.11); 8.4.3 (6.27); 9.1.1 (1.1); 9.2.2, 9.2.4, 9.2.5 (4.3); 9.2.7, 9.2.8, 9.2.9 (4.5); Tables 2.2.1 (6.1); 2.2.5 (6.4); 2.2.8 (6.5) from: Horowitz, A., 1979, The Quaternary of Israel.

Acta Botanica Neerlandica:

Fig. 8.1.3 (5) from Dupont, L. M., Hooghiemstra, H., 1989. The Saharan–Sahelian boundary during the Brunhes chron. 38(4): 405–415.

A.A. Balkema, Publishers:

Figs. 8.1.4, 8.1.5 (p. 445); 8.1.6 (p. 446); 8.1.7 (p. 440); 8.1.8 (p. 441) from papers by Schulz and by Schulz and Whitney from: Matheis, G. and H. Schandelmeyer (Eds.), Current research in African earth sciences — extended abstracts of the 14th colloquium on African geology, Berlin, 18–22 August 1987. 1987. 504 pp., Dfl. 135,00. A.A. Balkema, P.O. Box 1675, 3000 BR Rotterdam, Netherlands.

Bulletin of the Geological Survey of Israel:

Fig. 11.3.4 (12) from Eshet, Y., 1990. Palynological aspects of the Permo-Triassic sequence in the subsurface of Israel. 81: 1–57.

CNRS, Service des Publications:

Fig. 8.2.2 (14) from: Caratini, C., Bellet, J., Tissot, C., 1976. Étude microscopique de la matière organique: palynologie et palynofaciès. in: Arnould, M., Pelet, R. (Eds.), Géochimie organique des sédiments marins profonds, Orgon III, Mauritanie–Sénégal–Iles du Cap-Verts, 215–264.

Figs. 8.2.3 (16), 8.4.1 (15) from: Caratini, C., Bellet, J., Tissot, C., 1978. Étude microscopique de la matière organique: palynologie et palynofaciès. in: Arnould, M., Pelet, R. (Eds.), Géochimie organique des sédiments marins profonds, Orgon IV, Golfe d'Aden, Mer d'Oman, 265–307.

Elsevier Science Publishers BV:

Table 7.3.1 (I) from: El-Moslimany, A.P., 1990. Ecological significance of common non-arboreal pollen: examples from dry lands of the Middle East. Rev. Palaeobotan. Palynol., 64: 343–350.

Fig. 8.1.1 (3) from: Spaulding, W.G., 1990. Comparison of pollen and macrofossil based reconstructions of late Quaternary vegetation in western North America. Rev. Palaeobotan. Palynol., 64: 359–366.

Fig. 8.1.2 (9) from: Hooghiemstra, H., Agwu, C.O.C., 1988. Changes in the vegetation and trade winds in equatorial northwest Africa 140,000–70,000 YR B.P. as deduced from two marine pollen records. Palaeogeogr., Palaeoclimatol., Palaeoecol., 66: 173–213.

Fig. 8.2.1 (4); Table 8.2.1 (I) from: Scott, L., 1989. Climatic conditions in southern Africa since the last glacial maximum, inferred from pollen analysis. Palaeogeogr., Palaeoclimatol., Palaeoecol., 70: 345–353.

Fig. 8.3.1 (13) from: Hooghiemstra, H., 1988. Changes of major wind belts and vegetation zones in northwest Africa 20,000–5,000 YR B.P., as deduced from a marine pollen record near Cap Blanc. Rev. Palaeobotan. Palynol., 55: 101–140.

Fig. 8.4.5 (3) from: Vincens, A. 1986. Diagramme pollinique d'un sondage Pléistocène supérieur–Holocène du Lac Bogoria (Kenya). Rev. Palaeobotan. Palynol., 47: 169–192.

Fig. 8.5.2 (9) from: Horowitz, A., 1989. Palynological evidence for the Quaternary rates of accumulation along the Dead Sea Rift, and structural implications. Tectonophysics, 164: 63–71.

Fig. 9.3.1 (2) from: Fuji, N., Horowitz, A., 1989. Brunhes Epoch paleoclimates of Japan and Israel. Palaeogeogr., Palaeoclimatol., Palaeoecol., 72: 79–88.

Fig. 9.3.2 (2); tables 9.2.1; 9.3.1 (I; II) from: Horowitz, A., 1989. Continuous pollen diagrams for the last 3.5 M.Y. from Israel: vegetation, climate and correlation with the oxygen isotope record. Palaeogeogr., Palaeoclimatol., Palaeoecol., 72: 63–78.

Fig. 11.2.8 (I) from: Doubinger, J., Lòpez-Gòmez, J., Arche, A., 1990. Pollen and spores from the Permian and Triassic sediments of the southeastern Iberian ranges, Cueva de Hierro (Cuenca) to Chelva-Manzanera (Valencia–Teruel) region, Spain. Rev. Palaeobotan. Palynol., 66(1/2): 25–45.

Fig. 11.2.9 (7) from: Ouyang, S., Utting, J., 1990. Palynology of upper Permian and lower Triassic rocks, Meishan, Changxing County, Zhejiang Province, China. Rev. Palaeobotan. Palynol., 66(1/2): 65–103.

Geologische Rundschau:

Fig. 11.3.1 (3) from: Visscher, H., Van Der Zwan, C.J., 1981. Palynology of the circum-Mediterranean Triassic: phytogeographical and palaeoclimatological implications. 70(2): 625–634.

Israel Journal of Earth Sciences:

Figs. 9.4.2 (10); 9.4.3, 9.4.4, 9.4.5 (9) from: Horowitz, A., 1987. Subsurface palynostratigraphy and paleoclimates of the Quaternary Jordan Rift Valley fill, Israel. 36: 31–44.

Israel National Oil Company:

Figs. 8.4.4; 8.5.7 from an unpublished report by Horowitz, A. and Horowitz, M., 1990.

John Wiley & Sons Ltd., and Dr. John Imbrie:

Fig. 11.2.1 (data, Fig. 27) from: Walton, W.R., 1964. Recent foraminiferal ecology and paleoecology. In: Imbrie, J., Newell, N. (Eds.), Approaches to paleoecology, 151–237. John Wiley and Sons, New York.

Paléobiologie Continentale:

Fig. 9.4.1 (8) from: Horowitz, A., 1990. Neogene palynostratigraphy, paleoflora and paleoclimatic reconstruction in Israel. XVII: 245–257.

Palynodata:

Fig. 11.1.1 (8) from: Kremp, G.O.W., 1977. The positions and climatic changes of Pangaea and five southeast Asian plates during Permian and Triassic times. Paleo Data Banks, 7: 1–21.

Pollen et Spores:

Fig. 2.2.3 (8) from: Horowitz, A., 1969. Recent pollen sedimentation in Lake Kinneret, Israel. XI(2): 353–384.

Table 2.2.7 (I) from: Maley, J., 1972. La sédimentation pollinique actuelle dans la zone du Lac Tchad (Afrique Central). XIV(3): 263–307.

Table 2.2.6 (I); Figs. 2.2.4 (1), 2.2.5 (3) from: Horowitz, A., Weinstein, M., Ganor, E., 1975. Palynological determination of dust storms provenances in Israel. XVII(2): 223–231.

Tables 2.2.2 (2), 2.2.3 (1) from: Weinstein, M., 1981. The influence of slope direction on the pollen spectra. XXIII(3–4): 381–387.

Table 8.2.3 (2) from: Horowitz, A., Gat, J.R., 1984. Floral and isotopic indications for possible summer rains in Israel during wetter climates. XXVI(1): 61–68.

Princeton University Press:

Fig. 11.2.3 (3) from: Stehli, F.G., 1968. A paleoclimatic test of the hypothesis of an axial dipolar magnetic field. in: Phinney, R.A. (Ed.), The history of the Earth's crust, 195–205.

Quaternary Research:

Fig. 8.2.4 (4) from: Swain, A.M., Kutzbach, J.E., Hastenrath, S., 1983. Estimates of Holocene precipitation for Rajasthan, India, based on pollen and lake level data. 19: 1–17.

Fig. 8.4.2 (3) from: Van Campo, E., 1986. Monsoon fluctuations in two 20,000 Yr. B.P. Oxygen isotope/pollen records off southwest India. 26: 376–388.

Yale University Press:

Figs. 11.2.2, 11.2.4 (13) from: Stehli, F.G., 1968. Taxonomic diversity gradients in pole location: the recent model. in: Drake, E.T. (Ed.), Evolution and environment, 163–227.

Chapter 1

Introduction

Dry lands, either hot or cold, occupy a considerable share of the land surface of the Earth — extremely-arid and arid some 20%, semi-arid another 15% (Fig. 1.1). Their boundaries have constantly been changing through the ages and are still changing today, sometimes at a frightening rate. Dry habitats had, at least for the last several million years, played an important role in the global environments.

The main problem involved with studying the history and development of past arid environments is the scarcity of faunal and floral remains in rocks deposited in dry habitats. The rarity of fossils is a combined result of the poverty in both plant and animal life and the nature of erosion and deposition. The poverty of life has several causes, besides the obvious environmental stresses such as lack of water, strong winds and extreme temperatures. The period needed for evolutionary adaptation of plants and animals to extreme environments is quite long and may not have been sufficient in regions such as the Old World latitudinal deserts, which underwent frequent cyclicity of arid and non-arid conditions. Indeed, the New World or southern African longitudinal deserts, where dry refuges have almost always been available for considerably longer periods, thus enabling development of specialized adaptations, are much richer in their number of both plant and animal taxa.

Erosion processes are harsh and destructive in arid lands, chiefly caused by strong insolation, winds and occasional floods, which result in a rather rapid disintegration and break up of most biological remains. The continental deposits, which are hardly ever a favorable environment for fossilization, even in the temperate regions, are much less so in arid lands, where medium and coarse grained clastics are the predominant sediments and undisturbed continuous sequences are very rare. As E. S. Deevey (in Martin, 1963, p. V) had put it: "if the layers of silt in arroyos were like layers of peat in bogs, the prehistory of Arizona would be less a thing of shreds and patches, fitted together neatly but not without voids".

The combination of these factors makes it quite understandable why, when trying to study an arid land, most pollen analyses concentrate on the available, rather than desirable sequences. Such are wetter phases in the history of deserts, manifested by occurrences of temporary lakes (Van Campo, M., 1964, 1975; Van Campo, M., et al., 1968; Alimen et al., 1968; Faure, 1969; Maley et al., 1970; Maley, 1973; Spaulding

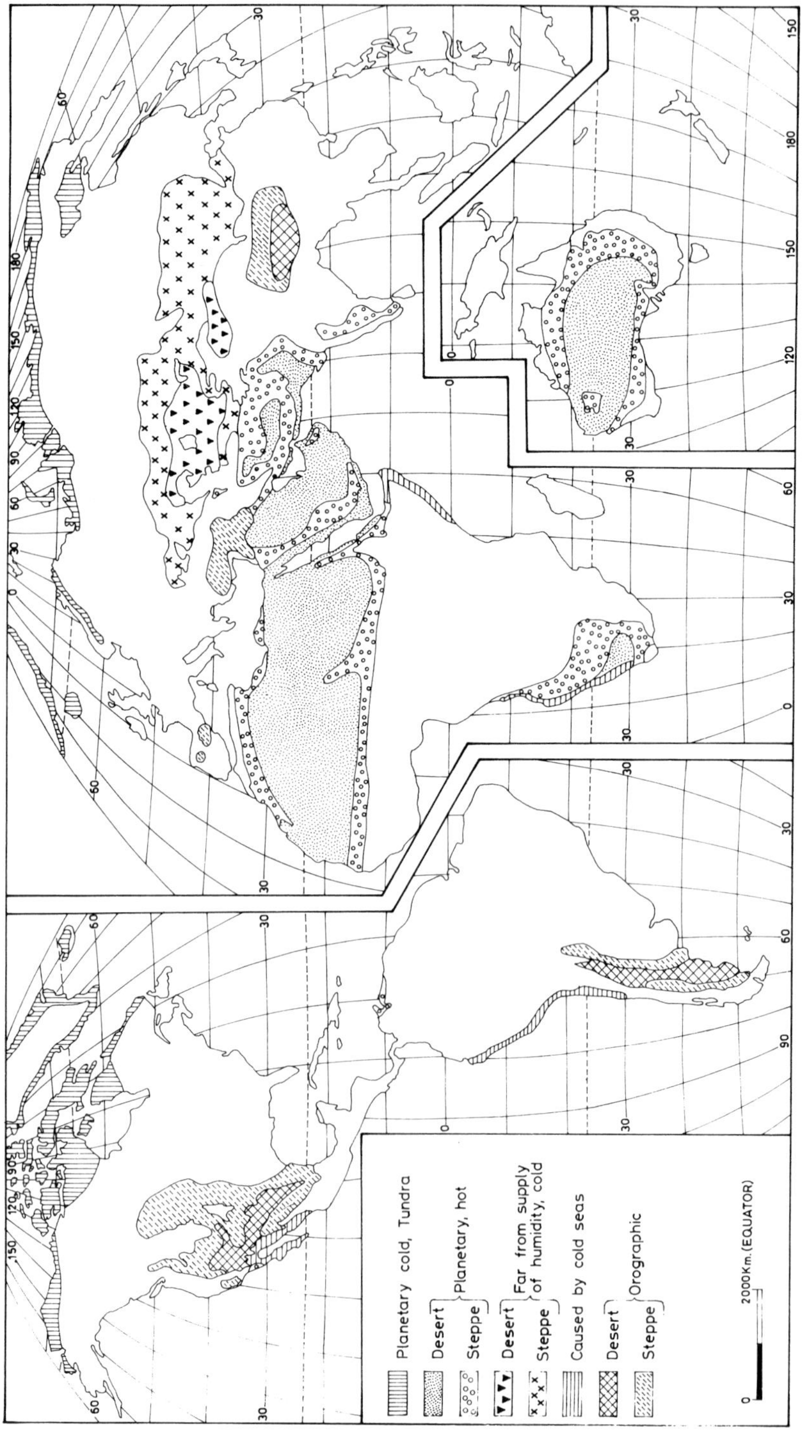

Fig. 1.1. Arid regions of the Earth.

et al., 1983; Van Zinderen Bakker, 1984; Schulz and Whitney, 1986, 1987; Schulz, 1987a); or by prehistoric and archeologic settlements in presently arid lands, which were usually only inhabited during humid climate phases (Mehringer, 1967; Bryant, 1975; Horowitz, 1976, 1977a; 1979; Bonnefille et al., 1987; and literally hundreds of references brought by Davis and Hall, 1988); or marine sediments adjacent to desert regions (Cross et al., 1966; Koreneva, 1971; Cheddadi, 1988; Hooghiemstra, 1988a). The following stage of palynological researches in arid lands, that takes place through the last years, centers on long continuous sequences, which could decipher the gradual developments in the deserts' history. Such are studies by Ritchie (1984), Horowitz (1987a), Lézine (1988, 1989) and others.

The importance of studying arid regions and especially their history and development, besides the pure scientific curiosity, mainly lies with the near future projects of land management and prevention of famine in extensive areas of the globe. Only by understanding the history of the desert, could we possibly predict such processes that may cause its expansion or contraction in the near future, which could directly affect the life of many people inhabiting these regions and their surrounding lands.

Climatology, by itself, can only be of limited help in studying the history and development of arid lands, chiefly because it is based on parameters which are not measurable for the past, such as amounts of rain, wind velocities, albedo, insolation, etc. The study of past climates has therefore to rely, at present, only on proxy-data recorded in geological sections. One of the best indications for past environments and thus paleoclimates, is the vegetation cover of a given region during a certain period and its development through time. To date, palynology seems to be the only practical way of studying continuous evolution of vegetation covers. This is because pollen grains, with their vast numbers and microscopic size, are incorporated in most sedimentary rocks, both in continental and marine environments.

Palynomorphs can be relatively easily extracted and studied in forested regions, due to their mostly good state of preservation and easy identification. The reconstruction of past plant communities from their pollen spectra is thus a common procedure in humid parts of the Earth, where pollen production is quite high by the extensive vegetation cover and rates of deposition are rather low, both making for a very high yield of pollen from small size samples.

This situation is essentially different in dry regions, where the vegetation cover is poor and rates of deposition are rather high, causing most samples macerated by the traditional methods to yield very little pollen, or sometimes none at all. To these, chiefly technical problems, some other difficulties are added in arid lands, mainly caused by the low masking factor of the local vegetation pollen. Such are the recognition and evaluation of processes as long distance transportation, redeposition and other natural agents which, for a rather long time held back palynological research in the drier parts of the Earth. This situation is manifested in many publications, such as by Spaulding (1977, p. 3) who maintains that:

"While palynology is the primary paleoecological tool in the temperate areas of the world, it lacks broad applicability in the (American) Southwest. In arid lands the drawbacks of palynology are the inability to identify most pollen grains to species level, combined with poor preservation and a general lack of suitable deposition basins".

Dodson (1983, p. 132) recognized the technical problems involved with pollen analysis in arid regions and stated that: "Drier environments, even though they may support intrinsically interesting and sensitive plant communities, are under-represented in the literature of pollen analysis. If collection and preparation techniques can be devised to make pollen analysis less difficult in these situations, then it is clear that significant advances can be made in the study of vegetation and environmental histories".

It is only during the last two or three decades that palynologists ventured into arid lands, to try and use pollen grains as evidence for environmental history. Pioneering studies, such as those by Martin (1963), Van Campo, M. (1964), Mehringer (1967) and others, had indicated beyond doubt that the possibilities are there, although some change in traditional attitudes is necessary for their exploitation. By developing new extraction methods and using much larger samples to start with, palynologists found out that pollen grains and spores, though tiny and made of seemingly perishable organic material, could withstand rather extreme transportation and deposition processes and be preserved within the continental rocks, from which they could be extracted and studied, bearing witness to the evolution of arid lands. Presently, palynologists are beginning to systematically study arid lands and the accumulating results are encouraging. One has only to leaf through the hundreds of references brought by Davis and Hall (1988), to realize the great extent of such studies. Symposia dealing with the topic of arid lands' palynology are becoming common in the last years, for instance Palynology et Climats (1980), Jacobs et al. (1985) and Singh (1988). Besides information obtained on the evolution of arid lands, the studies also reflect on global paleoclimates, since such fringe environments are very sensitive to even minor environmental changes, which are recorded in the flora and reflected by the ensuing pollen spectra in a distinctive way.

A fundamental distinction exists between tundras and other arid lands in terms of palynological research. Aridity involves, by its associated geological processes, many limitations placed for pollen analysis; while the tundras do not seem to bear most of these natural restrictions. The snow in winter and the low vegetation covering muddy surfaces in summer, help to protect against most of the extreme erosional and depositional processes. Thus, the typical sediments of tundras are peat and other low pH deposits, rates of deposition are relatively low and consequently pollen is abundant and mostly very well preserved. This book, therefore, will focus on arid regions, except for tundras that conform very well with methods applied by the "traditional" palynology.

1.1. Peculiarities of palynology in arid lands

Pre-Quaternary palynology became a powerful stratigraphic tool for almost the entire geologic time ever since plants, starting from the simplest unicellular forms, had appeared on Earth. Since palynology is not only restricted to pollen and spores of vascular plants, but also includes non-vascular, such as algae and bacteria, this information is presently dating far back, to about three billion years ago. Pre-Quaternary palynology had spread quite rapidly all over the globe, as an objective stratigraphic tool geared to both basic and applied research (*cf.* Tschudy and Scott, 1969; Traverse, 1988).

Concurrently, Quaternary palynology became a fundamental major contributor to our knowledge of environments during this period, which is limited to a very short span of the geologic history. Quaternary palynology, however, maintained its western European stigma for a long time and had therefore spread mainly in those regions enjoying a temperate climate, or occasionally the tropics, where plants, lakes and peat bogs are abundant ("The boggier parts of Earth", as Deevey had put it in Martin, 1963, p. V). Students from arid countries, who came to study the new discipline, were subject to the influence of the west European "cradle of palynology", an influence which took a long time to overcome, as it was a process which called for changes in both techniques and concepts.

Both European (*cf.* Faegri and Iversen, 1964; West, 1968) and North American (*cf.* Brown, 1960; Kummel and Raup, 1965) books dealing with palynological techniques, almost invariably suggest collecting samples for pollen analysis from lakes and peat bogs, discussing a variety of methods for doing so. Naturally, such water bodies are quite rare in the arid lands, which almost initially precludes these regions for possible palynological research by a potentially interested student. This limitation became so widely accepted that those rare palynological researches carried out in arid regions during the initial stages of study, were only on samples collected from the scarce water bodies, either extant, such as rift valley lakes in east Africa (Coetzee, 1967; Livingstone, 1975), or extinct, such as the fossil lakes of the Sahara (Beucher, 1975; Van Campo, M., 1975).

Instructions given in the above mentioned books and in other publications for collection of samples and extraction of pollen, are rarely applicable for sediments typical for arid lands. The sample size is only one example. Instructions usually begin by "take one gram of sediment", which almost invariably results, for samples collected in dry regions, in being barren under the microscope. The reasons are foreign to temperate regions, but quite common in the arid, namely the combination of poor pollen production due to the scarce, usually insect pollinated vegetation, accompanied by relatively high rates of accumulation of the sediments, due to strong erosional processes and predominantly clastic nature of the deposits.

As for extraction techniques, the problems initially resulted from the chemical treatments suggested for the samples, starting with dissolution of organic material and humic acids by using strong bases such as KOH. High organic content is not typical for sediments laid down under arid climates, thus maceration should initially involve dissolving mineral cements such as carbonates, gypsum and the like, which are soluble in acids, rather than in bases. This, again resulted in many samples being barren, while in fact, after a second run involving only acids, proved to contain sufficient pollen for analysis.

Another reason for discarding pollen spectra obtained from arid regions was the absolute number of pollen recovered from a sample, which was not thought sufficient for statistical treatment. For example, Van Campo, M. (1975), while discussing Quaternary samples from the Sahara, had omitted from the discussion all those which had yielded less than 200 pollen grains. This involves both technical and conceptual problems. The technical could relatively easily be solved, by macerating a larger sample. Indeed, Horowitz (1975), had obtained a decent spectrum of several thousand pollen grains, out of a breccia sample from Sterkfontein Cave in South Africa, only after treating some 2.5 kilograms of rock. This, of course, involved using different techniques and laboratory ware, which will be discussed in Chapter 5.

The conceptual problems concerned with a relatively low number of pollen grains per sample, are more complex only when larger samples cannot be obtained, such as when collecting from boreholes, where the amount of available material is limited. In such cases, for example the palynostratigraphy of a series of boreholes drilled in the Dead Sea Rift in Israel (Horowitz and Horowitz, 1985; Horowitz, 1987a; Levin and Horowitz, 1987), adjacent spectra, sometimes up to three or four, had to be combined to obtain a statistically significant spectrum. Evidently the stratigraphic resolution would be lower in that case, but still, the samples are of value and should, in our opinion, not be too hastily altogether discarded.

The problems of concentrating a large sample into a decent, statistically significant pollen spectrum should remain in the technical, rather than the conceptual part of the study. Unfortunately, it did remain quite frequently on the conceptual side and probably held back palynology in arid lands for a considerable time. Faegri and Iversen (1964), maintained that a pollen spectrum obtained by concentrating a large sediment sample could not be significant. To quote these authors (p. 73), “Such triumphs of preparation technique should be viewed with suspicion”. It should, though, be mentioned that the last edition of this book (Faegri et al., 1989) does not contain such constraints on the “poor” samples.

Another, a verily conceptual problem, in which “traditional” palynology differs from palynology of arid lands, is the use of arboreal pollen as the principal source of information. It is quite clear that in the temperate and tropical regions, where trees are the main constituent of the vegetation cover, their pollen is the best indicator for the environmental conditions. However, this is quite different in arid lands. Not only trees are quite scarce, but also many of those which do grow in deserts are insect pollinated, for understandable reasons, to ensure fertility.

Thus, the demand placed by Faegri and Iversen (1964, p. 124) "if ca. 150 arboreal pollen are counted in each spectrum, the diagram curves are comparatively safe", could hardly ever be fulfilled in arid environments. For this reason, palynologists working in arid lands have in the last years turned to non-arboreal pollen as the main source of environmental information (Vishnu-Mittre and Sharma, 1979; Brun, 1985; Davis et al., 1985a; Horowitz, 1987a; Cheddadi, 1988; El Moslimany, 1988, 1990; and others). This principle was recently accepted in a symposium dedicated to palynology of dry lands, convened by G. Singh, during the International Palynological Congress at Brisbane, Australia, 1988. It should however be noted that some trees had undergone physiological adaptations to various degrees of aridity, such as acacias, thus are quite common in many deserts. In these cases arboreal pollen originating in a local, arid environment can be found in the spectra.

It seems that one of the important characteristics of pollen spectra recovered from arid lands is their low masking factor. The poor vegetation and pollen production are prone to external interventions, much more than spectra obtained from richly vegetated regions and these "exotics" must be given thorough attention. Thus, the long distance transportation of pollen by winds, rivers and even sea currents, which only has a subordinate effect on pollen spectra from richly vegetated regions, is of prime importance in arid lands. A striking example, of the contribution of long distance transportation, could be seen in a study of recent sediments from the Bay of Elat (Aqaba), in southernmost Israel (Horowitz, 1966). The spectra obtained from sea bottom muds were quite rich in palynomorphs and several thousand were counted in each of the samples. This is not surprising by itself, because the rate of deposition of sediments in the deeper parts of the Bay is very low. However, the spectra contained very little pollen produced by the surrounding desert vegetation and were dominated by pollen and spores coming from two sources: the Lebanon and northern Israel on the one hand, some 400–500 kilometers to the north, represented by cedars, oaks and other typical Mediterranean elements; and on the other hand, spores and pollen derived from tropical ferns, palms and other plants, which had been carried by sea currents across some 25 to 30 degrees of latitude. These pollen and spores, which had arrived at the Bay of Elat from different parts of the world, across considerable distances, comprise up to 80% of the spectra and have no bearing whatsoever on the local environment, but rather on the activity of long distance transportation agents. Another example is of pollen, presently originating in the Sahara and Nile Delta, which are very common in silt arriving by dust storms to Israel, sometimes comprising up to 20–30% of the recent spectra in localities such as Jerusalem (Horowitz et al., 1975).

The exact location of the sampled site, especially in relation to wind and rain directions, may have a significant influence on the obtained pollen spectra. Again, this is much more important in arid regions, where rain facing slopes are covered by a much richer vegetation than those subject to a partial rain shadow. This was shown by Weinstein (1981), who had collected recent samples from north and south facing slopes of a wadi, located in the Carmel hills of Israel, a region enjoying a Mediterranean climate, where the rain and wind usually come from the northwest. The obtained pollen spectra considerably differed, across a distance of only some 400 meters. Those from the south facing slope were typical for a semi-arid environment, quite poor in arboreal pollen, while spectra from the north facing slope have been rich in arboreal pollen, typical for a wet Mediterranean domain. The differences between these pollen spectra are attributed to the considerably stronger insolation on the south facing slope together with the prevailing wind and rain direction, resulting in different vegetations on either of the slopes. Fossil spectra of these characters could easily be mistaken for representing interpluvial and pluvial climates, respectively.

These few examples only stress the importance, in arid regions even much more than in temperate or tropical, of studying pollen transportation and deposition processes, as well as local microecological variations, by evaluating recent spectra before any further step is taken into investigating the fossil ones. This, indeed, was the initial approach taken by many palynologists studying arid regions (Martin, 1963; Horowitz, 1969; Maley, 1972; Beucher, 1975; Van Campo, M., 1975; Weinstein, 1979; and many others).

The low masking factor of arid lands pollen can produce false results also due to local contributions, such as by vegetations intimately attached to springs, salt-pans or human settlements. Pollen recovered from travertines, in the extremely-arid region south of the Dead Sea in Israel (Weinstein-Evron, 1987a) contained many arboreal pollen, which seemingly indicated humid conditions. This, together with widely distributed travertines (Livnat and Kronfeld, 1985), led these authors to conclude that southern Israel had enjoyed a wet Mediterranean climate during the odd numbered oxygen isotope stages, with trees growing in the now desolate land. However, a closer study of nearby boreholes (Horowitz, 1987b) had shown that the time when the springs had been active, which is recorded by a considerable peak of Gramineae and Cyperaceae pollen in the drilled sequence, corresponds to a period of very low regional arboreal pollen values. It also corresponds to a period of accelerated subsidence of the rift valley, which was probably responsible for the appearance of fault scarp springs. These maintained local arboreal groups, whose pollen dominate the spectra recovered from the travertines, thus giving the false impression of a humid period.

Another difference between palynology in richly vegetated and dry lands lies with the degree of accuracy which could be presently assigned to the obtained results. There seem to be two main reasons for lower accuracy of conclusions drawn from pollen spectra in dry lands. The first involves the groups of pollen used for drawing conclusions, mostly of non-arboreal origin. The problem with these is, that in most cases identifications beyond the family level are very difficult or practically impossible, while in tree covered regions the arboreal genera and in many cases species are easy to identify. The other factor, which brings to the same drawback, is the state of preservation of palynomorphs which, in most cases is inferior in the predominantly coarse clastic and lime rich dry lands sediments. This is in contrast to the fine clastics and organic, generally low pH deposits in regions enjoying abundant rains and lush vegetation, which are much more favorable for preservation of the minute morphologic features of pollen, used to differentiate between species or closely related genera.

Extraction procedures applied to samples from arid regions necessarily contribute to difficulties in preserving the minute details of the sporomorphs, by the effects of mechanical disintegration and strong acids, which in most cases affect also parts of the exine. The acids applied to carbonate rich sediments cause effervescence and the gas bubbles help destroy the microfossils to a certain degree, especially those into which the cement had penetrated inside, during fossilization. This evidently increases the percentage of unidentified pollen in samples obtained from dry lands.

All these factors have a direct bearing on the significance of results, which has to be taken into consideration while attempting to draw conclusions. However, these do not seem good enough reasons to completely drop the subject. On the contrary, when more and more pollen sequences are collected from any dry region and the results start recurring, the ensuing conclusions concerning palynostratigraphy, paleoenvironments and paleoclimates become more and more sound. As an example, the Quaternary palynostratigraphy and paleoenvironments of the Dead Sea Rift, in Israel, had been quite safely established only after some twelve sequences yielded recurring results in the entire region (Horowitz, 1987a; Levin and Horowitz, 1987; and see also Chapter 9). Similar cases, of recurring results which led to a better established palynostratigraphy are also reported from other localities, such as southwestern United States (Martin, 1963) and southern Africa (Scott, 1989).

A somewhat different approach is thus needed for both techniques and concepts applied to palynological research in arid lands. Application of non "traditional" extraction techniques would most probably result in less samples being barren; while a different strategy in interpreting the significance of data, especially the non-arboreal components of the spectra and inclusions of allochthonous palynomorphs, could open new horizons in understanding paleoenvironments of arid lands and their development through time.

When more and more palynological data are collected from deserts of the world their history, development and controlling processes could be much better understood, to the benefit of all people concerned.

1.2. Arid lands — definitions and terminology

1.2.1. Definitions

The Oxford English Dictionary defines “arid” as “dry, without moisture, parched, withered: cause of the ground or climate. Hence, barren, bare”. The Webster Dictionary adds to it, “having insufficient rainfall to support agriculture without irrigation”. The Meteorological Glossary defines “arid” as a climate in which “the rainfall is insufficient to support vegetation”. It seems, however, quite difficult to exactly define an arid zone and even more so when it comes to its subdivisions into extremely- or semi-arid. Several different definitions for aridity are currently in use, mainly depending on profession, such as those favored by meteorologists or climatologists, by ecologists, botanists or zoologists, by farmers or economists and even by various politicians, each for his needs. Such necessities and definitions may change in time or following technological, genetic or physiological developments and are occasionally a matter of personal preference, so that a certain vegetation association could be considered as representing a “desert” in one locality, while in another it would be referred to as “fertile country”. The wealth of definitions only testifies to the lack of agreement among various students and to variability of needs and approaches.

A. Climatic

Climatologists have suggested indices for aridity which, although mostly expressed in meteorological terms, such as precipitation and temperature, also take into account the ensuing vegetation cover. Köppen (1923; and other publications), in his pioneering work, had defined an index of aridity by applying relations between average annual temperatures and precipitation, thus formulating the limits between the semi-humid, semi-arid and arid regions. Köppen also made a distinction between regions with predominantly summer rains and those in which winter precipitation applies. Köppen’s indices are only empirical, primarily based on regional plant associations. A “sliding” index of aridity was subsequently proposed, in order to define aridity in certain months or even days of the year, or indices which relate the climatic conditions to capabilities of the region to maintain vegetation. Thornthwaite (1948) took into consideration the efficiency of precipitation, as compared with the evapotranspiration of the vegetation cover and arrived at an index of aridity, defined by relations of water deficiency, surplus and the actual water need of the vegetation. Meigs (1953) divided the arid regions into three categories: extremely-arid (EA), occupying some 4% of the land surface of the Earth; arid (A), which take approximately 15%; and semi-arid (S), sharing 14.6%. Meigs took into consideration the rainy season, either winter, summer, or all year round and the winter and summer average temperatures. He also drew global distribution maps of these regions, which became the standard used by UNESCO. Meigs’ terms are also used in this book.

The one thing which seems common to all arid lands is their relatively high atmospheric pressure, which is the primary cause for lack of precipitation. Exceptions are only some restricted areas in otherwise fertile regions, which do not bear plants due to local edaphic conditions, such as dune sands or saline playas, or due to very strong winds. Another characteristic which accompanies low amounts of precipitation is the limiting temperature. In the hot deserts the high temperatures, sometimes up to 45–50 centigrade, are not really the limiting factor, as can be seen in these places where proper irrigation turns a region fertile, or in oases of the hot deserts. In the cold deserts and semi-deserts the low temperature is the most significant limiting factor, since throughout a considerable part of the year, when temperatures are below zero, no vegetative processes are at all possible. In the cold deserts, the length of the relatively short growing season dictates the type of vegetation which could eventually grow there.

B. Agricultural

In terms of agriculture, the extremely-arid and arid regions cannot be cultivated without artificial watering, while the semi-arid can maintain certain crops. It seems, however, that the main problem which agriculture faces in arid regions is the instability of rain regime. While in some years rains do occur in quantities which enable agriculture, other years may be almost completely dry, with very little or no rain recorded at all.

C. Botanical

The botanical definitions of aridity employ ecological argumentations, such as those presented by Zohary (1959), who defines aridity by the ensuing vegetation. Arid regions, subdivided into extremely-arid, arid and semi-arid, are primarily characterized by scarcity in trees, except for very limited areas. Extremely-arid regions (Fig. 1.2.1) are defined by an almost complete absence of vegetation whatsoever, although some plants would locally grow in places enjoying enrichment in water, usually wadi beds; arid regions (Fig. 1.2.2) are those in which the area covered by typical desert vegetation is considerably smaller than the barren land and plants are chiefly restricted to the lower wadi beds, where they are maintained by secondary enrichment in water; the semi-arid regions (Fig. 1.2.3) are characterized by non-arboreal steppe vegetation, which extends beyond the limits of the wadi beds and covers, however sparsely, the entire area.

Zohary's definitions had been proposed for the Near East, in which aridity is a relatively new factor (see Chapter 9). In localities where dry conditions persisted for longer periods, plant adaptations must also be considered. An extremely-arid region can thus maintain well adapted plants even outside wadi beds (Fig. 1.2.4); an arid biotope can carry arboreal vegetation, however sparse, all over (Fig. 1.2.5); while semi-arid environments can support a rather dense bush, with an almost total coverage (Fig. 1.2.6).

Fig. 1.2.1. Extremely-arid environment, with hardly any vegetation. Turfan depression, Xinjang, China.

Fig. 1.2.2. Arid environment with limited vegetation cover, trees only growing in wadis which enjoy secondary enrichment in water. Northwestern Negev, Israel.

Fig. 1.2.3. Semi-arid environment, with an overall cover of herbaceous vegetation, some shrubs and rare trees. Southwestern Israel.

These definitions hold for perennial vegetations of the three types of arid regions, but an important addition occurs following rare rains, a phenomenon especially characteristic for extremely-arid and arid zones. Such rains result in a sudden coverage of the entire land in these environments, with annual plants typical for each region (Fig. 1.2.7). Most of these plants live for a very short period, usually no more than two or three months. Most of them are wind pollinated, such as Gramineae (Poaceae) or Cyperaceae, or considerable pollen producers like Compositae. Although short living, these plants leave a marked impression on the pollen spectra by the great number of grains they produce and by the fact that they have no competitors which could mask their shares.

Such seasonal flourishing of annual vegetation is much more restricted in semi-arid regions, where climatic conditions are of a more stable nature and enable development of perennial plants associations which cover most of the area throughout the year, leaving no room for episodic, short living vegetation uprisings.

D. Palynological

When applying palynology for interpreting past floras, environments or climates, there is hardly any relevance for climatic or meteorological definitions in the way

(a)

(b)

Fig. 1.2.4. Extremely-arid environments, with adapted non-arboreal plants growing outside the wadis. a. *Welwitchia mirabilis*, Namib Desert. b. *Calligonum* sp. on dunes, southern Negev, Israel.

Fig. 1.2.5. Arid environment, with adapted *Acacia* and *Moringa* trees growing outside wadis. Southwestern Sahara, Senegal.

used for the present day, since the data on which the latter are based cannot be directly measured for the past. The botanical determinations are therefore much more adequate for the past and evidently preferable as a basis for palynological definitions and interpretations of paleoenvironments. The definitions of extremely-arid, arid and semi-arid environments by their vegetations seem to be the only ones which could be used in conjunction with palynological research and are therefore retained and used throughout this book.

I. Recent and late Cenozoic. The extremely-arid, arid and semi-arid environments are characterized in various parts of the world by typical plant communities, many of which are respectively represented in the pollen spectra. The principal palynological characteristic of these environments is the prevalence of regional non-arboreal pollen, or, in places where adaptation is at a progressive stage, occurrences of characteristic arboreal pollen. Exceptions to this rule may occur by intervention of transporting agents, which could bring arboreal pollen from considerable distances (Horowitz, 1966), or occurrences of trees growing in water enriched localities which do not represent regional conditions. Another representation of AP occasionally occurs in a basin and range country, where trees can grow at

Fig. 1.2.6. Semi-arid environments, with adapted trees forming a thick bush covering the entire area. a. Etosha, northern Namibia. b. Central Senegal, Baobab "forest".

Fig. 1.2.7. Extremely-arid region, covered with Gramineae following a rare rain. Namib Desert.

wetter higher elevations, while lowlands are barren or dominated by non-arboreal vegetation, as is the case in Southwestern USA (Martin, 1963). These "exotic" AP can take the upper hand in spectra from arid regions, due to the low masking factor of pollen produced by the poor dry land vegetation.

Another point which should be stressed especially when dealing with palynological definitions based on non-arboreal pollen is the possible level of identification. Due to difficulties in exact determinations it is usual to refer to pollen of a certain family by the family name, although some members of this family can easily be generically defined, thus not included under the wider term. Such is the case with Compositae, which may or not be subdivided into echinate and fenestrate, but do not include genera such as *Artemisia*, *Centaurea* or others. The exact use of family names may change from one region to another and should be clearly stated.

It seems that for arid lands' palynology the definition should be: "A land in which arboreal pollen are subordinate or derived from well adapted taxa, as compared with the non-arboreal, or where a different environmental provenance for the AP could be proved". Arid regions are frequently dominated by pollen of certain typical families and genera, such as Compositae, Chenopodiaceae, *Artemisia* and *Ephedra*. These plants also occur in other, more humid biotopes, but their prevalence can be used to define palynologically a dry land. Many plants which underwent adaptation to arid conditions, such as Cactaceae or *Acacia*, which are peculiar only to arid environments and could thus have served as excellent indicators, can hardly be used for a palynological definition of dry lands since unfortunately their pollen are not too common in the deposits.

Pollen spectra can be readily used to discriminate between semi-arid regions, with their typical steppe or bush vegetation on the one hand and desert plant associations on the other, since different families and genera usually occupy the two environments. Both extremely-arid and arid regions are palynologically characterized by a very limited number of taxa, which could however be used for differentiation between the two environments. Rare rains are the sole supporters for such regional vegetations, but the rain quantities may differ from extremely-arid to arid environments in a given region, resulting in developments of different plant associations. Thus for example extremely-arid regions in northern Africa and the Middle East are typified by associations of Cyperaceae and *Calligonum* pollen in sandy and Gramineae in rocky substrates, while arid environments are mainly characterized by Compositae grains.

Chenopodiaceae pollen are frequently used to define dry environments, since they are usually common in deposits from such regions. However, it seems that many members of this family only grow in wadis, where secondary enrichment in water is available (Fig. 2.2.11), while the elevated areas, which indeed reflect the regional conditions, are occupied by other plants. It seems therefore preferable to use pollen produced by the regional vegetation as general environmental characterizers, while leaving out the Chenopodiaceae, whose significance as indicators for regional conditions is questionable. They can however be used for defining local hydrologic conditions in restricted environments such as wadis, playas and so on.

II. Exceptions. As all definitions go, there are always exceptions to the rules, which should be taken into consideration to avoid possible pitfalls. When vegetation and consequently palynology of arid lands is concerned, such exceptions are usually the occurrences of "non-typical" plant associations. Relics of former floras, which only make a small part of the vegetation, are not included within such exceptions. It is only when the majority of the flora displays a nature seemingly alien to a given environment, that consideration should be given and care taken.

A few examples are the prevalence of *Artemisia* (Fig. 1.2.8) in arid regions of Southwestern USA such as Death Valley, or *Ephedra* in the arid Tarim basin and Taklamakan Desert, China (Fig. 1.2.9), that, if not noticed beforehand, could be erroneously considered typical semi-arid steppes by the pollen spectra; local adaptations of typical hydrophil trees such as *Populus euphratica* and *Tamarix* to arid conditions, again in the Taklamakan Desert (Fig. 1.2.10); or saline playas occupied by locally adapted variety of *Phragmites* in the Qaidam depression, China

Fig. 1.2.8. *Artemisia* in an arid environment, Death Valley, southwestern USA.

Fig. 1.2.9. *Ephedra* in the Taklamakan Desert, Xinjang, China.

Fig. 1.2.10. Aridity adapted *Populus euphratica*, Taklamakan Desert, Xinjang, China.

Fig. 1.2.11. Salinity adapted *Phragmites* (Gramineae), Qaidam depression, Qinghai, China.

(Fig. 1.2.11), which would, if ignored, give the false palynological impression of freshwater vegetation.

III. Pre-late Cenozoic. Palynological definition of pre-late Cenozoic arid lands is somewhat more complicated, since most of the palynomorphs have no well established, direct connections with parent plants whose environmental demands are sufficiently well known. Attempts at using index species to define late Paleozoic–early Mesozoic aridity (Visscher and Van Der Zwan, 1981; Eshet, 1990) are discussed in Chapter 11. In other cases, either in addition to the use of index species, or when these are not clearly recognized, another approach may be attempted, based upon a known phenomenon of the present day, namely the limited number of species in ecologically stressed environments, such as arid and polar regions (Stehli, 1968a; Horowitz, 1972; Prance, 1982), in comparison with the considerably greater number of taxa in tropical regions.

A contemporary example for this situation is brought in Zohary (1959), who indicated that the number of plant species in the entire area of the Sahara Desert is no more than 800–900, as compared with Israel, in which, within a much smaller area, more than 2,000 species are known for the Mediterranean environment alone. A possible explanation may be, that when ecological and environmental conditions are becoming more and more harsh, common to both hot and cold deserts, certain plants or animals whose adaptation is faster occupy all possible ecologic niches. This precludes other species from living in those regions, which results in a low species diversity for stressed ecosystems. This diversity is only relative, thus spatial paleoenvironmental gradients of pre-late Cenozoic periods can only be drawn by comparing the contemporary number of species from one region to another. Similarly, temporal changes can be studied along a given sequence. The low diversity of species can, therefore, be a very good indicator to past extreme environments, even when ecology of the species themselves is poorly known and even if identifications are based solely on morphology.

In terms of palynology, therefore, a pre-late Cenozoic desert could be defined by either its limited number of species as compared with other regions at the same period, or by the widespread occurrences of known xerophytic index species. Evidently the two methods, when combined, would contribute for sounder conclusions.

1.2.2. Classification of arid regions

Arid regions can be classified either by vegetational characteristics or by their climatic causes, depending on necessity and applications. The floral classifications are discussed above, while in terms of their origin and causes arid regions are divided into three main categories (Fig. 1.1): planetary, regional and edaphic.

The planetary include both the hot, subtropical arid lands and the subarctic, cold deserts and steppes. The planetary deserts, both cold and hot, are climatically typified by deficiency in absorbed solar energy. While this seems clear for the subarctic deserts, which are chiefly formed by cold, dry air masses coming from the polar regions, it may sound at first quite strange for regions like the Sahara or the Kalahari, known for their high diurnal temperatures. However, these high temperatures are a combination of very strong solar radiation, mainly due to the lack of any cloud cover and the descending dry, hot air masses. The absorption of solar radiation by the hot deserts surfaces is very low, due to their very high albedo, which precludes a thorough heating of the soil and subsoil. Such heating is essential for warming the air above, thus obtaining the low barometric pressures necessary for bringing rain into any region. One only has to remember the freezing nights in the deserts and the great difference in temperature between day and night, sometimes amounting to 25–30 degrees centigrade, to realize that even the hot deserts are indeed cold.

The regional deserts are also typified by high barometric pressures, chiefly caused by deficiency in absorbed solar energy. The three principal types are:

(A) Regional deserts in which aridity is caused by a cold ocean, either due to sea currents coming from the poles, such as the Atacama Desert in South America, or the Namib Desert in southern Africa, or due to upwelling of cold bottom waters, which happens, for example, at the Horn of Africa. The cold seas prevent formation of low barometric pressure cells and also considerably hinder evaporation, resulting in aridity, or sometimes even extreme aridity in the neighboring land areas. The main source of water for the scarce plant and animal life in such deserts is the frequent fog, which rises from the ocean almost daily and condenses into dew.

(B) Orographic deserts, in which a mountain range stands between the source of warm, humid air and the region itself, which is typified by descending dry, hot air, such as the Judean Desert in Israel, the arid regions of Patagonia and a part of the American Southwest.

(C) Arid regions such as central Asia, which are too far away from any possible source of rains.

Local edaphic conditions, usually quite restricted in area, are the cause of a third group of arid regions. The two most common types are: sandy deserts, where local aridity is caused by fast infiltration of water in the loose sands, which can thus hardly maintain any vegetation; and saline soils where due to local, endoreic drainage systems, playas develop, mostly barren except for adapted halophil plants. Local strong winds can also cause aridity by the very high rates of evaporation, sometimes combined with salt spraying, as is the case in many beaches or small islands in the oceans.

Edaphic aridity had been divided by Zohary (1959) into several types, according to the substrate:

(a) steppes or deserts characterized by gray, mostly calcareous soils;

(b) gypsum deserts, typified by concentrations of gypsum and occurrences of crusts;

(c) hammadah deserts, of an almost continuous cover of gravel and stones;

(d) sandy deserts;

(e) loess steppes;

(f) playa deserts;

(g) rocky deserts.

Each of these ecosystems maintains its typical plant associations, which can sometimes, if the area is large enough, be discerned by its pollen spectra.

A unique class of aridity may result from human activities, which in most cases involve over-exploitation of the vegetal resources, by extended grazing or by collecting plants for fire. This does not enable appropriate renewal of the natural vegetation, thus the barren surface, which has a high albedo due to exposure of the soil, prevents heating of the overlying air and formation of low pressure cells, essential for bringing rains. In this way such human induced desertification processes tend to amplify themselves by the positive feedback mechanism and, once started are very hard to control. Such processes are known for the historic periods in many parts of the Earth, where the initial climate was marginal and bear great significance for human habitation in boundary environments, such as the desert fringe. Human induced desertification is quite common in the present day (Fig. 1.2.12).

1.3. Concepts and aims of this book

"Palynology of Arid Lands" has been written following the great interest and attention given in the last decades to arid lands, due to their considerable economic importance as the only possible soil reserves on Earth. This interest, naturally, also sent palynologists to the field, who refused to accept the traditional limitations and ventured at opening new horizons for this profession. The book is aimed at bringing together the techniques and concepts which are different to those on which "traditional" palynology is based, in the hope that palynology would turn to be an important source of information concerning the history and development of arid lands and their future potentials. It is hoped that the book will bring forth our present state of knowledge in this growingly important branch of palynology, so that it could be much more widely applied.

The book is divided into 11 chapters, each dealing with one of the aspects of arid lands palynology. On its face, this subdivision approximately conforms that used in any other palynology textbook. However, the treatment of each of these topics, concerning arid lands palynology, is somewhat, or in some cases essentially, different.

Fig. 1.2.12. Human induced desertification. Satellite image of the northwestern Negev, taken in 1970, showing the international border of Israel and Egypt. The border is actually seen on the image due to overgrazing by Bedouin herds in the Egyptian side. The rectangle is a military camp, into which the Bedouins are not allowed.

Chapter 1 deals with differences between "traditional" palynology and the methods applied in pollen analysis of arid lands; with definitions and classifications of the latter discipline, with the terminology, concepts and aims of this book.

Chapter 2, "Recent pollen spectra in arid lands", concerns pollen production and transportation, both processes quite different as compared with temperate or tropical regions. Pollen production is quite poor, while transportation by winds, rivers, sea currents, groundwater or animals can take place over considerable distances, much more than is usual for the richly vegetated regions. The reason is not so much the difference in transportation, but in the fact that in forested countries the autochthonous pollen masking effect is much stronger than in arid regions, where the vegetation is quite poor, mostly insect pollinated. Deposition, preservation and initial destruction of recent pollen, until they become an integral part of the rock are also somewhat different in arid lands and discussed in this chapter.

Chapter 3, "Sediment–pollen relations in arid lands", discusses processes of weathering, sediment transport and depositional environments typical for the dry regions of Earth. Polyphase inclusion of pollen in sedimentary rocks, which is much more pronounced in arid regions, caused by redeposition, rainwater infiltration, groundwater movements, springs, human and animal activity, all are quite meaningful in the resulting pollen spectra. The quantitative relations of pollen and sediments are also quite different in arid lands, as compared with temperate and tropical regions, mainly because rates of deposition of most sediments, in relation to pollen production, are much higher in dry regions. This chapter is somewhat extended, accompanied by many illustrations, for the benefit of palynologists who are not too familiar with geologic and geomorphic processes in arid regions.

Chapter 4 presents sampling procedures, which again differ considerably from those exercised in the forested regions, mainly due to scarcity of pollen as compared with the sediment volume. Thus, sampling techniques and sample size must be quite different. Oxidation plays an important role in these regions, as do also differential preservation and destruction of the pollen grains. All these factors led to necessity of different sampling strategies.

Chapter 5, extraction techniques, brings forth various ways of macerating samples collected from arid lands. Here the problems are twofold: the relatively greater amount of deposit which has to be macerated for obtaining a significant pollen spectrum; and the nature of sediments, which necessitates appropriate physical and chemical treatments. These two factors are further complicated by the poor state of preservation of pollen, especially in clastic and carbonate rocks so common in dry regions, which implies toward as little mechanical and chemical treatments as possible, to obtain identifiable microfossils.

Chapter 6 treats the identification of pollen grains collected from arid lands, their counting and presentation. The first part mainly touches upon three topics: the state of preservation and differential destruction of the pollen, which necessarily lower the level of possible systematic identification; the vegetation, which is mostly non-arboreal, its pollen frequently hard to identify to the genus, let alone species; and the problem of identifying polyphase processes, such as redeposition and long range transportation of the pollen grains. A discussion is also added concerning the oldest problem in palynology, how many grains should be counted to make the spectrum significant, a problem which becomes more acute with the poverty in pollen of so many samples obtained from arid lands.

Counting and presentation of the results may also be quite different for pollen spectra collected from arid lands. The main reason being that, mostly, one is dealing with pollen of non-arboreal origin. Naturally, there is no point in drawing an arboreal pollen diagram, as is usual in forested regions. Rather, one may want to group together pollen produced by plants growing in similar ecologic conditions, to obtain some idea of the environment and paleoenvironment. Thus, the importance of non-arboreal pollen and the grouping of components by environmental affinities, by depositional processes, by landscape or by polyphase processes are discussed.

Various ways of presenting the results are the manifestation of different ways of interpretation, discussed in Chapter 7. Many factors alien to forested regions have to be taken into consideration in arid lands. Examples for possibilities involved in reconstructing paleoenvironments and past geologic, climatic and other processes are brought in Chapter 8.

For demonstration, a case study which summarizes the late Cenozoic palynology in Israel, together with its implications to the development of arid environments in this part of the world, is presented in Chapter 9.

Chapter 10 is dedicated to pollen analysis of archeological and prehistoric human occupation sites in arid lands. Pollen spectra recovered from such sites are an important source of information for understanding settlement locations, ways of life, hunting, gathering, agriculture, domestication, etc. For demonstration, a case study of palynological applications in understanding human settlements pattern and subsistence in Israel is further detailed here.

But arid lands are not only a matter of the late Cenozoic and undoubtedly existed on Earth during former periods. The possibilities of discerning such an environment from the palynomorph spectra are discussed in Chapter 11, based on species diversity and dominance, as well as on index species concepts. Again, case studies from the late Paleozoic and early Mesozoic deserts are presented.

The information contained in this book is only a beginning, with the hope to show some of the possibilities and to whet the appetite for new adventures in palynology of arid lands. Both technical, such as collection and maceration of samples and conceptual aspects as presentation, numerical treatment and drawing conclusions based on pollen spectra obtained from arid lands, need to be further developed. The same can be said for applications of palynology in arid lands, which could and should be widened. It is hoped that this book will serve as a basis for such developments and will encourage students and scientists to study arid lands, in the hope that accumulating knowledge will enable better land development and food production in the near future.

It is not known presently whether we could stop aridity or desertification of wide areas, literally lethal processes which tend to spread in the last years. But by better understanding arid environments we could, hopefully, at least partially turn these regions back into fertile lands, which some of them had actually been in the not so remote past.

It is hoped that this book will make its humble contribution toward the better future management of arid regions, which could only rely on better knowledge of the past — a past which could be revealed by further development of palynological research in dry lands.

Chapter 2

Recent pollen spectra in arid lands

Recent pollen spectra in arid lands and consequently the fossil ones, considerably differ from those obtained from the forested regions. Besides the obvious difference, of subordinate arboreal pollen, which results from the composition of the arid land flora, several other aspects have to be taken into consideration before drawing any conclusions. Two main factors make the pollen spectra obtained from arid regions different. The first is the overall relatively limited numbers of pollen grains produced by the local vegetation, which results in a very low masking factor for the obtained pollen spectra. This is of hardly any importance where only the local pollen are incorporated in the sediments, which in dry lands is a rather rare case. In most instances pollen are brought to the site by various transporting agents, depending on the sampled location and the geographical conditions.

Thus in studying pollen spectra in arid lands, consideration must be given to the ways by which pollen arrive at a certain location. Moreover, a distinction should be made between autochthonous pollen, which indicate the conditions around the site; and transported grains, that rather represent far away vegetations, together with the energy and possibilities of transporting agents, superimposed on the local environment characteristics. In many cases, when the local vegetation produces very little pollen and the transporting agents are potent enough to supply relatively large amounts of grains, the autochthonous pollen could easily be outnumbered or even almost totally masked by exotic ones.

Continental sites within an arid region are prone to pollen brought from great distances by winds, which could be quite strong in the desert. More than that, there is hardly, in many cases, any barrier on their way. Rivers and sea currents can carry pollen from their source areas, which are not necessarily dry and could frequently be located quite far away, to offshore and delta sediments. These pollen may occasionally almost completely mask the local, dry land vegetation.

Even groundwater may occasionally travel considerable distances under the desert, especially where the substrate comprises loose sands or gravel, sediments very common in these regions. The groundwater may thus appear as springs in localities far away from the intake area of the water. This phenomenon may at times be even more complicated, where desert springs are fed from a fossil aquifer, a situation common in the Sahara and southern Israel. These aquifers had been filled up with water during a former, more humid period. Such fossil water may transport and deposit in a site, pollen which have nothing to do with the present day environments. Groundwaters are also known to dissolve formerly deposited evaporites such as rocksalt, thus pollen trapped in these rocks are carried by the water and may be incorporated in younger sediments.

Pollen are also transported vertically along sections, by insects and other burrowing animals, as well as along roots and fissures, which may obscure the obtained results. Excrements and pellets of birds of prey, which may fly rather long distances, between the time they feed on rodents whose food contains abundant pollen grains, are also suspected to contaminate sediments by exotic pollen.

Being so common in arid lands, these factors almost always contribute to the pollen deposition at a certain site and may occasionally blur the local environment representation.

Pollen preservation in arid environments is subject to processes alien to the humid ones, due to the different types of sediments, being usually clastics and the nature of the cement, which is frequently rich in carbonates and sulphates. But even before incorporation within the sediments, some of the pollen cannot withstand destruction agents, both chemical and mechanical, which are much harsher in the arid environments. This results in a pronounced difference between airborne pollen spectra, obtained from pollen traps and those observed from the surface sediments.

These arguments serve to stress that, whenever a palynological study of an arid region is planned, it is essential to sample the recent surface spectra and understand their provenances and what they actually represent. For this purpose a study of the airborne pollen is of lesser importance, because these spectra have not yet been fully affected by processes causing differential destruction of the pollen grains, which may occur while they are being deposited and incorporated in their host rocks.

2.1. Pollen production

In terms of pollen production in extremely-arid and arid lands, initial distinction should be made between two groups of plants (Horowitz, 1989a). The first comprises those which are indeed adapted to aridity, such as cacti and acacias, that had developed adaptations to help pollination. The second consists of plants growing where water is available in dry regions, but are not typical of these, like reed and poplars; occurrences of many plants of this group depend on continuous import of seeds from less arid areas, where their survival is constantly ensured; wind pollinated plants of this group rely on their concentration around water bodies, where their close proximity makes wind pollination effective.

The situation is different in semi-arid regions, which maintain steppe vegetation or bush covering most of the area, where wind pollination is sufficient to ensure fertility, thus no special adaptations are required.

The greater majority of aridity adapted plants are insect (and other animals, especially birds) pollinated. The reason is the rarity of plants in dry environments, which would make wind pollination very risky. Insect pollinated plants growing in arid lands have either to produce greater numbers of pollen per plant than others, living under more favorable conditions, or develop special means for better attraction of pollinating agents, particularly insects and birds, in order to ensure fertility and survival.

Adaptations of insect pollinated plants for pollen production or better attraction of pollinating agents vary, but are fundamentally of several types: increased number of flowers per plant; increased number of anthers per flower; strong colors or odors; increased production of nectar; enlargement of the single flowers; timing of the flowering hours to the activity of the pollinating animals; and other, more specialized adaptations of certain plants, which are more of a curiosity.

Many of those plants which had adapted by increasing their pollen production became for palynology of arid lands as valuable as the wind pollinated plants of forested regions, since a great number of the pollen grains are liberated to the atmosphere after they had ripened, thus turning to be considerable contributors to the pollen rain and ensuing spectra.

A semi-arid region can readily be discerned from an arid by the main components of their pollen assemblages. The former is typified by prevalence of pollen produced by wind pollinated plants such as *Ephedra*, *Artemisia*, Gramineae and Chenopodiaceae, while the latter is characterized by higher percentages of pollen originating from drought adapted, primarily insect pollinated plants, of which the typical group for Israel and the Near East is Compositae (Horowitz, 1979; Weinstein-Evron, 1983; Horowitz and Horowitz, 1985; Horowitz, 1987a). Evidently, pollen of the above mentioned plant families do appear also in spectra obtained from forested areas in many parts of world; however, it is their prevalence, together with the low percentages of arboreal pollen, which characterizes the dry environments.

Adaptations by increasing the number of flowers per individual plant (Fig. 2.1.1) are very common in the dry lands. One has only to remember the beautiful sight of the desert, during the short blooming season, with millions of flowers covering every vegetated corner, to become easily convinced. Increased number of anthers by the individual flowers mostly come on account of the petals, which naturally makes some plant families more suitable for such adaptation. This can be seen with the Cactaceae and Compositae, for example, families whose flowers are typified by a considerable number of petals. Some of the petals may be retained for insects attraction, as is the case with Cactaceae (Fig. 2.1.2). The Compositae chose another way, enlarging and protruding their anthers on account of petals (Fig. 2.1.3), which made some of them practically wind pollinated. An extreme case of this type of adaptation can be seen in *Artemisia*. Some other plants had developed flower pods which grow high above the rest of the vegetation (Fig. 2.1.4), to better attract their

Fig. 2.1.1. Increased number of flowers in *Echinops spinosus*, Compositae, Negev Desert, Israel.

prospective pollinators, but in this way their pollen are easily discharged also to the wind.

Adaptations which are geared toward the struggle to better attract the attention of possible pollinators, are expressed by one or more of the following. Some plants had, instead of increasing the number of flowers, developed only a few, but much larger, like the irises (Fig. 2.1.5). Strong colors, especially the warm ones such as red, yellow or purple are typical for many desert flowers, (Fig. 2.1.6). Strong odors, some of them very pleasant to men, while other flowers have all kinds of what we would consider bad smells, but those apparently appeal to some

Fig. 2.1.2. *Cereus peruvianus*, Cactaceae, with many petals and anthers, northern Peru.

Fig. 2.1.3. *Centaurea*, Composite, with protruding anthers and reduced petals. Negev Desert, Israel.

Fig. 2.1.4. High flower pod in *Agave americana*, Amaryllidaceae, western Texas.

insects, such as flies, which are invariably attracted to flowers emitting smells of rotten meat, excrements and the like (Fig. 2.1.7). Increased, sometimes specialized nectar production (Fig. 2.1.8) is adapted by desert plants which cater for sucking animals such as bees, butterflies and birds. Many desert plants only flower at night time, early morning or early evening, hours in which their pollinators are active.

For the purposes of arid lands' palynology, there is a clear distinction between these plants which went through adaptational evolution, resulting in the practical increase of pollen production, especially those with enlarged number of flowers, anthers or both, or other plants which became adapted to special habits of their

Fig. 2.1.5. Enlargement of a single flower in *Iris mariae*, Iridaceae, central Negev, Israel.

pollinators. The former practically comprise the "wind pollinated" group of the dry regions, which is well represented in the pollen spectra, while the latter truly remained the "insect pollinated" and pollen of its plants rarely appear in the deposits.

Those plants which are primarily wind pollinated, either by nature such as Chenopodiaceae and Gramineae, or by adaptation like *Artemisia*, produce pollen in arid regions in a manner similar to that of any other environment, only occasionally in somewhat greater numbers.

The main difference in pollen production in arid regions is therefore the

Fig. 2.1.6. Deep red flower of *Tulipa amplyophylla* (Desert Tulip), Liliaceae, central Negev, Israel.

introduction of a new and important constituent of the pollen spectra, comprising grains produced by plants which had underwent evolutionary adaptations, in a way by which they became in practice "wind pollinated", due to the numerous anthers, flowers, or the combination of both. Some addition to this group of plants is made by those which developed high flowering pods, which facilitates the dispersal of their pollen in the air.

In spite of the occurrences of wind pollinated plants in dry and especially semi-arid regions; and in spite of various adaptations which result in a much higher pollen production by some insect pollinated plants, it should be clear when planning

Fig. 2.1.7. *Arum dioscoridis*, Araceae, smelling of rotten meat to attract insects. Northern Negev, Israel.

a palynological research in poorly vegetated lands, that the dominating factors are the much lower absolute number of pollen in the atmosphere and much higher amounts of mineral matter carried by the winds. This fact has a direct bearing on the recovery of pollen from the samples, which should therefore be larger than those used in the temperate and tropical regions. It also has significance when drawing conclusions, because of the considerably lower masking effect of pollen produced by the local vegetation, which in many cases may be under-represented due to pollen brought to the site by long distance transportation. Examples are the recent sediments of the Bay of Elat (Aqaba), in which pollen produced by

Fig. 2.1.8. Specialization for nectar sucking birds in *Nicotiana glauca*, Solanaceae, central Negev, Israel.

the local vegetation make only some 20% of the spectrum (Horowitz, 1966); the Mediterranean offshore sediments of Israel, where palynomorphs brought from Africa by the Nile (Rossignol, 1961) constitute up to 30% of the pollen sum; or the present day deposits of the Dead Sea, in which the desert plants pollen are considerably overridden by pollen originating from the Mediterranean vegetation and brought along by the Jordan River and northern winds (Rossignol, 1969a).

The relatively poor pollen production of arid lands vegetation, with its low masking factor, only stresses the need for studying the recent pollen spectra before any attempt is made to learn about the past.

2.2. Pollen transportation

The poor vegetation and pollen production in arid lands result primarily in a low masking effect of the local pollen flora. Various transporting agents, such as winds and rivers, which carry pollen over considerable distances, can therefore interfere with the local pollen spectra, by inserting components from far away provenances, sometimes thousands of kilometers from the analyzed site. These may quite frequently outnumber the local elements, to such a degree that may result in dubious conclusions.

Transportation of pollen can take place both in the horizontal and the vertical planes. Winds, rivers and sea currents carry pollen over great distances. Winds hardly face barriers in most places in the desert and may frequently be very strong; rivers, which more often than not originate in humid regions, may travel long distances through deserts and the numerous pollen carried from their provenance are hardly "diluted" by the much fewer local, desert or even river bank components; sea currents may act much in the same way. Groundwater may carry pollen from a humid intake area, through a porous aquifer to a spring in the desert, but may also be of fossil origin, or could have had dissolved previously laid evaporites; roots and fissures form paths along which pollen could be transported down the section.

It is therefore quite important to first establish the transportation paths of palynomorphs at a certain site, prior to any attempts at drawing conclusions from the obtained pollen spectra.

2.2.1. Eolian

Winds seem to be the most important transportation agent for pollen grains all around the globe, regardless of the environment. In the arid lands, however, their role becomes much more pronounced, chiefly because winds are very strong and can travel very long distances, up to several thousand kilometers, with very little barriers on their way. These strong winds play a threefold role in the obtained pollen spectra: pollen produced by the local, arid land vegetation, is mixed quite well and its part in the pollen spectra is representative of the regional, dry land flora; the winds can carry and transport pollen from desert areas into semi-arid and occasionally even into humid lands, thus intervening with pollen spectra obtained from the latter regions; winds arriving at the desert from more humid regions can carry pollen with them, which may be included within the arid lands sediments and may, due to the low masking factor of the dry land pollen flora, create a false impression of a much richer vegetation than practically grows in the studied dry land.

The pollen loads of winds are usually studied from dust, caught in a variety of traps smeared with sticky material, which are placed at some distance above ground. This was done in many localities, including arid lands such as the Sahara (Van Campo, M., 1975; Cour, 1974).

While this method is very good for studying the pollen loads of winds, it bears some drawbacks for palynlogists who are interested in the past. One is that such a study must be carried out throughout the entire year, because of different flowering periods for some of the plants and to account for winds coming from different directions, in various months of the year. The other, rather serious drawback for palynology of the past, is that such spectra, obtained directly from the winds, include many pollen types which never become fossilized, mainly because they cannot withstand the depositional processes involved, or are readily destroyed following sedimentation.

Another way of studying the recent pollen spectra is by collecting surface sediments. But in order to study wind transportation, these must be chosen so that no other agents are involved. Such sediments could to some extent be surface dust or barren soils, but not any deposits connected with water transport, such as those accumulated in playas, lakes, marshes, nor wadi beds.

The main problem involved with understanding the contribution of wind transportation of pollen grains in arid lands, is that most of the analyzed sites have at least some connection with flowing waters, thus a distinction must be made between the contribution of pollen brought by the water and pollen brought by the winds.

Two types of sedimentary environments are more frequently visited to collect samples for palynological research in dry regions, than others. These are either connected with water bodies, extant and extinct alike, or with human sites, which are naturally in some way or another tied with water, usually situated in wadis, near water sources. Consequently, while the present day wind transportation contribution of pollen can be relatively easily studied by using traps, there is a distinctive difference between results obtained from these traps and those from the study of recent sediments, to which water transportation and local vegetation also contribute pollen, quite significantly. Evidently, local pollen also affect the yield of traps, which should therefore be placed as far as possible from concentrations of plants.

Another essential difference between pollen spectra recovered from winds and those obtained from recent sediments is caused by the introduction, especially in the last years, of many cultivated trees, particularly to the arid and semi-arid regions. Such trees, many of which are wind pollinated, considerably contribute to the pollen load of the winds. But when subrecent sediments are analyzed, that in this respect may represent at least several tens of years, up to a hundred sometimes, the share of these trees is much diluted by the natural vegetation, which existed in the region before their introduction. It seems therefore, in many cases, better not to take these trees into consideration. On the other hand, pollen of such trees are very good indicators of provenances and transporting agents potentials and directions. The contribution of these introduced trees is much more prominent in areas which hardly maintain any arboreal vegetation. For example, in the northern Negev (Israel), near the city of Beer Sheva, the pollen load of winds includes some 60% arboreal pollen, as compared to values of between 6 and 20% AP obtained from recent soils of the same region. However, the pollen of introduced trees, such as *Eucalyptus*, *Casuarina*, rosaceous fruit trees and others, constitute some 90% of the arboreal pollen spectrum obtained from the winds (Weinstein, 1979). Thus the pollen spectra are essentially different when comparing the winds with the recent sediments.

Weinstein (1979) carried out a detailed study of pollen loads of winds in eighteen stations in Israel and analyzed the results monthly, over a period of one year. The results are summarized in Fig. 2.2.1, which shows the monthly distribution of pollen in the air and in Table 2.2.1, which gives the annual averages, from selected stations in Israel. It is clear from these data that blooming seasons are different for various groups of plants, which may later reflect in the sediments. Oaks and pines, which grow on the hilly region running along Israel, are represented in the pollen spectra of the recent sediments both east of the hills, in the Jordan Valley and west in the Mediterranean offshore, in approximately similar percentages (Horowitz, 1979). But in the Jordan Valley oak pollen make the bulk of the arboreal spectrum, while in the Mediterranean, west of the hills, pine is the dominant constituent. Oak flowers begin to appear already in January/February, while pine blossoms somewhat later. In January/February the prevailing winds are northwestern, while later on in the year eastern winds become predominant, which explains the difference between the recent pollen spectra obtained from east and west of the hilly range. Figure 2.2.2 shows the annual averages of pollen spectra over Israel, for arboreal and non-arboreal pollen groups; it can be seen that the mixing effect of winds makes the spectra quite similar almost all over the country.

The main differences between the various localities are in groups of pollen which are not so well transported by winds, being either produced by low plants, shrubs and herbs, or for some other, yet unknown reasons. Indeed there is a distinction between two pollen groups transported by winds. These which are easily lifted from their parent plants, thoroughly mixed in the atmosphere and consequently fall as a uniform pollen rain all over the region; and those pollen grains which do not yield so well to wind transportation and make for the local differences by their higher concentrations close to the places where the parent plants grow.

This phenomenon was also observed in a detailed study of recent, lake bottom sediments from Lake Kinneret (Sea of Galilee) in northern Israel (Horowitz, 1969, 1979), where the first group, of well transported pollen is more or less uniformly represented in the entire lake's area, while pollen which only submit to short distance transportation by the winds are more abundant closer to shores (Fig. 2.2.3).

Weinstein (1981) had analyzed pollen spectra from two closely located sites on Mount Carmel, Israel. One is situated on a north facing slope of a wadi and the other on the opposing, south facing one. The results are summarized in Table 2.2.2, which compares recent pollen spectra collected from dust, accumulated in these two sites. The vegetation cover of the two slopes in this area is given in Table 2.2.3.

The pollen spectra of the two slopes seem to differ significantly. The spectrum of the north facing slope is charaterized by high percentage of *Quercus calliprinos*

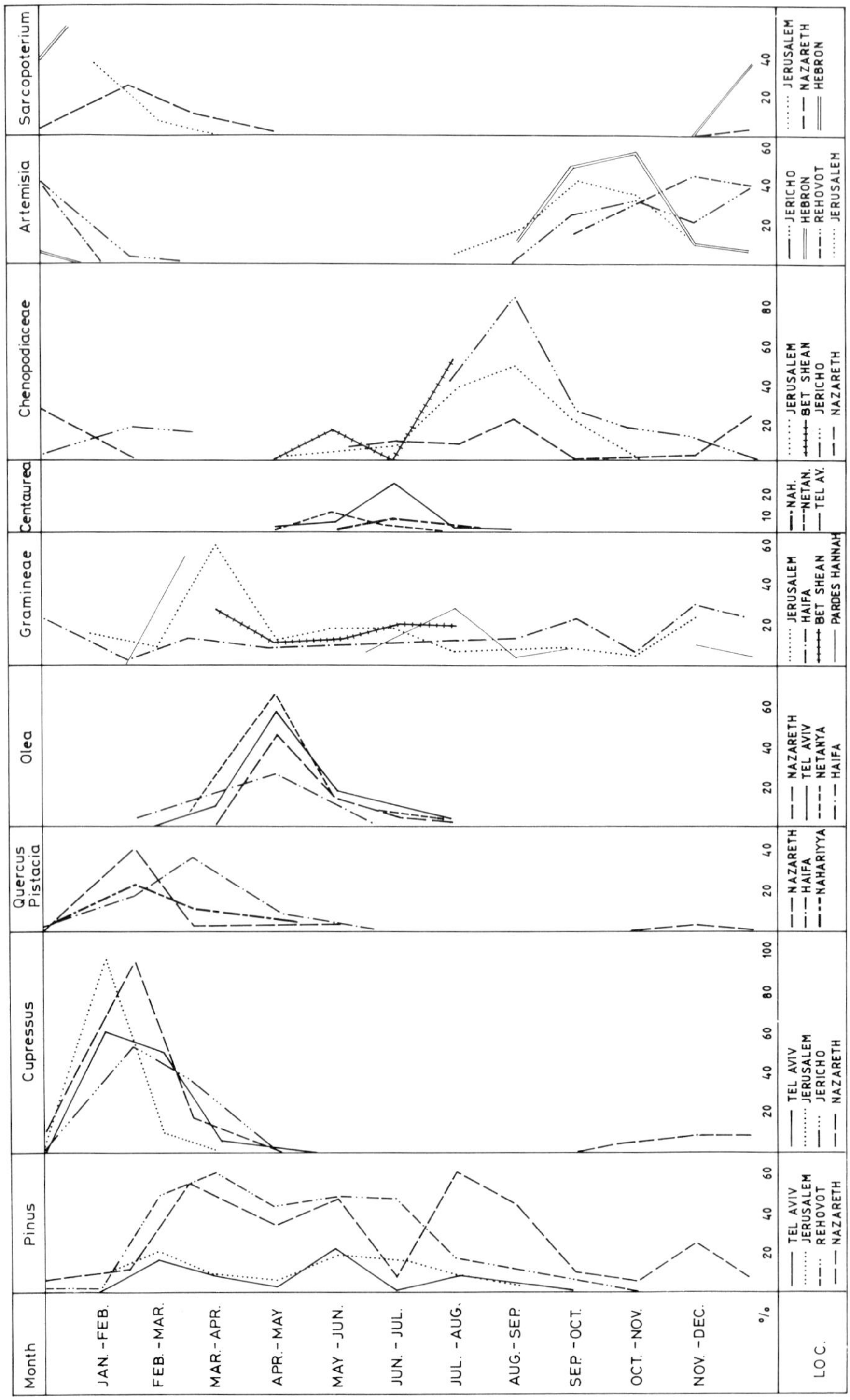

Fig. 2.2.1. Monthly distribution of airborne pollen in selected stations in Israel. Reprinted with permission from Weinstein (1979).

TABLE 2.2.1

Annual averages of airborne pollen spectra from selected localities in Israel (total number of counted pollen = 100%)

Pollen of	Locality							
	Qiryat Shemona	Zefat	Nazareth	Haifa	Afula	Tel-Aviv	Jerusalem	Jericho
Quercus spp.	6.4	4.3	4.9	5.6	2.7	2.2	3.8	1.7
Pinus halepensis	2.6	4.2	20.2	7.5	9.5	5.6	8.7	1.4
Olea europaea	8.0	17.0	6.9	5.4	3.9	8.3	7.4	2.6
Pistacia spp.	0.4	0.3	0.8	1.0	0.3	0.1	0.1	0.1
Cupressus semperv.	6.3	12.7	4.1	9.3	8.5	11.0	1.3	9.7
Eucalyptus spp.	9.0	9.0	14.9	20.0	19.7	12.8	5.0	0.8
Casuarina spp.	4.6	0.7	4.2	8.0	4.6	4.9	5.3	9.6
Other AP	1.9	1.3	1.9	2.7	4.6	2.9	3.3	12.0
Total AP	39.2	49.5	56.9	59.5	53.8	47.8	35.0	37.9
Cereals	2.5	1.5	0.7	0.7	1.4	0.9	2.9	0.7
Gramineae	13.9	12.9	8.5	13.3	10.5	11.2	15.2	7.6
Compositae	7.4	1.8	2.9	2.7	1.2	3.4	5.4	1.7
Cruciferae	2.3	1.5	2.0	0.7	2.1	1.4	3.6	2.5
Chenopodiaceae	6.2	3.5	7.8	8.0	6.2	10.3	11.6	25.7
Artemisia spp.	14.3	14.5	5.9	6.8	3.8	8.5	9.7	12.9
Other NAP	14.2	14.8	13.7	7.6	20.7	15.1	14.7	10.2
Total NAP	60.8	50.0	42.1	40.5	46.2	52.2	65.0	62.1
Total counted	300	300	300	300	300	300	300	300

Reprinted with permission from Weinstein (1979).

and *Pistacia*, typical for the association of plants covering this slope. These species are accompanied by some pollen of *Ceratonia siliqua*, which also grows there. Pine pollen are mostly derived from introduced trees, planted all over. On the south facing slope *Q. calliprinos* is only represented by less than 25% of it's share on the northern, while *Pistacia* is much better represented. The differences in the non-arboreal pollen spectra are much less pronounced, but notable is the higher representation, on the south facing slope of Compositae, *Sarcopoterium spinosum* and Cruciferae.

Each of the pollen spectra seems to correlate quite well with the vegetation composition of its relevant slope, thus chiefly representing local conditions. Again, as is the case with the coastal plain and the Jordan Valley recent sediments, the arboreal pollen percentages do not considerably differ from one slope to another. However, the internal composition of the arboreal pollen group does differ, which shows that the arboreal/non-arboreal pollen ratios are indicative of the regional environmental conditions, while the internal composition of the spectra could be used for specifying local conditions, in this case the direction in which the slopes face.

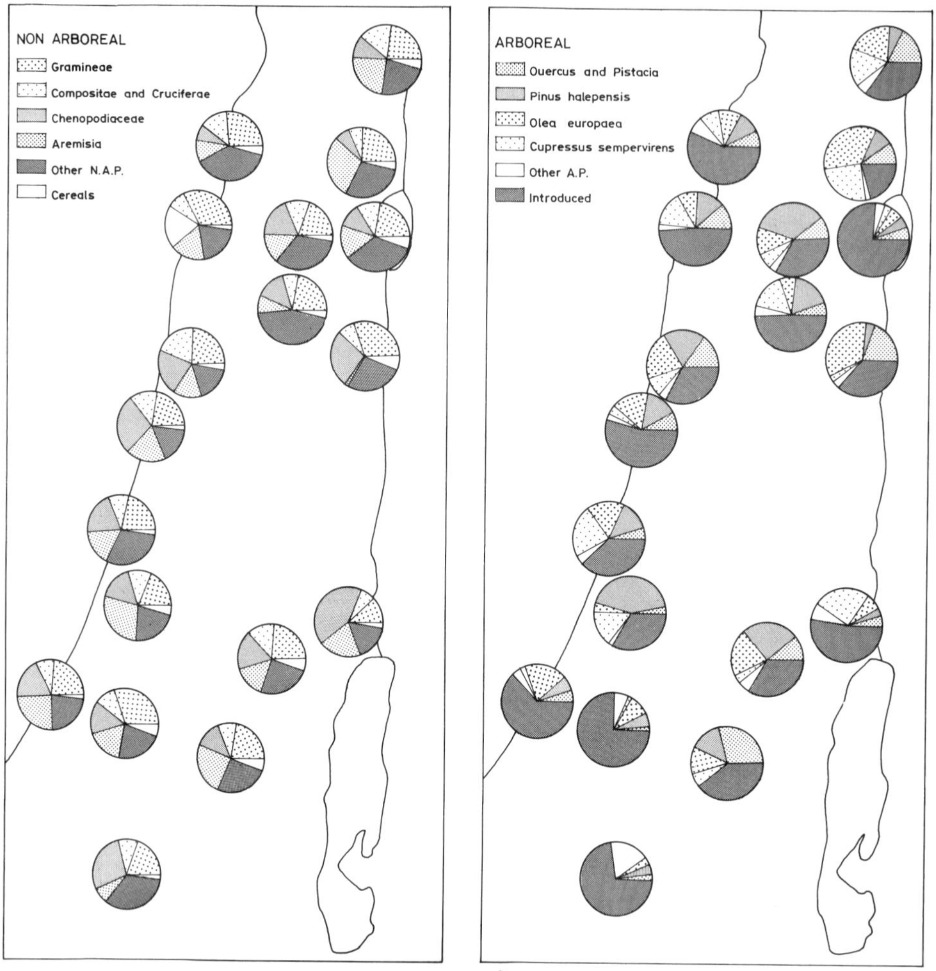

Fig. 2.2.2. Annual averages of AP and NAP from Israel. Reprinted with permission from Weinstein (1979).

Attention, however, should be especially given while studying archeologic or prehistoric sites, even more so to comparisons of samples collected from single horizons (as opposed to continuous sequences) from different sites. Such comparisons between two sites situated on oppositely facing slopes may be quite misleading, if the slope direction is not taken into consideration.

The yearly distribution of pollen recovered by traps from the in air in the Beni Abbes region (Van Campo, M., 1975; Beucher, 1975), northwest Sahara, is displayed in Table 2.2.4. Besides the obvious seasonality of different pollen groups brought to the site, Van Campo divided the obtained spectra by provenances of the pollen grains into several groups: those originating from taxa not represented in the

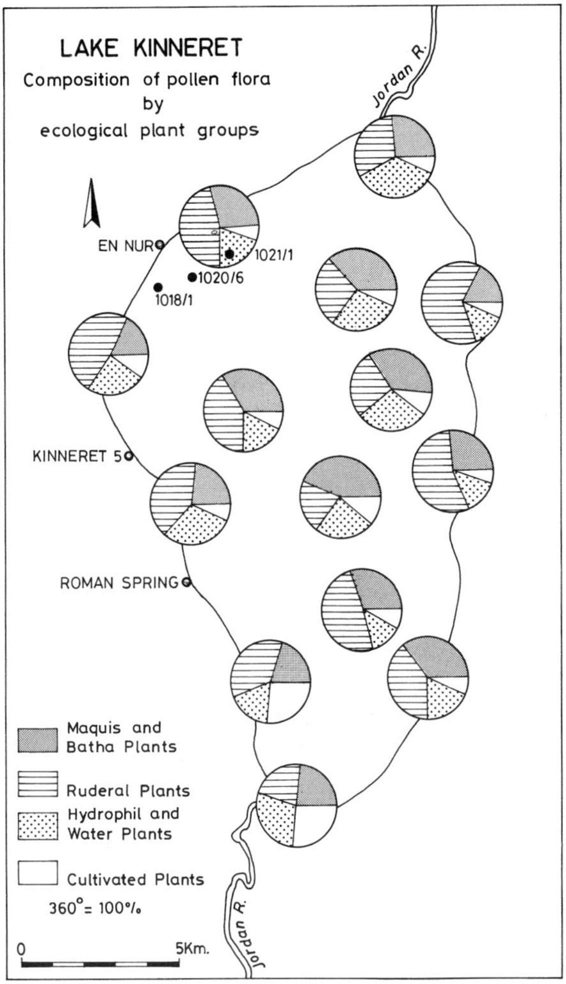

Fig. 2.2.3. Recent pollen spectra and locations of palynologically analyzed water sources, Lake Kinneret, Israel. Open circles-springs; black circles-drillings. Reprinted with permission from Horowitz (1969).

Sahara, brought by the winds over considerable distances, from the Mediterranean and even from the Pontic regions, which comprise more than 20% of the spectra; exotic pollen, brought by winds from the tropical regions to the south, which hardly appear at all in the traps exposed to the wind, but do occur in very small numbers, about 1/4%, in a dust sample collected from the same locality and may have been brought by occasional floods; taxa growing exclusively in the Sahara are only represented at Beni Abbes by some 2.5% of the spectrum; those which grow in the Sahara, but not solely there, comprise some 70%; while pollen of introduced plants only make 1.25% of the spectra.

TABLE 2.2.2

Pollen spectra (percentages out of total counted) of dust samples collected from north and south facing slopes on Mount Carmel, Israel

Pollen of	Locality	
	North facing slope, Abu Usba Cave	South facing slope, upper Oren Cave
Quercus calliprinos	34.4	7.3
Pistacia spp.	10.7	27.4
Pinus halepnesis	9.8	2.1
Ceratonia siliqua	3.5	11.4
Olea europaea	1.4	2.1
Eucalyptus spp.	3.8	0.2
Other AP	1.7	2.2
Total AP	65.3	52.72
Gramineae	7.3	9.1
Compositae	2.0	8.3
Chenopodiaceae	0.7	1.9
Umbelliferae	2.4	2.1
Cruciferae	0.7	3.8
Labiatae	4.4	3.8
Artemisia monosperma	1.4	–
Sarcopoterium spinosum	1.9	7.5
Plantago spp.	4.9	–
Other NAP	8.3	11.1
Total NAP	35.0	47.4
Total counted	572	519

Modified from Weinstein (1981).

The one thing which is indeed striking in these results is, that during certain months, June for instance, 25% of the total pollen is from arboreal origin, from trees none of which grows in the region. This only shows the considerable contribution of allochthonous pollen, brought by the winds, to spectra obtained from the desert.

In places where local vegetation is very poor, such as the extremely-arid region around the Dead Sea, recent sediments yielded considerable shares of arboreal pollen, brought from the Lebanon and northern Israel both by winds prevailing in the region and the Jordan River (Rossignol, 1969a). Results of the analyses are summarized in Table 2.2.5.

It should be noted that such arboreal/non-arboreal pollen ratios are also quite similar in the northern rift valley lakes of Israel, Lake Kinneret and the Hula (Horowitz, 1979), in which the arboreal pollen takes some 20 to 25% of the total spectrum. This influx, especially of those arboreal pollen which are easily carried by the winds, into remote arid regions from faraway provenances, must be taken into consideration when discussing results obtained from dry regions. This

TABLE 2.2.3

Vegetation cover percentages of the north and south facing slopes on Mount Carmel

	North facing slope	South facing slope
Total coverage	120%	40%
Perennials coverage	100%	50%
Main taxa		
Quercus calliprinos	50%	–
Ceratonia siliqua	10%	8%
Pistacia spp.	5%	6%
Olea europaea	5%	–
Calycotome villosa	5%	1%
Salvia fructicosa	20%	1%
Cistus spp.	4%	–
Stachys distans	1%	2%
Phagnalon rupestre	0.5%	1%
Ruta chalepensis	0.5%	1%
Number of perennial taxa	19	12
Annuals coverage	20%	50%
Number of annual taxa	25	47

Reprinted with permission from Weinstein (1981).

TABLE 2.2.4

Recent pollen spectra from Beni Abbes region, NW Sahara, grouped by provenances, comparing ground dust flow and horizontal filter sampling

Pollen of	Ground dust flow sampling		Horizontal filter
Locality:	High Algerian Plateau (33°N)	Beni Abbes (30°N)	Beni Abbes (30°N)
Taxa not growing in the Sahara			
Temperate and Mediterranean	18.85	20.24	24.05
Tropical	–	0.26	–
Taxa growing in the Sahara	0.94	5.82	2.56
Taxa not exclusive to the Sahara	77.16	71.80	69.41
Planted	0.17	0.34	1.25
Unidentified	2.72	1.21	2.41
Total counted	2853	3533	2894

Data from Van Campo, M. (1975), Beucher (1975).

phenomenon is even more pronounced in depositional basins such as lakes or seas, where the local vegetation pollen, which are not carried to great distances by the winds, are usually under-represented. Thus, when comparing pollen diagrams obtained in Israel from the rift valley lakes, the arboreal/non-arboreal ratios and their fluctuations along the entire late Cenozoic are almost identical in the Hula

TABLE 2.2.5

Pollen spectra of recent sediments from the Dead Sea, Israel (total counted = 100%)

Pollen of	Northern basin		Southern basin		
Water depth:	326 m	350 m	3.3 m	5.0 m	3.8 m
Quercus spp.	2	2	2	1	5
Olea europaea	5	5	4	–	4
Pistacia spp.	4	1	2	–	0.5
Pinus halepensis	5	4	3	11	9
Other AP	3	3	1.5	1	1.5
Total AP	19	13	12.5	13	20
Gramineae	19	15	5	6	13
Cyperaceae	3	1	0.5	2	1
Compositae	6	7	13	11	9
Chenopodiaceae	17	28	32	56	22
Cruciferae	2	1	2	1	3
Artemisia spp.	11	10	15	–	16
Plantago spp.	10	10	5	1	3
Sacropoterium spinosum	6	6	2	1	4
Ephedra spp.	0.5	1	2	1	2
Other NAP	7.5	11.5	11.5	6	8.5
Reworked spores	1	0.5	–	2	0.5
Total NAP	81	87	87.5	87	80
Total counted	243	334	424	287	293

Data from Rossignol (1969a), Horowitz (1979).

basin to the north, through to the southern Dead Sea basin, some 250 km to the south (Horowitz, 1987a), reflecting the mixing and long distance transportation by the winds.

Pollen grains are also carried from the desert regions into the vegetated ones by strong winds, especially by those involved with dust storms. Horowitz et al. (1975) made a study of dust collected in Jerusalem, brought by both southeasterly and southwesterly dust storms. The paths of these dust storms are given in Fig. 2.2.4; Table 2.2.6 summarizes the results and Fig. 2.2.5 shows the provenances of pollen grains arriving in Jerusalem with the dust storms. Figure 2.2.6a is a satellite image of one of these storms. The contribution of pollen grains brought by the dust storms is quite obvious in the pollen spectra from Jerusalem, which are otherwise typically Mediterranean. It should however be noted that although the storms and dust arrive from considerable distances, most pollen are derived from the closer vicinity. This seems to result from the scarce vegetation in the storm provenance, which cannot contribute much pollen to the winds' load.

Pollen transported by dust storms and strong winds from the Sahara to southern Europe, across the Mediterranean, are reported by Van Campo M. and Quet (1982). Figure 2.2.6b is a satellite image of such a storm. Moar (1969) reports of

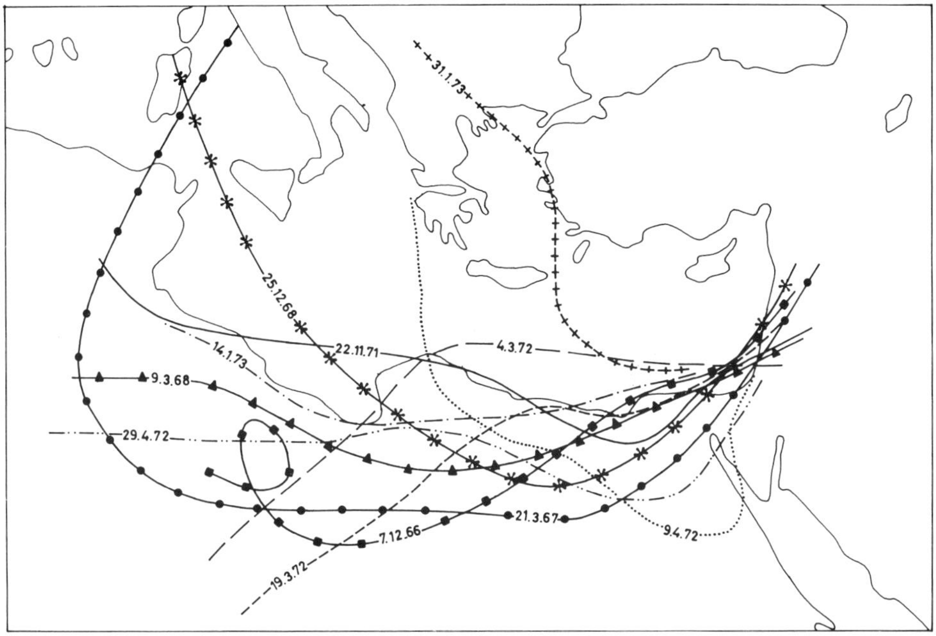

Fig. 2.2.4. Trajectories of dust storms arriving in Israel. Reprinted with permission from Horowitz et al. (1975).

Casuarina pollen in New Zealand, brought there from Australia over the Tasman Sea, together with red dust, a distance of some 1,600 kilometers, while Straka and Kappen (1988) report pollen carried by winds to Antarctica over distances of up to 3,300 kilometers. These cases are rather curiosities, which at any rate only contribute minor shares to the pollen spectra.

Not all pollen grains are transported by the winds in the same manner, which causes some disadvantage when studying pollen from arid lands. For instance, one of the typical trees in arid lands, in many places in the world, is the thorn tree (*Acacia*), but its pollen are hardly carried by the winds for reasons unknown. *Acacia* is therefore always under-represented in the winds (Van Campo, M., 1975) and also in the sediments (Horowitz, 1966, 1979; Maley, 1972, 1981).

Hooghiemstra et al. (1986) and Hooghiemstra and Agwu (1986) made thorough, detailed studies of pollen and spore distribution in recent marine sediments off the northwestern African coast, from which they could deduce on the seasonal wind patterns and their relations to vegetation belts in the western Sahara. The general outline of interpretation of the pollen signal from the palynological records of the northwest African marine sediments is discussed by Hooghiemstra (1988a). Hooghiemstra and his collaborators based their study on the fact that the climatic gradient in northwest Africa is reflected in the vegetation by a series of distinct

TABLE 2.2.6

Pollen spectra from dust samples collected in Jerusalem (total counted = 100%)

Pollen of	Date of storm								
	9-4 1968	31-3 1969	29-4 1972	12-3 1968	29-3 1971	13-1 1968	25-11 1968	7-5 1973	17-1 1969
Quercus spp.	8	15	2	3	–	2	3	15	18
Olea europaea	11	1	3	14	2	3	5	11	26
Eucalyptus spp.	3	2	1	1	3	9	4	1	1
Cupressus semperv.	11	9	11	3	–	18	1	4	2
Pinus halepensis	2	1	1	1	–	2	2	1	2
Other AP	1	3	6	3	2	6	6	2	3
Total AP	36	31	24	25	7	40	21	34	52
Gram. + Cyp.	17	37	31	30	27	2	5	12	28
Cereals	33	1	1	16	1	3	9	4	10
Plantago spp.	1	1	6	12	2	–	0.5	1	2
Chenopodiaceae	2	10	10	9	34	7	47	21	1
Umbelliferae	1	1	1	2	1	–	1	2	1
Compositae	2	3	2	2	10	2	11	6	2
Artemisia spp.	2	4	2	1	7	14	–	9	–
Cruciferae	1	1	16	1	4	2	5	2	2
Other NAP	5	11	7	2	7	30	0.5	9	4
Total NAP	64	69	76	75	93	60	79	66	48
Total counted	200	200	254	200	131	177	188	210	255

Reprinted with permission from Horowitz et al. (1975).

belts. Pollen and spores from this area are transported to the Atlantic mainly by the trade winds and African Easterly Jet. They have defined seven pollen assemblages in the recent surface sediments from the eastern Atlantic, offshore northwest Africa, suggesting that pollen distribution in the marine sediments can be explained by their provenance, the timing of pollen release for transport and the major transport systems (winds) effective at that time of the year. They concluded that in this way the distribution of selected assemblages and taxa could trace the positioning of the major wind belts.

Hooghiemstra (1988a) brings forth isopollen maps of spectra recovered from the recent marine sediments west of the African coast, maintaining that the main contributor of pollen to these sediments is the wind. This, in opposition to some other localities, such as the Orinoco Delta (Muller, 1959), the Great Bahama Bank (Traverse and Ginsburg, 1966), the Gulf of California (Cross et al., 1966) and the Mediterranean (Koreneva, 1971; Rossignol, 1969b; Cheddadi, 1988), which mainly benefit from pollen brought by rivers. It seems, however, that some contribution by the Senegal River cannot be overlooked in this area, as shown by Hooghiemstra.

Following the studies by Hooghiemstra and his collaborators on the recent sediments, wind directions for the past had also been established. Hooghiemstra

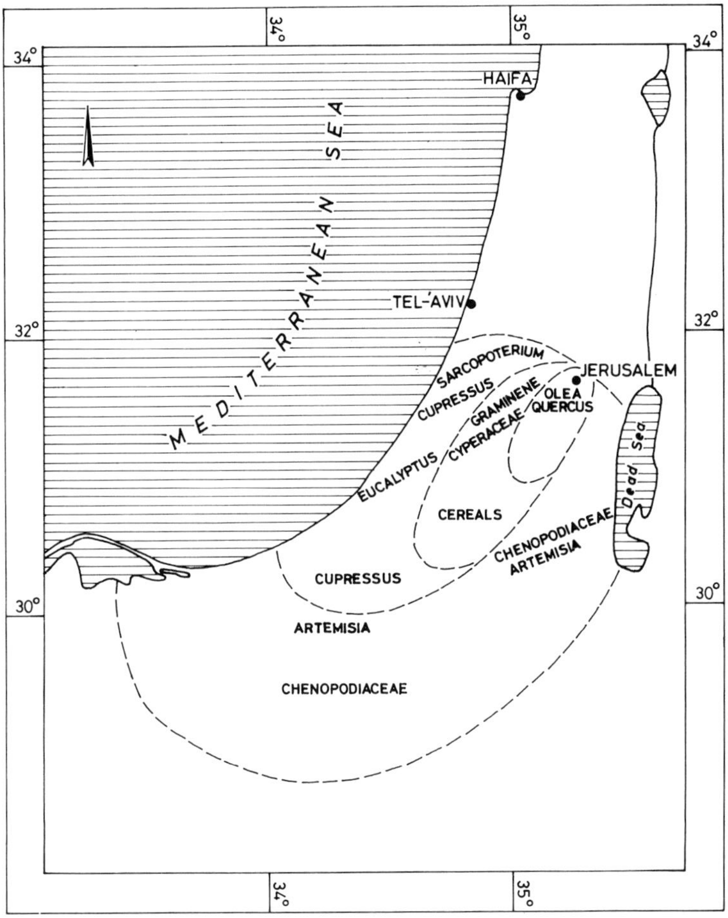

Fig. 2.2.5. Sources of pollen grains arriving in Jerusalem during dust storms. Reprinted with permission from Horowitz et al. (1975).

(1988a) studied paleowind directions for the period of 9 ky BP. Hooghiemstra et al. (1987) did the same for sediments dated 18 ky BP. The following step was to establish continuous changes in the wind directions and vegetation zones of the northwest African coast, between 20,000 and 5,000 years BP (Hooghiemstra, 1988b) and between 140 and 70 ky BP (Hooghiemstra and Agwu, 1988; and see Fig. 8.1.2).

These studies show that once the mechanism of wind transportation of pollen is established, one can proceed and study paleowind directions. Marine sediments seem to be one of the best sources for such studies, since continuous cores are quite easy to obtain, but similar studies can be carried out in continental sediments as well. Indeed, correlations of the marine and continental sequences would probably be the best source of information (Lézine and Hooghiemstra, 1990).

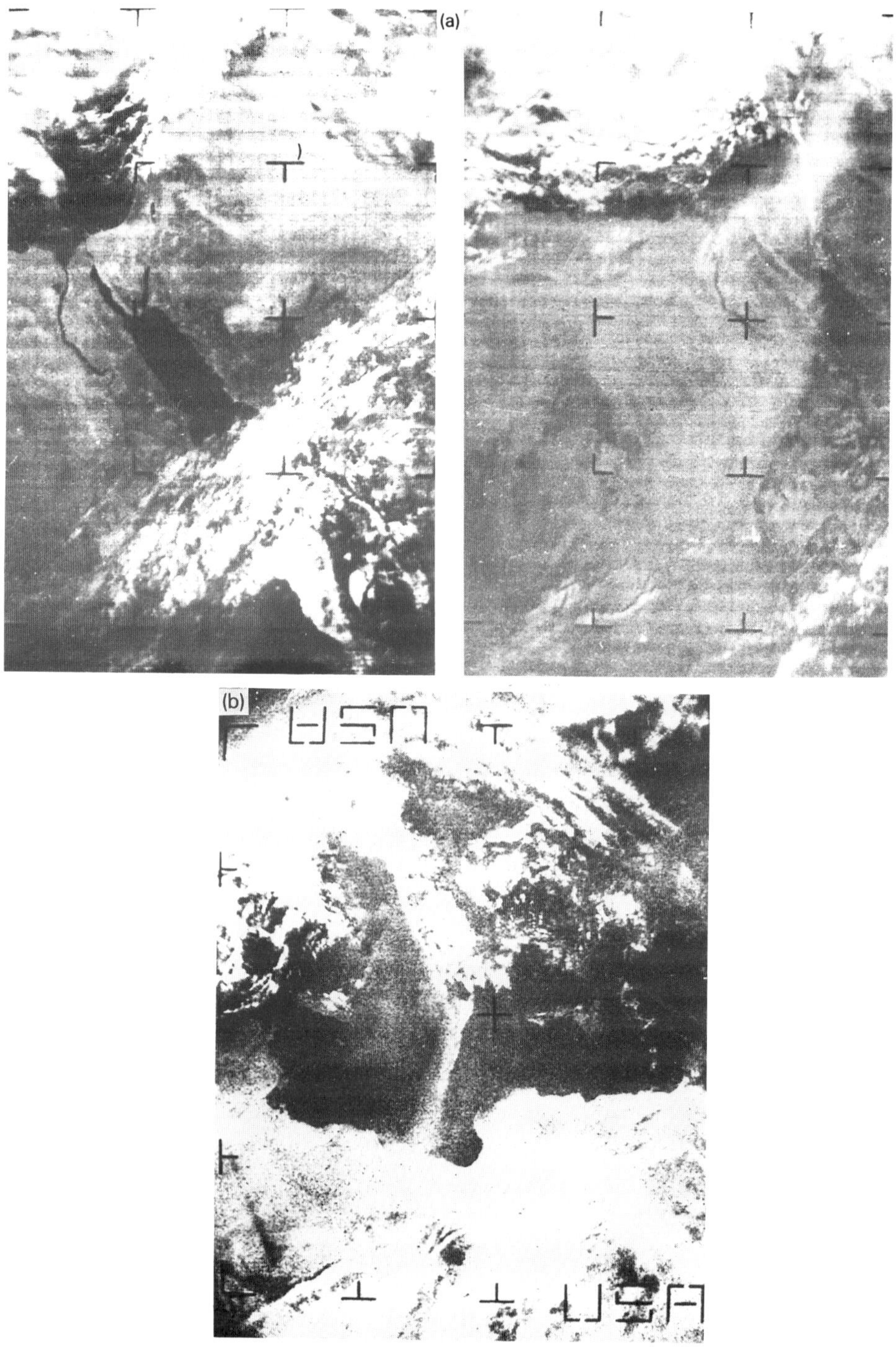

Fig. 2.2.6. Satellite images of dust storms originating in the Sahara. Courtesy of E. Ganor: a. 21.3.67, storm carrying dust to the Near East. Left, before the storm; right, the storm. b. 10.11.71, storm funneling dust to Italy.

One of the problems involved with wind transportation, as discussed above, is the occurrence of exotic pollen, arriving from considerable distances to a certain site. The importance of studying the present day processes cannot, therefore, be overly stressed, because these seem to serve as the only reference to what was happening in the past. In this way, climatic changes can be concluded from occurrences of arboreal pollen in the desert, even for periods when trees had apparently not been growing in this region. As an example, in the central Negev, in Israel (Horowitz, 1979), recent sediments yielded between 2% and 4% arboreal pollen, mainly oaks and olives. Samples collected from prehistoric sites in this area yielded arboreal pollen in the order of 10% to 15%. Trees could either grow in the vicinity of the site, or brought by winds from the north, where these trees grew, without the trees themselves actually growing near the site itself. However, the different percentages must reflect some climatic change, because a stronger activity of northern or northwestern winds at a site which is presently arid, implies that these winds must have also brought some more rain to the place. Trees had not necessarily grown near the site during the prehistoric settlement times, but the increase in percentage of arboreal pollen undoubtedly reflects a change in the winds regime, which must have been connected with some degree of climate change.

2.2.2. Fluviatile

Fluviatile transportation of pollen and spores, originating in continental plants, both terrestrial and aquatic, is a well known phenomenon, discussed by many authors. Most studies are concerned with transportation from continental environments to the sea, inclusion of palynomorphs in marine and especially in deltaic sediments. Such are studies by Traverse and Ginsburg (1966), who discussed the palynology of surface sediments of the Great Bahama Bank, as related to water movement and sedimentation; Muller (1959), who dealt with palynology of the recent Orinoco delta and shelf sediments; Cross et al. (1966), who studied the source and distribution of palynomorphs in bottom sediments of the southern part of the Gulf of California; and Koreneva (1971) and Cheddadi (1988), who studied the palynomorphs in the Mediterranean bottom sediments. These studies, among others, deal with rather long rivers, sometimes several thousand kilometers, which originate in forested areas, such as the tropics (Muller, 1959; Koreneva, 1971), or high mountain ranges (Cross et al., 1966) and which travel through an arid land, to pour their load to the sea. Since marine sediments deposited close to arid lands are one of the best sources of information for the dry regions, the contribution of such rivers must be understood before an attempt is made to draw any conclusions about the arid land itself.

Transportation by rivers in arid lands has two other aspects, which should also be taken into consideration. Certain rivers, quite commonly occurring in arid lands, may frequently not end their way in oceans or seas, but rather in an endoreic

Fig. 2.2.7. The Dead Sea. Satellite image of an inland, endoreic basin in an extremely-arid region.

basin, within the arid land. For example, the Jordan River in Israel which originates in the Mediterranean domain, north of the country and ends in the Dead Sea internal, hypersaline lake (Fig. 2.2.7); or the Okawango River, whose headwater is in the subtropical regions of Angola and pours its water, through an inland delta in northwestern Botswana, to support subterranean aquifers running under the Kalahari Desert sands.

The third aspect of fluviatile transportation of pollen in arid lands is by intermittent rivers (wadis, arroyos), which are only active during seasonal floods (Fig. 2.2.8). This process seems to be one of the most important agents of pollen

Fig. 2.2.8. Flood in wadi. Judean Desert, Israel.

transportation in arid lands, mainly because most samples collected from these regions are, in some way or another, connected with intermittent streams. Such are samples collected from wadi beds, which are sometimes the only locality where sediments are accumulated; most archeologic and prehistoric sites, which are usually situated along intermittent streams in the desert, frequently embedded within terraces; or playas, salt lakes and sometimes perennial lakes, receiving most of their waters from intermittent floods. Such seasonal floods are not necessarily connected only with wadi (arroyo) beds, but can sometimes develop into flush floods, which, for a while, may wash great areas in arid lands (Fig. 2.2.9).

Fig. 2.2.9. Flush flood. Central Negev, Israel.

The floods, which are common to most dry regions, have a considerable transporting capacity due to their large, almost instantaneous volumes of turbulent waters and are consequently involved in two types of transportation. On the one hand, transporting pollen which were produced concurrently, or close to the time of the flood; it must be mentioned in this respect that in arid lands the flowering season usually coincides with the time of rains and floods; while another factor, which cannot be overlooked in the desert, is that such floods are responsible for considerable removal of older debris and its redeposition at a different locality downstream, which also involves redeposition of pollen grains, included in the removed and redeposited sediments.

A. Rivers flowing to the sea

The River Nile is a good example for long distance transportation of pollen and spores, originating mainly in the tropical and subtropical regions of Africa and brought through the Sahara to the southeastern Mediterranean, close to the desert and semi-desert areas of northern Sinai and southern Israel. The recent pollen deposition in the Mediterranean offshore Israel was studied by Rossignol. The preliminary stage of this study dealt with the southern sector of the offshore

(Rossignol, 1961), while a more comprehensive study of the entire offshore along Israel's coast, was published later (Rossignol, 1969b).

Pollen and spores are brought to the Sinai and Israeli coasts by the combined action of the Nile and a longshore sea current, which flows east through the Nile outlet and northward along the Israeli coast. The absolute frequencies of pollen and spores are highest along an area parallel to the coastline, situated about ten kilometers seaward. The frequency diminishes toward the coast, most probably as a result of the sandy type of sediment and also farther seaward, due to loss of energy of the transporting agents (Fig. 2.2.10).

Recent sediments of the Mediterranean offshore are dominated by Gramineae and Cyperaceae, sometimes accompanied by *Sparganium*. The second most common group comprises pollen grains of conifers, of which *Pinus halepensis* is the dominant element. Some grains of *Cedrus libani* also appear, but are quite rare. Another conifer, which is quite abundant somewhat more in the northern locations, is *Podocarpus qracilior*. Next in frequency are pollen grains of Compositae, including some *Artemisia* and *Centaurea*, accompanied sometimes also by others such as *Asphodelus microcarpus*. Pollen grains of Chenopodiaceae and *Ephedra* are quite rare. A distinct group, which is quite abundant in the recent pollen spectra of the offshore Mediterranean, comprises spores of cryptogamous plants. Of these, more than 50 species were encountered in the various samples, mainly representing such families as the Gleicheniaceae, Cyatheaceae, Anthocerotaceae, Polypodiaceae, Lycopodiaceae, Osmundaceae, Schizeaceae and others. This group is occasionally accompanied by pollen of Combretaceae.

Horowitz (1979), while discussing the results obtained by Rossignol, divided the recent pollen spectra of the offshore Mediterranean into three groups: pollen representing plants which grow exclusively in Israel; grains clearly arriving from the tropical regions of Africa, such as fern spores, Combretaceae and *Podocarpus*; and those which do occur in Israel, but also grow along the Nile and in the Nile Delta, of which Gramineae and Cyperaceae are the most important.

Rossignol regarded all pollen derived from plants which grow in Israel as autochthonous, but a comparison of the airborne and recent sediments' pollen spectra indicates that Gramineae and Cyperaceae, the most common group in the recent sediments, cannot be safely regarded as coming solely from autochthonous origin. In the airborne pollen spectra (Table 2.2.1) collected at the coastal plain (Weinstein, 1979), Cyperaceae are very rare and in most cases completely absent, whereas Gramineae are quite common. It seems therefore, that most Cyperaceae and probably some of the Gramineae are allochthonous at the coastal plain, having been brought by the Nile and derived from its bank and delta vegetations, which is probably the reason for their over-representation in the Mediterranean surface sediments, as compared with such sediments collected in other localities in Israel (Horowitz 1979).

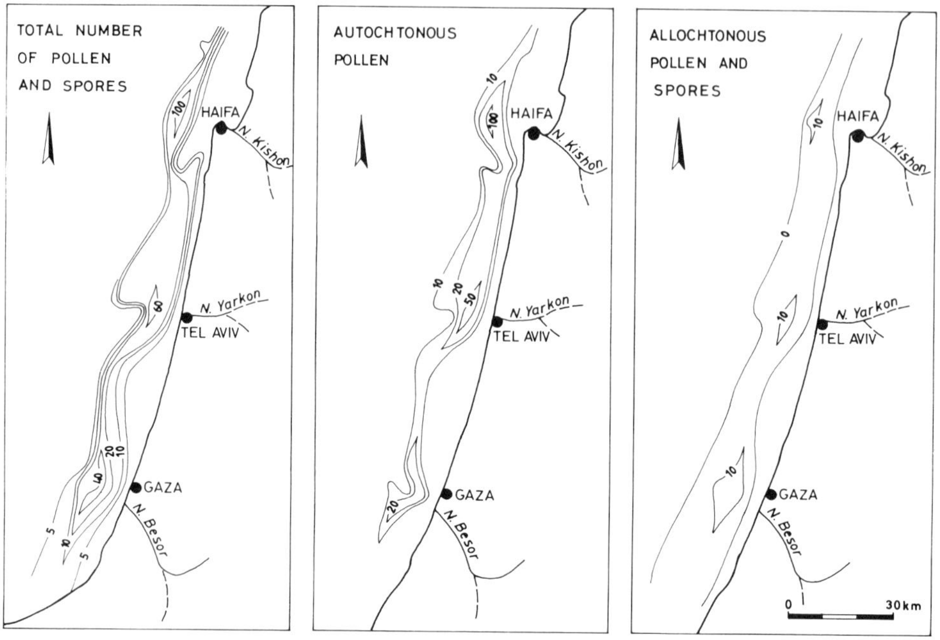

Fig. 2.2.10. Pollen deposition offshore Israel. Numbers are grains per gram sediment. Data from Rossignol (1961, 1962, 1964, 1969b).

Occurrences of tropical palynomorphs in the offshore sediments of the southeastern Mediterranean could be used for delineating the evolution of the River Nile (Horowitz, 1974a, 1979, 1989e; Sneh et al., 1986; Horowitz and Derin, 1987). These palynomorphs however, do not bear any direct evidence to the evolution of the deserts which surround the southeastern Mediterranean.

Local rivers also contribute pollen grains to the offshore environments, as may well be seen in studies by Rossignol (op. cit.). The total number of pollen and spores (Fig. 2.2.10) is significantly higher in marine sediments located in front of outlets of the main rivers. To the south, a peak can be seen in front of the outlet of Nahal (river) Besor; in the central part of the country in front of Nahal Yarkon; and to the north, in front of Nahal Kishon. Pollen rain from winds is considered more or less uniform, while these superimposed peaks are caused by the rivers.

B. Inland rivers

Pollen and spores brought by rivers into inland, endoreic lakes (or other types of sedimentary basins), comprise an important part of spectra obtained from the basin's sediments. It is not always easy to distinguish between contributions made by the rivers, as opposed to winds, especially when both flow in a similar direction, as is the case, for instance, in the Jordan Rift Valley in Israel (Rossignol, 1969a;

TABLE 2.2.7

Recent pollen spectra from Chari River, entering Lake Tchad from the south, compared with the lake and the surrounding marshes (spectra grouped by provenance; total counted = 100%)

Pollen of	Location		
	Malalie (Chari R.)	Baga-Sola (Lake, center)	Koukou (Marsh, NE)
Sahelian elements	1.13	1.64	2.25
Sahelo-Sudanian elements	2.93	0.82	1.35
Southern, allochthonous	4.0	3.6	–
Northern, allochthonous	–	0.2	–
Poly-regional elements	3.61	1.42	1.08
Aquatic plants	2.3	9.93	40.68
Combretaceae	1.12	0.69	0.27
Gramineae	67.6	23.84	23.94
Cyperaceae	14.2	54.61	28.17
Others	3.16	3.25	2.25
Regional trees	3.83	1.36	2.52
Submerged trees	–	3.0	–
Total AP	8.8	8.63	2.8
Total NAP	88.0	88.1	94.9
Undetermined	3.1	3.2	2.2
Total counted	443	3162	1111

Reprinted with permission from Maley (1972).

Horowitz, 1969, 1979). In such cases, only palynomorphs carried almost exclusively by water can be taken as indicators for river transportation, such as spores of ferns growing at the river bank (*Thelypteris palustris* in northern Israel, for example), or submerged plants, which discharge most of their pollen grains to the water, such as *Myriophyllum* and *Lemna*. Although undoubtedly some of these plants also grow in and around lakes, these pollen and spores could be used to follow the path of the Jordan River into Lake Kinneret (Horowitz, 1969). Certainly, the river also brought some other palynomorphs, which could not be distinguished from those arriving by winds.

A different example comes from Lake Tchad, in central Africa, which receives water, among other sources, by the Chari River coming from the south, across a stretch of several hundred kilometers. Maley (1972), in a study of recent sediments in Lake Tchad, concluded that the allochthonous southern group, which is represented by pollen grains of plants growing in the Sudanian domain and Sudanian-Guinean sector, have only been encountered in the river's sediments and in deposits of the southern area of the lake, which is under direct influence of the waters of the Chari River (Table 2.2.7). The fact that none of these pollen grains were found in the ponds nearby, led Maley to conclude that the river alone is responsible for transport from the south of the basin. Long distance arrival of pollen grains by winds to Lake Tchad is quite negligible from the south, because southern winds in this region do not have great strength on the whole, as compared with winds coming from the north to Lake Tchad, which contribute considerably more pollen to the lake's sediments.

Notably, pollen arrive at Lake Tchad by winds from a different provenance as those brought by the river. This differs from the situation in the Jordan Valley, where the prevailing winds blow in the same direction as the River Jordan flows, thus both agents carry almost similar constituents. Except for the minor truly fluviatile indicators mentioned above, no distinction could be made between the winds and river's contributions.

It should be noted in this respect, that pollen transport by rivers is chiefly limited to those parts of the depositional basin which are directly affected by the rivers' waters and are hardly found in other places in the lake. Similar situation is also observed in marine sediments (Fig. 2.2.10), as discussed above. Thus such typical fluviatile palynomorphs are very good indicators of river paths and depositional processes, rather than of the regional vegetation composition.

C. Wadis

Wadis, (or arroyos as they are known in the American Southwest; dongas in southern Africa; and a host of other terms), are typified by their intermittent activity, following the rare but rather strong rains and thunderstorms, mostly in the form of violent floods, which among all other debris also carry along pollen grains. This way of pollen transportation is very common, especially in arid and extremely-arid domains and is responsible for the composition of pollen spectra obtained from most analyzed sites in these regions. The floods transport pollen from two sources: the wadi vegetation, most of it flowering during the short rainy season, so that water will still be available to make fruit and seed; and pollen derived from sediments previously accumulated along the flood's path, which are lifted by the water and redeposited downstream. Some pollen from the regional vegetation are almost always included, as may also those derived from anthropogenic sources, especially if settlements are located upstream.

It is very difficult to assess the share of such redeposited pollen and their numerical relations with those produced at the very year of the flood. It therefore seems much more useful to use the regional vegetation pollen for drawing conclusions, rather than the wadi vegetation pollen (Horowitz and Horowitz, 1985; Horowitz, 1987a). This could only be done by omitting the wadi vegetation pollen, after its establishment, from the total spectrum and analyzing the rest for concluding on regional environments. On the other hand, the wadi vegetation pollen could be used to discern stages in the development of the wadi itself and its drainage system. An example which demonstrates the essential differences in the recent pollen spectra collected from wadi deposits and those obtained from sediments on nearby highlands, can be seen in the central Negev, southern Israel, which is presently arid to extremely-arid. A recent pollen spectrum, obtained from wadi sediments at Nahal Zin (Horowitz, 1976) reads as follows:

Total arboreal pollen 1%, derived of olive.

Total non-arboreal 99%, comprising:

- Gramineae and Cyperaceae 19%
- Chenopodiaceae 77%
- Other NAP 2%

A recent pollen spectrum obtained from dust collected from the surface on nearby highlands at Har Harif (Horowitz, 1977a):

Arboreal pollen, none at all

- Gramineae and Cyperaceae 91%
- Chenopodiaceae 6%
- Papilionaceae 2%
- Occurrences of *Artemisia*, *Zyqophyllum*, Compositae.

The striking difference is between relations of Gramineae and Cyperaceae pollen on the one hand, which dominate the highlands following rare rains (*cf.* Fig. 1.2.7) and Chenopodiaceae, which prevail in the wadi sediments. The dominant plant growing in the wadis of the central Negev is *Atriplex*, which is probably responsible for most of the Chenopodiaceae pollen obtained from the recent sediments. Gramineae and Cyperaceae also grow in the wadis, but are generally outnumbered by *Atriplex* (Fig. 2.2.11).

These differences mainly depend on the degree of coverage and composition of the wadi vegetation. In extremely-arid regions, where wadi vegetation itself is poor, the differences between recent pollen spectra obtained from the wadi bed and those from airborne dust traps are much less prominent, as demonstrated by Van Campo, M. (1975) in the Sahara (*cf.* Table 2.2.4). These differences again stress the necessity of studying recent pollen spectra, in as many as possible locations in an arid land and not to restrict the analyses to those localities which are more convenient for sampling, such as the wadi beds themselves alone. Special attention must be given to those places in which the wadis develop a local, much richer vegetation, which does not represent the regional conditions, as might be the case with oases (Fig. 2.2.12).

The problem of a well developed local vegetation in oases must be given special attention when analyzing samples collected from archeological sites, which in most cases are intimately connected with such environments.

2.2.3. Sea currents

Sea currents can carry pollen over considerable distances, up to several thousand kilometers, thus introducing exotic components to pollen spectra from far away localities. Pollen and spores transported by sea currents are mostly those brought to the sea by rivers, which in many cases causes an enrichment in spores over

Fig. 2.2.11. Wadi vegetation in the Negev Desert, Israel, with dominating *Atriplex*, Chenopodiaceae.

Fig. 2.2.12. Oasis made by a perennial spring in the central Negev highlands, Israel, supporting well developed hydrophil vegetation, including trees such as poplars and tamarisk.

pollen in the ensuing spectra (*cf.* Fig. 2.2.10 and the accompanying discussion). The reason for this enrichment is the fact that most spore producing pteridophytes grow along rivers, ecologically connected to perennial waters. Thus fern spores travel almost exclusively by water, while pollen produced by wind pollinated plants travel chiefly by air. Sea currents then act as a secondary enrichment agent for those palynomorphs essentially subject to fluviatile transportation. It may also be possible that spores can withstand water transportation better than pollen, but this assumption still needs proof.

Not much information is presently at hand on such transportation processes, but two examples are brought here from Israel. A longshore current sweeps eastward the southern coast of the Mediterranean and turns north when it approaches the Israeli sector. This current carries palynomorphs initially brought by the Nile, which originate in tropical Africa, from the Nile Delta to the Israeli coast. These are incorporated in the recent offshore Mediterranean sediments in considerable quantities, up to 20–25% (Rossignol, 1961, 1969b). These palynomorphs mostly comprise fern spores, discussed above (see Chapter 2.2.2), but some pollen grains are also brought, in subordinate quantities, notably those produced by *Podocarpus gracilior* and Combretaceae.

Another sea current is known to sweep the east African coast, northward into the Red Sea and, through the Tiran Straits, to the Bay of Elat (Aqaba). The velocity of this current was measured at Tiran Straits and amounts to some 3 knots (Red Sea Charts, 1848). This current flows on the sea surface until it reaches the northernmost end of the bay and there turns to the bottom, where it deposits mainly spores, accompanied by minor amounts of pollen grains, which seem to have been transported from as far as Madagascar (Horowitz, 1966). The spectra comprise some 25–30 spore genera from families such as Cyatheaceae, Polypodiaceae, Vittariaceae, Hymenophyllaceae and Lycopodiaceae, (identified by D. Welman, University of the Orange Free State, Bloemfontein, South Africa, 1968). The common accompanying pollen grains are again, as in the Mediterranean, of *Podocarpus gracilior* and Combretaceae. These exotic, tropical sporomorphs make up to some 10% of the spectra obtained from the recent sediments of the Bay of Elat. The paths of sea currents bringing exotic palynomorphs to the Israeli coasts are shown in Fig. 2.2.13.

2.2.4. Groundwater

Pollen are washed from the air by rainwater and, at least in part, are incorporated in groundwater either immediately or following some period of surface runoff, by infiltration. Groundwater can travel considerable distances in aquifers and the grains are transported, especially in coarse grained beds such as pebbles and sandstones or karstic horizons with very little filtration. Transportation in fine grained aquifers may involve filtration, thus only smaller grains would be carried

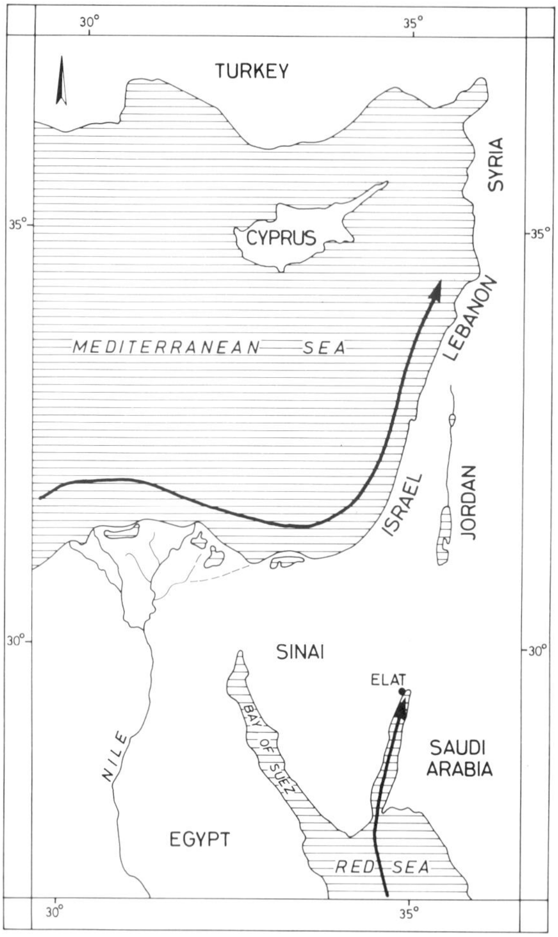

Fig. 2.2.13. Sea currents transporting pollen and spores originating in Africa to the Israeli coasts.

on, if not altogether blocked. On their way, groundwater can also dissolve some of the rock in which they travel and pollen incorporated in such rock can thus comprise part of the spectrum carried by the water (Horowitz and Langozky, 1965; Horowitz, 1968).

Pollen grains can be transported by groundwater in a variety of time planes: palynomorphs are transported horizontally, from the intake area to the discharge locality, which may sometimes be quite remote, located in a different environment; grains from dissolved aquifer rocks, especially evaporites, can be brought up the section and incorporated in younger rocks; pollen grains collected from the intake area can sometimes be trapped in the aquifers, especially in fine grained ones in which the water is filtered, thus moving pollen down the section, into older strata.

The last, but not least factor, which seems to be peculiar to arid lands, is the occasional occurrences of large reservoirs of groundwater, which had accumulated in deserts such as the Sahara, Sinai and southern Israel long ago, in times of a different climate regime (Sonntag et al., 1978; Gat and Galai, 1982; Horowitz and Gat, 1984). Such fossil waters, when discharged by springs at the present day, may carry pollen which had been incorporated during their intake time. Reference to the present day is just an example, while dealing with the recent sediments, but such processes of transportation by groundwater could have practically taken place at any time.

These processes are of considerable significance in dry regions, by incorporation of exotic pollen in spectra, for several reasons: it is quite common for groundwater, emerging as springs in arid regions, to be incorporated in the aquifers at an intake area in which rains are more abundant; highly porous aquifers, especially sandstones and pebbles, are quite common in dry lands; rocks which tend to be more susceptible to dissolution processes by groundwater, such as rocksalt or gypsum, are again more abundant in the arid regions; and the phenomenon of fossil groundwater, which had accumulated during a previous, humid period, seems to be peculiar to some of the deserts. In addition, the local vegetation low masking factor makes the exotic grains quite important quantitatively. Since it is a frequent practice to sample spring deposits or closely associated sediments for palynological studies, groundwater transportation should not be overlooked.

In order to follow pollen transportation processes by groundwater in Israel, six water sources, located inside and around Lake Kinneret, northeastern Israel (Fig. 2.2.3), have been sampled and palynologically analyzed (Horowitz 1970, 1979). About 100 liters of water were processed from each of these, taken directly from the source head and pollen extracted by using a continuous flow centrifuge. Five of the sources are hot and saline, while the sixth discharges freshwater and was taken for reference. The aim of the study was to check whether groundwater, emerging as springs, significantly contribute to recent pollen deposition in Lake Kinneret; and to see whether the saline solutions, incorporated in these springs, carry pollen grains which could be detected in the water, which may give clue to the source of salinity. It was also thought that the age of the rocks from which the saline solutions originate could be suggested by the pollen spectra of the spring water.

Results of these analyzes are summarized in Table 2.2.8, as percentages of the various constituents of the spectra, based on the total sum of counted pollen and spores. The spectra have been divided into three groups: arboreal pollen, non-arboreal and fern spores. The prevailing arboreal pollen, except in samples from the most saline Roman Spring and Kinneret 5, are *Pinus halepensis* and *Eucalyptus*. This is quite different from the results obtained by analysis of recent sediments from Lake Kinneret (Horowitz, 1969), where oak pollen predominate. The main reason seems to be that pine and eucalyptus were extensively planted in the last years, while the recent samples from Lake Kinneret constitute the

TABLE 2.2.8

Pollen spectra from saline sources around Lake Kinneret, Israel, compared with a freshwater well (1018/1) in the same area (samples comprise 100 liters of water; total counted = 100%)

Pollen of	Source					
	Roman Spring	Kinneret 5 well	1020/6 well	En Nur Spring	1021/1 well	1018/1 well
Salinity (mg Cl/l):	18,200	5,930	3,150	3,000	3,000	130
Quercus spp.	8	–	4	3	8	3
Olea europaea	2	5	5	–	–	5
Pistacia spp.	2	–	2	–	–	2
Ceratonia siliqua	2	–	2	–	–	1
Pinus halepensis	2	5	10	14	34	47
Introduced trees	6	–	17	23	16	8
Other AP	–	5	2	7	16	6
Total AP	22	15	42	47	64	72
Gram. + Cyp.	57	27	23	14	24	6
Chenopodiaceae	–	–	10	24	4	7
Compositae	4	–	7	–	–	2
Plantago spp.	2	9	5	7	–	1
Ephedra spp.	2	5	–	–	4	–
Artemisia spp.	2	–	9	–	–	2
Other NAP	1	–	4	8	4	10
Fern spores	10	44	?	–	–	–
Total NAP	78	86	58	53	36	28
AP/NAP Ratio	0.28	0.17	0.72	0.89	1.78	2.57
Total counted	135	180	97	126	140	200

For locations please refer to Fig. 2.2.3. Reprinted with permission from Horowitz (1979).

uppermost several centimeters of mud, representing approximately the last 100 years, thus pollen produced by the natural vegetation greatly dilutes those from the recently introduced plants. The rest of the arboreal pollen extracted from the springs' waters display shares generally similar to spectra recovered from the recent sediments.

Centrospermae, *Plantago* and Gramineae are much more abundant within the non-arboreal pollen group in the spring waters, as compared with their shares in the pollen spectra obtained from the recent sediments. In the two samples collected from the Roman Spring and Kinneret 5, some fern spores were also found. These were not identified taxonomically, but were seen to be identical to some of the taxa recovered from rocksalt of Pliocene age of the Sedom formation (Horowitz and Zak, 1968) and from rocks of the coastal plain of Israel, assigned to the early through middle Pliocene Pa palynozone (Horowitz, 1974a; Horowitz and Derin, 1987).

It is especially noteworthy to observe that arboreal/non-arboreal pollen ratios increase as salinity of the source decreases. In the freshwater well, 1018/1, AP/NAP ratio is as high as 2.47, whereas in the most saline, Roman Spring, it only amounts to 0.31.

Pollen and spores found in the water samples could be subdivided into three categories, by their stratigraphic characteristics: pollen produced by plants growing in the region only today and especially introduced trees; those which are exclusively fossil, of Pliocene age; and taxa occurring in both recent and Pliocene sediments. It is very difficult to distinguish between pollen grains of similar taxa of the latter group, so these could not be taken as stratigraphic guides.

The first group includes pollen of *Eucalyptus*, *Casuarina* and orange tree (*Citrus aurantium*), which have been introduced to the country in the last decades. A considerable part of the pine pollen also seems to belong to this group. Only the fern spores can be attributed to the second group with certainty, since no fern producing such spores is presently known from the Near East.

The influence of mixing recently circulating groundwater with the fossil, saline solutions (by the dissolution of a buried salt bed, as possibly is the case), seemingly of Pliocene age, is quite conspicuous in the arboreal/non-arboreal ratios in the pollen spectra obtained from the water samples. In the freshwater spring this ratio (2.47) is much higher than in the recent sediments of Lake Kinneret, around 0.4. The main reason for this is, most probably, because the water intake area of the aquifer is in the hilly regions of the Galilee, which is quite well forested by both natural and introduced trees. AP/NAP ratios of pollen spectra from the saline springs, which decrease with increasing salinity, indicate that some of the NAP are contributed by the source of salinity, which is probably rich in pollen of Centrospermae and Gramineae. This, indeed, is typical for pollen assemblages recovered from the Pliocene evaporites of Pa palynozone (Horowitz and Horowitz, 1985), where AP/NAP ratios are in the order of 0.25 or less.

The study of the springs around Lake Kinneret clearly showed that the present day pollen assemblages, incorporated in the groundwater, only represent the vegetation of the intake area, which may be quite different from the regional average or the discharge location. The results also indicate that older rocks, in this case Pliocene rocksalt, can be dissolved and their pollen brought up the section, to be incorporated in much younger sediments. A verification of the assumption that the pollen spectra of the saline springs represent a buried salt bed came 15 years after this study was made, when a borehole drilled just south of Lake Kinneret, Zemah 1, hit rocksalt of Pliocene age at depth of 1350–1730 m (Marcus and Slager, 1985).

Since springs contribute much of the water to certain arid land water bodies, these factors should be taken into consideration. The best way, of course, for studying recent processes is to analyze water of the spring itself, to define its contribution to pollen spectra. Analyzes of spring waters from the Arava, which is the sector of the Dead Sea rift to the east of the Negev desert, presently arid to extremely-arid, yielded pollen spectra with oak, pistachio, olive and other trees which do not grow in the region today (unpublished personal observation, 1980). These waters had been dated by radiocarbon to some 12,000 to 15,000 years ago, which proved their fossil origin. Such pollen are not known from recent sediments of this area not intimately connected with springs (Horowitz, 1976), which indicates that the pollen recovered from springs were being brought by groundwater, up the section.

Such cases may have also occurred in the past, possibly recorded in travertines from the northern Arava. These yielded uranium series ages corresponding to dry periods in Israel (Livnat and Kronfeld, 1985), but pollen spectra (Weinstein-Evron, 1987a) rich in arboreal constituents. These arboreal pollen, if assumed autochthonous, may have been included within the travertines either due to a better developed local vegetation, maintained by the springwater, or from regional trees, growing in a more humid climate. The AP could also be of allochthonous provenance, either temporal such older grains carried by paleowaters, or spatial, by the springs having their intake area in a more humid region. The springs are situated on a faultline which was active at that time (Horowitz, 1987b), thus may have exposed to the surface an aquifer containing fossil water, a situation also known for this region at present (Gat and Galai, 1982).

Spring deposits, when not contaminated by allochthonous components, mostly contain pollen grains originating from the nearby vegetation, rather than the regional, as is indicated in the Sahara (Ballais and Cohen, 1985). Scott (1975), in an analysis of pan sediments from Alexandersfontein (near Kimberley, South Africa), noted prevalence of local constituents in pollen spectra obtained from spring eye deposits, as opposed to concurrent sediments collected adjacent to the spring eye. Such differences have subsequently been found in other localities in the Orange Free State (Scott, 1988) and were also noted (Scott, 1989) between pollen contents of contemporaneous spring deposits and coprolites collected nearby.

A possible indication to pollen transportation by groundwater down the sequence is given in Horowitz (1975). The study involved pollen analysis of the Middle Breccia from Sterkfontein, west of Johannesburg, South Africa, a site which has yielded remains of some of the earliest hominids. The processed samples comprised breccias made of Precambrian dolomite fragments, interbedded within red clays and possibly laterites, cemented by carbonate travertine into a hard rock. The pollen spectrum, of about 400 grains, was divided into two groups on the basis of the state of preservation of the grains: the first assemblage comprised heavily corroded and carbonized palynomorphs, while the second included only slightly corroded, hardly carbonized grains. A pollen assemblage from recent sediments, taken for comparison, was much better preserved than the two obtained from the breccia.

A suggestion was made, that the badly preserved, assumingly older material, represents the environment at the time of accumulation of the cave sediments, bones and artifacts; whereas the second, better preserved material, was brought into the sediments by percolating water at a later stage, a process that may have been responsible also for the travertine deposition and subsequent cementation.

The badly preserved part of the pollen spectrum obtained from the Middle Breccia of Sterkfontein indicates an open grassland, an environment slightly drier than the present day. Vrba (1974) had analyzed mammal bones from the same horizon and arrived at similar environmental implications, which seems to justify the separation of the pollen spectrum into the two groups. The better preserved part of the pollen spectrum indicates a humid climate, more than at present. No age could be attributed to the better preserved sector since no continuous sequences are known for this region, but due to its essential difference from the recent one, it was suggested that the groundwater activity took place some time in the past, during a humid phase, between the formation of the breccia and the present day.

The conclusions arrived at by the study of the Middle Breccia from Sterkfontein and the possible contribution of groundwater to the pollen spectrum, are by no way sound as yet and are contradicted by other scientists (see Scott and Bonnefille, 1986). They are brought here only to show the possibilities involved with pollen transportation by groundwater.

Rainwater can carry pollen down the section, already in the first stage of downward infiltration, when percolating commences through the soil, sands, porous or fractured rocks. It was found out that in the Middle East, some types of pollen are carried (or can withstand destruction) in this way better than others, such as those produced by Compositae and *Pinus* (Bottema, 1975; Weinstein-Evron, 1986). The infiltration of such pollen was particularly noted in sediments connected with prehistoric sites in the Levant, but there is no reason why similar phenomena could not occur in other types of sediments.

2.2.5. Animals

Animals can transport pollen grains both temporally and spatially. Some, especially several genera of bees and wasps do so by collecting pollen from flowers and depositing them in their burrowings, as food for their larvae. Other may do it by their food reservoirs or other activities, not necessarily limited to pollen grains, such as the packrat (*Neotoma* spp.) middens; feces, which become coprolites, of both herbivores like *Hyrax*, or carnivores, such as *Hyaena*; or pellets of birds of prey such as the barn owl, particularly those which feed on rodents.

Burrowing animals are very abundant in arid lands, since this is one of the best ways of behavioral adaptation to the extreme environmental conditions on the surface (Fig. 2.2.14). The subsoil is cooler and more humid than the outside air and serves as shelter for many animals in the deserts. For the same reasons, incidentally, so many desert animals are nocturnal. The burrowing animals of arid lands represent almost all phylla of the animal kingdom, from worms to mammals.

Fig. 2.2.14. Burrowing striped ground squirrel (*Xerus erythropus*), Kalahari Desert, Botswana.

Some of them make their own burrows, while others occupy holes made by a different animal, but the common thing to almost all of them, is that they transport pollen up and down the section. Some mixing of pollen grains can occur in soils by activity of earthworms (Walch et al., 1970). This is quite rare in deserts, but may occur in semi-arid regions, such as parts of Israel, where earthworms are rather common. Similarly, mole rats cause soil mixing in semi-arid lands where they are quite common, which may affect pollen as well (Weinstein-Evron, 1986).

Transporting pollen up the section is primarily done by the burrowing activity itself, when subsoil is brought to the surface and scattered among the more

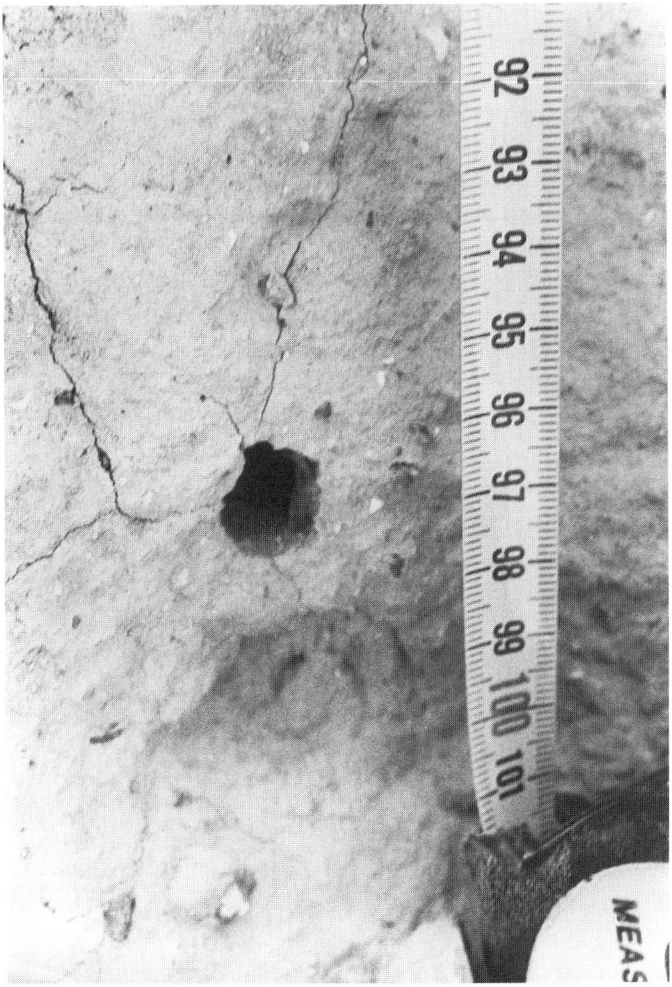

Fig. 2.2.15. Wasps burrows in an archeological site, reaching almost a meter below the surface. Tell Ira, northern Negev, Israel.

recent sediment. In most cases analyzed in Israel, this way of transport is almost meaningless concerning amounts of pollen included in the younger sediments, mainly due to the rather rapid oxidation of the dug materials. However, attention should be given to signs of burrowing, while collecting samples.

The more meaningful animal activity concerns pollen transported down the sequence, especially by those insects which primarily collect pollen for food, such as bees, some kinds of wasps (Bottema, 1975) and ants (Fig. 2.2.15). These make reservoirs of pollen in the subsoil, which may cause quite a lot of trouble when not detected in time. The burrowings can go quite deep into the subsoil, where the animals seek humidity and a depth in the order of one meter or more is not exceptional even for insects, worms, isopods and some other animals. Larger reptiles and mammals, such as *Varanus*, or porcupines, can even dig much deeper holes, which may in some cases become traps to pollen transported by winds, especially if they face the prominent wind direction.

While it seems easy and straightforward to discern burrowing activity of the present day, it may turn quite difficult when sampling older sediments, where the hole had already been filled up long ago. Such a hole is usually filled up with sediment from the close vicinity and the similarity in material makes it sometimes very difficult to detect. This problem becomes more acute when samples from drillings are taken for pollen analysis. It was also found out that such burrowing activity may especially become problematic when archeologic, or prehistoric sites are being sampled, thus special care should be taken while trying to collect from such localities.

It appears quite rare to find any burrowing activity in lakes or other water bodies situated in arid lands, because most of these are, to some degree, saline and their bottom fauna is usually very poor. An exception may occasionally be observed in lagoons or playas directly connected with the sea, where burrowing animals frequently dwell. Burrowing animals can transport pollen up and down the section, but not over any horizontal distances, because the areal distribution and radius of activity of the individual animals is quite restricted.

The only possible way of long distance transportation of pollen by animals, was found to be connected with the pellets and excrements of birds of prey (Fig. 2.2.16). Many of these are migratory and may cover distances of several hundred kilometers in a single day, during the migration season. Although they do not feed directly on pollen or plants, many of them do feed on rodents or other animals, which eat quite a lot of vegetal material, including the pollen grains. When the remains of these animals are excreted by the bird, either as pellets or as excrements, they were found to contain considerable amounts of pollen originating from the region in which the prey was living prior to migration. This process particularly occurs during the spring migration, when most plants are in bloom. Such pellets and feces are not a common contaminant of most types of deposits in the arid lands, but they can contaminate sediments connected with ruins or rockshelters, in which the birds rest for the night, or sometimes the day, depending on the particular time of activity of a certain bird. Such localities may have intimate connections with archeologic and prehistoric sites, or places in which sediments are being trapped — caves for instance. It is sometimes very difficult to differentiate pellets, especially coprolites of migratory birds, from those produced by local birds of prey, which are a very good source of information for the regional, arid land vegetation, again through their prey.

It may occasionally be possible to discern such transportation by analyzing not only the pollen grains from pellets, but also rodents or other prey bones, which

Fig. 2.2.16. Pellets of barn owl (*Tito alba*). Courtesy of E. Tchernov.

may give some clue to the primary environment. However, this seems to be quite a tricky situation, because it may happen that the paleoenvironment is figured out from pollen grains and strengthening came by analysis of the bones in the pellet, or vice versa, while both are in fact allochthonous.

Some pollen transportation, but apparently never to a great distance, is by actions connected with feeding and other activities of animals and men. Both herbivores and predators consume pollen, adhered to their food, which is not digested in the predominantly acid environment of the intestinal system and reappear in their feces. These feces, while fossilized and turned to coprolites (Fig. 2.2.17), may therefore serve as indication of the local vegetation and also of the diets of those animals or men. Such studies are mostly connected with archeologic and prehistoric sites, or caves and rockshelters (Fig. 2.2.18), localities in which coprolites can easily be found and identified as such. O'Rourke and Mead (1985) concluded, while studying pollen from two caves in the Grand Canyon, that "The major input of pollen to Grand Canyon caves comes from animal dung and from the local and regional vegetation".

Pollen have been found in coprolites from deserts in many localities, such as the Sahara (Pons and Quezel, 1958), South Africa (Scott, 1987; Scott and Bousman,

Fig. 2.2.17. Fossil hyena coprolites, Equus Cave, South Africa. Courtesy of L. Scott.

Fig. 2.2.18. Hyrax midden, in a rockshelter near Clarence, Orange Free State, South Africa. L. Scott, who kindly took me there, is posing for scale.

TABLE 2.2.9

Comparison of pollen spectra from synchronous coprolites at Equus Cave and sediments from nearby pan, Taung, southern Kalahari (total counted = 100%)

Pollen of	Sample from					
	Coprolites			Pan floor, spring area	South Section	
	A	B	C		Shore veg.	Karoid veg.
AP						
Pinus spp.	–	–	–	–	–	2
Myrtaceae	–	–	–	2	2	2
Podocarpus spp.	–	–	1	–	–	–
Olea spp.	–	1	–	–	1	2
NAP						
Gramineae	76	67	66	9	20	4
Cyperaceae	2	8	3	73	–	3
Anthospermum spp.	9	4	3	1	2	–
Trilete spores	1	–	1	–	–	–
Compositae	3	11	9	6	5	60
Cheno-Ams	1	3	9	6	67	18
Aizoaceae	3	1	3	1	–	6
Other NAP	5	5	5	2	3	3
Total counted	250	250	250	250	250	245

Data from Scott (1987).

1990), the Near East (Darmon, 1989; Fall et al., 1990), Patagonia (Markgraf, 1985) and other places. It was found out that spectra recovered from coprolites are not identical to those obtained from contemporary sediments, at the same locality. This is because pollen in coprolites represent, except for the environment, dietary and other various habits of the animals or man. Scott (1987) demonstrated differences in pollen spectra obtained from modern surface samples and subrecent coprolites from a salt-pan in the Orange Free State grassland (Table 2.2.9).

Several studies, made on human coprolites from the American Southwest (Bryant, 1974a, b, 1975; Williams-Dean and Bryant, 1975), served to assess the diets of prehistoric men in this region. Trevor-Deutsch and Bryant (1978) made a similar study on suspected human coprolites from Terra Amata Paleolithic site in Nice, southern France. Unfortunately, in this case, the presumed coprolites proved to be an unreliable source of information for the diet.

Coprolites, although containing pollen eaten by the animal, are exposed to the air and, being a rather sticky material in their fresh stage, make an excellent trap for pollen grains travelling by winds, at the same time. However, the differences seen between pollen spectra recovered from coprolites and those obtained from synchronous sediments, seem to indicate that coprolites should be better used to study diets, rather than regional vegetation or environment. This is stressed by

Moe (1983), who studied pollen contained within sheep's feces in Norway and concluded (p. 105): "Pollen analysis of feces provide virtually no information about the quantitative composition of the vegetation of the sheep's environment, but do provide valuable information about their diet" Caulton and Gibson (1988) who analyzed camel feces from Kuwait, among other localities, fully agree with Moe's conclusions.

In contrast, Thompson et al. (1980), while analyzing pollen from coprolites of the Shasta ground sloth and comparing the results with vegetal remains in the same material, concluded that spectra from the coprolites have high representations of anemophilous plants and low representations of the dietary items shown in the cuticle analysis.

Most researchers who studied coprolites do not fully agree with Moe's conclusion, thus paleoenvironmental studies are being conducted on fossil feces (Martin et al., 1961; D'Antoni, 1978; Scott and Klein, 1981; Scott, 1987, 1990; Darmon, 1989). There may well be a difference between the environmental signal recorded by various animals' feces, however this was not studied in detail, since for such a study coprolites of different animals from the same locality are needed. While this can easily be found with the recent material, it is rarely the case for the past.

A rather similar controversy can be seen when middens are concerned. These are made of vegetal material collected by various mammals and mixed with their excretions, usually urine, such as the packrat in the American Southwest (Fig. 2.2.19), the rock wallaby in Australia and some mammals in the old world. The vegetal material represents the immediate vicinity and the great advantage of the middens is, that one can study both the vegetation and pollen grains from the same sample. It seems, however, that the pollen grains recovered from middens represent both the collecting activity of the animals and the pollen load of winds, deposited on the sticky surfaces of these middens. The stickyness is chiefly due to the urine, apparently secreted by the animals to define territories. The middens also have the advantage of being easily dated by radiocarbon. A detailed discussion of the significance of packrat middens from the American Southwest and hyrax from South Africa and the Near East, is given by Betancourt et al. (1990). It should however be stressed that Davis (1990) found out that (p. 331) "Pollen percentages in cave sediment differ from those in packrat middens due to additional transport mechanisms (on plant tissues and on the packrats themselves) for middens".

Thompson (1985) found out that pollen spectra from packrat middens exhibit little variability within a midden, though samples from different middens of the same age and even from the same site, may have disparate palynological assemblages. A detailed account of pollen transport by packrat and its relation to the vegetation is given by Davis and Anderson (1987) for the American Southwest, whose conclusions are somewhat opposing to Thompson's. They indicated that variability among and within middens is high and the accuracy with which midden pollen reflects vegetation is lessened by this variability. Also, that small size

Fig. 2.2.19. Packrat midden dated at 13,400 ± 250 yr BP (A-1357) from the Kofa Mountains, southwest Arizona, USA. The conifer needles are from the pinyon pine taxon that was widespread in the Sonoran Desert during the last glacial, but now has a limited distribution. Courtesy of Thomas R. Van Devender. Scale bar 1 cm.

samples are inadequate for drawing conclusions. They maintain that a sample size of approximately one kilogram should much better reflect the vegetation, by overcoming most of the variabilities. Van Devender (1988), in a discussion of Davis and Anderson's paper, prefers subsampling of the large sample, which should be divided into 20–30 small ones; he also states that analysis of midden pollen to the exclusion of other data provides inadequate and misleading paleovegetation records, thus should not stand alone as a new analytical method, suggesting that always, pollen analysis of packrat middens should be accompanied by studies of the vegetal remains and pollen from other contemporaneous deposits.

It seems that, concerning these opinions, only further study would better establish the relations of pollen assemblages obtained from packrat middens with the environment and transport. It is interesting to note that this discussion is centered on packrat, rather than other animals middens, most probably because the former was subject to a greater number of studies. This, naturally, increased the amount of information and also exposed the differences in ways of thinking of the many researchers.

The hyrax, known from the Near East and Africa, has a somewhat different strategy for creating middens, by a peculiar behavior, whence a group of these rock dwellers always excrete at a certain place for a long period of time. The accumulations of sticky excretions are quite large (Fig. 2.2.18) and make an excellent trap for windborne pollen. Thus, although grains are undoubtedly also added by the food, studies by Scott (1990) and Scott and Bousman (1990), on hyrax and dassie-rat middens from South Africa seem to indicate a marked similarity between pollen spectra recovered from these and assemblages from synchronous sediments. These middens, therefore, act more like pollen traps for grains carried by the winds and less for those transported by the animals, which is in contrast to the cases presented by coprolites and packrat middens. Fall et al. (1990), while analyzing fossil hyrax middens from southern Jordan, could conclude on both records of paleovegetation and of human disturbance from the pollen spectra, by separating the natural from anthropogenic components.

2.3. Pollen deposition, preservation and differential destruction

Pollen deposition in arid lands follows the same rules as elsewhere and behaves similarly to other particles deposited from the transporting agents. The main difference from the mineral particles, namely the better adherence of pollen to the water and air, which delays its deposition in comparison with the mineral matter, is again not peculiar to dry lands. One of the principal pollen deposition agents in the deserts is, not surprisingly, the rare rains. These help to sweep the air clean of its pollen contents, along with the other finest dust particles, which makes the ensuing muds one of the best pollen traps of the arid regions. This process of deposition may have a single or multiple stages, when secondary transport and deposition further downstream is caused by floods, which accompany the rains in the desert quite frequently.

Processes of pollen deposition in water bodies in and around a desert, such as lakes, marshes or a surrounding sea, are again not peculiar to the dry land. Here too, the better adhesion makes pollen travel farther and be deposited at a slower rate as compared with most of the mineral matter.

Different types of pollen may be subject to variability in both the transportation distances and the time necessary for settling down. Those grains which are not easily carried by the transporting agents are naturally the ones laid down first. Thus, a study of the settling times of different grains can yield results which may inflict on the distance of the parent plants from the locality of deposition. Since filtration by dense vegetation, known for the temperate lands (Tauber, 1967), plays a very little role in arid regions, such studies could indeed be useful. Pollen grains of *Acacia*, for example, travel very little and are to be found in any significant numbers only very close to the growing localities (Rossignol, 1969a; Horowitz, 1976, 1977a; Weinstein, 1979; Daoud, 1984).

Preservation of pollen in the sediments, unless totally destructed (or entirely preserved, which is very rare, if at all) is always a function of differential destruction. Destruction of pollen grains and other associated microfossils takes place in several stages, starting from the minute they are produced by parent plants and exposed to various transporting agents, ending only when they finally find their way to the microscope slide. Most of these processes occur anywhere and are not peculiar to arid lands. However, some of these are more pronounced in the dry regions, due to the different composition of the sediments and their cements, mainly the predominantly clastic nature, high pH and Eh (oxidation potential) and low to extremely low organic matter contents.

The first stage involves palynomorphs which are destructed, mostly by oxidation, almost as soon as they are outside their parents' domain. These delicate grains almost never form a part of any pollen spectrum, not even airborne, thus in practice are of no interest to the palynologist.

The second stage affects those palynomorphs which are destructed in much the same way, while still in the air or water, but are less prone to the process, which thus takes a longer time. This group of pollen is of importance for arid lands palynology, due to the longer transportation and deposition times normally involved in this environment, as compared with the forested regions. The longer time thus necessary involves inclusion of more pollen types in this group, making the airborne pollen quite different to the one deposited on the surface, both in the number of individual grains and in their species diversity. Cour et al. (1973) and Cour and Duzer (1976) demonstrated, in samples collected from the Sahara, that the number of pollen per square centimeter of a trap placed perpendicular to the wind is up to four times larger than, at the same time and locality, the number caught by horizontal traps. Percentages of the various constituents also vary between the two positions of the traps, indicating the essential difference between the airborne and the surface pollen spectra. The results of this study are summarized in Table 2.3.1.

Thus, in the tropical and temperate regions much can be learned from airborne pollen spectra on the possible significance of fossil ones, while in dry lands this method may sometimes be misleading, has to be considered much more carefully and preferably replaced with study of the recent or subrecent sediments.

The third stage of destruction involves palynomorphs which are finally deposited, but cannot withstand the surface conditions for a long time. Many of these types may be present in recent sediments for a while (Weinstein-Evron, 1986), the length

TABLE 2.3.1

Airborne pollen (%) flux in Beni Abbes (NW Sahara), April 1968, comparing trapping on vertical and on horizontal filters

Pollen of	Vertical filter	Horizontal filter
Taxa present in the Sahara		
Cheno-Ams	33.12	26.95
Gramineae	11.80	14.58
Cyperaceae	1.78	1.92
Compositae	5.35	8.71
Cruciferae	3.00	4.48
Plantago spp.	3.72	4.92
Ephedra spp.	1.70	2.71
Tamarix spp.	0.22	0.18
Phoenix dactilifera	0.06	0.35
Acacia spp.	0.02	0.04
Others	2.84	4.87
Total Saharan	64.03	70.51
Taxa of Mediterranean provenance		
Cupressaceae	5.30	2.93
Oleaceae	4.64	5.35
Quercoides	2.49	5.68
Abietinae	0.75	0.82
Umbelliferae	1.45	1.84
Myrtaceae	0.66	0.04
Artemisia spp. (Saharan in part)	11.68	6.30
Castanea sativa	0.69	0.07
Others	0.45	0.74
Total Mediterranean	28.12	23.39
Taxa of tropical provenance		
Sapotaceae	0.26	0.15
Combretaceae	0.04	0.01
Total tropical	0.30	0.16
Undetermined	7.46	5.40
Total counted (same filter area)	21,260	9,459

Data from Cour et al. (1973).

of time depending on conditions such as oxidation, humidity, bacterial activity, pH, rates of deposition or how long it takes for the topmost layer to be buried and other factors. The common result is, that in a given region different recent deposits would yield different pollen spectra. As an example, the topmost muds of the Dead Sea yielded (Rossignol, 1969a) 92 taxa, while recent wadi sediments not far away (Weinstein-Evron, 1987a) only contained 18 identifiable types of pollen.

Again, in forested regions this phenomenon is of a lesser importance, chiefly because the depositional domains are not as variable as in deserts and destruction agents are less active. Also, usually one would analyze a sequence collected from a basin which has a recent counterpart. This is not necessarily the case in arid lands and sometimes impossible to do, for example, when extinct lakes' sediments are being studied in a desert, in which no lakes exist at present; or, when samples were analyzed from prehistoric sites in Israel, results had to be compared with recent pollen spectra taken from a variety of environments, because none could be paralleled exactly to the site (Horowitz, 1979). The only way to deal with the problem is then, to collect reference samples from as many environments of deposition as possible, preferably such that seem similar to the fossil ones being studied.

The fourth stage of differential destruction occurs in the sediments, following their burial. This could be divided into two substages, the first which involves destruction before compaction and the second, which occurs during and after lithification and diagenesis. Unfortunately, not much is known about these destruction processes, although some experimental work was done on this subject.

The first substage involves physical processes, such as temperature and humidity changes; chemical reactions, mostly oxidation; and deterioration through biological activity, principally carried out by bacteria and fungi. Holloway (1989) made a series of experiments with pollen grains and subjected them to both temperature and humidity changes. It was found out that temperature changes, in the range common to temperate North America, had virtually no effect on the pollen preservation; while wetting and drying the pollen caused alteration of the exines, sometimes even up to a severe damage. This process should be regarded carefully with pollen collected in arid lands, which are almost always exposed, before final burial, to rather extreme humidity fluctuations. Holloway further brings several examples of pollen destruction in various sites and explains the loss of pollen as a result of wet and dry periods which had affected the analyzed sequences.

Chemical processes which cause deterioration of palynomorphs are mostly attributed to oxidation. Experiments done by Havinga (1984; and earlier publications), Erdtman (1971) and others on palynomorphs preservation in various soils, indicate that the grains are subject to differential destruction, depending on both the character of the soil and the type of palynomorph. A study of lake sediments from Minnesota (Cushing, 1967) also showed differential destruction of palynomorphs in various types of deposits. Bacterial and fungal activity was found to be responsible for degradation of pollen and spores (Elsik, 1966a, b). Unfortunately, little was done in this respect concerning desert sediments, but indications from preliminary studies (Hall, 1981; Holloway, 1981a, b; Bryant and Holloway, 1988) are along similar lines.

The second substage also involves both mechanical and chemical processes. Among the mechanical, probably the most common is the flattening of palynomorphs by the compaction during burial, which becomes more and more progressive as the sediment is buried deeper. This flattening usually affects all palynomorphs in a similar fashion, but subsequently some types are more difficult to identify that others, particularly those prone to crushing, which makes for the same effect as differential preservation, in terms of the pollen analyst. Compaction can become serious and even destroy completely the palynomorphs, throughout the stages of regional metamorphism. Another mechanical damage, which may even be more common than compaction, especially with relatively young arid lands deposits, is caused by crystal growth in and around the grains, throughout the lithification processes. Carbonates, anhydrite and gypsum are the more common minerals which may grow inside palynomorphs in the dry lands. Pyrite is a common mineral in sediments deposited in reductive environments, well known to palynologists in the temperate regions, but seems subordinate in drier domains.

A common chemical process which can deteriorate, or even completely destroy palynomorphs is carbonization, which can gradually wipe off all their morphologic characteristics. This process takes a long time and requires high temperatures usually only achieved through deep burial, except where deposits are intruded by magmas, or are situated close to subsequent intrusions, as was found in some of the sequences penetrated by boreholes drilled in the Dead Sea Rift fill (Bein and Feinstein, 1988). Some chemical changes can also affect palynomorphs during metasomatosis, by the action of groundwater or by contact with hydrocarbons.

The fifth stage of possible differential destruction does not represent any natural causes, but rather the chemical and physical processes involved in macerating the samples, for the extraction of palynomorphs. Not all palynomorphs are resistant in a similar way to preparation techniques and some are destroyed more easily then others. This again is not peculiar to the dry regions, but may be of much more significance there, due to the usually initial bad state of preservation of palynomorphs in these environments. The state of preservation is a function, among other things, of the time involved in the transportation and deposition processes, which is longer in the desert environments; the type of sediments, being mostly clastics, more often than not quite coarse; and the chemical environments, which are usually of a rather high pH and oxidation potential. All these cause a much greater susceptibility to destruction during the maceration. It is not always possible to overcome this phase and sometimes impossible, thus the only way seems to treat the samples as little as possible, both chemically and mechanically (see details in Chapter 5) and to try collecting, macerating and comparing synchronous samples from different depositional domains. Sometimes, several different treatments should be attempted with a certain sample and may indicate which is the better way to avoid damage with a particular type of deposit in a given environment.

A discussion of differential pollen preservation in two types of marine sediments from the Mediterranean, north of the Sahara coast, is given in Cheddadi (1988), who concluded that grains are much better preserved in black, highly organic sapropels, than in yellow-buff, oxidized silts. This conclusion is based, besides the better state of preservation of the palynomorphs ornamentation, on the much higher frequency of pollen and spores in the sapropels, sometimes by three orders of magnitude. This may as well be a result of different rates of deposition and does not necessarily apply to preservation. Cheddadi further states that vesiculate gymnosperms pollen, together with spores, are more resistant to destruction in the oxidized yellow-buff silts, as compared with other palynomorphs, which should explain their higher percentages in these deposits. While this may be true, it is very difficult to prove beyond doubt that the differences in spectra do not mark a veritable floral change, thus the paleoclimatic implications may consequently prove doubtful.

While analyzing borehole sequences from the Dead Sea region, in southern Israel, it was found out (Horowitz, 1979, 1987a) that many of the species reported for recent sediments (Rossignol, 1969a) of the Dead Sea do not appear in the borehole samples. Two possible explanations were suggested, that either the poor state of preservation does not permit detailed identifications, as could be made for the recent sediments; or that some of the pollen taxa have been completely, or at least partially, destroyed while being buried, or following burial. It was found out that for the Quaternary of the Dead Sea Rift, oak and some non-arboreal pollen were the least affected by burial and their ratios could therefore be used as a paleoclimatic indicator. It is however clear that this would not do as a general rule, thus for every region and type of deposits the more resistant palynomorphs should be studied, realizing that some information may have been lost by the differential pollen destruction. Another approach (Bottema, 1975) calls for applying correction factors for over-represented pollen, such as Liguliflorae in prehistoric sites of southern Europe.

Chapter 3

Sediment–pollen relations in arid lands

Weathering, transportation and deposition processes of pollen and sediments in deserts are in most cases entirely different than in other environments on Earth. In this respect, there is practically very little difference between the hot (Fig. 3.1) and cold (Fig. 3.2) deserts, which consequently result in quite similar landscapes.

Weathering processes disintegrate the parent rocks, creating sediments available for subsequent transport and deposition. In arid lands these processes mostly produce medium to coarse clastic sediments because, due to paucity in water, chemical and biological weathering are subordinate and mechanical disintegration is the major agent. This also affects pollen grains, which may have originally been embedded in those parent rocks, as they also may entirely disintegrate. In itself,

Fig. 3.1. Hot desert, central Negev, Israel.

Fig. 3.2. Cold desert, Tibetan plateau.

this is a welcome process to palynologists in arid lands, since it greatly reduces the chances of redeposition of older pollen in younger rocks. However, since these mechanical processes also affect rocks in whose pollen contents the palynologist may occasionally be interested, such phenomena may be viewed with mixed feelings. The mechanical disintegration also facilitates intrusion of air (oxygen) into the rocks, thus necessitating much more labor in digging deeper into outcrops while samples are sought for, to collect from unaffected rock.

Some, but certainly not all of the sediments formed by mechanical disintegration are medium or coarse grained clastics. When these are being transported by powerful floods or strong winds, they tend to crush and destroy pollen which are concurrently carried by those agents. When added to the initially poor vegetation and pollen production, the results are that, usually, the coarser fractions of desert sediments are very poor in pollen grains. This is somewhat different with finer fractions, which have a considerably lesser mechanical destructive capability, therefore frequently prospective for palynologists. Rates of deposition vary, but in most cases fine clastics accumulate quite slowly, thus giving time enough for the rare pollen to be enclosed in the deposits. These pollen are frequently quite well preserved, since the close packing of the fine clastics hinders, to a certain degree, oxidation processes so common in coarser clastics, such as sands and gravel which are much more permeable.

Chemical deposits, notably evaporites, are frequently formed in dry lands, in confined or temporary water bodies. The water surfaces make quite good pollen traps, while the usually closely packed evaporites help protect from subsequent oxidation. On the other hand, pollen in such deposits could be damaged by crystallization effects and internal crystals growth, which may cause mechanical damages to the trapped grains. It should be noted that these crystallization processes are quite rapid and may affect palynomorphs in a matter of weeks following deposition.

Deposits connected with anthropogenic and animal activities, such as archeologic accumulations, dung heaps, feces or midden, which may contain more organic compounds than others common to dry lands, seem to make very good pollen traps and preservation media. However, the origin of such deposits may cause initial preferences in their pollen contents, which should be carefully considered.

Marine sediments deposited adjacent to arid lands may prove a very good source of information about their neighboring regions. These are not essentially different from marine environments in other localities in terms of their sediments, with the exception of certain coastal features, but their pollen contents is affected by processes on land, which should be studied beforehand.

Sediment–pollen relations in arid lands are therefore a combined result of a variety of processes, some of which differ from those known for other parts of the world, notably the weathering, transportation and deposition of both sediments and pollen.

3.1. Weathering

Weathering processes of rocks in arid lands are mainly mechanical, caused by the extreme diurnal temperature and sometimes humidity changes, strong winds and floods. Added to that are subordinate chemical processes, aided by the scarce humidity brought about by rare rains, or by the more frequent dew in some deserts, especially the coastal ones (Glennie, 1970). Biologic weathering did not attract much attention in the arid lands, but lately more and more indications of considerable activity of lichen and blue-green algae are gathered (Danin et al., 1982, 1983). Crystal growth seems also to be responsible for some weathering of rocks cropping out in deserts, especially in extremely-arid zones (Wellman and Wilson, 1965; Amit, 1990). Oxidation processes have not drawn enough attention in the dry lands, mainly because most of the mineral matter is hardly affected by these. Organic materials, prone to oxidation, are never too common in the deserts, thus oxidation which is so cardinal for palynologists, was generally neglected by both geologists and geomorphologists. Consequently, many rocks collected in deserts which have been macerated for pollen, proved barren at first try. Only when digging deep into the outcrops, sometimes even several meters to arrive at rocks unaffected by oxidation, samples which contained rich pollen spectra had been recovered. This, incidentally, is true not only for rocks initially deposited under desert conditions, but for almost any rock sample collected from outcrops in the present day deserts.

3.1.1. Insolation

The widespread phenomenon of rock splitting (Fig. 3.1.1.1) has drawn much attention among arid land geologists and geomorphologists and consequently, bitter disputes. The two main schools explain the splitting by either being a result of rapid diurnal temperature changes, or by mechanical impact (or maybe both?). Glennie (1970) brings a short summary of the disputes and concludes that (p.16) "Insolation seems the most likely reason for the splitting of larger boulders and is also thought to play a part in the formation of much finer fragments".

Personally, I do tend to accept Glennie's postulation since, while staying overnight in the desert, one can actually hear the rocks cracking and splitting. This cannot have anything to do with mechanical impact, since it is heard also on flatlands, far from any rolling boulders which could have broken the rocks, particularly in nights of calm air, with no wind at all.

Be the reason what it may, the phenomenon is very common in all deserts of the world, hot and cold alike, where the ensuing rock fragments seem to form a considerable part of raw materials for the slopes, wadis and windborne sediments. The rock splitting, probably accompanied by crystal growth and some other microchemical and possibly biological processes, is also responsible for the exfoliation of rock surfaces (Fig. 3.1.1.2), which frequently ends with formation, in situ, of rounded gravels. These are easily carried further on by floods or gravitational creeping down slopes. It seems that in deserts, many if not the majority of rounded wadi gravels, are not formed by water polishing and transport, but as a result of exfoliation which takes place in situ, before the gravel is supplied, "ready made", to the wadis.

3.1.2. Winds

Winds are strong and frequent in the dry regions and considerably affect the landscape, especially so because there is hardly any vegetation cover to protect from their activity. The winds' erosion is manifested in creation of deflation surfaces (Fig. 3.1.2.1) and hollows, caused chiefly by the lifting up of finer clastics, usually accompanied by polishing action of the windborne sand grains, which attack exposed rock surfaces (Fig. 3.1.2.2). Winds may considerably affect archeologic sites in dry regions, which had usually been constructed on tells higher than the neighborhood, thus very much prone to the process (Fig. 3.1.2.3). Deflation

Fig. 3.1.1.1. A split pebble, probably due to temperature changes. Central Negev, Israel. Scale bar 2 cm.

Fig. 3.1.1.2. Exfoliation in granite, forming "pebbles" *in situ*. Namib Desert.

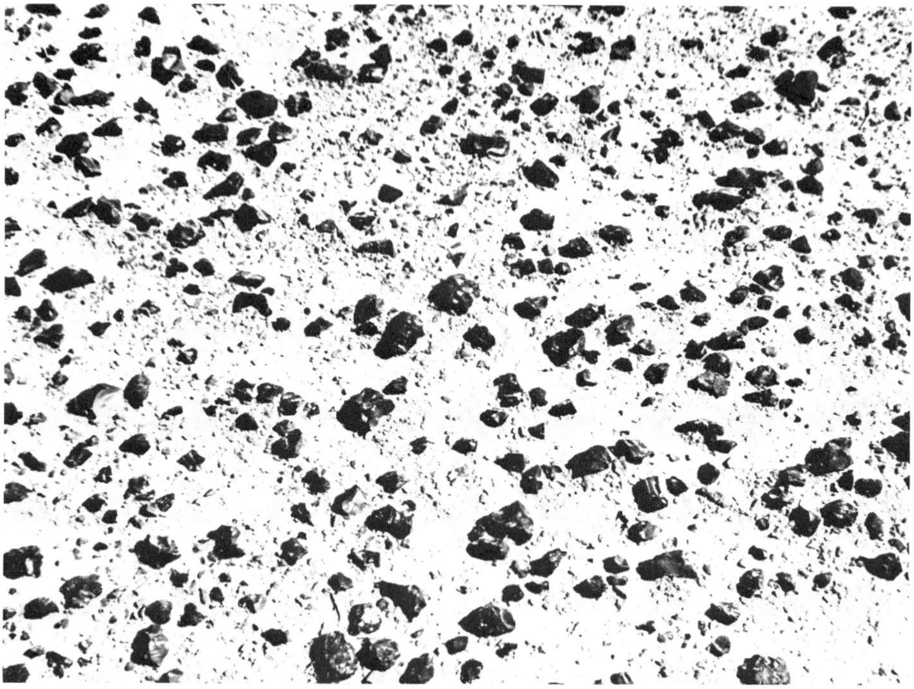

Fig. 3.1.2.1. Wind deflated surface, with flint implements of several Late Upper Paleolithic and Epipaleolithic cultures. Northern Sinai, Egypt.

Fig. 3.1.2.2. Wind polished pebble ("Dreikanter"), some 20 cm long. Namib Desert.

Fig. 3.1.2.3. Wind swept archeological site. Tell Ira, Negev, Israel.

supplies most of the finer clastics for loess and sand dunes, consequently laid down as windborne deposits. Sorting is a result of winds velocities, while deposition, of particularly the finer clastics, is usually a function of occurrences of suitable dust traps such as water bodies, vegetation cover, caves or rockshelters, or is caused by rains.

Deflation processes are important for palynological research, for two reasons: they expose to the surface older strata, thus in most cases, particularly on flat or higher areas the seemingly present day deposits may be quite older, which calls for attention while sampling; they can produce deflation hollows (Fig. 3.3.2.2) sometimes rather deep and many kilometers across. When wind or hydrologic regime changes, these can become depositional basins in which finer clastics accumulate, a process which is frequently aided by the outflow of groundwater into the deflation hollows. Such depressions are therefore prospective for obtaining sections suitable for pollen analysis, which would testify to the history of the region since the last major change in wind regime. Usually such sections are not too long, but nevertheless form an important source of information. It seems that in some regions the deflation hollows may turn from the erosional to the depositional phase following the exposure of groundwater, rather than the change in wind regime.

The groundwater maintain vegetation in most cases, if not too saline. Considerable salinity can also stop the deflation processes, at least to a certain degree, by the formation of durable evaporitic crusts such as rocksalt, gypsum, dolomite and the like. In many cases, prospective sections were found underlying the groundwater level at a certain stage. The deposits are occasionally also rich in organic material and in many instances provide suitable substances for radiocarbon dating.

The polishing action of winds depends on their load. The best polishing agent seems to be quartz sand grains, but since these are never travelling high, the polishing is particularly prominent at surface and near surface levels. This activity, besides polishing the rocks, also produces considerable amounts of very fine dust, which is added to the windborne deposits, in the same manner as dust originating by other processes. Polishing processes do not seem to have any direct bearing on palynological research, although they may become helpful, as an additional source of information on paleowind directions.

3.1.3. Floods

Although rain is rare in the arid regions, when it comes it may frequently take the form of thunderstorms, pouring considerable amounts of rainwater for a very short time, which almost invariably results in floods. These may appear, depending on the terrain and the amount of rain, as flush floods covering large areas (Fig. 2.2.9), or confined to the wadis (Fig. 2.2.8). In both cases, floods are responsible for mechanical weathering, by crushing the rocks and gravel on their way, not so much by the action of flowing water, but rather by their clastic load. Floods also rearrange the formerly laid clastics, by sorting and deposition downstream. Surprisingly, it was found out that pollen grains can, in many cases, withstand this mechanical process of weathering and transport, thus wadi sediments usually produce decent spectra (Martin, 1963; and many others). It may be argued that pollen can withstand flood erosion because the grains are included in the finest clastics fraction, which is transported separately from the coarser one. This is because the finer fraction, whose particles have a much larger surface area per weight as compared with the coarse, is carried by the water in a layer which overlies the one in which the coarse sand and gravel are transported. The finer fractions are transported farther away downstream relatively to the coarse, in any given stage of the flood, but then, when the flood fades away, are deposited among the coarser gravel (Fig. 3.3.1.1). This phase of deposition is a very quiet one and the delicate pollen grains are embedded within fine matrix, among the coarser sands and gravel. It is therefore possible, by mechanically separating the coarser from finer fractions, to obtain pollen grains from the latter.

3.1.4. Chemical

Chemical and biological weathering, although seemingly subordinate in arid lands, may considerably affect pollen preservation. Both groups of processes primarily depend on at least some supply of humidity, for which a common source is the frequent dew, occasionally the rare rains, or occurrences of groundwater seepages. Humidity bounded by physiological activities of some organisms, especially lichen and blue-green algae which are quite common in arid lands, may also play an important part in biological and chemical degradation processes.

The most frequent products of chemical weathering are clays and soluble salts, such as common salt, gypsum and carbonates. The fine clastics and frequently also the crystals of soluble salts, are usually blown off the rock's surface and subsequently deposited with the rest of the fine windblown dust. Concurrently, water percolating through such sediments dissolve and when evaporate, recrystallize the solubles, to form the abundant subsoil crusts of arid regions. While the chemical weathering products are blown off the rock's face, the latter becomes "scarred" (Fig. 3.1.4.1), which is a typical appearance of so many rock surfaces in the deserts. When such scars become deep enough, they are termed "tafoni" (Fig. 3.1.4.2), which can

Fig. 3.1.4.1. Scarred rock, as a result of chemical weathering. Central Negev, Israel. Scale bar 10 cm.

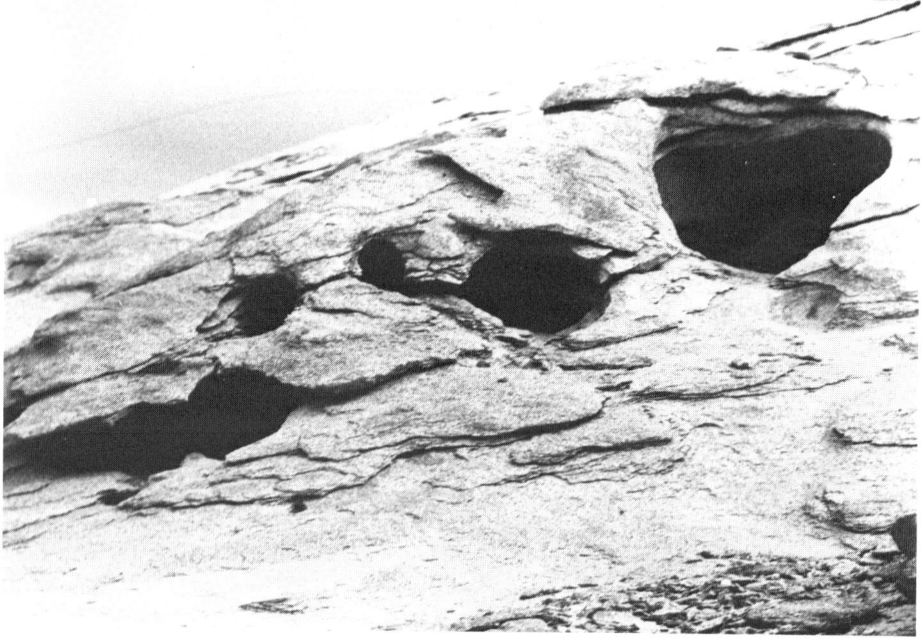

Fig. 3.1.4.2. Tafoni, developed in granite. Namib Desert.

turn into rockshelters for desert animals or men, whose trash, excrements and food reservoirs, usually accompanied by trapped windblown fine sediments, make a prospective source for palynological material (see Chapter 2.2.5). Tafoni are quite common in deserts, occurring on almost any type of rock.

The soluble salts formed by chemical weathering, while crystallizing, can further cause mechanical destruction of the rocks (Beaumont, 1968; Amit, 1990), a phenomenon which is apparently more common in regions with alternating humid and dry conditions of the ground, a possible outcome of daily dew. Mechanical breakage is also thought to result from differential expansion of the crystals, with the changes in temperature. Such changes in volume are also attributed to variations in the percentage of water molecules in the crystal's lattice, due to hydration and dehydration reactions, which can result in disintegration of rocks into smaller particles. The various crystallization processes affect pollen very badly and this is especially visible in subsoil crusts, which frequently contain deformed and crushed grains, occasionally hopeless for identification.

The most harmful process of chemical weathering is, however, oxidation. Many of the rocks formed in dry lands are friable and devoid of almost any organic material. When air gets into the rock's hollows, it only needs a slight humidity to become a strong oxidizing agent, especially in relatively high temperatures. The rarity of any other organic materials make pollen grains the first and almost only candidates for oxidation. Being composed of organic material, however resistant, most pollen grains cannot withstand oxidation and are literally turned to ashes.

3.1.5. Biological

Biological weathering can be either mechanical or biochemical. Mechanical mainly involves activity of burrowing animals (Figs. 2.2.14, 2.2.15), which is very prominent in arid environments, as are penetrating roots, both of which can reach down to considerable depths (Fig. 3.1.5.1). These processes do not seem to affect

Fig. 3.1.5.1. Roots penetrating to a depth of more than 1 m at an archeological site. Tell Ira, Negev, Israel.

pollen directly, but may cause stratigraphic mixing, to which attention must be paid while collecting samples (see Chapter 2.2.5).

"Lower" organisms, notably lichen and blue-green algae, seem to participate in processes of rock weathering, mostly by dissolution. These are given more attention in the last years (Danin et al., 1982, 1983; Danin, 1989) and such dissolution patterns are beginning to be used for dating. The blue-green algae are very common endolithic organisms in desert environments and seem to help disintegration of the rocks. On the other hand, it was also found out that these algae can, in some cases, protect finer sediments from wind deflation (Danin, 1990).

A biologic agent which may dissolve mineral matter, but seems to protect pollen grains, is animal excrements and decay products of plants. Both help turn their close microenvironment into acid and reducing, which is favorable for pollen preservation. However, too much of these materials may form a cradle for bacteria and fungi, which can again be harmful to the pollen.

It seems that the biological factors do not take a major part in the weathering processes of the arid lands, mainly due to the scarcity of life, combined with the strong and almost immediate oxidation of the biological products.

3.2. Sediment transport

The two principal agents for sediment transportation in arid regions are winds and water. However, due to the difference in climatic conditions between arid and humid regions, transport activities in both are quite different too. Except for winds and water several other agents, although having only a subordinate role in transportation, are important for palynology of arid lands. Such are sediment transport by animals, man or groundwater, which are intimately connected with many sites from which samples are collected for pollen analysis in dry regions, sites which seem to comprise an important source of information concerning the history of these areas.

3.2.1. Winds

Light winds are very common in dry lands, but are hardly ever connected with the transportation of sediments, which is readily evidenced by the clear air and excellent visibility, typical for these regions. But occasionally, wind velocities are considerably increased and may result in dust or sand storms. Winds of high velocities are not too rare also in the humid regions, but sediment transport by these is only confined to unprotected areas, such as coasts, where dunes are widespread; periglacial regions, mostly characterized by loess transport; and some other, rather rare natural combinations, such as wind funneling which had formed the Great Sand Dunes National Monument in the middle of the Rocky Mountains, Colorado.

The combination of two interconnected factors, the lack of protective vegetation cover and the dryness of the ground, makes the stronger winds an important transportation agent of desert sediments. The scarce flora permits a direct contact of the wind and weathered rock surfaces, which supply sediments ready for transport; wind transport essentially affects only dry sediments (Glennie, 1970), but these are evidently very common in arid lands, especially since the strong winds considerably help desiccation.

Eolian transport affects the particles in several ways, depending on their size and on the wind velocity. The finer particles, up to an average size of 200 μm, are easily lifted by the winds, sometimes up to considerable elevations above ground level, which may even attain several kilometers (Fig. 2.2.6). These particles can stay in suspension over large distances and are quite frequently deposited outside the desert margins. Usually, such dust is carried from drier regions into the less arid ones, where it is being trapped by vegetation, forming loess beds (Schultz and Frye, 1968; Horowitz, 1979; Liu, 1987, 1991). These loess deposits may occasionally turn into aridic soils (Dan et al., 1981).

Sand grains, up to a size of 1–2 mm, are usually transported by the wind by way of saltation. Bagnold (1941) stated that the sand size particles, having received a supply of energy from the forward pressure of the wind, have acquired horizontal velocity which is converted into an upward movement, following impact with the ground. The pressure of the wind pushes the grains onward until, when sinking to the ground, they would bounce upward once again, in a parabolic trajectory. Since the density of the sand grains is considerably higher, some 2,000 times, than that of the air, such movement cannot take place much higher than several centimeters above ground level. Bagnold further states that on impact with the ground, such saltating sand grains can cause a creeping movement of larger particles, up to some six mm in diameter.

It seems however, that at least in some deserts winds are more powerful, causing a continuous transport of sand by lifting it much higher above ground level, thus carrying the grains over considerable distances at a short time. Such sand storms have been personally observed in the Negev, Israel and in the Mojave Desert, SW USA.

The saltating way of movement of sand grains has a considerable destructive activity on pollen grains, which seem, consequently, to be very rare in sand dunes. In addition, the loose packing of sands facilitates ventilation and water percolation, which considerably hasten oxidation.

Dust, being consisted of a finer clastic fraction, mainly silt and clay size particles, is transported by winds in a manner essentially different to sand grains. Once the fine dust is lifted it can stay in suspension for a long time, up to several days. It is deposited when the wind has significantly quietened, if the storm is accompanied by rain or if a suitable trap or barrier is met. The dust particles, being suspended in the wind, have a very low abrasive capability, thus hardly damaging pollen which are carried concurrently. Consequently, loess deposits are a good source for palynological studies.

3.2.2. Water

Transport by water has three principal facets in arid lands. The first involves sediments carried by rivers which flow perennially, especially those which originate in other regions and only cross the desert on their way to the sea, such as the Nile or the Senegal, commencing in the tropics and crossing the Sahara (Fig. 3.3.11.1). The second, which is much more important in the formation of desert sediments, involves the wadis, usually dry, which are filled up with water during floods, carrying considerable loads of sediments over a short period of time (Fig. 2.2.8); these may sometimes develop into flush floods, wiping the entire area (Fig. 2.2.9), that occasionally deposit a blanket of usually fine grain sediments over a large surface. The third involves transport by groundwater, mainly of dissolved salts, but occasionally also pollen, discussed in Chapter 2.2.4.

Perennial rivers are quite rare in arid lands, but do occur occasionally. Some may take their water from springs and are truly desert streams. However there are not too many of these, since most rivers or streams flowing in arid and extremely-arid regions originate outside these. The two main provenances for such rivers are tropical regions as for the Nile, or higher mountains close to, or within the deserts, where rains are more frequent, such as the Colorado originating in the Rocky Mountains. Perennial rivers transport sediments in much the same way as such rivers do elsewhere, thus have nothing in particular in dry lands. Some of them transport deposits from their initial provenances, which are occasionally laid down on their way, usually on floodplains within the arid zones; or may at times redistribute former desert sediments, which they erode on their way and deposit downstream. These two processes may occur concurrently. A special case are those rivers which originate in humid regions and pass through arid ones, to be terminated in internal basins in dry lands (Fig. 2.2.7), such as the Jordan River. Sediments transported by such rivers become part of sequences accumulated in the endoreic basins and are a very important source of information, since in most cases continuous sequences can be collected for study. However, most pollen thus deposited originate outside the arid land and should be treated accordingly. These rivers frequently carry very fine clastics and dissolved minerals, which are usually laid down in their final destination (Horowitz, 1987a).

Floods, a common phenomenon in the dry lands, can also affect the perennial rivers, which at that time behave, in terms of sediment transport as wadis, depositing the coarser load whenever their higher than usual energy diminishes. However, since they continue their perennial activity after the flood had subsided, the finest clastics and dissolved minerals may be carried all the way to the river's final destination. Such floods are usually connected with sudden rains in arid regions, but may also develop from stronger rains in the rivers' provenances. The River Nile is famous for such floods, which occur at the peaks of monsoon activity in tropical Africa. At this time the Nile acts like a wadi, in terms of sediment transport and deposition.

Wadis are very common in all arid regions around the globe. They are usually dry and may occasionally remain so for years. They become active only when the rare thunderstorms hit the arid region, filling them with water. The most remarkable characteristic of the wadis at that time is the floods, which can carry considerable volumes of water during a very short time, usually several hours. Such floods are quite violent and can carry almost anything with them. Large volumes of clastics, coarse and fine alike, are redistributed by the wadis during the floods, consequently deposited downstream. These clastics may arrive at the final base level of the wadi, which can be either the sea or an internal basin and be deposited there; or, the sediments may travel downstream as long as the floods' energy is still sufficient and be laid down somewhere along the wadi channel. In both cases, it is more frequent that not that the energy of such a flood is diminishing gradually, thus depositing a mixture of coarse and fine clastics. Such a mixture is a good prospect for pollen grains, which are always laid down toward the end of the flood and embedded within the finer matrix, however not at the same location for different floods, which makes long continuous sequences quite rare along wadis.

Groundwater, although not too common in deserts, can transport dissolved minerals, sometimes even along rather long routes, especially in karstic aquifers, which frequently occur in deserts (Frumkin, 1990), although not formed under such dry climates but during a former, more humid phase (Horowitz, 1979), or in the frequent coarse clastic aquifers. The most common minerals thus transported are carbonates and sulphates. Finest clastics, such as clay minerals, are also transported, but their quantities are usually subordinate. Pollen grains are known to be transported by groundwater (Horowitz, 1979). The dissolved minerals are precipitated when groundwater emerge as springs, or sometimes even within the subsurface, as subsoil crusts (Lavee and Wieder, 1990).

3.2.3. Animals and man

Animals and man can transport both sediments and pollen, although hardly ever to any great distance or in considerable quantities. Greater distances are only covered by migratory birds, but these never carry considerable quantities. Some of the present day anthropogenic activities also carry sediments and pollen, sometimes even in significant quantities, over considerable distances. This transport is caused both by digging and burrowing activities and by feeding habits and excrements. These transport processes are not of any significant importance when the sedimentology of a dry land is concerned, but may be quite meaningful while analyzing human settlements, a site which seems to be favored by many arid lands palynologists, especially those who team with archeologists. The sediments thus transported can be either mineral, as is the case with digging, or organic, such as excrements, pellets and middens. Both excrements and pellets can occasionally be transported over considerable distances, especially by migratory birds. A more detailed discussion of these transportation processes is given in Chapter 2.2.5.

3.3. Depositional environments

Depositional environments can be divided in arid lands into these situated within the arid land proper, such as wadis and streams, salt-pans, lakes and inland sebkhas, springs and oases, dune fields, loess accumulations, soils, caves, rockshelters and archeological sites; those which neighbor arid lands such as coasts, coastal sebkhas, deltas and seas; and environments which are there only due to past geological events such as rivers crossing deserts. In addition, many rock types found in arid lands had been formed in depositional environments which existed in the region during former geologic periods, sometimes under a completely different climate, like the ancient glacial deposits of the Sahara and the Kalahari.

3.3.1. Wadis and streams

Wadis are typified by their intermittent floods (Fig. 2.2.8) and ensuing deposits depend on water energy and on volume relations of water and carried sediments. If the water/sediment ratio is low, the resulting deposits would form an unsorted muddy gravel, or mudflow conglomerate, which is laid down from a slurry of sand and mud, sometimes comprising rather sizable boulders (Fig. 3.3.1.1). A higher water/sediment ratio would result in better sorting, with occasional grading of the deposits. When the water energy of the flood decreases, usually due to absorption of the water in the dry wadi bed, the entire load of sediment is deposited, which may occasionally block the wadi channel. The following flood would then have to find itself a different route, causing formation of a typical braided pattern. At the termination of the floods' activity, the finer clastics are deposited, sometimes covering large areas with mud, which dries up and forms typical mud cracks (Fig. 3.3.1.2). These muds, before being dried up by the sun, form an ideal trap for concurrent airborne pollen, which are subsequently sealed in the mud when it completely dries up. Such sediments and landforms are characteristic of wadis, controlled by their intermittent activity. When no floods affect the wadi, it is frequently being filled up with eolian sediments, mainly sands (Fig. 3.3.2.1), which would at a later stage appear in outcrops as a mixture of water and wind laid sequences (Glennie, 1970). Such wadis could, when the sealing is extensive, turn to inland sebkhas or lakes.

It should be noted that many wadis follow channels cut in the substrate during a former, more humid period, when these have actually been streams, of

Fig. 3.3.1.1. Unsorted wadi deposits, with large pebbles and fine clastics. Central Negev, Israel. Scale bar 10 cm.

Fig. 3.3.1.2. Cracked mud in a wadi, following flood and subsequent drying up. Central Nagev, Israel.

Fig. 3.3.1.3. A terrace consisting of fluviatile deposits, with sedimentary structures indicating perennial flow during a former period. Golmud River, Qinghai, China.

Fig. 3.3.1.4. Alluvial fan, at the transition of the Edom mountains to the Arava Rift Valley. Southern Jordan.

perennial activity. Such processes would result in good sorting of the sediments (Fig. 3.3.1.3). It was found out that, in the Near East, alternation of dry and humid periods during the Quaternary caused formation of terraces in wadis, where sediments have been laid down during the humid phases in perennial streams, while downcutting occurred at a subsequent, dry period typified by powerful floods (Glennie, 1970; Horowitz, 1979; Goldberg, P., 1986). This process is even evidenced in the extremely-arid region of Elat, southern Israel (Gerson, 1981).

Floods can occasionally, when water amount exceeds the capability of the channel, go over the banks and form a flush flood (Fig. 2.2.9), depositing typical sediments on the floodplains. McKee et al. (1967) described these deposits as a combination of steeply foresetted sands, horizontally laminated sands and sets of climbing ripples. The occurrences of clear sedimentary structures in floodplain deposits seem to result from a considerable lowering in the water's energy, following spreading of the large volume of water and sediment on a flat surface, which is much less confined than the wadi channel itself.

A somewhat similar situation may occur when wadis, flowing in a confined channel of a hilly or mountainous terrain, where the flood is strong, arrive onto a flat country downhill, where water energy considerably diminishes, to form alluvial fans (Fig. 3.3.1.4). Sediments of these fans are not as well sorted as those on floodplains, since in many cases boulders do reach alluvial fans.

Wadi deposits (and it should again be stressed that by the term "wadi" reference is only made to intermittent activity) are therefore characterized by their being exclusively clastic, with poor or no sorting and hardly any development of fine sedimentary structures. They are frequently interbedded with eolian sands and silts. Pollen grains are deposited in wadis, sometimes in considerable quantities, during the last phases of a flood, when the finer clastic fractions are laid down. These fine clastics are thoroughly dried up by the sun, usually not more than several days after the flood. During these days, when the mud is still wet, it can trap pollen from the air. These dry mud cakes make excellent pollen reservoirs due to their sealing and occasionally subsequent burial by the next flood or by windborne deposits. Indeed, many researchers found wadi sediments to be a very good prospect for pollen analysis, from which spectra of quite well preserved pollen, in sufficient quantities, had been obtained. The only drawback is the inclusion, in great numbers, of pollen produced by the wadi vegetation, which, if a general climatic picture is to be drawn, should be subtracted from the total pollen sum (see Chapter 9).

Streams, in which water flow all year round, are not so common in extremely-arid and in arid lands (which is also true when they originate outside the region), but are quite frequent in the semi-arid. Their sediments, again characteristically clastics, are typified by good sorting and clear sedimentary structures. Such streams are also characterized by finer clastics, as compared with the wadis. These streams are prone, particularly in the arid and extremely-arid regions, to occasional floods, in which case the final sedimentary outcome would constitute alternations of well and poorly sorted deposits. Such stream deposits and especially the finer ones, were found to contain well preserved pollen, in sufficient quantities for analysis.

3.3.2. Inland sebkhas and salt-pans

Depressions lacking outgoing drainage can turn in arid lands into temporary, or sometimes even perennial shallow, saline water bodies. These are termed differently in various languages, like sebkhas (a term used for both inland and coastal ones), salt-pans, marshes, or occasionally even "lakes". The term "inland sebkha", used by Glennie (1970) seems appropriate, as is "salt-pan", used in the English speaking countries of the southern hemisphere. It seems that "marsh" should rather be reserved for shallow accumulations of relatively low salinity water, while the term "lake" involves a permanent, somewhat deeper water body, usually continuously fed by at least one perennial river, by springs, or by high level groundwater.

The origin of such depressions may involve several possible causes. Structural, either by downfaulting, or by uplifting which makes a barrier over a previous drainage path, or occasionally the combination of those two processes; sedimentary, by flood deposits or sand dunes blocking a channel (Fig. 3.3.2.1); or erosional, such as the shallow, sometimes large basins formed by wind deflation on otherwise flat

Fig. 3.3.2.1. Sossusvlei, an inland sebkha formed by dunes blocking the drainage. Namib Desert.

Fig. 3.3.2.2. Salt-pan, in a depression probably first carved by wind deflation, subsequently filled up with sediments. Near Oodnadatta, South Australia.

regions (Fig. 3.3.2.2). These depressions can get their water supply from wadi floods or from groundwater, when its level is high enough. The inland sebkhas maintained by groundwater tend to be perennial, in many cases.

The one thing common to all internal drainage basins in arid lands, including sizable lakes fed by perennial rivers, is their salinity. The degree of salinity depends on supply of salts by water, or occasionally also by winds; the rate of evaporation as compared with water supply; and time — the longer the history of an internal drainage basin, the more saline it may become. Considerable salinities are found when water dissolve salt bodies on its way, as is the case with the Dead Sea, in addition to other causes.

The typical sediments deposited in inland sebkhas are of three main types: fine clastics, which seem to be of eolian origin in most cases, except where sebkhas are fed by wadis, which could supply also some coarser clastics, usually confined to the margins; most inland sebkhas can maintain at least some vegetation; even if water is too saline for a well developed plant life, bacteria still flourish, which add a significant amount of organic materials to the deposits; a third compound is evaporites, which are almost always present in some amount, usually comprising gypsum or rocksalt. When the inland sebkhas temporarily dry up, which in some

Fig. 3.3.2.3. Alternations of evaporites and eolian sediments in an exposed salt-pan sequence. Sossusvlei, Namib Desert. Scale bar 10 cm.

Fig. 3.3.2.4. Black mud of an inland sebkha, covered by an evaporitic crust. Southwestern Sahara, Senegal.

regions may happen almost every year, an evaporitic crust is formed, which protects the sediments from oxidizing, at least to a certain degree (Fig. 3.3.2.3). The evaporites, being hygroscopic, also hold water, thus below the seemingly hard crust one frequently finds black muds (Fig. 3.3.2.4), even in extremely-arid regions such as the Sahara.

A somewhat more complicated situation sometimes occurs with relatively large depressions, in which freshwater seepages are situated around their rims. The water flows to the central, somewhat deeper part of the depression, where evaporation may form impervious layers. These do not permit water seepages at the deeper part of the depression, which are therefore confined to the rims. The types of sediments associated with such a situation are travertines at the pan's rims and fine clastics entrapped by the freshwater vegetation, usually with less organic material. Rarely, when vegetation is rich enough, local peat bogs may develop in such localities (Scott, 1982).

Pollen grains are usually well preserved in inland sebkha and salt-pan sediments and can easily be extracted in sufficient quantities in most cases, which accounts for many palynological studies carried out on such deposits. The low rates of deposition of predominantly fine clastics, the wet, usually spacious sticky surface which acts as a pollen trap and the good sealing caused by frequent periods of drying up, make the sediments a very good source for pollen. This may change occasionally, when deposition of evaporites exceeds fine clastics, which makes extraction laborious and pollen preservation poor, chiefly due to crystal growth. Pollen spectra obtained from sebkhas and salt-pans reflect the regional vegetation, but are also prone to the influence of long distance eolian pollen transportation and occasionally to over representation of the local, hydrophil and halophil floras.

3.3.3. Lakes and marshes

As stressed above, the term "lake" applied here does not include water bodies such as sebkhas and salt-pans, even if these are perennial, in contrast to the use of this term in some other publications. Lakes are quite rare in arid lands and mostly occur in tectonic depressions (Fig. 2.2.7), fed by a river (or rivers) originating in a more humid environment. The lacustrine sediments appear in two distinct facies, either coarse clastics of the lakes' margins, or the fine clastics and chemical precipitates of the deeper parts.

Lakes in arid regions are fed, besides their main perennial supply of water, also by floods which arrive by the numerous wadis surrounding such depressions. The coarser clastic fraction brought by wadis is deposited at the lakes' margins and hardly ever travels far into the lake itself, which usually involves the buildup of alluvial fans. The finer clastics are carried in suspension and distributed all over the lake's bottom. These are also accompanied by the finer fractions transported to the lake by its principal water supplier. The combination of these two sources makes fine clays and shales the typical lacustrine sediments and in this respect, arid zone lakes are not much different than others, occurring in more humid environments.

What makes the difference is dissolved minerals which at a certain stage are deposited due to evaporation. This deposition depends on the overall salinity and the relations between various ions in the solution, thus the ensuing precipitates vary from lake to lake and even in different regions of a certain lake, if large enough. Deposits can occasionally vary with the seasons. The rainy season brings water and clastics, while during the dry evaporation takes the upper hand. The result is that many lakes in arid regions tend to form varved sediments, of alternating clastics and precipitates (Fig. 3.3.3.1).

A third aspect of lakes involves the interfluves. These regions, separating between wadi inlets into the lake, are prone to geomorphic processes which result in the formation of lakeshore terraces. Mostly these are erosional, but often sediments also accumulate there, usually a mixture of medium to fine clastics, with occasional evaporites. Such lakeshore terraces may occasionally be of interest to palynologists, especially those involved with archeologists, since many sites are located on these terraces. Another importance of the lakeshore terraces lies with the study of tectonic lakes which, due to continuous subsidence, do not leave outcrops of their truly lacustrine sediments. Before deep boreholes have been drilled, these terraces had been almost the sole source of information for the history and development of the Dead Sea Rift in Israel (Horowitz, 1979). Pollen analyses carried out on their deposits helped reconstruct the stratigraphy and paleoenvironments of this region. At a later stage, when deep boreholes had been drilled and palynologically analyzed, the spectra obtained from lakeshore terraces were correlated with the continuous palynostratigraphy (Horowitz, 1988a).

Freshwater, or nearly freshwater marshes, same as lakes, are not too common in arid lands. They sometimes accompany oases, or formed on the floodplains of large rivers crossing deserts. These marshes only rarely develop into peat bogs, but are typified by highly organic sediments, formed from a mixture of the decaying rich marsh vegetation and additional fine clastics, arriving by the winds and sometimes the floods (Fig. 3.3.3.2). The sequences laid down in such marshes are almost never continuous and usually do not attain any considerable thicknesses.

Pollen is generally abundant and well preserved within the deeper lake sediments, notably in the fine clastics. Chemical precipitates may cause mechanical deterioration of the pollen, chiefly by crystals growth; however, this is not too common in lacustrine deposits and chemical sediments analyzed from ancient lakes in the Dead Sea region, Israel, proved productive. Bacterial activity on the lake's bottom can also sometimes adversely affect pollen preservation. It should be noted in this respect that even in the hypersaline, usually saturated waters of the Dead Sea, bacteria do exist, however quite rare. Since most of the desert lakes are saline, pollen destruction by bacterial activity is not too common. The rather slow rates of deposition of sediments in most arid region lakes, the relatively long sequences

Fig. 3.3.3.1. Lakeshore terrace sediments, a mixture of clastics and varved sediments. The latter composed of dark fine winter clastics and white summer evaporites. Dead Sea, Israel. a. General view, scale bar 20 cm. b. Detail of varves, scale mm.

Fig. 3.3.3.2. Dark, organic rich marsh sediments in an arid region, mixed with pebbles brought by floods. Dead Sea, Israel. Scale bar 10 cm.

which are in many cases continuous, especially in terminal basins and the generally good state of pollen preservation, make lacustrine sections a very good source for palynologic information.

The lakeshore terraces are more problematic, but could be prospective when the sediments consist of considerable shares of fine clastics, even if the latter only comprise the matrix for the coarse gravels.

Pollen spectra obtained from lake sediments are generally considered a reliable source of information for the regional vegetation, but also with these, care should be taken with occasional over representations of the local, hydrophil and sometimes halophil vegetations.

3.3.4. Springs and oases

Springs and oases are not too common in arid regions, but when they do occur, they are of considerable importance for both plant life and human habitation, which makes their deposits a significant source of samples for palynological studies. Spring waters in arid lands are frequently characterized by high concentrations of dissolved minerals, most of which are precipitated upon discharge. These deposits can take several forms, which frequently fall under the terms "travertine" or "tufa". The most common minerals of these deposits are carbonates, either in the form of calcite or aragonite, occasionally quite rich in magnesium. Springs in volcanic regions may contain considerable amounts of dissolved silica, which is usually deposited as a kind of opal.

Since springs support plants in their close vicinity, some of these are usually engulfed within the mineral matter, occasionally forming highly organic travertines. In other instances the plant material becomes oxidized, or sometimes may be entirely replaced by mineral matter.

Another kind of spring deposit is formed when the water flow, upon discharge, through a porous rock such as sands or gravel. In this case (Fig. 3.3.4.1) the dissolved mineral matter supplies the substance for cementation of the older, porous rock. This process is responsible for the formation of multi-age rocks, whose clastic particles are of an earlier age than their cementation. The differences in age between the matrix and the cement could sometimes stretch over considerable geological ages. Such rocks may contain pollen grains representing either of the two generations, or in some instances both, thus should be treated accordingly.

Fig. 3.3.4.1. Older gravel, cemented by recent spring carbonate precipitates. Central Negev, Israel. Scale bar 10 cm.

Fig. 3.3.4.2. Oasis made by a spring on a fault scarp, with well developed hydrophil vegetation. En Feshkha, Dead Sea.

Oases are of two main kinds: either connected with springs, whose discharge is sufficient for watering a surrounding area in which hydrophil vegetation can develop (Fig. 3.3.4.2); or those enjoying a high groundwater table, although without actual seepages (Fig. 3.3.4.3). Both types can support trees, although only locally, evidently also human and animal life. Oases connected with springs usually develop some kind of marshes, which deposit highly organic sediments composed of plant material and fine mineral matter, largely of eolian origin, entrapped by the dense vegetation. Oases of high groundwater table can be connected with various types of deposits, depending on the groundwater level. If it is high, some seasonal marshes can be formed, with their typical organic sediments, which may sometimes interfinger with purely eolian deposits, laid down in the drier period. Somewhat deeper groundwater levels, which would still support grasses, would be typified by loess type deposits, as the windborne dust is trapped by the vegetation. These deposits may occasionally contain some plant debris. When groundwater level is quite deep, usually only trees such as date palms can grow, typical of many localities where dune sands cover the area. No particular sediments are usually connected with this type of oasis, although the organic matter contents of associated deposits could be high and help pollen preservation.

Fig. 3.3.4.3. Oasis formed by high level groundwater table. En Evrona, southern Arava Valley, Israel.

Many palynological studies in arid lands are geared toward understanding human settlements in these marginal environments, which are almost always connected with springs or wells, fed by groundwater. Pollen is frequently well preserved in spring deposits of various kinds and may occur in considerable quantities. Consequently, although spring deposits are not too common in arid lands, they constitute an important source for palynological and paleobotanical information, if caution is taken in separating the local signal from the regional, as well as for datable material.

3.3.5. Eolian

Sand dunes are quite common in extremely-arid and in arid environments, while loess is more characteristic of the semi-arid regions (Fig. 3.3.5.1). This primarily reflects the generally lower winds energy in the latter, combined with the more extensive vegetation cover. Dunes are formed wherever a suitable source of sand is available, their great abundance in deserts resulting from extensive exposure of bare rocks and the agents of weathering, which supply sufficient quantities of sand grains. Dune sands occur naturally, as in other regions, along coasts, but are common also inland and even on highlands in extremely-arid and arid regions

Fig. 3.3.5.1. A 12 m high continuous loess sequence in the northern Negev, Israel. The loess was deposited in this region during a former, pluvial period, when the area enjoyed a semi-arid climate and is now being eroded by floods.

Fig. 3.3.5.2. Dunes of the Namib Desert.

(Fig. 3.3.5.2). Dunes are characterized both by their shapes and sedimentary structures, especially cross bedding, depending on wind velocities and directions. Sand dunes include, beside the sand grains, almost always some finer clastics. These may be included within the sand almost contemporaneously with its deposition, following the drop in wind velocity, but may also be deposited at a much later stage, while the dunes are being stabilized (Pye and Tsoar, 1987). Only a careful study can reveal the cause and timing of inclusion of the finer fraction. This may turn to be of significance when collecting samples for pollen analysis, since pollen deposition accompanies the fine clastics, rather than the sand by which it is usually destroyed. If the time difference between deposition of the two fractions is considerable, pollen spectra would have no bearing on environmental conditions during the sand dunes accumulation period.

Dust is being deposited almost everywhere in and around the deserts. In semi-arid regions situated on the boundary of deserts it is trapped mainly by the vegetation (Pye and Tsoar, 1987), to become loess. In the arid regions, where there is no sufficient plant cover for this process, the sinking dust is enclosed within the existing deposits, thus enriching sediments, such as sand dunes, playa deposits and wadi gravel with finer particles, including the pollen grains which may have been transported together with the dust. It was however found out (Danin and Ganor, 1991) that even the rare plants, including such groups as mosses, serve as dust traps in the desert. This finer matrix is in many cases almost synchronous with the other eolian deposits in which it is embedded, except sometimes for a delay of several days, due to the slower rate to deposition of the fine particles, if they have not been washed down by a rain. The inclusion of finer dust particles within the coarser desert eolian clastics renders most wind laid desert sediments suitable for pollen analysis, by a simple separation of the fine matrix from the coarse clastics. It is quite rare to find coarse eolian clastic sediments without any trace of finer particles and it is the pollen deposited together with the latter, that can serve for analysis.

Some caution must be taken, especially when analyzing pollen from sand dunes. Although in many cases the dust trapped within the sand is synchronous, it may occasionally be transported and deposited at a later stage. Such may be the case with fossil or stabilized dunes, which are not buried, in which dust may be incorporated after thousands of years, under a different climatic regime. These processes are known from Israel, where dust from the Sahara is captured at present within ancient sand dunes of the coastal plain, as well as other non sealed deposits (Yaalon and Dan, 1967).

Rare pollen are occasionally found in dune sands, but are usually in bad shape. Better spectra are sometimes obtained from dunes when a fine clastic fraction also occurs. As mentioned above, it is advisable to initially ascertain the time relations between sand and finer clastics deposition, since pollen had most probably arrived and incorporated with the latter.

3.3.6. Soils

Soils are usually quite poorly developed in arid regions in terms of pedogenetic processes, due to the lack of water and vegetation, the principal agents necessary for soil formation. Distinction should however be made between various degrees of aridity, since the semi-arid regions, which maintain vegetation all over, though not usually arboreal, do support pedogenesis, thus well developed soils are found in many such regions. Soils in arid lands are significant contributors of samples for palynological research, since in many cases archeologic and more often prehistoric sites are embedded within them.

The predominantly mechanical disintegration of rocks in arid climates, the poor leaching, insufficient vegetation cover and the relief, more than composition of the bedrock, control the types of soils which are formed. Pieces of bedrock comprise a significant part in the soils formed on hilly or mountainous regions, while alluvium is the main constituent in the valleys. The restricted water infiltration accompanied by rapid drying up, causes formation of various shallow subsoil crusts and concretions, the most common of which are composed of carbonates, gypsum, or iron oxides. The sparse vegetation does not support significant chemical and biological processes, thus the soils are usually poor in organic compounds, their profiles rather shallow.

Dan (1981) summarized the types of soils in the extremely-arid, arid and semi-arid regions of Israel, in relation to physiography:

Mountainous soils, in the extremely-arid regions, vary from bare rocky slopes and lithosolic regs (Fig. 3.3.6.1) on the plateaus, through saline calcareous desert lithosols developed on chalks and marls and coarse stony and gravelly desert alluvium in the gullies. The arid regions are typified by brown lithosols, usually saline and calcareous; loess and coarse desert alluvium occur in gullies. The semi-arid regions are characterized by shallow brown soils and lithosols and some natric grumosols; loessial stony brown soils and various colluvial–alluvial soils occur in gullies (Fig. 3.3.6.2).

Extensive valleys are typified, in the extremely-arid regions, by various kinds of alluvium, which can comprise coarse to fine gravel, some loam, or sand. Some soils can be fine, consisting of alluvial or eolian sands, occasionally forming solonchaks. The arid zones maintain various serozems, mainly loessial and stony, highly calcareous (Fig. 3.3.6.3) and various calcareous alluvial soils. The semi-arid soils are mostly highly calcareous brown grumosols and brown, silty alluvial soils and alluvial loess.

Plains and undulating regions, in the extremely-arid zones, maintain mostly regs, while in the depressions, coarse desert alluvium, sometimes loamy, accumulates. In the arid regions, stony, sometimes loessial serozems occur, together with alluvial loess in the lower places. The semi-arid plains are characterized by loessial, light brown soils (Fig. 3.3.6.4).

Fig. 3.3.6.1. Lithosolic reg, on an extremely-arid plateau. Atacama Desert, northern Chile.

Fig. 3.3.6.2. Colluvial-alluvial "soil", in a wadi, in an extremely-arid zone. Central Negev, Israel.

Fig. 3.3.6.3. Highly calcareous soil, in an arid region. Western Negev, Israel.

Fig. 3.3.6.4. Light colored loessial soil, in a semi-arid region. Northwestern Negev, Israel.

Fig. 3.3.6.5. Exposed calcrete, Etosha, Namibia. Scale bar 10 cm.

Sand dunes do not develop any significant soils under extremely-arid and arid conditions, while in the semi-arid zones they may occasionally turn into sandy regosols, but in most cases remain unchanged.

Subsoil crusts are common in many arid regions, usually composed of carbonates, termed calcrete (Fig. 3.3.6.5). These can occasionally attain up to several meters in thickness and since frequently the covering soil is blown away by winds, make the landscape surface. Crusts which are less common may be composed of silica, iron oxides or gypsum.

Pollen grains are sometimes found in arid lands soils and subsoil crusts, depending mainly on their oxidation potential. Red soils hardly ever yield pollen, yellow and buff soils occasionally do. Grains are usually corroded, either due to oxidation or bacterial and fungi activities, or both. Caution must be applied with pollen spectra obtained from soils, especially when the grains are relatively well preserved, because such pollen may have been enclosed within the soils at a later date, by percolating rainwater or by subsequent fine dust deposition. The main problem with pollen obtained from soils and subsoil crusts seems however to be caused by the downward infiltration of rainwater, however rare the rains are, that may wash the grains down the profile. Mostly these pollen grains are trapped within the mineral concretions of the B horizon and again, their exact stratigraphic connection with the soil must be ascertained.

3.3.7. Caves and rockshelters

Caves and rockshelters could either be formed in the subsurface, mostly as dissolution cavities in a karstic system, sometimes as lava cavities, subsequently exposed by uplift and erosion; or subaerially, as erosional features, by coastal abrasion, river bank erosion and the like. Karstic caves are not formed in arid regions, except for those carved in easily soluble rocks, such as rocksalt (Fig. 3.3.7.1).

Fig. 3.3.7.1. Karstic cave in an extremely-arid region, carved in rocksalt, presently active. Mount Sedom, southern Dead Sea, Israel.

Fig. 3.3.7.2. Karstic cave in an extremely-arid region, a relic from former, more humid times, presently inactive. Central Negev, Israel.

However, many karstic caves do occur in arid lands, inherited from former, more humid periods (Fig. 3.3.7.2).

As long as caves and rockshelters are in the process of being carved, they usually do not accumulate deposits. When these processes slow down, caves and rockshelters become sediment traps for four principal types of deposits: chemical precipitates, crystallizing from dripping water; breccias, formed from mechanical disintegration of the cave ceiling and walls; windblown sediments, sands and silts, aerially transported into the cave or rockshelter; and organic sediments, the type of which depends on the animals (or man) which chose the place for dwelling. Some caves and rockshelters are dominated by a single type of deposit, but a mixture of up to all four can occur. Occasionally, when the climate and occupation change, sediments also change their character, or the caves and rockshelters can be affected again by accelerated carving, erosional processes.

Under extremely-arid and arid conditions, cementation due to activity of dripping water is quite rare, but is much more common in semi-arid regions. The more frequent occurrences of cave and rockshelter sediments in arid lands are those comprising a mixture of windblown and organic beds, mixed with rock debris (Fig. 3.3.7.3).

Fig. 3.3.7.3. Cave in an arid region, where eolian sediments are mixed with rock debris and some organic matter (darker). Dead Sea, Israel.

Pollen is quite frequently well preserved in cave and rockshelter sediments and may occur in considerable concentrations, since rates of deposition in these environments could be rather low. Leaching and cementation can occasionally account for corrosion of the pollen grains, as can bacteria and fungi, which sometimes flourish in the usually damp, organic rich sediments. Care should be taken in ascertaining wind directions or the origin of organic deposits produced by different animals, for defining and understanding any preferences in the pollen spectra.

3.3.8. Archeologic and prehistoric sites

Archeologic and prehistoric sites in arid lands can either be embedded within an otherwise natural sedimentary environment, such as soils, wadi deposits, loess, caves and rockshelters, etc., or act as an independent environment, in which characteristic deposits are accumulated. The former case does not necessitate further discussion here and reference is made to descriptions of the appropriate sedimentary environments. An exception, quite common in open, unbuilt sites, is the fires made by the people, which leave characteristic ashes, usually mixed with windblown dust, which forms a fine grained, highly organic deposit. Pollen are usually also burnt out by the fire, but when the heat subsides airborne grains could be entrapped together with fine dust, thus the highly organic microenvironment may be favorable for preservation of the grains.

The environment of deposition of an archeologic site, in which building activities had been carried out by men, calls for some attention when palynological research is planned. Deposits connected with human sites are of two main categories: either formed deliberately, by man's activities, or accumulated through natural processes, in traps supplied by the anthropogenic environment. In most sites these two types of deposits occur together and it is sometimes quite difficult to discern one from the other. Man made sediments are of various kinds, connected either with building, living, or destruction activities. All of these serve for palynological studies, thus a thorough understanding of modes of their formation and stratigraphic context would render the results a higher level of significance.

Building activities involve building materials, of which stones, wood and mud (adobe) are principally used. Gravel and sand are used for pavement and foundations, due to their drainage virtues and may contain, although rarely, pollen from their source; stones and wood are less problematic, since these are not usually used for palynology; mud bricks (Fig. 3.3.8.1) are a different story: they are made of raw material collected in the nearby vicinity, in most cases muds of the last flood in a nearby wadi; usually water and plant debris are added and this process is carried out in a pool, open to the air (Fig. 3.3.8.2). The bricks are cut and being dried up in the sun. Usually this activity takes place at the end of the rainy season, so that the bricks would have the opportunity to thoroughly dry up. But at the same time, incidentally, is the flowering season of many arid lands and cultivated plants, thus their pollen are trapped by the mud, both at the end of floods, while still in the wadi and while lying exposed in the pool. When the bricks are cut and dried, they contain plenty of pollen from the very season in which the construction activity took place. It was found out that usually these concurrent pollen make the majority of grains in the mud bricks, while pollen coming with the source material are subordinate (Drori and Horowitz, 1988–89). Furthermore, the drying process seals the mud bricks from any later contamination, which also protects from oxidation, thus making them one of the better deposits for pollen analysis of archeologic sites.

Mortar (Fig. 3.3.8.3) and plaster (Fig. 3.3.8.4), which are prepared in essentially similar way as mud bricks, can also be used for pollen analysis. Pavement is sometimes laid in much the same way and could then also be used. Levelling of floors is either done by digging, or filling up a certain piece of land. Bedrock may be used in this process, dug on one location and used for filling up another and care should be taken not to sample it, because it may represent a former habitation period due to the frequent recycling applied while building. It should, in this respect be remembered that a secondary use of mud bricks may also occur, but this seems to be less common. Living activities involve formation of deposits

Fig. 3.3.8.1. Mud bricks. Budhist temple, 6th Century AD, near Kuche, Xinjang, China.

Fig. 3.3.8.2. Making mud bricks. The mud is prepared in a pool which acts as a pollen trap. Turfan, Xinjang, China.

Fig. 3.3.8.3. Mortar (4 cm thick), cementing flint blocks. Byzantine, Tell Ira, northern Negev, Israel.

Fig. 3.3.8.4. Plaster. Budhist temple, 6th Century AD, near Kuche, Xinjang, China.

Fig. 3.3.8.5. Trash accumulation (or ash layer?). Byzantine, Tell Ira, northern Negev, Israel.

such as dung, trash and the like, which are highly organic. Usually these are mixed with windblown dust. Fire, which is part of daily activities, for cooking or heating, produces ashes quite common in many sites. These, too, are usually mixed with windblown dust, producing a fine grained, highly organic deposit which in many cases includes plenty of well preserved pollen. It is often quite difficult to discern between the origin of these grayish-blackish deposits which, in an excavation, tend to look quite similar (Fig. 3.3.8.5).

Deposits connected with destruction activities involve accumulations of stone rubble, broken pottery and ash. Burning of a conquered place seems to have been quite popular with certain people and cultures, thus such ash layers are very common in the stratigraphic columns of many archeological sites. They should not be mixed up, although very similar, with ashes of cooking or heating, corpse burning rituals, or other daily activities. Site burning ashes were found to contain pollen grains, which had, most probably, been included within the deposits following the destruction. It is however quite difficult to define the time span in this case.

Natural sediments which are involved with archeological sites are mostly windblown silts, which may fill up lower recesses in buildings and other structures and may yield good pollen assemblages. This type of sediment resembles very much loess. There is however a stratigraphic problem involved with such deposits, since their significant accumulations usually only commence after a site is abandoned.

3.3.9. Organic

Depositional environments in which vegetal material is the major component of the sediments, such as peat bogs, hardly ever exist in arid lands, for understandable reasons. Rarely, such environments do exist in semi-arid regions, most frequently as a result of blocked drainage, creating small scale basins in which peat can accumulate. Such peats are usually typified by their high pH and carbonate contents, which may be harmful for pollen preservation. It was found out that in the Hula basin, northern Israel, pollen are usually quite damaged in peat and lignite horizons, while within intercalating silty and clayey deposits, even though also of a high pH, pollen is much better preserved (Horowitz, 1979; Weinstein-Evron, 1983). Bacterial and fungal activity may, in addition to the basic environment, contribute toward the destruction of pollen grains in such organic sediments.

Accumulations of purely vegetal material are known from the arid regions, resulting from animals activities, such as nest building, midden collections and excrements. Animals' and occasionally also vegetal remains can sometimes be found in pellets of birds of prey. This may not be a proper environment in terms of sedimentology as understood by geologists, but is sometimes quite useful for the palynologist, since many such deposits are quite rich in well preserved pollen. The processes under which these bio-accumulations are formed are discussed in Chapter 2.2.5.

3.3.10. Coastal, deltaic and marine

Some of the coastal, fluvio-marine and marine depositional environments bordering arid lands are an important source of samples for pollen analysis, since in many cases these would be the only localities where continuous sequences could be obtained, pollen preservation is relatively good and extraction is an easy, straightforward process. The more promising, in this respect, are marine sediments, deltaic and coastal sebkhas. The first two environments could often provide rather long sequences, which however deserve a much more sophisticated and costly equipment for sampling. Most of the sebkhas are short living geological episodes, but can more readily be sampled. Other coastal deposits, such as dune sands and alluvial fans, are more problematic both in evaluating their chronostratigraphy and in the state of preservation of pollen grains.

A. Coastal

I. Alluvial fans. Wadis pouring their flood loads at the desert beaches are a common phenomenon. Such alluvial fans are usually well preserved on the

Fig. 3.3.10.1. Older alluvial fans in an extremely-arid region, cut by subsequent erosion. Namib Desert.

continent side, but are quickly dispersed in the sea by currents (Glennie, 1970), so that hardly ever real deltas are formed. The rare streams, those which originate in the desert, do not ever carry any considerable amounts of water and sediments, except for times of floods when they behave like wadis and therefore do not build any deltas. Again, deposits brought by these ephemeral streams are very quickly dispersed by sea currents.

Two main stratigraphic problems are connected with the continental alluvial fans. The processes of repeated deposition and erosion, which can be concurrent during a single flood in different parts of the same fan, make its internal stratigraphy very complicated, thus the assessment of a continuous sequence quite difficult. Also, many alluvial fans or "fan deltas" which occur at deserts coasts have been formed in older times, under a different, more humid climate and are only cut today by erosion (Fig. 3.3.10.1). Thus, alluvial fans can hardly ever be used for a very detailed continuous information, but rather for sampling single horizons, such as those connected with archeologic sites, or with any other particular occurrences of interest that can be adequately dated.

Pollen can be recovered from alluvial fan deposits, in much the same way as from wadi sediments. The state of preservation varies, as does the frequency of pollen in the different samples. It is very difficult to tell ahead what the prospects for obtaining a decent pollen spectrum are, but it was generally found out, when the stratigraphy is evaluated, that these sediments are worth trying.

II. Coastal sebkhas. Coastal sebkhas are formed landward (Fig. 3.3.10.2), behind some kind of a bar, usually originating on the seaside. The more common are either sand bars, forming from the redistribution of sands by longshore currents, or bars made by coral reefs, which grow up to the upper tide level and can enclose lagoons landward. When this closure becomes tight enough and does not facilitate a free water exchange with the sea, a sebkha is formed. In some localities such sebkhas are formed by mangroves, which trap sand and silt within their root system. During periods of rapid sea level changes, either due to eustasy or to local structural disturbances, sebkhas may be created as a result of former, deeper topography on land, which is blocked by the coastal sediments. Sebkhas can also be formed on the landward side of the coast, following blockage of streams or wadis by coastal sand dunes.

Salinity of coastal sebkhas is determined by an interplay of freshwater coming from the hinterland, mostly as percolating groundwater which rise to the surface at the coastline, floating on the coastal groundwater interface; seawater percolating through the bar; and evaporation. This intricate interplay is different from one sebkha to another, changing throughout the year, thus making a variety of salinities, from hypersaline through hyposaline sebkhas, which may even occasionally completely dry up through the drier months of the year.

Sediments accumulated in coastal sebkhas are usually fine to very fine eolian sands, silts and clays, which may alternate with evaporites such as rocksalt, gypsum and dolomite, either deposited from the hypersaline water, or by total evaporation, when the sebkha dries up. These evaporitic crusts are quite common and in most cases protect the sediments from oxidation, thus preserving the organic contents, which frequently makes their dominating color black. The amount of organic material in sebkha sediments only depends on local production, which can be very high in places such as mangrove stands, or where algal mats are common. This organic contents may prove beneficial for the preservation of pollen on one hand, but may also form a very good substrate for bacteria of various kinds, some of which could entirely destroy the exines. Quite frequently the excess organic matter would deserve special treatments while macerating for pollen, such as partial oxidation or acetolysis. In most cases, however, coastal sebkhas proved to be a very good source for pollen, which are abundant and usually well preserved. The rich organic contents is also quite useful for radiocarbon datings, in the younger sequences.

III. Coastal dunes. Coastal dunes are very common in arid environments and may sometimes travel quite far inland, climbing considerable elevations (Fig. 3.3.5.2). The sand grains are mostly quartz or carbonates, depending on

(a)

(b)

Fig. 3.3.10.2. Coastal sebkhas of the northern Red Sea. a. Behind a coral reef, as the poor captain discovered too late. b. Enclosed by mangroves.

source materials, but may occasionally be gypsum, especially where coastal sebkhas are subject to wind deflation. As is the case with alluvial fans, the rapid movements of such sands make the assessment of a detailed stratigraphy very difficult. Also, the nature of deposits hinders, to a great degree, pollen preservation. The main difference between coastal and inland dunes is that the former are deposited by winds coming from the sea, whose pollen load is very poor to begin with. Also, for the same reason, coastal dunes comprise very little finer fractions, which makes any prospect for pollen quite doubtful.

IV. Other coastal environments. Desert coastal and intratidal environments can accumulate a variety of sediments, mostly clastics or bioclastics, from coarse through very fine, depending on supply and energy. Some of these, like subtidal fine silts, can yield well preserved pollen assemblages, but most of the coarser clastics seem to be a rather poor prospect.

B. Deltaic

Deltas are developed at deserts margins only when a large river crosses the arid region and pours its load to the sea. Such rivers never originate in the desert itself, but in some remote, humid region, typically the tropics, like the Nile (Fig. 3.3.10.3). Although the water come from a far away provenance, some amounts of sediments are usually added to such rivers on their way, to be also included in the deltaic deposits. These, however, are not different than other deltaic formations deposited elsewhere, characterized by medium to fine clastics, with some organic compounds added from the well developed delta vegetation and deposited at a very high rate. The high rates of deposition, coupled with the relatively steep relief of the continental slope on which sediments are deposited, aided by large amounts of water enclosed within these sediments, make delta deposits prone to slides, slumps and turbidity currents, which may often blur the stratigraphy. Thus, although this environment is favorable for pollen deposition and preservation, measures must be taken to ensure slumps had not affected the sampled sequences. Occurrences of turbidites should warn against regarding a section as continuous. Incidentally, slumps and turbidites can be easily detected by X-ray photographs (Fig. 3.3.10.4) of a collected core, while still in its plastic tube, an advisable check before any further labor is invested.

Provenances of sediments and palynomorphs carried by a large river crossing a desert can vary in time, following climatic or structural changes in its headwater region or on the way, which may be reflected in its deltaic deposits. Such is, for example, the situation with the Nile, which, during the dry Quaternary interglacials takes most of its water and sediments from the tropics, depositing fine silts and clays in the delta; while during the wetter glacial periods, when water was more abundant in the Sahara (Faure, 1969; Van Campo, M., 1975; and others) and at the same time monsoons have been less active in the tropics (Rossignol-Strick, 1983),

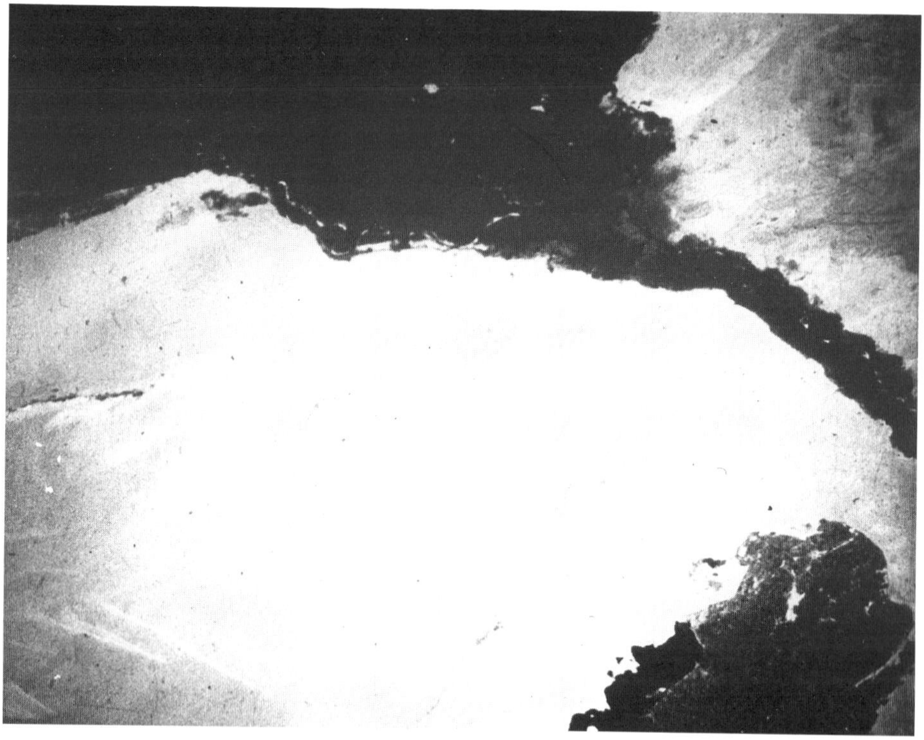

Fig. 3.3.10.3. The Nile Delta, as seen from space.

Saharan sands comprised the main load of the Nile and were deposited in its delta (Horowitz, 1979).

A rare case is the occurrence of inland deltas in arid regions, such as the Okawango Delta in northern Botswana, which disappears under the Kalahari sands. In terms of deposits, these are not essentially different from deltaic sediments formed at the contact of a large river and the sea, but since the gradients are less steep no turbidites are known from such an inland delta. Eolian contributions to both sediments and pollen are greater in such cases.

Rich, well preserved pollen spectra had been recovered from many deltas, including those at desert borders, proving to be a valuable source of information on both their provenance region and the traversed arid land. Two factors which need to be established beforehand, prior to drawing conclusions, are the stratigraphic significance of collected sequences and provenances of the pollen grains.

C. Marine

Shallow marine environments are characterized by predominantly biogenic carbonate deposits in the hot regions, originating from the well developed fauna

Fig. 3.3.10.4. X-ray of a sea bottom sediment core, showing distortions in bedding due to slides and slumps. Core diameter 10 cm. Courtesy of G. Almagor.

and flora. Thus the typical shelf sediments are various kinds of carbonate sands, which are subsequently cemented into carbonate rocks. These are somewhat disturbed where large wadis enter the sea and clastic deposits are accumulated (Fig. 3.3.10.5). In localities close to large deltas, the longshore currents spread deltaic sediments on their way and these may extend for hundreds of kilometers, as is the case with the Nile sediments, which reach all the way through the Sinai coast and up north, to northern Israel (Horowitz, 1979).

The shelf sediments, which are mostly coarse to medium clastics, whether originating from the fauna and flora or from the wadis and deltas, are usually

Fig. 3.3.10.5. Shelf sediments in an extremely-arid region, showing carbonates cementing clastics brought by a wadi, forming beachrock. Scale bar 10 cm. Bay of Elat (Aqaba).

rich in oxygen, due to the perpetual water movements, burrowing activities and the flourishing algae. As a result, pollen preservation in these formations is quite poor (Rossignol, 1969b; and compare also Fig. 2.2.10), although occasionally decent spectra are obtained, probably where pollen was deposited directly from the air and buried quite rapidly.

Slope sediments are finer, comprising sometimes considerable amounts of fine clastics, mixed with plankton. Benthic life is less developed as compared with the shelf and sediments are usually pollen productive. However, the same warning which pertains to deltas holds true also with the slope sediments, namely, they are prone to slumps. Indeed, a detailed survey along the Israeli continental slope indicated that slump scars and heaps make most of its topography (Almagor, 1976). Thus, care should be exercised with the stratigraphy before relying on sections collected from such environments. Deep sea sediments constitute very fine, well bedded clastics, not too rich in oxygen or burrowing benthos, thus seem to be ideal for pollen analysis. Their rates of accumulation are slow to very slow and cores collected may represent a considerable time span. In some cases, where seas are confined by narrow straits, such as the Mediterranean and the Red Sea, stratification of the entire seawater body may occur under certain oceanographic conditions, causing stagnation of the sea bottom, which results in the formation of black, highly organic sapropels. Many researchers have tried to connect these sapropels directly with paleoclimates, either glacial or interglacial (*cf.* Rossignol-Strick, 1983), while others (*cf.* Horowitz, 1987a) thought that the stagnation is a result of the water balance and showed that sapropels have been deposited in the Mediterranean in times of a positive water balance, when the sea received excess amounts of freshwater either from melting glaciers, during and particularly at the termination of glacial periods, or from the Nile, in interglacial times. A further discussion of these processes is given in Chapter 9. Both the "normal" deep sea sediments and sapropels yield pollen, but it seems that preservation and differential destruction processes are different for those two types of deposits (Cheddadi, 1988). It seems that pollen is much more abundant in the sapropels, up to three orders of magnitude more than in the yellow-buff, oxidized silts. Pollen recovered from the sapropels seem also to be better preserved. It cannot however be ruled out that the higher pollen concentrations in sapropels are not merely a result of their slower rates of accumulation, but of better preservation, or a combination of both.

Another problem which should be given a considerable attention while dealing with deep sea sediments, is that since they are being laid down quite far away from the arid land's coasts, they may include, in addition to pollen originating from the xerophytic vegetation, grains transported by winds and sea currents from yet other, far away environments. If these remote sources are high pollen producers, the local, desert ones, may be masked to a great extent (compare also discussions in Chapter 2), thus the ensuing picture can be subject to different interpretations, some of which may subsequently, following further studies, turn entirely wrong.

Marine environments around cold deserts are quite rare, the latter usually located deep within continents. Marine sediments of cold seas are usually dominated by fine clastic deposits.

3.3.11. Environments of deposition, accidentally located in arid lands

By this heading, reference is made to environments of deposition which are found in arid lands due to heritage of geological processes, but do not really belong there.

This case is quite clear with extremities, such as large rivers crossing otherwise barren lands (Fig. 3.3.11.1), or ancient glacial landscapes found in a hot desert. It becomes, however, more problematic with sedimentary formations which are usually found only in deserts, thus frequently thought typical for this environment. Due to the fact that most arid lands are subject to erosional processes, more than to depositional, such features can become very common surface phenomena.

One such example is the huge fan deltas found in almost any dry land, especially

Fig. 3.3.11.1. Space view of the River Nile, crossing the Sahara (Red Sea at the top).

where relief is prominent. It was found out (Horowitz, 1979) that most of these fan deltas in Israel (Figs. 3.3.1.4, 3.3.10.1) had been formed under rather humid climates, in the past, but are indeed mostly eroding today, with very little sediments added during the present day floods.

Another example, which seems to need some more consideration, is the occurrence of red beds. Glennie (1970) brings forth a variety of opinions about red beds and their applicability as typical desert sediments. According to Glennie (it seems that opinions have not changed much until present), the only subjects on which all geologists agree are that most red bed sequences are terrestrial in origin, while the causes of color are the presence of red stained clays and a coating of red iron oxide on grains of sand and silt. Other than that, most researchers think that humidity is necessary for the oxidation of iron-rich minerals such as hornblende, augite and pyroxene, to form the red coloration.

Walker (1967) claimed that the red color had been formed after deposition, in an arid environment and defined six critical factors for this process:

(1) Presence of iron-bearing grains in the original sediment.

(2) Post-depositional conditions favoring intrastratal alteration of the iron-bearing grains.

(3) Eh–pH interstitial environment that favors formation of the ferric oxide.

(4) Absence of subsequent reduction of ferric iron.

(5) Enough time for alteration of the iron-bearing minerals.

(6) Possibly an elevated temperature.

Since these critical factors are widely accepted, I would like to make some comments on the issue. Factors 1 through 4 seem straightforward. The fifth factor, however, needs clarification as to how long is "enough time"; as an example, a red paleosol was formed on early Holocene sands in Israel during the Atlantic period, which was no longer than 2,500 years (Horowitz, 1979); moreover, presently agriculture is practiced also on sand dunes in Israel, where plastic covers are used to keep the crops in a warm, humid microenvironment; such plastic covered sands began reddening after only some 3–4 years. Factor 6 is open to a similar question, of how elevated should the temperature be, is it sufficient to have warm summers, or high temperatures should be all year round? since red soils are formed today in southern Europe, this factor seems of little significance in connection with deserts. Noteworthy, aridity is not mentioned as a mandatory condition.

Presently, red coloration is typical for warm, humid climates such as the tropics and wetter subtropics with their typical laterites, or the northern Mediterranean region, where such climate is typical for the summers. Therefore, I do not accept the arid environment as suitable for formation of red beds. Rather, these had been formed whenever a humid climate, of a subtropical or a wet Mediterranean character, took over the desert. When the climate changed to arid the red beds are preserved, under the desert conditions, from being reduced again. It should be noted in this respect that red loams and soils had only been formed in Israel (Horowitz, 1979) during past humid phases, not under the dry conditions at the present day, neither during the dry interpluvials.

It thus seems that two sources can be viewed for red beds in arid lands: either oxidation in situ under temporary more humid phases, or the erosion of previously red rock formations. The red beds which crop out are only preserved in the desert environment, having in fact been formed and therefore typical for more humid conditions. Under a temperate climate, or when they are buried in marine environments, red beds are reduced and loose their typical colors (Glennie, 1970).

The prospect of obtaining pollen from red beds depends, to a great extent, on their origin. Red beds which had been formed in situ under a warm, humid climate, such as the many horizons of paleosols in Israel, usually do not contain any pollen due to oxidation in time of their formation. But these soils may also be enriched with secondary eolian silts and clays, which are deposited together with pollen grains at a later date. In such a case pollen were found to be included in red beds, but their stratigraphic relation with the reddening processes is dubious.

The other possibility, of red beds being eroded from a former red formation and only being preserved under the dry conditions in the arid land, calls for more attention of the palynologist. Since these red beds had not been oxidized within the desert environment in which they are now found, there is no reason why they should not contain pollen, which were enclosed within the secondary deposition of these beds. Indeed, such secondary red beds proved to yield quite rich, sometimes even well preserved pollen spectra (Horowitz and Horowitz, 1985).

3.4. Cementation

Cementation processes affect almost any kind of sediment and are quite common in arid environments. These have two main implications for palynology: the mechanical degradation of pollen grains caused by dissolution and crystallization processes; and the techniques which should thus be adopted for macerating samples for pollen analysis. Cementation is very common in coastal sediments, due to the high amounts of carbonates in these deposits; in sebkhas, both inland and coastal, due to the abundant evaporites; in most argillaceous sediments, due to recrystallization of clay minerals; in soils, by the leaching and formation of subsoil crusts; and in almost any other type of deposit which is in contact with groundwater.

Pollen grains, usually fairly to badly preserved, had been obtained from almost any type of cemented rock, using various acids to remove the cement.

In contrast to most inland dunes, the coastal ones bordering hot deserts almost always include carbonate grains, produced by the rich marine fauna (and some algae) of the nearby warm sea. These provide the main substance for subsequent cementation of the dunes (Fig. 3.4.1), which become calcareous sandstones. There is, however, a stratigraphic problem, as yet not satisfactorily solved, with the cementation, which may imply on the accuracy of pollen spectra that could occasionally be extracted from such calcareous sandstones. The cementation process is believed by some researchers to be contemporaneous with the dune formation, or insignificantly subsequent (Yaalon, 1967; Glennie, 1970), while others (Horowitz, 1979) suggested that the principal part of the process takes place under a different, more humid climate. The cementation involves percolation of meteoric waters through the porous sands, dissolution of carbonates and precipitation, which takes place during the drier time of the year. This process can contribute pollen grains carried within the percolating rainwater to the initial sandy deposits, thus their stratigraphic significance may be doubtful.

Fig. 3.4.1. Alternations of cemented coastal dunes and paleosol. Near Tel Aviv, Israel.

Coastal deposits are frequently well cemented, mostly from dissolution and precipitation of carbonates, originating from either the remnants of marine fauna and flora, or from seepagas of groundwater, floating on the seawater/freshwater interface and surfacing at the beaches. These make beachrocks, typical for many desert coasts (Fig. 3.3.10.5), which again, may bear the stratigraphic uncertainty similar to other cemented rocks such as sand dunes, which had originally been deposited as loose clastics.

Primary cementation of fluviatile clastic sediments in the arid zones is usually quite poor. Where no subsequent cementation occurred, even sands and gravel as old as Paleozoic may remain loose, a rather common phenomenon with the Nubian sandstones of North Africa. Secondary cementation usually occurs when water seepages saturate the clastics, which may be a result of a following, more humid period, a spring which commenced activity due to faulting and so on. In most cases the cementation is carbonate (Fig. 3.3.4.1). When, for some reason, a wadi ceases to act like one in terms of transport and deposition, soil formation processes may cause cementation by down leached carbonates, iron oxides, gypsum, or other minerals. The cement of clastic desert sediments should therefore be viewed with some care, since it may represent a subsequent period, not necessarily maintaining

Fig. 3.4.2. Cemented cave breccia, with flint and bone fragments. Hayonim Cave, Israel. Scale mm.

similar environmental characteristics as compared with the time of accumulation of the initial clastics. Indeed, it was found that in most such cases cementation was formed under at least somewhat different conditions.

Cave sediments are very often cemented by hard carbonates and form typical cave "breccias" (Fig. 3.4.2). These are sometimes very important sources of information, especially when caves had been inhabited by man, however the time relations between the deposits and the cementation should be established before any attempt at drawing conclusions is done. These breccias should not be confused with the typical stalactites and stalagmites of an active cave, which are indeed formed at the very time of the groundwater activity. Pollen obtained from the latter type of deposits usually represents the load of the groundwater at that time.

3.5. Polyphase inclusion of pollen in sediments

Pollen grains can arrive from more than a single source and be included within a given sedimentary environment, by a variety of processes. This phenomenon can be very useful when the various sources can be accurately traced, to obtain additional information from the spectra; it can, on the other hand, lead to false conclusions if care is not being taken in dealing with each of the provenances separately. Polyphase inclusion of pollen within the sediments could occur everywhere, but is much more pronounced in pollen spectra obtained from arid lands, due to their lower masking factor as compared with the forested regions and the frequently considerable lengths of pollen transportation and supply lines. The different sources donating pollen to a certain rock may be contemporaneous, causing deposition of grains from more than a single environment at a given period; or may be asynchronous, mixing pollen from different ages in a certain sedimentary horizon. Occasionally, pollen from such different ages may appear at different levels of preservation. Such cases may be easier to discern, but unfortunately this is not always the case and even when it is, convincing evidence should support assignment of certain states of preservation to assumed ages or processes.

The descriptions which follow are largely based on studies carried out on recent pollen spectra. It is fairly easy to study these, but seems much more complex with the fossil assemblages, especially when the sediments had been laid down in environments which differ from the present day ones, as is frequently the case with older rocks.

3.5.1. Multiple transporting agents

Pollen grains are brought to many environments of deposition by more than a single transporting agent. This may bring to a single site grains originating in different ecological environments, which, if not presented and studied separately, may blur the obtained picture. This process can be discerned quite straightforwardly while dealing with recent sediments, when transporting paths can be easily followed, but may be much more difficult in the past. For example, middle Holocene sediments in the Negev, southern Israel, yielded several percents of oak pollen, as compared with the recent ones, in which none are recorded. This may have resulted from either oaks growing far more south than today, or due to stronger northern winds during the mid-Holocene. It may even be a combination of both, as it seems (see Chapter 9). Similar problems of interpretation are also discussed for the late Pleistocene deposits of the Sahara.

The problem may become more complicated when the provenances of the pollen are uncertain, due to objective difficulties in their identification. Such is the case with changes in the shares of *Artemisia* in late Quaternary deposits of the eastern Mediterranean (Cheddadi, 1988). Since it is almost impossible to identify the species, it remains uncertain whether the increase in *Artemisia* is due to expansion of the desert species or the steppe populations, each of which grows in a different region.

One of the extreme cases of multiple provenances for pollen brought to a site is the marine sediments in Israel (see Chapter 2.2.3). In the Mediterranean offshore, a considerable contribution is made by the Nile from tropical lands, subtropical regions, the Sahara and the Nile Delta, besides pollen arriving from Israel; in the Bay of Elat, some 70% of the recent spectra constituents are from palynomorphs brought by sea currents from equatorial Africa, mixed with wind transported pollen from as far north as the Lebanon, with a minor share of pollen produced by the desert vegetation which surrounds the Bay.

These few examples only stress the necessity of exact identification of pollen provenances and transporting paths, which could contribute to polyphase inclusion of grains in a single sample.

3.5.2. Dust

Detailed studies of dust accretion and deposition in the Negev, southern Israel (Gerson and Amit, 1987; Pye and Tsoar, 1987), had assessed two main types of dust traps. The first involves areas with low, steppe vegetation, in which dust is added to the surface or the near subsurface of the sediment or soil, forming loess. Only a small amount of the dust penetrates deeper into the deposits. The second comprises gravelly and sandy surfaces, usually devoid of any significant vegetation cover, which trap the dust in the voids. In this case the dust penetrates deep into the preexisting deposits and forms a part of the rock. A third type of dust trap is also mentioned, hardly important for geologists but meaningful to archeologists and palynologists, which is roofless buildings and ruins. Dust is also trapped by mosses in the desert (Danin and Ganor, 1991), also of little significance to geologists, but samples for pollen analysis collected from previously wetter localities may be affected by this process.

Since pollen almost always accompanies dust, it would be quite important to follow the processes involved with dust trapping. The first type of trap only involves processes of long distance transportation, discussed above (Chapter 3.5.1). The second type involves both long distance transportation and inclusion of younger dust (and consequently pollen grains) within older rocks. Since these are gravelly and sandy and the coarser fraction always discarded during the maceration (see Chapter 5), the ensuing pollen may well represent a different period than that during which the gravel was accumulated. This may occasionally be only a later season in the same year, but may also be much later and lead to false conclusions. These two processes are also involved with ruins, which are not so important when natural conditions are concerned, but may become significant when pollen is studied in connection with archeological sites, when again, precaution should be exercised.

3.5.3. Redeposition

Of all difficulties involved with palynology of arid lands, redeposition seems the least significant from the practical point of view. Although redeposition of sediments is a very common process in dry regions, especially due to floods and strong winds, palynomorphs seem to be less prone to the process, most probably due to their fragility, which does not leave much resistance to the harsh mechanical weathering processes. The susceptibility of palynomorphs to oxidation is also a welcome virtue in cases of possible redeposition. Since the process of reworking sediments usually takes place by floods, the combination of humidity and turbulence seem to help in oxidizing the would be redeposited pollen quite easily. Consequently, very little is known of redeposited pollen in wadi deposits and the problem seems subordinate in this environment of deposition, from which a considerable share of the samples are collected for palynological studies in arid lands.

Redeposition of pollen may however occur in such environments where slumps are common. These are chiefly marine deposits of the continental slope, deltaic sediments and archeological sites, which call for some attention in this respect. Redeposition of palynomorphs is also known from such localities where older evaporites are dissolved in rainwater, groundwater, or where an evaporitic body forms a diapir which intrudes a lake, as is the case in the Dead Sea. The dissolution of evaporites keeps the grains in solution, from which they are deposited and enclosed in younger sediments. Since the entire process is carried out in solution, chances for oxidation and mechanical damage are greatly reduced. Clearly, the possibility of identifying such mixtures of older and younger palynomorphs depends on the differences between the two assemblages, in state of preservation, typical taxa and age. when the different sources are closely related, for instance in age, distinction becomes very difficult.

3.5.4. Water infiltration and cementation

Water, infiltrating through porous, permeable rocks and soils, can carry pollen and deposit the grains in another layer by a variety of ways. Rainwater can wash pollen from the surface down the sequence, sometimes into appreciable depths. This can happen synchronously, with the deposition process, which is of less importance since it does not produce false stratigraphy, but may also occur at a much later date, thus incorporating younger pollen within older deposits. This is a very frequent phenomenon in arid lands, where permeable rocks are very often laid down, such as in wadis and sand dunes. Water thus percolating can dissolve various types of rocks on their way, depositing the pollen in other horizons, younger and older alike.

Cementation processes, which also include the formation of mineral horizons within soils, are not necessarily contemporaneous with the sedimentation period. This can bring younger pollen into older rocks, as was seen with breccias in the Sterkfontein hominid site, South Africa (Horowitz, 1975; and see Chapter 2.2.4 for details), or include older pollen within younger rocks, as is seen with Quaternary clastic sediments in the Dead Sea region, into which Pliocene grains are added, through the dissolution of rocksalt of that age (Horowitz and Horowitz, 1990).

The percolating water with its load of pollen may become significant especially when analyzing soils in arid regions, which are particularly prone to such processes. It is sometimes quite difficult to assess, when obtaining pollen from a soil profile, what period it exactly represents. A detailed discussion of the problem, however in the temperate regions, is given in Dimbleby (1985). The problem in arid regions is more acute, due to the low local pollen masking factor, its high rate of initial oxidation and the frequent better sediment permeability, as compared with the temperate regions.

3.5.5. The role of springs

Springs are indeed a special case of pollen transportation by groundwater, which deserves attention in arid lands because spring deposits are frequently sampled for pollen analysis. Polyphase inclusion of pollen in spring deposits can result from several sources, such as a remote intake area of the water, dissolution of older rocks, or the spring being fed from a fossil water reservoir.

Another point that needs attention with springs concerns the problem of local vegetation which, being sometimes well developed, constituting chiefly of high pollen producing wind pollinated plants, can almost completely mask the regional, rather poor pollen producing vegetation.

A frequent case with extremely-arid and arid regions springs, somewhat less in the semi-arid, is that the intake area of their waters is quite far away from the spring itself and may be located within a different environment, in which the climate is more humid and rainfall more frequent. Consequently, the spring water may contain pollen grains which would be included in the spring deposits, indicating a more humid environment than the one in which the spring is actually located. This phenomenon depends to a great extent on the nature of the aquifer and is especially common with karstic ones, where, due to the large cavities in which the water flow, filtration and loss is minimal. It is a common case with springs in the Judean Desert, east of Jerusalem, which are fed by karstic aquifers whose intake area is to the west, in the Mediterranean domain. Dissolution of older rocks by younger groundwater flowing in the aquifers can also bring exotic pollen to the spring deposits, as is the case with springs in Lake Kinneret region, northeast Israel, where aquifers which are active today dissolve at depth Pliocene evaporites, thus analysis of water taken from the springhead and from adjacent sediments revealed a wealth, sometimes up to 40% of the spectra, of Pliocene pollen and spores (see Chapter 2.2.4 for details).

Another phenomenon, which is quite abundant in deserts such as the Sahara, Sinai and southern Israel, is the occurrence of fossil aquifers, in which water had been accumulated during the wetter late Pleistocene. These are chiefly sandstone aquifers, in which filtration is somewhat better than in karstic ones, but the fossil water still contain some pollen grains evidencing the former, more humid period of water intake. These aquifers are at present hardly ever replenished.

All these factors, if not properly taken into consideration, can easily lead to incorrect conclusions.

3.5.6. Human and animal activity

Polyphase inclusion of pollen by human and animal activity is mainly due to burrowing and building operations, which can move sediments up or down the section. Burrowing animals are quite common in dry lands, belonging to almost all terrestrial phylla. Human activities are usually confined to prehistoric and archeologic sites and are relatively easily discerned, while burrowing by animals can occur anywhere, sometimes quite difficult to observe. It is especially difficult when sampling is done by boring, which is occasionally necessary in order to obtain non-oxidized material.

Inclusion of pollen from far away provenances can be a result of feeding and migration habits of birds of prey, discussed in Chapter 2.2.5.

3.5.7. Artificial contaminations

However evident, artificial contaminations should not be neglected, especially in arid lands palynology, where, due to the initial low pollen contents may override the spectra. Two "environments" are particularly prone to contaminations: drilling activities, producing borehole samples and laboratory procedures.

Drillings in which cores are being taken seem to be quite safe as long as samples are taken from inside the core, as compared to those where only well-cuttings are available. Unfortunately, it so happens that coring is very expensive and when samples are collected from water, oil, gas, or any other commercial drillings, usually only cuttings are available for the palynologist. Provided care is being taken, cuttings are almost always a valuable source of information, especially when no other choice exists. Probably the best check of the results would be, whether these recur in several drilled sequences. An example is the sedimentary fill of the Dead Sea Rift, which attains several kilometers in thickness and was penetrated by more than ten drillings. Only when results started recurring, conclusions could be drawn with some confidence (see a detailed discussion in Chapter 9).

Two main sources of contamination can occur in cuttings: the more common is cavings, which are pieces of rock enclosed within a deeper sample on its way up through uncased drill shaft, contaminating it with younger palynomorphs. A thorough petrographic analysis of the cuttings, by an experienced field geologist or well-sitter, can avoid most of the troubles. A less common cause which can introduce exotic pollen into cuttings is the mud or water used for cooling and lubrication of the drilling bit. The mud usually comprises some proportion of clays, which could have originally contained palynomorphs that may contaminate the cuttings. As a measure of precaution, a sample of the mud should be processed for its pollen contents before the cuttings are being studied. If the mud is indeed palynologically exotic to the drilling site and stratigraphy, being imported from another region, or differing significantly in age, the problem is minimal. In drillings where pollen bearing mud of local origin is used, it is sometimes very difficult to ascertain what the amount of contamination is.

Water used throughout the drilling activities is almost always of local origin, but could also be sampled and studied beforehand. Except for rare cases, however, water would contain only present day contaminants, which are frequently discernible by their state of preservation, susceptibility to staining or the occurrence of protoplasm. A first warning for possible contamination of borehole samples by water is the unmistakable appearance of pollen produced by introduced plants, such as, for example, *Eucalyptus* in Israel.

Laboratory procedures can introduce exotic pollen in two main ways. If the laboratory is not cleanly sealed and its air filtered, pollen from nearby plants may be included in the slides, which again, due to the low masking factor, can be troublesome. This can be solved by using a clean laboratory, or, if the processing of samples is made far away from the arid lands, the contamination is readily seen. For example, a good indicator for such contamination in Israel, as mentioned above, is the occurrence of *Eucalyptus* pollen in the slides, since this tree was introduced only less than a hundred years ago to the region. It is however, useless when recent spectra are studied. The problem may turn more serious when the contamination consists of pollen that are otherwise also present in the spectra, such as cypress and pines in Israel. In these cases only the state of preservation may become helpful for distinction.

The second possible source of foreign pollen is the water and chemicals used throughout the extraction processes, which can also contain exotics. A method to discern both contamination sources is to run a sample known to be sterile, as a control. If pollen is found, it indicates probable contamination, the source and nature of which should then be ascertained.

3.6. Quantitative relations of pollen and sediments in various arid lands environments

It is quite difficult to assess ahead the quantitative relations of pollen and sediments in various arid lands environments, or, in other words, what would be the expected yield of a given sample. Three main factors contribute to the pollen concentration in a given sample: the initial pollen production; the rates of deposition of both pollen and sediments; and the subsequent preservation and differential destruction of the grains. A fourth, artificial factor, does not affect the natural ones, but may be of importance to the final results — the methods used for extracting palynomorphs from the samples.

If pollen production is taken as a constant (which in practice is not the case, since various plant communities produce different absolute annual amounts of pollen), then the rate of deposition of sediments would be the principal factor in determining the expected yield. This has also some bearing on the principles of arid lands palynology — success in obtaining a decent pollen spectrum from a relatively large sample may only be the result of a high rate of sediment accumulation and is therefore significant in terms of stratigraphy and reconstruction of past environments.

Leafing through the various environments of deposition discussed above, high rates of sediment accumulation and consequently "poor" pollen yield, are usually found in wadis, sand dunes, some archeologic sites, alluvial fans, lakeshore terraces and some parts of deltas. Low rates of sediment accumulation are typical for inland sebkhas and salt-pans, marshes, deeper parts of lakes, loess blankets, springs, some oases, some caves and rockshelters, some biogenic and most deeper marine environments. However, not all environments in which rates of deposition are low act as good palynologic prospects, since the differential destruction and preservation may completely alter expected results.

Unfortunately, though, when a palynological study is planned for an arid land, the choice of sampling sites is never extensive and one has to do with what is at hand. Since each environment has its preferences in pollen accumulation, it seems best to try and process as many samples from as many different environments of deposition as possible. However, while this principle is always true, compromise must usually be made between the desired and the available, to make studies practical.

Chapter 4

Sampling

Sampling procedures in arid regions differ from those applied in richly vegetated ones, for several reasons. The first, apparently most important, is the relative scarcity of pollen in arid regions sediments, which initially involves processing a much larger amount of material from each sample intended for pollen analysis. To that, logistic problems may also be added, as in many cases samples are collected from an arid region during a relatively short field season, to be later on processed in a laboratory which might be quite far away from the field. Larger samples are therefore needed both because of the probable scarcity of pollen grains in the sediments and, if a further run of extraction is found indispensable, it may be time consuming and sometimes very expensive to get back to the sampled locality. It would thus be helpful to have the material at hand. Another reason for collecting a large amount of material is that sometimes, especially in arid regions, the prevailing elements of the flora, such as *Artemisia*, produce considerable quantities of pollen grains, dominating the spectra. It may occasionally be found necessary to use minor accessory elements to distinguish, for example, between a cold and a warm *Artemisia* steppe. Such accompanying palynomorphs are usually quite rare, thus considerable numbers of pollen grains have to be identified and counted in slides before a proper solution could be established to such questions.

Another problem, of a similar nature, lies with the local vegetation. If sediments are collected from a salt-pan, a spring or a temporary marsh in the desert, the local, halophil or hydrophil plants which grow close to the water body, not representing general climatic or environmental conditions, would frequently mask the regional information. In such cases an additional extraction may be needed to provide a larger number of palynomorphs, so that enough regional pollen can be counted. For these reasons, the amount of material collected must be much greater than in forested regions, where in most cases a single gram or two of sediment is sufficient for any possible need.

The size of a sample to be collected would depend on the natural conditions which had determined the expected absolute frequency of pollen per gram sediment; on distance of the sampled site from the research facility, which determines the ease with which resampling could be done; on laboratory techniques applied for the pollen extraction; and also on the individual palynologist's personal preferences. Mehringer (1967) collected some 200 grams from arroyo outcrops;

I found out by experience, that at least one kilogram of sediment should be collected, wherever possible even more than that, to allow for all above mentioned limitations.

Needless to stress that the general rule for collecting samples for pollen analysis, namely the cleanliness of the samples, is self evident and is even more important in arid lands, due to the initial low pollen contents of most samples.

Since pollen production is usually poor in arid lands, it is quite important to try and find out, or at least estimate, rates of deposition of the sampled sediments, so that the collected samples would yield significant pollen spectra. Clastic sediments, that are the most common type of deposits in arid regions, are generally deposited at a rate which is proportionate to the grain size. Fine silts are accumulated quite slowly, while sands and gravel accumulate at a higher rate. However, this rule dose not always hold true, thus a thorough study of the stratigraphy of the sediments and their accumulation rates should be carried out for a good estimate of the samples size and also for defining the required vertical sampling intervals along a section.

Besides mechanical destruction, the main factors involved with the preservation of pollen grains in arid lands sediments are connected with oxidation and cementation processes, which are much more potent in these regions as compared with the forested. Bearing in mind the problems of oxidation and cementation, a palynologist would usually prefer to collect loose, fine grained, dark gray to blackish sediments, over the yellowish and certainly the reddish-brownish coarse clastics, well cemented rocks or evaporites. However, some precautions should be taken also in this respect. Firstly, it was found out that dark gray to blackish sediments indeed usually contained pollen grains, but on the other hand, may also contain considerable quantities of other organic substances, which may be difficult to get rid of during the extraction processes. It was found out by experience that, while trying to oxidize the excess organic material, most of the pollen grains are also destroyed. On the other hand, if oxidation is not applied, slides prepared from dark gray sediments are often very difficult to study, because of the visual masking made by the organic material. Another pitfall frequently occurring in arid lands is manganese oxide, which renders the rock a blackish color, although entirely oxidized.

It seems that yellow-buff, fine grain sediments are probably the most promising. However, one should not refrain from collecting even reddish or brownish deposits, which on a first glance would look completely oxidized. Many of these proved to have been oxidized prior to their last phase of deposition in the desert, carrying along their oxidation colors on the one hand, but containing also pollen grains on the other. The red breccias of Sterkfontein Cave in South Africa (Horowitz, 1975) and many paleosol horizons in Israel (Horowitz, 1979) are examples of red sediments which do contain pollen. In other words, if one works in the desert, one has to collect almost any available sample, regardless of first impressions, to give it a try, as long as more suitable materials are not available.

It is unfortunate that sequences ideal for pollen analysis are quite rare in deserts; if only these would have been looked for, the amount of available information could be minimal. Therefore, any kind of sediment, provided it is within stratigraphic context, may prove worthwhile for examination and experiments of extraction in various possible ways, which of course involves a larger sample to start with, to try and find out whether it does contain pollen or not.

A note about those pollen "poor" samples. Personally, I do not believe in pollen poor samples, because if a sample yielded several pollen grains, a larger sample from the same horizon would yield more, until a significant pollen spectrum could be obtained. It involves much more laboratory work, it calls for collecting bigger samples, but it frequently pays off with the information yielded by its pollen contents, concerning the environment and other data of the sampled horizon. It is therefore worthwhile to try once and again, because in most cases we found out that pollen spectra could be obtained from almost any sample collected in an arid region.

Another factor, which is not unique to dry lands but should be taken into consideration while sampling for pollen analysis, is the vertical interval between samples in a given sequence. Here again, the rate of deposition of the sampled section plays a very important role. It seems, for example, of no use to sample at intervals of five centimeters a sequence in which one meter may have been deposited during a yearly flood. This of course is an extreme case, but it has happened more than once in the deserts. The sampling intervals have to be decided upon considering two factors: one is, as said above, the rate of deposition, for which a good dating framework of the sequences is indispensable and must be available; the other concerns the type of information one would like to obtain from the studied sequence. It is quite clear for example, that general environmental trends necessary for regional palynostratigraphic studies, or for economic purposes such as drillings, require much less counting and consequently a smaller sample than, for instance when trying to reconstruct plant communities in detail, or to delineate ancient agriculture.

One should also bear in mind the practical capabilities of the palynological laboratory and the analyzing palynologist. As an example, Quaternary sequences in Israel attain sometimes up to 3–4 kilometers in thickness (Horowitz, 1987a). It does not seem practical to sample such sequences every five or ten centimeters, because to analyze these would take more than a lifetime, while the general tendencies of the Quaternary environments and the general stratigraphy could be studied by analyzing samples collected every ten or twenty meters. A compromise, at any rate, must be agreed upon between the theoretical and practical considerations and stated clearly.

A rather significant and interesting discussion of the statistics, techniques and implications used for sampling and subsampling of palynological sequences is given in Weinstein-Evron (1987b), who shows that for different sequences, different sampling intervals are needed, which depend on rates of deposition and the required information. Thus subsampling of palynological sequences can be an efficient means of achieving reliable results, while avoiding disturbing noise. In addition, such procedures can save considerable time and effort, particularly in the study of boreholes which penetrate long sequences. Weinstein-Evron therefore recommends that one should first sample the sequence, study in detail a few sections from various depths, statistically smooth their graphs, evaluate the desirable amounts of subsampling and only then proceed with analysis of the remaining sections, each with its appropriate subsampling fraction. It is quite clear that such a procedure is not unique to cores or drillings, but applies as well to exposed sequences.

Weinstein-Evron suggested that during this process occasional, more detailed checking will be done on the subsampled parts of the section, in order to validate the subsampling rate. This could especially be significant for transitional or problematic sections, such as when a change in the rate of deposition is indicated by the datings. The rate of subsampling should be determined individually for each and every site, but the general procedure seems to be applicable to different locations. When a series of sequences is collected from a single site, subsampling should preferably be performed independently for each of them.

The stratigraphy of sampled localities in arid lands has to be studied in detail, because temporally alternating environments can have different manifestations in erosion and deposition. Wadi sediments in southern Israel (Goldberg, P., 1976, 1986; Horowitz, 1979) have only been accumulated during more humid climates, in times of perennial activity (Fig. 3.3.1.3), while erosion prevailed in the dryer intervals. Thus, a palynological study of seemingly continuous wadi sequence in this region may result in obtaining a rather smooth curve, indicating an environment which would seem to have been continuously more humid than the present day. Only a detailed geological study, which would discern possible hiata in such sequences, could thus provide information on their applicability in deciphering continuous trends.

An opposite situation exists in the coastal plain of Israel where the sediments, especially sand dunes and coastal deposits, have been laid down during periods of high sea levels, which correspond to dry environments correlative with the European interglacials. Erosion prevailed in times of regression and low sea levels, during wet climates synchronous with glacials. Thus, the sequence obtained from boreholes in the coastal plain (Rossignol, 1969b) represents seemingly continuous dry environments for the entire Quaternary.

The probability of such processes should evidently be taken into consideration while sampling sequences in arid lands and it must be carefully evaluated whether a sequence is continuous or interrupted, that may involve depositional lacunae or erosional stages. Depressions which are usually filled up with sediments seem to be a very good source for continuous sequences in temperate and tropical environments. But in arid lands some of these may also prove problematic, such as salt-pans (Fig. 3.3.2.2), in which at certain times winds blew out sediments from the depressions, or the water level was lowered, creating lacunae in a seemingly continuous sequence. Care with this respect should particularly be practiced while sampling drillings, especially those from which only well cuttings are available, since in these such missing horizons are much more difficult to detect.

To conclude, sampling in arid lands should follow a detailed stratigraphic study of the region; involve collecting rather large quantities of material for each sample; not avoid collecting samples even if they do not look primarily promising by their color or texture; should involve sampling all available sequences and not only the ideal ones (which may be more often than not altogether absent from a given region); and sampling intervals should represent a compromise between the desired level of information and the practicability of processing and analyzing the material.

4.1. Airborne pollen

Sampling airborne pollen in arid lands is not essentially different as compared with other environments thus techniques applied are quite similar. The problems involved with sampling in arid regions are logistic rather than technical, the usual case being insufficient sampling, either in quantity of material per sample or in the spacing of samples, vertical and horizontal. This is because most arid lands researches are carried out by expeditions rather than by people who practically live there and can sample on a regular basis. It is quite rare that such an expedition would be able to stay an entire year, which is the least significant period for establishing a clear picture of the airborne pollen distribution.

Various kinds of pollen traps are described in detail by Ogden et al. (1974) and need not be reproduced here. The principle of most traps is quite straightforward — a sticky surface is exposed to the air for a certain amount of time and is then taken to the laboratory for maceration. Most such pollen traps are also protected in one way or another from the rain, which may wash the trapped grains — a precaution which may sometimes seem superfluous in the desert. However, it better be there, chiefly to protect against the strong solar radiation which may deteriorate the pollen grains even before these are collected and brought to the laboratory. Weinstein (1979) had very good pollen yields from very simple traps, made from two sheet metal boards. The lower, 20 × 20 cm, on which a glass plate of almost the same size, covered with glycerin jelly was placed, while the upper protecting it 20 cm above, measured 30 × 30 cm. The two sheets have been connected by metal rods, welded to both and protruding below the lower, to serve as a stand.

Several researchers had used gauze filters wetted by a sticky, non evaporating liquid instead of plain surfaces, which serve to trap the pollen load by filtering it from the passing wind. Usually both horizontal and vertical filters were placed, so that the wind load could be compared with the actual vertical pollen rain at each location. This technique was suggested by Cour (1974) and subsequently used by many students of arid lands (Schulz, 1984; Van Zinderen Bakker, 1984; and others). Cour also used an air vane, to increase the concentration of trapped pollen by facing the trap perpendicular to the wind direction at all times. Such a trap would naturally only give the average airborne load, while if more detailed information is needed on the provenances of the pollen, it would be best to locate the traps each across the path of a different wind, preferably only at a time during which each blows.

The type of sticky substance used by various pollen hunters is usually either silicone oil or glycerin jelly. It seems that the first is preferable in arid lands, due to its better resistance to drying up, although the maceration is somewhat lengthier, unless the mounting is made with the same substance.

4.2. Recent sediments

Collection of recent sediments can be performed in every environment in arid lands in a manner similar to other regions, provided that several points are considered. The first and most important is, one must make sure that the samples are indeed recent. Two factors are involved with terrestrial deposits: erosion, especially wind deflation (Fig. 3.1.2.1), can expose older beds to the surface, which may thus create a false impression of being recent; and, very low rates of deposition in certain environments, may result in samples representing a much longer time than initially desired by the palynologist.

When collecting underwater recent sediments special care must be taken not to sample slumped layers, which may quite frequently occur in deltas and continental slopes, or in deep tectonic lakes. Most of these factors could usually be checked while analyzing the slides, when occurrences of pollen derived from recently introduced plants could be used for indicating the recency of samples, when they occur in sufficient quantities. However, for practical reasons this check could be too late, for it may sometimes be difficult and quite costly to revisit the sampling site for a more suitable set of samples.

As a rule, when sampling wet deposits it is advisable to add to each of the samples a measure of formaldehyde before the bag or jar is sealed, especially if some time is expected to pass between collection and maceration of the samples, even in the order of several days. This would deter bacterial activity and help preserve the pollen intact; also, formaldehyde being a reducing agent, would protect the sample against further possible oxidation, to which wet sediments are particularly prone.

Sampling underwater recent sediments is usually done by dredging, or the use of the topmost layer from a core or drilling. It is important to make sure that dredging does not penetrate too deeply into the top layer of sediments, as may be the case with some types of dredgers, or, if it does, that the age of penetrated deposits is well established. Such procedures apply to lakes, some of the deeper marshes, pools connected with springs and waterfalls and most marine and deltaic environments. Mehringer (1967) suggested that for obtaining an approximation of the pollen contents of soils, it would be preferable to divide the sample into subsamples. He collected in each area between 10 and 40 such subsamples, mixing them together before proceeding with maceration. He also suggested that, in order to minimize contamination, it is preferable to collect the samples in winter time, when most flowers are not yet in blossom.

Similar results to those obtained by the subsampling procedure have apparently been also reached by a technique used in the Sahara by several French palynologists (Cour et al., 1973; Van Campo, M., 1975). A filter, coated with some sticky substance, was placed on a rod connected to the back of a moving vehicle. The dust raised by the rolling wheels was trapped on the filter, which seems statistically equivalent to the subsampling described above. When applying such methods it is important to make sure the vehicle indeed moves only on recently laid deposits and not, for instance on wind deflated surfaces, so common in deserts.

It is sometimes desired, however, to study recent sediments in a somewhat different manner, to establish the natural processes before disturbances were made by man's activities. In such a case, it would be preferable to sample subrecent sediments, only care should be taken to define how old the "subrecent" would be in any given environment. This could, for example, be risky with arid lands soils in such cases when these are not forming today, where a seemingly "subrecent" sample may represent an ancient period, frequently up to several thousand years old.

Wadis and arroyos seem to be the most common sources of recent samples in arid regions. It seems that the best samples for recent sediments are the mud cakes which are deposited at the end of each flood, separated by typical mud cracks into "sample size" portions (Fig. 3.3.1.2). These seem to best represent pollen deposition processes in the wadi and are easy to collect and macerate, since they only comprise the finer fraction of the sediments. Drying up by the sun seals these mud cakes from further contamination and also helps protect against subsequent oxidation. Recent muds from perennial streams, where these occur in arid lands, can also be easily collected and again, seem to represent fairly well the contemporaneous pollen deposition processes. It is also possible to sample the sediment load of the floods' or streams' water, by using sediment traps, as was shown by Peck (1973). This may be important in a region which is subject to numerous annual floods, but it should be warned that the technique is quite costly and implies extensive labor.

Sampling sebkhas, both inland and coastal, as well as salt-pans, marshes and oases is a straightforward process, where the uppermost layer is collected. Here too it is important to first decide whether recent or subrecent spectra are desired, which would determine the maximum required depth below surface of the collected sample. Some drops of formaldehyde would be very useful to conserve the samples.

Sampling recent sediments from eolian environments may sometimes be quite tricky. It is usually not advisable to sample sand dunes, first because their pollen content is very poor and second, since the sand grains and consequently the pollen, if they are there at all, are in a continuous state of movement and redeposition. Older, stabilized dunes, in which these processes had already been ceased, do not in most cases represent the recent environments and should therefore be avoided. On the other hand, loess makes a very good source for recent and subrecent samples, usually containing numerous pollen grains, which are quite abundant and well preserved in this type of deposit. But the main limitation with loess is its abundance in only the semi-arid regions, while it hardly ever accumulates in the arid and extremely-arid domains. Loess, or rather windblown dust, can occasionally be sampled in the latter environments, from hollows in the rocks and similar traps, but the age of such samples has first to be very carefully established.

The problem of the exact age of the sample is more complex when soils are being collected, both due to the almost free vertical movements of water, minerals and pollen within the soil profiles and the fact that many of the soils presently situated in arid lands are not indeed of recent age. Paleosols had frequently been formed in arid regions under different environmental conditions, usually a wetter climate and richer vegetation cover. these paleosols have for some reasons been protected, at least in part, from subsequent erosion or deflation, thus constituting the topmost present day layer, seemingly recent.

Spring deposits are easy to sample, although extraction could be tedious due to the high amounts of carbonates and organic remains. A problem may arise there, since some of the desert springs make themselves mounds of travertines, in which the water may change outlet quite frequently, thus necessitating detailed datings and stratigraphic studies to establish which part of the mound indeed represents the recent conditions. Results obtained from spring deposits should be treated with considerable care, since in most cases there is a marked over representation of pollen produced by the local flora, which may occasionally completely mask the regional elements. Similar precautions should be applied with the study of sediments collected from oases, in which hydrophil vegetation is well developed.

Caves and rockshelters can be very good sources for recent sediments in the arid regions. This is especially significant when older sequences are planned to be studied from these environments of deposition. Both eolian and biogenic sediments accumulate in these sediment traps, but erosion can also occur. Removal of top layers, especially by water currents, is a well known feature in many caves in the Near East, particularly those connected in one way or another with karstic activity. It should therefore be established, before sampling, that the topmost layer is indeed of recent age and not exposed as a result of erosion. Caves and rockshelters which mainly accumulate eolian deposits may have done so under a certain wind regime, but hardly trapped any dust under a different climate. This may also be the reason why, in many such environments, the topmost layers are not of recent age, as can frequently be seen when analyzing their human remains.

It is also important to note, in this connection, the direction in which the opening of the cave or rockshelter is facing, in relation to the directions of prevalent winds and the distribution pattern of the vegetation in the region. It would be very risky to sample recent sediments from one cave and apply the results for conclusions drawn from a sequence obtained from another, facing a different direction. Davis (1990) suggests collecting samples from the center of the cave's chamber, where eolian deposition is most rapid. He also indicates that pollen spectra differ in various parts of a cave or rockshelter, thus it is advisable to collect from more than a single locality and mix, to obtain an average spectrum. Weinstein (1981) reported good results from samples collected by sweeping dust off the cave's walls.

One of the more complex problems in arid and especially extremely-arid regions, is the appropriate sampling of recent sediments which would later be used as a basis for interpretations pertaining to archeologic or prehistoric sites. The problem is that most of these sites are presently abandoned, thus it is very difficult to find exact recent counterparts whose pollen deposition processes would parallel those of the past settlements. It is therefore, in most cases, only possible to sample recent sediments from the vicinity of a site, which would represent the natural conditions of the present day, or any recent changes resulting from man's activity, which would be superimposed on the natural processes. Comparisons with pollen spectra obtained from the archeologic and especially older, prehistoric sites should therefore be viewed with great care.

Recent samples from biogenic environments, such as excrements, pellets, food reservoirs and midden, can easily be obtained by checking that the organic material comprising the sample is relatively fresh. It should not, however, be too fresh, because as long as it is still wet and sticky it acts as a pollen trap, absorbing grains that would consequently have formed a part of a fossil sample, such as a coprolite. The best recent representatives seem to be organic accumulations of feces, midden, pellets and the like which are several months old and had time enough to trap all the pollen they could and become thoroughly dry. Needless to say, this also makes for a more pleasant collection. Davis and Anderson (1987) report that packrat midden are not a homogeneous deposit in terms of its pollen contents, thus various parts of the midden may yield different spectra (due to collection by the animal at different seasons?). It is therefore necessary to collect several subsamples from any particular midden, to obtain an average pollen representation.

Coastal environments in arid lands are not essentially different from such localities elsewhere in terms of their sampling procedures, except for dunes, which are usually avoided and alluvial fans, in which the age of the samples should be very carefully assessed, due to the repeated depositional and erosional processes common to these environments of deposition.

Special attention must be paid while sampling recent sediments from environments which are not indeed typical for arid lands, but are only situated there due to some inheritance of past geologic histories. This implies to large rivers crossing deserts, such as the Nile (Fig. 3.3.11.1) or the Senegal, in which the recent river muds would usually yield pollen originating in their source area, in numbers which may exceed those produced within the surrounding desert. Red beds, frequently occurring in dry regions, can serve as sources for recent samples provided it was established beyond doubt that these are redeposited, recent accumulations of formerly red materials. In such a case, when the reddening processes had long been terminated, the red sediments proved to contain unoxidized pollen. Red beds are therefore worth sampling and analyzing for pollen, especially when no other sediments are available.

Water tanks, which are very common in farms situated in arid and especially semi-arid regions, can be an excellent source for recent samples, as suggested by Martin (1963) and other palynologists, especially in the American Southwest. The sludge collected at the bottom of these water tanks can easily be obtained and processed. It is again suggested to add some formaldehyde to these samples, for better protection of the pollen. The only problem which is apparent with such samples is that they cannot be free from human intervention in the pollen spectra, by the introduced and cultivated plants. To these, a "noise" is added by ruderal weeds, especially the Chenopodiaceae and Amaranthaceae ("Cheno-Ams" in SW USA). It would be useful, when sampling water tanks, to collect some other recent deposits as well, further away from any settlement or farm, to establish the human intervention for a significant distinction and better understanding of the natural part of the recent pollen spectra.

4.3. Water

It may be found necessary to sample water, to establish its contribution to recent pollen spectra obtained from sediments which are formed in close relations with seepages, such as spring or oasis deposits. Springs in arid lands may be fed by aquifers taking their water from far away provenances, which could be represented in the arid region deposits by allochthonous pollen, signifying a different environment; or from fossil water reservoirs, that may contain grains which had been enclosed in former periods under different climates and vegetations, thus adding exotic, redeposited pollen to the deposits (*cf.* Table 2.2.8).

It was also proved useful to sample water from rivers crossing deserts, especially those originating in an entirely different environment. This may help to establish relations between pollen produced by plants growing in the humid region, transported by the water and those contributed by desert or bank vegetations, while the river flows through the arid land. In the case of such rivers it would be advisable to sample the water as close to the source area as possible, the ideal locality would be in the transition zone from the humid source region to the arid land. Care should be taken with the occurrence of dams, as is frequently the case in arid lands.

When considering sampling springwater, or water from high groundwater tables which maintain oases, the best procedure is to do so directly from the springhead. Better still, but this is not always possible, is to sample the water from a drilling which penetrates and pumps the aquifer. In oases supplied by high groundwater tables it may prove necessary to dig into the aquifer level, which usually is not too deep, wait some time for water to accumulate in the small thus formed pool and go on with sampling by pumping the water with an electric pump (car battery driven are available, not too expensive), or manually. Sampling of groundwater may only prove difficult in oases where the water table is quite deep, which would necessitate extended digging. Such oases are characterized by growth of trees, usually poor in grasses.

Sampling is a straightforward procedure, in which water is pumped and collected into suitable containers. It is only necessary to estimate the expected pollen concentration, so that the amount of water would yield a sufficient quantity of grains. Experience shows that usually 100 liters of water would constitute a sufficiently large sample, yielding a significant amount of pollen. As with sampling of wet sediments, formaldehyde is added to the water, to prevent pollen deterioration by bacterial activity and oxidation.

It may prove useful, especially when several samples of water are to be collected, to take along to the field a continuous-flow centrifuge, a pump and a generator. This way separation between the water and its contained pollen bearing sediment can be carried out on the spot, thus saving hauling, acquiring containers and a large enough vehicle (please remember that each water sample may weigh 100 kilograms!). Separation carried out on location may also prove useful for determining the quantity of water necessary to obtain a large enough pollen spectrum, by judging from the amount of sediment collected in the centrifuge tubes. The process of centrifuging could go on until a sufficient amount is collected. A continuous-flow centrifuge, as well as an electric pump, does not require too much current, which could be obtained from the car battery with the aid of a simple 12 VDC/AC (110 or 220 volts as the case may be) converter. An alternative, also not too expensive, is to obtain a small gasoline operated AC generator. It is also possible, in this context, to make an initial microscopic examination while still in the field, which does not necessitate much more instrumentation. This way seems preferable to organizing another sampling trip.

Traverse (1990) used silica filters for water, while in the field, with the aid of a gasoline pump for driving the water through the filters. The filters had been brought to the laboratory, the silica dissolved and the palynomorphs extracted. This way of sampling can again save hauling and seems advisable when water is to be sampled. Previously, Traverse used to evaporate the water and use the remaining sludge for extracting the pollen. Such evaporation could be carried out in the field, following sampling, provided that care is taken to avoid contamination.

4.4. Outcrops

Outcrops could either be natural or man made. Natural outcrops are quite common in arid lands due to both the dominating strong erosion, especially where floods and winds are frequent, the poor soil formation and the scarce protective vegetation cover. A special case, which is typical for arid lands is wind deflation, which creates superficial outcrops even on flat lands where the underlying layers are exposed, not necessarily connected with a strong relief. Artificial outcrops may result from various activities such as roadworks and diggings for foundations, or intentionally done by the sampling team for the collection of material. This last sort of activity also includes archeological excavations, from which samples are subsequently collected for pollen analysis.

The usual measures of caution should also be observed when sampling outcrops, namely avoiding unnecessary contamination, establishing the stratigraphy of the sampled outcrop and, in the case of moist or wet samples, adding some formaldehyde to prevent deterioration of the pollen by bacteria, fungi and oxidation. Sample size should naturally be considerably larger than usually collected in humid environments. Samples should be appropriately labelled, but it was also found useful, for future reference, to label the outcrop from which they had been collected. It seems that the easiest way is to stick label tags with the sample number written in clear, big lettering, on the outcrop itself and take a photograph. The use of a large frame Polaroid camera is quite handy, because notes could be written directly on the photograph, while actually observing and sampling the sequence. This procedure is especially important in arid lands, since many of the sampled sites are remote and not likely to be visited again. Such outcrops could also easily be destroyed by a subsequent flood or covered by windborne deposits, processes which act quite rapidly in the dry regions. It is needless to stress the importance of adequately labelling artificial diggings, many of which would subsequently be covered by the digger himself, or by the elements.

Most natural outcrops in arid lands occur along wadi (arroyo, donga, or as the local term may be) channels, where sampling is a straightforward process, which usually does not require much labor. It is advisable to clean the sampled outcrop from any recent deposits, which might have been washed down from higher horizons, or from the eolian dust, which could have penetrated inside the sampled sequence, especially if the latter comprises medium or coarse clastics, which is a usual case in wadis, or if the deposits are cracked. Sampling of cliffs is done from the base upward, to avoid unnecessary cleaning of spilled material.

A point which should be considered, but could evidently be only tried out with an experimental extraction run, is how deep oxidation had penetrated inside the layers which crop out at a certain locality. The rule of thumb is, the deeper you dig the lesser are chances of oxidation and contamination of the samples. The problem is that in some cases this depth was found out to be several meters, which may make outcrop sampling impractical, or require the use of a compact drilling equipment. A preliminary extraction could establish the desired depth for sampling and could help obtaining pollen productive samples with the minimal required labor. Unfortunately this is not always possible, especially where the laboratory might be far away from the area in which sampling is carried out. It is suggested in such cases to consider the possibility of applying a basic extraction in the field, a method which would necessitate minimal equipment and power supply.

Detailed instructions for sampling wadi sequences are given by Mehringer (1967), who states that a depth of 60–80 centimeters would be sufficient to obtain non contaminated, pollen productive samples. He also suggests using composite sequences along a section, with each of the stratigraphic units sampled at the locality where it is best developed. Sampling intervals, as in other cases, are the compromise between desired information and practical considerations. Mehringer advises that vertical intervals in fine grain sections should be closer than in coarse clastics, as the rate of deposition of the latter is more rapid.

Outcrops are usually very common in regions subject to recent or subrecent tectonic activity, where differential block movements are common. In such localities a great variety of layers could be exposed, either by the tectonic movements along fault scarps, or due to subsequent erosion into a newly formed lower base level, which exposes older strata. Ancient lake sediments, lakeshore terraces (Fig. 3.3.3.1), older wadi accumulations (Fig. 3.3.1.1), alluvial fans (Fig. 3.3.10.1) and other deposits may thus crop out, serving as excellent localities for sampling, thus help extending the length of sampled sequences.

Outcrops can also be found within other sedimentary environments, such as springs, internal sebkhas, marshes or lakes, due to lowering of the groundwater table (Fig. 3.3.2.3), thus exposing older strata through erosion. Changing sea levels can cause erosion of near coast formations (Fig. 3.4.1) such as sand dunes, coastal sebkhas and the like. Stream activity may expose sections of loess deposits (Fig. 3.3.6.4), sand dunes and older wadi beds. However, when sampling sediments accumulated in these environments, it is also advisable to make artificial diggings wherever possible, to obtain clean, less oxidized samples. This is especially suggested for sampling younger sediments, of late Quaternary and Holocene age, which are frequently very porous and friable. Digging may be limited by the available labor and occasionally also by the depth of groundwater. Heavy equipment is quite welcome for such diggings, but is usually costly. Unfortunately it is not always at hand, especially in far away localities. The use of hand held drilling equipment was found useful.

One of the more frequent questions asked while sampling outcrops in arid lands is, which is the preferred type of sediment for collection to obtain the best results in terms of pollen contents. Experience indicates that windblown dust is probably the best source for pollen grains, being also one of the easiest materials for sampling and extraction. The close packing of the fine grains comprising the dust also helps, to a certain degree, to protect from subsequent oxidation and contamination. However, such material is not always available and even when it is there to sample, it rarely comprises continuously the entire sequence to be studied. It was thus found out that practically every kind of deposit should be collected and given a chance, to be tried for its pollen contents.

As discussed above, highly organic sediments, seemingly best for pollen analysis, may occasionally prove barren or very poor in pollen, while such sediments as yellow-buff, or sometimes even red beds may contain enough pollen. Thus, the general rule is to sample every type of available sediment in sufficient quantity to allow for several runs of extraction.

4.5. Boreholes

Boreholes, when samples availability for palynological research is concerned, are of three main types. The first comprises hand drillings, which could be performed by the palynological study team in various sedimentary environments, usually those in which fine grained, sometimes organic, waterlogged sediments are accumulated, or as described above for outcrop material. Suitable environments are sebkhas, both inland and coastal, salt-pans, marshes, shallow lakes and oases. Many types of hand driven augers had been specially designed for such purposes and the only general recommendation for arid lands is, to use the rotary ones, made to penetrate more consolidated sediments. The pushing augers are only good for soft deposits such as peat and gyttia, which are quite rare in arid lands. The main drawback of such hand drillings is that the depth of penetration of the auger is limited, thus not allowing collection of long sequences. Sampling is usually limited to late Quaternary and Holocene sequences, especially in environments of deposition which are presently subsiding. The use of mechanical hand held drilling equipment is evidently preferred in terms of convenience and labor, but is more costly and may require some skill. The depth of penetration is usually restricted also with this type of auger.

Sampling intervals can be planned according to necessity and possibilities, but the amount of material which can be thus collected for each horizon is limited by the diameter of the available drilling equipment. It is advisable to consider the possibility of drilling several holes, very close to each another, in order to obtain samples large enough, if a necessity arises for further extraction runs.

Gravity cores collected from deeper lake and marine sediments, usually by coring devices operated from a boat, also belong to this category which, in most cases, is only suitable for sampling geologically young deposits.

As usual, formaldehyde should be added to the wet samples, to prevent deterioration. It was found out that cores, both gravity and those obtained by manual augers, when stored, tend to deteriorate in a short time even if they remain in their original plastic casings, well sealed and under refrigeration. Sampling the cores on the spot and immediate addition of formaldehyde help prevent this process, keeping samples as good as fresh for pollen extraction even after a long time (personally tried on peat, which kept some five years and sea bottom sediments which kept at least ten).

The second group of boreholes useful for palynological studies are those specially made for the study using heavy machinery, from which continuous cores are usually obtained. Such boreholes are ideal for studying continuous sequences of considerable lengths and duration, much longer than could eventually be obtained from the hand driven equipment. The same precautions for sampling and handling the material described above also hold true for these cores, although palynomorphs contained in deeply buried sediments, which have been subject to considerable compaction, have a much lesser tendency for deterioration. The only drawback of this method of sample collection seems to be the high costs involved with the process of drilling and continuous coring, either on land or underwater.

The third group of boreholes involves commercial ones, drilled in search of water, oil and natural gas, various reconnaissance studies, foundations and so on. Unfortunately, with such boreholes palynological study is frequently of only a secondary priority, although used by exploration teams for defining stratigraphy. These boreholes are usually made by rotary drilling, where the samples are brought up by the drilling mud or water, as well-cuttings. Naturally, such sampling procedure would not provide samples which are as good and reliable for palynological study, as compared to those collected from continuous cores. However, frequently this would be the only available source of samples for studying deeply buried sequences and should not, by all means, be hastily discarded.

Several precautions should be taken when sampling well-cuttings, some of which applied concurrently while drilling activity is going on, at the borehole site, which usually concern the vertical spacing of samples and their state of cleanliness, as well as their reliability. It was found out that occasionally, routine procedures at the time of drilling such as washing samples from the drilling mud and drying them up in an oven, could be detrimental to subsequent palynological studies. Washing could remove much of the finer fraction, a case which may especially be detrimental when coarse clastics are penetrated by the drill. However, even though washing the mud from the sample may seem a welcome activity, much of the unconsolidated, fine grained sediments risk to be thus washed as well (Fig. 4.5.1). These usually include the finer matrix of sands, gravel and conglomerates. Thus the sample would only contain the coarser fraction, while the finer, in which pollen is usually contained,

Fig. 4.5.1. a. Well-cuttings of gravel, washed from their finer matrix by the well sitter, devoid of pollen. En Gedi 2 borehole, near the Dead Sea, Israel, 25–30 m. b. Well-cuttings of gravel from which the finer matrix was not washed, that yielded abundant pollen. En Gedi 2 borehole, near the Dead Sea, Israel, 90–95 m.

would be lost. It is therefore preferable to refrain from washing the samples and try to get the "virgin material" to the laboratory for processing. The samples should naturally be sealed as soon as possible, to avoid contamination. Again, a measure of formaldehyde proved very useful with wet cuttings.

Samples are usually also dried up in an oven during normal drilling activities, to enable their packing. Excessive heating may occasionally be applied for drying up such samples following their washing, sometimes exceeding comfortable temperatures for the palynomorphs, thus the delicate organic material constituting exines can be damaged or even completely perish, practically burned to ashes. Exceedingly high temperatures may also result in fine clay-rich sediments turning to ceramic-like material, from which extraction is hopeless. In addition most pollen have probably been destroyed by the high temperature anyway.

Problems concerned with the drilling mud and water or other liquids used in its composition are that they may contain contaminations of exotic or even local palynomorphs; or the mud could contain various chemicals which may react with reagents used for extraction, resulting in possible formation of insoluble salts or gels. The type of drilling mud used and its added chemicals may change along the penetrated sequence, particularly when it is quite long, of varied lithologies or includes evaporitic horizons. Consequently, a thorough discussion with the mud engineer may prove useful. Samples of the mud or different muds should be collected and processed for control.

In many such boreholes samples are only collected every several meters, rather than continuously. A palynologist who accompanies the drilling, on location, can obtain much closer sampling intervals if willing to actually collect the material himself (or herself). An important problem involved with well-cuttings is the contamination by cavings, which are rock pieces, usually derived from younger strata, falling into the rising drilling mud from the borehole walls (see discussion in Chapter 3.5.7). It is therefore advisable that for collection of commercial borehole samples, the palynologist would act as his own well-sitter, to ensure availability and reliability of the material.

4.6. Archeologic and prehistoric sites

The most important problem, when sampling archeologic and prehistoric sites, is to establish the contemporaneity of sampled sediments with the periods that are to be studied. Too many factors are involved here, which may add to confusion in the stratigraphy. Ancient habitations are always ruined to some degree, which implies that much of the sedimentary material presently comprising the archeological sites may have been deposited in the place only after destruction or abandonment by the inhabitants took place. Thus, such later sediments do not represent the processes which took place during the settlement period. Windborne dust accumulations in many archeologic and prehistoric sites may frequently postdate occupation times, thus their stratigraphy should be studied in detail to establish synchroneity, prior to sampling.

Another problem especially pertaining to arid lands, is the wind deflation, which may be quite active in certain regions. Wind deflation removes the finer fraction of the sediments, leaving behind the coarser as relics, thus exposing to the surface older strata, which could even be considerably older than the site itself. Since artifacts are not removed by the wind, being too large, a false impression is common that they are embedded in a layer which seems synchronous, but may in fact be considerably older. It is quite common in places such as the Negev, southern Israel, to find prehistoric sites which had been wind deflated even down to the bedrock, thus comprising artifacts representing several cultural stages, all mixed up (Fig. 3.1.2.1). It is usually no problem when the situation is that no deposit worth sampling is left, or where a mixture of artifacts is evidence of deflation. However, in many cases a single-culture site can be found, embedded within a relic of fine material which is indeed much older than the assemblage of artifacts due to wind deflation. In such cases artifacts had merely travelled down to be seemingly a part of this layer. Extra care should be taken when sampling regions subject to wind deflation, that affects both prehistoric and archeologic sites, such as tells, which are usually elevated above the vicinity, therefore more prone to deflation processes.

The most significant prehistoric sites to be sampled are those buried within a sequence of continuous deposition and protected from further erosion. This could be ascertained by a careful analysis of the artifacts, making sure that each of the horizons only contains a single cultural assemblage, with no apparent mixtures. In this respect, "transition cultural phases" which contain artifacts typical of two successive cultures should be primarily viewed with reservation, although in some cases they may turn to be truly so.

In archeologic sites, the most significant samples for palynological studies, which would be contemporaneously meaningful, were found to be mud (or adobe) bricks (Horowitz, 1979; Horowitz et al., 1981; O'Rourke, 1982; Drori and Horowitz, 1988–89). The mud, at time of building, acted as a pollen trap, sealed from consequent contamination when being dried up. The possibility of older contamination from the raw material of the mud should not be ruled out, but it was found that usually (op. cit.) this was not the case.

All other types of deposits should be regarded with care, to make sure no asynchronous samples are collected. Similar care should be applied while collecting samples from jar fillings, or burial sites. The time of ultimate sealing of such dust traps has to be safely established. Floors can supply adequate samples when sealed by an overlying floor of a known age. Plaster and mortar, made while building the site, can also make reliable samples, for reasons similar to mud bricks. However, due to the usually large amounts of lime used for plaster and mortar production, it was found out that the state of preservation of pollen can be quite poor in these materials.

Exotic or "strange" materials which are occasionally found in archeological sites may be of some interest, especially regarding particular activities. Sites in Israel had yielded material proved to be honey by its pollen contents, asphalt whose provenance could be traced according to the enclosed palynomorphs, wine jar linings which proved to be pine resin, again by their pollen content, indicating that Romans, at least in this part of their empire, probably enjoyed drinking Retsina with their meals or parties. Most of these "strange" materials, incidentally, look like small, black lumps. Suggestion is made here not to discard such occasional finds, but rather give them a chance with palynology (see also Chapter 5.4.5).

4.7. Biogenic substances

Biogenic substances comprise coprolites, which are fossilized feces, pellets of undigested parts of food vomited by birds, food reservoirs, or middens excreted or collected by various kinds of animals. Some of these, like packrat or hyrax middens, are readily recognizable and very easily sampled, while others, such as some coprolites, may prove quite difficult to positively identify as such in the field. It had already occurred that mineral concretions had been mistaken for coprolites.

Again, the main problem with these biogenic substances is their stratigraphic context. It is quite straightforward with coprolites and pellets found embedded within sequences, such as cave sediments, or any other sedimentary layer. The only stratigraphic problem with middens may occur when younger material gets spilled over, or infiltrates inside an older accumulation. The stratigraphy of middens can very easily be established by radiocarbon dating provided no internal mixing as described above had occurred, thus these substances form an important source of samples in many arid regions around the world. To the best of my knowledge no midden whose age is beyond radiocarbon range is reported, so that this possible stratigraphic problem was not raised as yet. The suggestion made by Davis and Anderson (1987), that several subsamples should be collected from the recent material, should also be followed with fossil middens, to ensure a better homogeneity of the pollen spectra. Naturally, it must first be established that such subsamples are indeed all of the same age.

Food reservoirs may prove more difficult to assign exact stratigraphic position, except of course when they are still within radiocarbon range. This is because many of the reservoirs are stored by burrowing animals, especially some kinds of bees and wasps. These could be embedded within sediments which have no time connection with the age of storing. Such food reservoirs can also contaminate the sediments in which they are contained and care should be exercised in this respect.

Chapter 5

Extraction

Essentially, all extraction techniques applied in palynology regardless of region, type, or age of the samples, aim at obtaining a microscope slide, or a scanning electron microscope (SEM) stub, on which palynomorphs would comprise the majority of components. Ideally, all other material should not be present, to enable easy identification and counting. However, even with the best of samples, this aim is practically impossible to achieve, except with single grain or reference collections that are not practically pertinent to sediment samples.

This ideal is even more remote when samples collected from arid lands are processed, due to the relative scarcity of pollen in the sediments and their usually poor state of preservation. The first factor calls for leaving some "carrier" in the processed sample, so that the final deposit would have enough volume to be treated; this carrier would in most cases be natural, comprising some other organic particles which are almost impossible to get rid of anyway, or it can be artificially added while processing the sample, constituting any kind of preferable organic substance. *Lycopodium* spores, contained in calcium tablets in a known quantity (Stockmarr, 1971), have been found useful for this purpose.

The usually poor state of preservation in most arid lands deposits calls for some caution while applying various methods of extraction techniques. The main principle here is, to treat the sample as little as possible. This, naturally, deserves compromises to be made between the ideal, clean, pollen rich slide or stub and the probability that when such a preparation is finally achieved, no palynomorphs are left there to be studied.

The descriptions of various extraction techniques which follow in this chapter, are specifically geared toward the maceration of pollen poor sediments. It is assumed that the reader is familiar with the standard techniques, described in various classical works, such as Brown (1960), Faegri and Iversen (1964 and later editions), Kummel and Raup (1965), or lately in Traverse (1988), Faegri et al. (1989) and Moore et al. (1991).

The extraction techniques described in this chapter aim primarily at obtaining a slide or stub, with palynomorph material suitable for study. Usually, these would contain a considerable quantity of "dirt" on the first microscopic scan, when occasionally only a careful observation will reveal the microfossils (Fig. 5.1); in most cases, if these could be thus identified, extraction was stopped at this point.

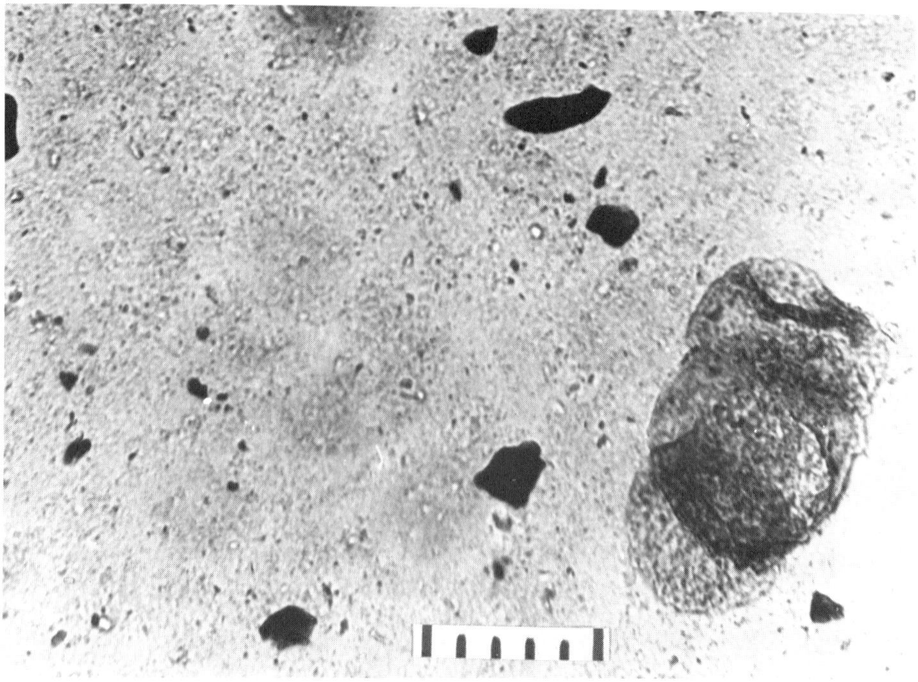

Fig. 5.1. Microscope slide, typical for arid lands deposits, with some pollen and plenty of clay. Ami'az East 1 borehole, near the Dead Sea, Israel, 1,098–1,103 m. Scale units: 10 μm.

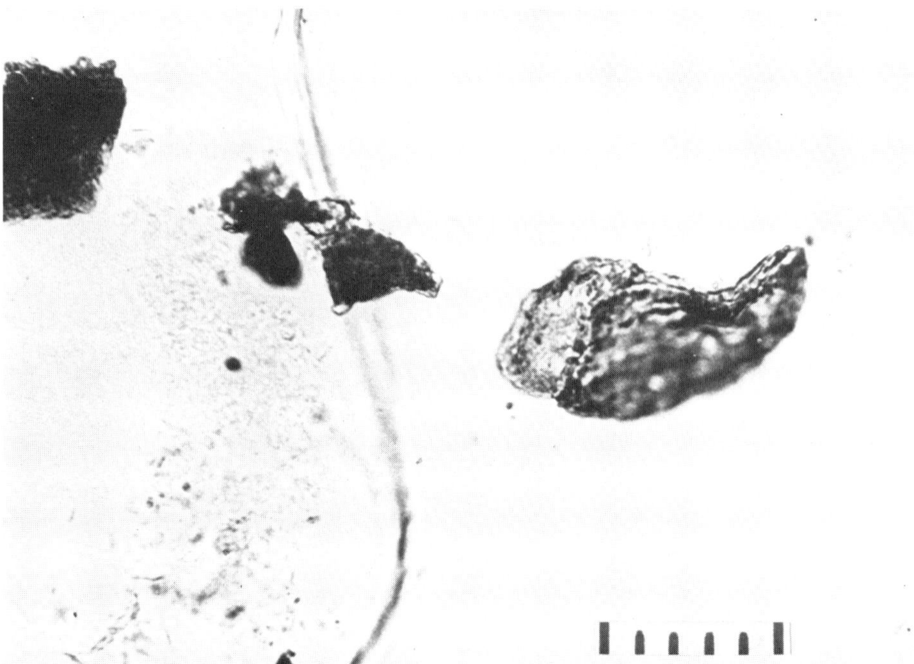

Fig. 5.2. The same sample as in Fig. 5.1, after HF cleaning. Scale units: 10 μm.

Whenever further cleaning of the sample was attempted (Fig. 5.2), it was not before a safe amount of "dirty" material was kept aside, in case the cleaning would destroy the pollen as well.

5.1. General

5.1.1. Sample size

The size of the samples must be considerably larger than of those collected from pollen rich deposits. The actual quantity of processed material is frequently limited by the available laboratory equipment and this point should be taken care of ahead, by securing larger glassware and especially a large capacity centrifuge. An optimal size for a centrifuge is such that can accommodate 1,000 cm^3 swingout bottles, but a 500 cm^3 would usually do. Needless to say, the larger the capacity of the centrifuge, the better are the chances of obtaining pollen, even from the more stubborn samples.

Usually, the sample size macerated by various authors is in the order of 50–100 grams (Martin, 1963; Mehringer, 1967; Horowitz, 1979; Weinstein-Evron, 1983; and many others). Samples smaller than 50 grams proved disappointing in many cases. Bonnefille et al. (1980) had tried to extract pollen from samples weighing only 20 grams from a borehole in Lake Abhe, Djibouti, but only three out of 63 samples yielded significant spectra; another 13 yielded pollen, but not in sufficient numbers. It is expected that if larger samples would have been processed, many more would yield significant spectra. Addition of carriers may also be of help in such cases (see above). Dumait (1964) indicates that some oil companies processed samples of up to a kilogram; Horowitz (1975) could not obtain pollen until some 2.5 kg were treated. In fact, more and more evidence indicates that the chances of success with larger samples are considerably higher.

The first problem involved with the treatment of large samples, except for the equipment, is the cost of chemicals. It is therefore suggested to use only technical grade reagents and refrain from using the laboratory grade, which is much more expensive. Such technical grade reagents, especially HCl, may contain pollen and an experimental run should be performed every so often, to ensure knowledge of such possible contaminations. However, it is not advisable to use tap water, which is almost sure to contain pollen. Distilled water from a colona, which are filtered, are the best choice, but filtered tap water can safely be used if filtration is done with a 1 μm filter, which is now easy to get, since it is manufactured for daily household and can be permanently installed at the water inlet to the laboratory. An experimental run may prove important also here.

Samples should be washed before the beginning of maceration, to rid of atmospheric or packing contaminations, preferably in 10% HCl, but occasionally distilled (or filtered) water would do. Scraping the outer surface may occasionally be necessary, especially with friable sediments. The extent of scraping, or how thick a layer should be removed, is determined by experience.

5.1.2. The influence of chemicals on palynomorphs

Various chemicals used during the maceration processes may adversely affect the palynomorphs. No particular study was made in this respect with sediments collected from arid lands, but it seems safe to assume that, considering the usual poor state of preservation of pollen in these, the effects of chemical treatments could only be worse. The rule of thumb is therefore, to avoid as much as possible, any chemical treatment which is not vitally important. I have personally found out, over long periods of experiments, that most chemical processes designed to clean the slides from unnecessary debris result in partial, or occasionally total, obliteration of the palynomorphs. It was also found by experience that it may take much longer to clean a sample, even when the palynomorphs are not ruined through the process, than to count a "dirty" slide. Counting such slides is in a way a matter of getting used to. Once the mental barrier is removed, routine work becomes quite easy. Naturally, this would depend on personal preferences of each palynologist, occasionally also on availability and cost of help, but the possibility of losing some of the pollen should always be considered when cleaning samples is attempted.

Commonly acceptable methods in "classical" palynology, such as bleaching, HF treatment, acetolysis and especially their combination, may prove hazardous and cause partial or even total destruction of the grains (Hafsten, 1959). De Vernal et al. (1983) report that the application of Lüber oxidation agent (a mixture of 50% nitric acid and concentrated hydrochloric acid) for 2–3 minutes, caused differential destruction of pollen obtained from organic sediments. These authors, prior to the application of the oxidation agent, also used hot 10% KOH, 48% HF, 10% HCl (twice) and acetolysis. It is not clear whether the destruction of the grains took place by the oxidation alone, or was considerably initiated by the previous treatments.

Brown (1960) summarized the action of chemicals on pollen and spores and arrived at several conclusions, which may have significance especially for the maceration of samples containing poorly preserved palynomorphs. Oxidation agents which are in common use, such as the Schulze reagent, sodium perborate, potassium permanganate, nitric acid and hydrogen peroxide are very destructive to palynomorphs, thus the important trick involves the knowledge of when to stop their action. Boiling in 10% KOH also caused deterioration of certain palynomorphs and Brown suggests not to use it, except when no other choice is possible. Acetolysis is safer, but has also caused deterioration of pollen, especially if applied for a long time, as also discussed by Erdtman (1952). Brown even raised the question whether the better sculpturing seen after acetolysis treatment, may not be a result of a partial dissolution of the exine.

We found out that almost any oxidation treatment, even for a very short duration, may destroy some of the more delicate exines, thus we avoid use of such procedures altogether. When no other choice remains, oxidation of arid lands' samples should be performed with the utmost care, under constant supervision. Every step must be followed by sampling the residue and microscopic observation, to determine when would be the best time to stop the procedure. This procedure would also ensure that, if pollen is digested at a certain stage, a residue from the former is still available for examination. KOH treatment is usually unnecessary with arid lands deposits, which are in most cases very poor in humic acids anyway. Acetolysis was occasionally found useful, but should be performed with care, again taking control samples.

Several chemical treatments, which do not seem to cause any visible deterioration in well preserved pollen and spores, can however cause changes in their size. Michel (1955) studied these changes in detail and concluded that enlargement of grains, in ascending order, is due to application of KOH; KOH + perborate; KOH + acetolysis; and oxidation + acetolysis. The last procedure caused expansion of up to 150%, as compared with KOH alone, which also causes some expansion of the palynomorphs (Brown, 1960). It is therefore quite important to make a clear recording of the chemical treatment applied for each sample, especially in these cases where identification is also based on the size of grains, such as the differentiation between natural Gramineae and cereal pollen.

5.1.3. Filtration

Filtration is a common procedure with pollen poor samples and could take place at various stages of the extraction process. Filtration is done for obtaining cleaner final residues, performed either to get rid of coarse material, such as vegetal debris, or of very fine substances, such as clays. The first stage of filtration involves using filters which have mesh openings larger than the largest expected palynomorphs. Usually 210 μm (65 mesh) would do, but in some cases larger mesh diameters (corresponding to smaller mesh numbers) should be used. This stage is a simple, straightforward procedure and usually does not require nothing more than the filter itself, placed on top a proper beaker or test tube. Gentle shaking, or light mixing, may be necessary at this stage. The samples should be well dispersed, either mechanically, by mixing, or, if this proves insufficient, by using various dispersing agents (Kummel and Raup, 1965). As will be seen later, in most cases this stage is carried out while the sediment is placed in 10% HCl, thus care should be taken to use appropriate laboratory ware, like plastic or stainless steel screens. It was found out that brass, or stainless steel filters can withstand quite a long period of use with diluted acid. It should be washed well in water immediately after use, to prolong its life.

The second stage of filtration, aimed at getting rid of very fine particles, is usually more complex and involves use of a vacuum pump of some sort. It is almost always carried out at the final stages of the extraction, when only very little material was left, usually immersed in water. An elaborate way of doing so is described in detail by Vidal (1988). Filtration in one way or another was however applied by many authors before, who used a simple device made by a filter seated on a funnel, which is connected to a vacuum pump. It may seem preferred to use the more elaborate equipment, but the results tend to be identical.

The technique for an easy, quick filtration, is as follows: a funnel, large enough to hold the filter (10–18 μm opening, according to necessity and expectations), is connected to a vacuum pump by a piece of rather rigid plastic tubing, which would not collapse when vacuum is applied. The vacuum pump (I usually use the simple type, which is connected to the water tap, but any other type could as well be used) is activated and the filter is placed on top of the funnel. Some sealing may be necessary, if the filter does not fit the funnel properly, for which any kind of household silicone sealant, or a wide rubber band is appropriate. The residue, immersed in water (or any other liquid, depending on the procedure applied) is mixed well and poured onto the filter. Gentle sprinkling of water (or the other liquid), with circular movements covering the entire surface of the filter, should help the filtration. In case the filter is clogged, the best way we found out is to mix the residue gently, for which a fine, good quality painters brush proved ideal. When filtration is accomplished to satisfaction, the residue is poured to a centrifuge tube, using a fine sprinkle of water on the reverse side of the filter. The brush should also be washed in the same way, so as not to loose any material. Care must be taken not to use too much water here, which may cause overflow of the centrifuge tube. Size of the tube should be chosen by experience, but since this stage may be the final one of the extraction and the amount of residue may be minimal, it is advisable to use as small a tube as practical, preferably a conical edge one. A simple 500 cm^3 plastic spray bottle is ideal for sprinkling during all stages of filtration. Its nozzle should be cut or adjusted to obtain a proper spray.

5.1.4. Mechanical separation

Most samples, except for loose substances, have to be crushed before extraction procedures begin, to facilitate and hasten the chemical reactions by enlarging the surface area of the treated material. This is a straightforward process, which is not peculiar to deposits collected from arid lands. However, since such samples are usually larger, crushing could save considerable time of processing. The only precautions here are to keep the sample from being contaminated and to refrain from entirely pulverizing the material. Bits in the order of several millimeters across are usually the best. Crushing is done after the sample was put into a strong bag, of either paper or plastic, to avoid contamination and loss. Washing, with 10% HCl or water, may be necessary following the crushing, especially if the bag tears apart, or leaves paper or plastic pieces mixed with the bits of sample.

Ultrasonic treatment, for mechanical cleaning of the sample by invoking a better dispersion, was found quite hazardous even with well preserved pollen (Dodson, 1983; Heusser and Stock, 1984; and others). Its first sign of destruction is usually seen in vesiculate (saccate) conifers pollen, which loose their bladders. After experimenting with this method, we found out that it is best to avoid using ultrasonic altogether, since the chances of destruction of poorly preserved palynomorphs are far superior to any possible success.

Mechanical disaggregation of rocks, especially shales and some kinds of clays, can be aided by the introduction of organic solvents. The methods, discussed in Brown (1960) and Kummel and Raup (1965) are usually of no use with sediments of arid regions, in which cementation more often than not is caused by crystallization of carbonates, while pure shales are indeed rare. In most cases, disaggregation is achieved by dissolving the carbonates with the aid of HCl and no further treatment is necessary.

Various devices, which combine mechanical and physical treatments, were developed by palynologists over the years, for extraction of pollen from poor samples. However, these never became widely used, probably because they involve considerable labor. Such is the "Vibroseparator", presented by Dumait (1962, 1964), which combines vibrations with flotation in a heavy liquid, to separate the organic fraction from the mineral; or various ways of frothing a liquid in which the disaggregated sample is placed, by the use of a gas stream, where pollen is floated and collected (Kummel and Raup, 1965, p. 581).

5.1.5. Gravity separation

Gravity separation involves two sets of methods, which are essentially different. The first consists of separating the various particles in a liquid environment, which only depends on relative cohesions of the different substances, while the second applies heavy liquids in which the desired, organic material floats, while the heavier, mineral particles, sink down to the bottom.

With the first set of methods, the bulk of disaggregated material is mixed well within the liquid and the mixture is let to settle down. Smaller or lighter particles will take longer to settle than heavier or larger. The differences in settling velocity of particles having similar specific gravity mainly depend on their size and to some extent on their shape. The larger particles have a lesser surface area relative to their mass and cohesion is therefore weaker, making settling time shorter. In the same manner, flat particles will settle slower than round ones. Organic material, which is lighter than the mineral, will take longer to settle. The method could be used for two purposes: either to get rid of large, heavy particles which are of no interest to the palynologist, such as gravel and sand, or to get rid of the finest fraction, only several micrometers in size, both organic and mineral, which is annoying due to the haziness it renders the final microscope slide.

The first method is commonly known as the "swirling" technique, developed by joint efforts of several palynologists in the American Southwest and brought to perfection by Mehringer (1967; for details see Chapter 5.3). The second is better known as the controlled centrifugation technique.

Controlled centrifugation, destined particularly to get rid of the finest mineral and organic particles, is discussed by several authors, summarized by Brown (1960) and by Kummel and Raup (1965). The principle is quite simple: at a certain stage of the extraction procedure, centrifugation is carried out at a lower than usual total radial velocity, which is a combination of the centrifuge speed and the time the sample is subject to the process, so that palynomorphs would settle down, while the finest particles would still remain in suspension. Practical applications of controlled centrifugation are far more complex, since it is necessary to establish, by trial and error, the right speed and duration of centrifugation for each type of sample. This is done by checking both the residue which had settled down and the suspension, which is further centrifuged until the liquid is clear. A common problem here is that when the suspension contains considerably more fine particles than pollen, the latter may be masked in the slide. Thus a false impression that nothing is lost is quite common, when in fact some of the pollen are still in suspension. Similar results could also be achieved by decantation, after letting the sample containing liquid to settle for a certain length of time. Centrifugation only hastens the process, but the usual limitation is the centrifuge tubes size, while with decantation larger samples can be processed.

Controlled centrifugation, or decantation, is even more problematic with samples collected from arid lands, where even adjacent horizons may considerably differ in lithology from each another, thus involving quite a lot of labor in determining centrifugation speeds and times. Filtration, under a partial vacuum (see above), seems to produce better, safer and faster results than controlled centrifugation and is therefore preferable for any practical use.

Heavy liquids are widely used for separating the organic from the mineral material, by flotation. A variety of liquids are in use, both organic such as bromoform/alcohol mixtures, or mineral, such as zinc halogenides and others. Most palynologists tend to believe in the liquid they personally use, though practically it does not essentially matter. We found out that a saturated solution of zinc chloride in concentrated HCl is the most convenient and cost effective. Its convenience lies in the fact that almost the entire extraction procedure, except for the final mounting, is done in an HCl environment, which eliminates many stages of repeated washing. Details are found below.

Preparation of the mixture is quite simple: one kilogram of zinc chloride crystals is slowly dissolved in half a liter of concentrated HCl, both of technical grade. The heat produced by the process only helps the dissolution of the crystals. Needless to say, it is advisable to mix the two under the fumehood. The mixture has a rather long shelf life and would only need some shaking if it stood without use for some time (weeks, or even months). If contamination by pollen of any of the reagents is suspected, filtration through any kind of ceramic or glass filter would solve the problem. Verification of the specific gravity by a suitable hydrometer (1.500–2.000 range is ideal) should be performed after the solution has cooled.

The use of organic compounds for gravity separation has the advantages of being less corrosive and easier for mixing. On the other hand, organic compounds are more expensive, no less smelly, which necessitates a fumehood anyway. their use requires many more stages of washing, dehydrating and hydrating afterward. Results of using either zinc halogenides or organic compounds are similar when slides are judged, provided the procedure is carried out meticulously (in both cases).

5.1.6. Centrifugation

Centrifugation is the most common procedure used in almost any method of extraction and its only aim is to hasten gravitational separation. Usually the sediment is used and the supernatant liquid discarded, except for a single stage during the separation by heavy liquids, in which the liquid, containing the desired floating organic matter is kept, while the mineral deposit discarded.

The most common centrifuge used in palynological laboratories is the swing head. It seems that separation is better with this type than with the angle head (Brown, 1960). In the following descriptions of extraction processes, when "centrifugation" is mentioned with no extra details, it is assumed that a swing head centrifuge is used, with a speed of 3,000 RPM, for a duration of five minutes. Whenever different times or speeds are required, these will be specifically mentioned.

5.1.7. Washing

The term "washing", which is very common in the descriptions of various extraction techniques, summarizes the following procedure:

(1) Pour off the supernatant liquid of the last centrifugation.

(2) Let the test tube stand upside down for a while, to get rid of the last drops of the previous liquid.

(3) Fill the test tube with the washing agent mentioned in the process (e.g. water, alcohol, 10% HCl, etc.).

(4) Mix thoroughly, until the deposit is entirely in suspension.

(5) Centrifuge (3,000 RPM swing head, five minutes unless otherwise specified).
(6) Pour off the supernatant liquid, stand the test tube upside down.

5.1.8. Mixing

When this term appears in the extraction process, it means a thorough mixing, until the entire sediment is in suspension and no lumps remain in the liquid, or sticking to the test tube walls. This is especially important during gravity separation using heavy liquids, since such lumps may, more often than not, float, thus the included clays would cause haze in the slide, their enclosed pollen practically lost for identification. This would necessitate repeating the separation procedure, or the unnecessary application of hydrofluoric acid treatment.

5.1.9. Acetolysis

Acetolysis is a common procedure in the preparation of palynological material. Its main purpose is to dissolve cellulose, thus increasing pollen concentration within the organic fraction of the sample. The process also has some influence on the palynomorphs, that usually become somewhat larger and darker in color, which makes them easy to discern without staining. Acetolysis, probably by dissolving also some parts of the exine (Brown, 1960), makes the morphological features of palynomorphs more prominent and easy to distinguish. Many versions of this method have been published, but it seems that the principle is similar with most of them. A sample method is as follows:

(1) Dehydrate the sample with glacial acetic acid. This is done by adding the acid to the sample containing test tube, mixing well, centrifuging and pouring the liquid. WARNING: Please note that dehydration is essential, because if the acetolysis mixture is added to water, or to a wet sample, explosion may result.

(2) Add the acetolysis mixture up to about two thirds of the height of the test tube and insert a thin glass rod into the tube. The mixture is made from 90% acetic anhydride, mixed with 10% concentrated sulphuric acid. It should be mixed just before use, because it does not keep for more than several hours.

(3) Heat the tube in a water bath, raising the temperature gently until the bath water is boiling, while mixing the contents of the tube with the glass rod. When the water bath has commenced boiling, remove the test tube, centrifuge and decant. WARNING: do not pour the acetolysis mixture into water as explosion may result.

(4) Wash in glacial acetic acid, centrifuge and decant.

(5) Wash with water to remove acid, preferably twice.

Some authors combine acetolysis with bleaching or oxidation, but this may be destructive even to well preserved pollen and spores (Faegri and Iversen, 1964), thus should not be performed on samples with pollen in a bad state of preservation, as are most of those collected in arid lands.

5.1.10. Staining

Staining has its believers and non-believers. Some use it always, regardless of the type of processed material, some hardly ever, while each claims this or that to be the best. Various stains and staining techniques are given both by Brown (1960) and by Kummel and Raup (1965) and will not be repeated here, particularly because, when dealing with samples from arid lands, we found out that it is better not to use stains. The main reason for this seemingly "negative" approach is that the slides are quite poor in pollen and many other organic substances which are present may tend to be stained in the same manner that pollen are. Consequently, we found out that the entire slide may turn red or green, depending on the stain used, thus the rare pollen were only more difficult to detect. Another reason is, that poorly preserved pollen occasionally tend to absorb the stain quite strongly, thus their finer details cannot be discerned anymore. A suitable mounting medium, with a considerable contrast in refractive index, makes the pollen seen much more clearly than while using staining.

5.1.11. Picking the last drop from the test tube

A common problem, when extracting from pollen poor samples, is the extremely small quantity of residue left in the last stage of the procedure. More often than not, the quantity would only suffice for a single slide or stub, thus it is vital to get every drop of it out of the test tube. We found out that one of the ways is to use 15 ml conical edge tubes in the final stages. After the last stage of centrifugation, when the residue has properly settled down, to wipe dry the inside of the tube with an absorbing paper, rolled to a cylinder. A small piece of glycerin jelly, the size which would be used for a slide (or bigger, if more residue is seen at this stage, again only a matter of experience), is inserted into the tube down to the bottom and mixed well with the residue, while being gently heated. Heating should only be to the degree of melting the glycerin jelly, but not let it bubble, so care should be taken here. To be on the safe side, this could be performed in a hot water bath, but this way the tube takes some time to heat. A quicker way is to heat over a small flame, but be careful not to boil the mixture. Despite this warning, it may occasionally be found necessary to boil the residue at this stage, if for some reason it contains too much liquid. This may result from either inserting too large a piece of jelly that would dilute pollen on the slide, or if the last centrifugation did not remove all the liquid. The latter case is occasionally a result of the shape of the test tube, conical more prone to holding the liquid. If you seem to run into troubles with excess liquid at this stage, try using different tubes, or boil very gently.

The mixture of residue and glycerin jelly, when uniform, is let to cool down, which could be hastened under running water, if necessary, or the tube put in a glass of ice water. The mixture is then being taken out of the tube with the aid of a small spatula and placed on the slide. It is again heated, melted, mixed and a cover slip is applied. Sealing is done in many ways and no special suggestion is being made here. We are using paraffin for that purpose, which is melted and penetrates under the cover slip. The excess is wiped out with xylene (see detailed description below).

Heusser and Stock (1984) have used, when the final pollen residue is smaller than 0.1 ml, a modified disposable plastic syringe, to obtain the material. They cut the plastic syringe, which is used as a centrifuge tube, to about two centimeters. After final centrifugation the syringe is again cut, to provide easy access to the pollen concentrate at the bottom of the tube.

It is a matter of experience, or personal preference, to find out which way is more convenient, also depending on the mounting medium which is used. A general suggestion, if not using glycerin jelly, is to pipette the residue out with the little of liquid left in the tube and dry it on a slide. Any type of mounting medium could then be added.

5.1.12. Mounting media

Various mounting media are in use and it seems that every palynologist has its own favorite. Again, for detailed information the reviews of both Brown (1960) and Kummel and Raup (1965) are suggested. Choosing the right mounting medium does not have any specific preference when samples from arid lands are treated, so that no recommendations are given here. The two important factors are the keeping quality and refractive index of the mounting medium, which should have a pronounced contrast with that of the palynomorphs, especially when no staining is practiced. Considering the first factor, glycerin jelly seems to be a very good answer, but its keeping quality is far from satisfactory, if sealing is not properly done. Proper sealing, preferably with paraffin which creeps under the cover slip, seems to help keep the slides for a long time. To achieve such sealing, both the slide and the cover slip should be perfectly clean from any greasy substance. We have at our collection slides thus made which kept for at least 28 years, still in a good shape.

The procedure for proper sealing of slides in which glycerin jelly is used for mounting, is as follows (before mounting, both slides and cover slips are kept in alcohol and wiped dry with a clean cloth just before mounting is carried out):

(1) Take some residue out of the test tube, either with the aid of a spatula or a small piece of glycerin jelly held at the tip of a needle. The size of the piece of glycerin jelly, in both cases, should be such that, when put on the slide and flattened under the cover slip, will not protrude out of the latter, but would rather leave some free space from its margins. It takes some practice to estimate the right size, so it is suggested to make the first batch of slides with no residue, until the feeling of the right amount is acquired.

(2) Put the piece of glycerin jelly containing residue at the center of the slide, heat gently until it melts. Avoid boiling.

(3) Pass a cover slip over a small flame and place it on the drop of melted glycerin jelly, on the slide. Flatten with a gentle pressure of your fingertip. Some additional, however careful heating may be required at this stage.

(4) Put several scrapes of good quality high melting point paraffin at the edge of the cover slip. Heat gently until paraffin melts and penetrates below the cover slip, encroaching the flattened glycerin jelly drop. Some "guidance", with the aid of a very small flame, is useful here.

(5) Wipe the excess paraffin with a fine blade. Clean the rest of the paraffin with a cloth, soaked with xylene.

Synthetic resins, which are being constantly developed and improved, seem to combine both the necessary refractive index contrast and the keeping qualities. Most of these however, do not permit any rotation of the grains after the residue was mounted, as opposed to glycerin jelly, pure glycerin or silicone oil. We did not find this limitation too serious, since optical sections could in almost all cases replace such procedures as turning the grains around. These mounting media also have the advantage that grains could be easily located, even after a long time, by their coordinates.

When choosing a mounting medium, it should also be remembered that various media have an influence on the mounted palynomorphs, affecting either their shape or size. Jacobs (1985), for example, indicates that "pollen grains mounted in glycerin jelly or glycerin were significantly larger than those mounted in silicone oil". Attention should especially be given when comparisons of grain sizes are made in slides prepared in different laboratories.

5.1.13. Keeping the residue

It may happen even with samples which are relatively poor in pollen that at the end of the extraction process, the amount of residue left is more than necessary for preparing the slide, or slides needed for a certain type of study. In such cases it is advisable to keep the rest of the residue for future uses, in case it is needed. This is done by storing the extra amount of residue in small, 5 cm^3 vials with sealed caps, in a solution of glycerin to which several crystals of phenol are added, in silicone oil or any other suitable substance with adequate keeping properties. Properly labelled and stored, slides can be prepared after many years. It is suggested that labelling is done with the aid of a diamond pen, rather than stickers which can fade or fall off after some time.

5.1.14. Electron microscopy

The preparation of polliniferous material for electron microscopy has nothing in particular when arid lands are considered, therefore not discussed here in any detail. Electron microscopy can either be of the transmitted electron beam type, which is seldom of interest to a palynologist dealing with fossil material, or with the scanning electron microscope (SEM), which may be more significant, by helping to reveal the minute morphologic features of the grains. It should be however noted that the routine work, of identifying and counting the palynomorphs, is usually done under a transmitted light microscope and not the SEM.

The latter may prove of great help occasionally, especially when reference material is studied, either extant or fossil. A detailed account of techniques for preparing fossil palynomorphs for study with both the transmission and the scanning electron microscopes is given by Leffingwell and Hodgkin (1970). Preparation of modern palynomorphs for scanning electron microscopy by the critical point drying method is discussed by Garner and Bryant (1973). The interested reader is referred to these papers.

Practically, we found out that preparing material for the scanning electron microscope is a simple, straightforward procedure and does not require adhesives used by most investigators. Some of the final residue is washed in plain water, centrifuged, again mixed with very little water to achieve the required dilution, picked using a fine pipette and sparsely smeared on a stub which is left to dry up, protected from dust. When extraction is complete, some sticky substance always remains in the liquid, which, when dries up on the stub, proved stable enough to withstand gold or carbon coating under vacuum.

5.2. A basic extraction method

This method was primarily developed to check whether a prospective pollen poor sample is worth a further treatment and what are the chances of obtaining a sufficient number of grains. Also, this method is useful for determining the type of process or further treatment which may be required. Since the procedure is relatively short and not too expensive, it was found useful to apply to almost any type of sediment, before tedious and lengthy preparation processes are attempted.

Luck was frequently on our side while applying this basic technique, so that more often than not it was found out that the preliminary thus prepared slide contained enough pollen grains for meaningful results, in the order of several hundred. This preparation method was therefore adopted as an initial standard in our laboratory. The slide photographed in Fig. 5.1 is a result of such extraction, while Fig. 5.2 shows the same sample after HF cleaning (see below).

The described steps must be followed very carefully, since sloppiness in any of them would result in an almost sure failure. Each of the steps seems critically essential and we could not find any more shortcuts. It is nevertheless hoped, that even with that some progress could be made in the future.

(1) Put 25–30 grams of sediment in a 1,000 ml beaker. If particles are too

large, crush beforehand to quicken the reaction. The ideal size of the particles seems to be 2–4 mm across. Add 50 ml concentrated HCl (the fuming, 36% variety gives excellent results, but an effective fumehood is necessary) and mix well. If effervescence is too wild, add some ethyl alcohol, preferably using a plastic spray bottle. The nozzle of such a bottle should be cut or adjusted, according to type, so that a fine spray will take the entire diameter of the beaker, from a convenient spraying point. When effervescence ceases, or becomes very slow, add 25 ml of concentrated HCl, to make sure all cement has been dissolved. Some types of cement, like gypsum, anhydrite and dolomite, take longer to dissolve and at this stage, after having all samples for the day lying peacefully in the acid, the best thing to do is take a lunch break. Before the break, however, mix the samples again and, with the aid of a glass rod, scrape the walls of the beaker to return all the froth into the mixture at the bottom. If it is suspected that the sample is dominated by hardly dissolving cement types, it could be put for the lunch break in a hot water bath, which hastens the reactions.

(2) After an hour or two mix well, let insoluble coarse particles settle for a few seconds and pour the mixture through a 210 μm screen, via a funnel, into a 100 ml centrifuge tube. If the sediment clots the screen, wash it down with a thin spray of 10% HCl, but watch that the tube does not overflow. Balance with 10% HCl, centrifuge three minutes at 3,000 RPM (A swing head centrifuge is used for this process; if only an angle head is available, times and speeds of centrifugation have to be determined by experiments). Pour the supernatant liquid and place the tube upside down for five minutes. This step should be performed in a single movement of the hand, not to disturb the sediment. While still upside down, lift the tube and wipe dry its inside from below, without touching the sediment, with absorbent tissue. We found it advisable to use paper towels colored by an acid sensitive pigment, usually blue, which changes color when in contact with any harmful acids used throughout this process. This gives a first warning that some was spilled, before your skin itches or pants (skirts, too) need replacement.

(3) Add zinc chloride solution, of a specific gravity 1.85 g/cm^3 (see above for instruction how to prepare), approximately twice the amount of sediment in the tube, never more than 45 ml, otherwise subsequent dilution would not be sufficient for settling of the organic materials. Mix very well, until there is no lumps and the mixture is homogeneous. This step requires patience, because even minute lumps would make a slide full of clay. Good mixing is usually achieved by using a vortex creating instrument, such as "Vortex Genie" and the like. A rubber stopper pushed firmly into the tube is helpful at this stage, to avoid spilling. A thick glass rod is handy for mixing, but attention should be drawn to its tip, which should not be bulging beyond the perimeter of the rod. The rod is held parallel to and touching the tube's wall, mixing is done by circular motions, while rubbing the mud lumps against the tube wall. Some effervescence may still occur at this stage, so one must be careful especially when the rubber stopper is removed, which should be done

away from the eyes. Centrifuge two minutes at 800 RPM, than raise the speed to 3,000 RPM, for further ten minutes.

(4) Pour the supernatant liquid into another tube (we use several sets of tubes and beakers bearing similar numbers, so that no change is made during the process, thus a list made at the beginning, which refers the samples to ware numbers does not need updating during the process). Fill the tube with 10% HCl, mix well. This step is easy with a plastic spray bottle, whose nozzle is removed and the jet of 10% HCl is aimed at the side of the tube, at the height of the zinc chloride solution. Some practice may be necessary, after which the changing sound of the jet will indicate when the tube is nearly full. Centrifuge six minutes at 3,000 RPM.

(5) Discard the liquid (or save it for reuse of the zinc chloride). This step must be performed with care, since the amount of deposit may be minimal. Do it in a single continuous gentle movement, without any shaking; if for any reason the liquid was not completely discarded at the first movement, do not attempt to tip the tube again, but rather repeat the centrifugation. When the tube is upright, wipe its tip dry.

(6) Add 14 ml distilled water, mix well (again, a plastic spray bottle with an appropriately cut nozzle to give a very thin jet is helpful at this stage). Transfer into a 15 ml cone edged tube. Centrifuge three minutes at 3,000 RPM and make a slide, using whichever mounting medium you prefer. A wet slide with water can be made, but make sure you have enough residue left for a permanent slide, if you are lucky and it contains sufficient pollen grains. Clays are almost always present in the slide at this stage (Fig. 5.1) and some experience will show whether further cleaning is required. It is usually at the end of stage 5 of the process when clay occurrence is apparent, by the color of the residue. If the color is gray to black, clear, further cleaning is usually not necessary; if it is yellowish or brownish, opaque, a short hydrofluoric acid treatment usually makes the miracle (Fig. 5.2) and leaves only the organic particles. The simple way to do it is to use plastic tubes at least from stage 4 onward and, at the end of stage 5, to add some 20 ml 48% hydrofluoric acid. THIS STAGE MUST BE CARRIED OUT IN THE FUMEHOOD AND IF YOU HAD NOT USED THEM BEFORE, IT'S TIME FOR GLOVES! Mix well, carefully, let stand some ten minutes, centrifuge six minutes at 3,000 RPM. Discard the HF (see step 5 for details), add 10% HCl almost to the top of the tube, centrifuge six minutes at 3,000 RPM, after which refer again to steps 5 and 6.

Clays and fine organic particles can also be got rid of at this stage by filtration through a fine sieve, in partial vacuum, as described above.

When the slide is observed under the microscope, decision must be made as to further treatment, if at all. In many cases it will be found out that this process produces a decent spectrum, sufficient for most stratigraphic purposes. If the slide contains pollen, but is masked with clay, an HF treatment is advisable. If not enough residue is at hand, the process could be repeated, this time with HF prior to making any slide. If the slide is very poor in pollen, but however even a small number is there, it could be considered to process a much larger sample, using a procedure described below. If no pollen are present at all, following a detailed scrutiny, the sample may be considered barren. However, it is worth sometimes to try a larger sample before final decision is made.

5.3. Clastic sediments

These are the more frequent type of deposits treated in a palynological laboratory geared to study arid lands, being the most common rocks formed in these regions. Since the majority of clastic sediments, especially those of continental origin in arid lands, are deposited at considerable rates, the first and foremost goal of extraction is to concentrate palynomorphs to an acceptable level for microscopic examination and numeric treatment.

Very fine, clay-size clastics are deposited in arid lands almost only in aquatic environments, both marine and continental. The rates of deposition are usually slow, so that pollen concentration can be relatively high, thus in most cases extraction is performed with a 10–20 grams sample. The main difference between waterlaid fine clastics and the unconsolidated sediments described below, is that the former may have gone through some cementation processes, thus requiring additional stages of extraction.

Almost all other clastic rocks found in arid lands have been deposited at higher rates, therefore the first stage would be to get rid of the coarser particles and treat the rest as one would deal with fine clastics. The sample size, accordingly, would have to be considerably larger, in the order of 50–250 grams. Even larger samples, up to several kilograms each, could be treated in the same way, only using larger laboratory ware, until reduced to a manageable quantity of fine grained residue.

5.3.1. Unconsolidated sediments

These include most types of samples collected for recent pollen studies. Dust captured by airborne pollen traps, that comprises both fine mineral and organic particles; and all kinds of fine grain sediments, as long as no cementation processes took place. The reason for including both dust and sediments in this category is, because essentially the only activity involved in extracting pollen is their mechanical separation from the mineral matter. Any chemical processes involved are only aimed at treating the palynomorphs themselves, for examination under the microscope, or getting rid of excess organic material. In most cases no oxidation would be necessary and acetolysis seems to take care of most organic "dirt". It is advisable not to use KOH, since it may affect the grains in adverse ways (Brown, 1960). Some authors use hydrofluoric acid for dissolving clays, but this seems not necessary with unconsolidated sediments, from which clays can be removed by other, simpler methods, such as gravity separation.

A typical procedure for such preparation, of both dry and wet samples, including those collected by traps smeared with glycerin or glycerin jelly, is as follows:

(1) Add to the sample 10% hydrochloric acid up to ten times its volume and pass through a 210 μm sieve.

(2) Centrifuge, pour the supernatant acid with all that floats, which would mostly comprise plant debris. Turn the test tube upside down over a filter paper for about ten minutes.

(3) With the aid of filter paper rolled into a cylinder, wipe the inside of the test tube dry. Add zinc chloride solution (sp. gr. 1.85), no more than half the height of the test tube. Mix very well. Centrifuge at 3,000 RPM, eight minutes.

(4) Pour the supernatant liquid into a test tube, no smaller than the former. Top up with 10% HCl, mix well. Discard the sediment left in the former tube.

(5) Centrifuge at 3,000 RPM, 15 minutes, pour the liquid.

(6) Wash with 10% HCl.

(7) Wash twice with distilled water.

(8) Acetolyze the residue.

(9) Mount the residue.

Dust samples which were collected by traps using silicone oil should first be washed with xylene, after which two ways are open: either go on using organic liquids (recipe follows), or wash the sample several times with absolute alcohol, following which it can be transferred into an aqueous medium and extraction continues as detailed above.

In case organic compounds are preferred (more expensive, no less smelly, but non-corrosive and easier to handle), the procedure is as follows:

(1) Wash the sample in absolute alcohol, centrifuge, decant.

(2) Add a mixture of bromoform and alcohol (sp. gr. 1.85), no more than half the height of the test tube. Mix very well. Centrifuge at 3,000 RPM, eight minutes.

(3) Pour the supernatant liquid into another test tube. Discard the sediment from the former.

(4) Fill the test tube with absolute alcohol, mix well. Centrifuge at 3,000 RPM, 15 minutes.

(5) Wash the residue with absolute alcohol.

(6) Wash with 70% alcohol.

(7) Wash with distilled water.

(8) Acetolyze the residue.

(9) Mount the residue.

Either of these techniques, which are essentially similar, would usually result in good quality slides, especially with recent material. A haze, which may appear in the slide, usually results from very fine clays, which are sometimes difficult to rid of with the heavy liquids, due to the extreme cohesion of the finest particles. The occurrence of clays in the final stages is also evident from the color of the residue, which turns yellowish. In this case two possibilities are open, if one prefers a cleaner slide: either filtration, through a very fine (12–20 μm) sieve, or application of hydrofluoric acid treatment for some time. The time cannot be specified, since it may change from one sample to another. Usually 20–30 minutes are sufficient to remove the fine clays.

Fine organic particles, if very abundant, can also be annoying, but would not cause yellowish coloration of the residue, which will look black. These can be removed by sieving or by oxidation. Sieving proved very useful for cleaning the residue and is, in our laboratory, a routine procedure. Oxidation, if at all attempted, should be done with great caution, preferably leaving some non-oxidized residue aside, just in case the process turns too successful, destroying the palynomorphs as well as the dirt.

Several palynologists, especially French researchers working in the Sahara such as Cour (1974), Daoud (1984) and others, have used a rather complex method for separating pollen from airborne dust, trapped on gauze filters. The first stage involved dissolving the gauze in concentrated sulphuric acid for 24 hours; decantation, which took another 24 hours; treatment with 20% hydrochloric acid; acetolysis; 10% KOH; filtration through a 200 μm filter; and in some cases gravity separation was done at this stage, after which the palynomorphs were stained and mounted.

It seems that most stages are not really necessary and a much simpler method, such as described above, would bring the same results. It is of no use to dissolve the gauze filter, which could easily be washed and the grains collected. If it is suspected that some pollen are left on the gauze, this can be checked by dissolving one or two and comparing the results with washing. Needless to say, the simpler methods are less expensive in terms of materials and labor and take a much shorter time. We found out (Weinstein, 1979) that the results are indeed similar, thus the shortcut was welcome.

5.3.2. Fine clastics

It is sometimes quite difficult to tell, without an X-ray analysis, whether fine clays have undergone cementation and if so, of what type, or are only compacted. Cementation of fine clays can be of two main types: either by precipitates, such as carbonates and sulphates, or by secondary intergrowth of the clay minerals. The first type of cementation could be treated by 10% HCl, hot in the case of sulphates or dolomites, while with the second type hydrofluoric acid seems inevitable. At any rate, since carbonates and sulphates are quite common in arid lands and also in surrounding water bodies, the HCl treatment should precede any other, so that no calcium remains in the sample before applying HF. If it does remain, insoluble deposits can form which are very difficult to get rid of.

A basic procedure for extraction from pollen poor fine clastics, following Caratini et al. (1976) and others, is as follows:

(1) Disaggregate the sample in 10% hydrochloric acid. The size of the sample would depend on expectation and availability. Caratini et al. (1976) used 25 grams, but some of their samples yielded poor slides, which indicates that a larger quantity was in fact needed. Stir the sample in the HCl until effervescence terminates. Add some concentrated HCl to make sure there is no more carbonates. If dolomite or sulphates are suspected, the mixture should be heated over a hot water bath, while occasionally stirring, for at least 15 minutes. Centrifuge and decant.

(2) Wash with 10% HCl.

(3) Add 70% hydrofluoric acid, leave overnight in room temperature, stirring every once in a while, or boil for one hour in a hot water bath. Centrifuge and decant.

(4) Wash with 10% HCl.

(5) Float organic material in zinc chloride solution. Centrifuge 15 minutes at 3,000 RPM. Pour liquid into another test tube, discard sediment. Caratini et al. suggested centrifuging in two stages: at 200 RPM for ten minutes, followed by 1,800 RPM for half an hour.

(6) Dilute with 10% HCl, mix well, centrifuge ten minutes at 3,000 RPM, discard liquid, stand test tubes upside down.

(7) Wash with 10% HCl.

(8) Wash with distilled water.

(9) Mount.

This method is quite simple, not time consuming and would take care of at least 90% of the samples. Attention should be paid after the first stage, because in many cases, when carbonates are the only cement, HF treatment could altogether be avoided, which makes the process even shorter. Experience shows that this would be the case with a large part of the fine grained samples collected from arid lands, which have not been buried too deeply. Sediments older than the Neogene would almost surely require HF treatment.

A general procedure for macerating pollen poor fine clastics is brought in Heusser and Stock (1984), who summarize many years of experiments. The size of sample they suggest is 5 grams, or 5 cm^3, for which they also give exact amounts of chemicals and volumes of glassware. This method, described below, would need some modifications for sediments from arid lands, due to their usually higher carbonate contents. Unfortunately, Heusser and Stock found it impossible to use HCl at the first stage, because for some reason it deters the activity of sodium pyrophosphate, which is used as a deflocculating agent in the process. This method is therefore more suitable for clayey sediments. The process is rather lengthy and labor consuming, thus it is suggested here to first try the simpler one, described above, going through this one only when there was no success with the former. In case it is found necessary to process larger samples, quantities of chemicals and sizes of ware should correspondingly be increased.

(1) Boil briefly five grams of disaggregated sediment in 200 ml 1% KOH. Centrifuge and decant.

(2) Add 200 ml 0.001 M sodium pyrophosphate solution, mix well, sieve through a 150 μm nylon mesh screen, held on a two-piece large plastic Buchner funnel. (Some manual stirring may help while the sample is poured through the filter, to prevent clogging. If you expect larger palynomorphs, use an appropriate screen.)

(3) Sieve–decant with sodium pyrophosphate, through a 7 μm micron screen. This is done by suspending the sample fraction smaller than 150 μm in five liters of cold pyrophosphate solution and letting it settle. Sieve gradually through a 7 μm nylon sieve, supported by a metal mesh held in a ring stand, adding more of the solution while manually stirring to prevent clogging. Add pyrophosphate and continue sieving until the liquid coming out of the sieve is clear.

(4) Wash the residue on the screen, with water, into a 50 ml centrifuge tube. Stir, centrifuge and decant.

(5) Wash in 10% HCl, to rid of carbonates.

(6) Add 52% HF, stir, boil in a hot water bath for an hour, or leave at room temperature overnight. Centrifuge and decant.

(7) Add 10% HCl, stir well, boil shortly, centrifuge and decant.

(8) Wash with water. Check sample for occurrence of siliceous materials. If such are present, repeat from stage 3.

(9) Acetolyze. Check the sample for occurrence of amorphous organic matter or charcoal-like particles. If these occur, oxidize the residue with a mixture of 25 cm^3 80% sodium chlorate, 75 cm^3 glacial acetic acid and 6 cm^3 concentrated sulphuric acid. The oxidation is done at room temperature while occasionally stirring, monitoring continuously until the residue turns pale yellow. In highly organic samples this process may take up to 12 hours. Constant monitoring is necessary, because with some types of sediments no color change will occur and oxidation must be stopped before pollen are destroyed. Centrifuge and decant.

(10) Wash with water.

(11) Wash with 10% NaOH, in a hot water bath, for ten minutes.

(12) Sieve through a 7 μm nylon filter.

(13) Stain and mount residue.

This procedure gives very good results with samples in which palynomorphs are well preserved, but note that it may be hazardous when preservation is not as good. Especially, as said before, I would refrain from oxidizing, if not vital.

5.3.3. Coarse clastics

These seem to be the most common type of sediments treated while studying palynology in arid lands, especially because they are so common in wadis, from which samples are being frequently collected, either in connection with stratigraphic or archeologic studies (Martin, 1963; Mehringer, 1967; Beucher, 1975; Horowitz, 1979; and many others). The main problem involved with extracting pollen from such sediments is their relative poverty in grains, resulting from combination of the usual poor pollen production in arid lands and the high rates of deposition of coarse clastics. On top of that, Mehringer (1967) reported that pollen are sometimes difficult to extract from some alluvial deposits due to the formation of resistant colloids and the presence of obscuring inorganic debris. The method described below seems to take care of almost all these limitations.

Coarse clastics are very rarely composed of only coarse particles, usually containing some amount of finer components. It is with the latter that pollen are associated, thus consequently the extraction process first aims at discarding the coarse fraction, only then treating the finer in the same manner as fine grained clastics would be macerated.

Methods for doing so have been developed by groups of palynologists working in the American Southwest and summarized by Mehringer (1967), who gave a detailed description of the process, which follows:

General: This procedure is only an outline. Steps may be added or deleted as one gains experience with the sediment types of a particular region and age. Sample size is approximately 50 grams. The reagents and distilled water are stirred well with the sediment, when added, at each stage. Following each step, the sample is centrifuged for 2–3 minutes and the supernatant liquid decanted. Approximately 40 cm^3 of reagent or distilled water is added at each stage, unless otherwise specified.

(1) Mix the sample with 200 ml water, in a 500 ml plastic beaker. Add concentrated HCl, 25 ml at a time and stir slowly. This is continued until the material can be vigorously stirred with no further reaction. This step removes much of the carbonate cement and assures the release of pollen from the calcareous matrix.

(2) Stir the mixture with a circular motion, until a strong vortex is formed and all material in the beaker is in motion. Make sure the sediment is also raised from the center of the beaker!

(3) Let the beaker stand 45–60 seconds, until the heavier fraction such as sand and gravel start to settle out. While the suspended material is still in motion, pour it slowly through a #100 mesh screen into a 250 ml beaker. Wash any sediment on the screen with a strong jet of water from a plastic spray bottle.

(4) Stir the material again in the same manner as above, allow to stand some 90 seconds and pour suspension into 50 ml nalgene tubes. Centrifuge and decant the supernatant liquid.

(5) Repeat the last step, until only 2–4 cm^3 of sediment remain in the bottom of the test tube.

(Steps 1–5 are termed "The HCl swirl method".)

(6) Wash the 2–4 cm^3 residue with 10–14% HCl.

(7) Wash with 25–30% HCl.

(8) Wash with distilled water.

The HCl washing steps remove the remaining carbonates and generally eliminate any vigorous reaction when, next, hydrofluoric acid is added.

(9) Add 50% HF, starting with 10 ml, stir gently. If no reaction starts within two minutes the remainder of HF is added and stirred well. If reaction occurs, the remainder of HF is added in two or three steps, gently stirring following each addition. Leave the sample in the 50% HF, unstoppered, but protected from contamination by a piece of paper towel, in a well ventilated fumehood, for 6–24 hours. Mix occasionally. Centrifuge and decant.

(10) Add 70% HF, allow to stand (as above) for 6–24 hours. Place in a gently boiling water bath for 20–30 minutes.

(11) Wash twice, with boiling water.

(12) Add 25 ml of 20% nitric acid, allow to stand for ten minutes.

(13) Wash twice with hot water.

(14) Add 30 ml of concentrated HCl and place the tube in a boiling water bath for two minutes.

(15) Wash twice with hot water.

The nitric acid treatment oxidizes some of the organic colloids, which become water soluble after the HCl treatment.

(16) If it seems necessary, acetolyze.

(17) Wash with water.

(18) Add 5–7% KOH or NaOH solution. Place in a boiling water bath for 2–3 minutes.

(19) Wash with water, until the decanted solution comes clear. This last step may require from two to as many as 12 repetitions, depending on the humic acids contents of the sample.

(20) Transfer the remaining sediment into 5 ml shell vials. Centrifuge.

(21) Remove most of the water from the vial with a capillary pipette.

(22) Place a drop of the pollen containing residue on a microscope slide, add a drop of glycerol and stain with basic fuchsin. Cover the preparation with a No. O cover glass.

(23) Store the remaining residue in the shell vial, in a mixture of two parts glacial acetic acid, two parts glycerol and three parts water.

This procedure is quite lengthy and labor consuming, but served as a sound basis for modifications by various investigators (Fish, 1984; and others). The two principal modifications concerned enlargement of the sample size and replacement of the HF treatment by zinc chloride (or bromide) flotation. Fifty grams of sediment would usually be adequate for regions such as the American Southwest, in which the desert vegetation is quite well developed and pollen production, in relation to average rates of deposition would be sufficient for a decent spectrum. However, in deserts where vegetation is much poorer, such as the Sahara, the Negev, some parts of the Kalahari, Namib, Atacama, etc., samples of only 50 grams are far from being sufficient for yielding a significant amount of pollen, thus must be considerably enlarged, quite frequently by an order of magnitude.

The standard procedure applied in our laboratory for extracting pollen from coarse clastics shows good results with most samples, up to 90–95% of those macerated. Only if necessary, a second, more elaborate run is performed on the stubborn samples, along the lines described above. It goes as follows:

(1) Place 250–500 grams of sediment in a large plastic container (2–5 liters laboratory beaker, depending on the sample size, or a household bucket from which the metal handle has been removed). Add concentrated HCl slowly, stirring constantly. If effervescence is too strong, keep at hand a spray bottle with ethyl alcohol and spray some until the reaction quietens down. This would also result in a much nicer odor, since some of the alcohol would turn into an esther. Concentrated HCl is used, rather than the usually recommended 10% solution, in order to keep the amount of liquid manageable. The acid is quickly neutralized by the abundant carbonates right at the beginning, thus its actual concentration would soon drop to 10–15%. Keep adding slowly the concentrated HCl, until effervescence stops. Add some more, to make sure it had not stopped only because all acid has been neutralized.

Note: certain samples may contain large limestone pebbles and it is not necessary to waste time and acid to consume them. In this case, do not let the reaction go all the way, but stir well, let stand 5–10 seconds until only the pebbles settle down and pour the mixture into another plastic container. Make sure what remains in the first only comprises coarse, unnecessary particles (with no cement, for instance) and discard. It may be helpful to pour through a 210 μm sieve, since some of the carbonate particles may float due to attached gas bubbles.

(2) Stir vigorously with a circular motion, using a large plastic spoon, making sure all the sediment is in suspension. Let stand for 45 seconds, then pour into another plastic container. Discard what was left in the first one.

(3) Repeat step 2, but this time let the material settle for 60 seconds, before pouring the suspension into another container.

(4) Repeat, with standing time of 90 seconds.

(5) Some samples, which would still comprise too much material to be handled by the centrifuge, may require several repetitions of this "HCl swirl", with longer standing time, which should be determined by experience. Usually standing times longer than 120 seconds are not recommended, since some of the pollen may settle down. Occasional monitoring of the coarse part of the sample, before being discarded, may help tell at what stage pollen are also wasted.

Stages 1 through 4 (and 5, if necessary) are usually done in the afternoon, while the rest of the procedure is carried out the next day.

(6) When the last stage of swirling has been completed to satisfaction the containers are left covered, overnight, in the ventilated fumehood, until all sediment has settled down. Next morning, the liquid is pumped out of the

container, with the aid of a vacuum pump. This process should be done very carefully, because some of the organic material, including pollen, may float on the surface of the liquid. The sucking end of the vacuum pump tube is inserted into the container *before* the pump is activated, about 1 cm below the surface of the liquid. When the pump is turned on, the tip should be sunk slowly into the container, following the descending liquid level, always keeping it some 1 cm below the surface. Care should be applied when most of the liquid is sucked out, not to disturb the uppermost layer of the settled deposit. Slight tipping of the container can help get rid of some more liquid, but should be done very gently. A little practice would make this operation an easy routine. Turn the vacuum pump off *before* the tube is taken out of the liquid and *do not* make any second attempts to pump liquid. If too much liquid is left, the container must stand for at least another 12 hours before resuming pumping. Another possibility, if large enough centrifuge bottles are not available, is to get rid of the excess liquid by repeated centrifugation.

(7) Stir well the sediment in the remaining liquid, transfer into 500–1,000 ml centrifuge bottle. The amount of remaining sediment should not fill more than a quarter of the bottle after centrifugation. If the quantity is larger, repeat swirling until it is reduced to manageable quantity. Centrifuge eight minutes, 3,000 RPM, decant the liquid, stand the bottle upside down for at least 10–15 minutes on a filter paper. If the amount of sediment is smaller, so could accordingly be the size of the centrifuge bottle or tube.

(8) Wipe dry the inside of the bottle (or tube) with a clean filter paper, taking care not to touch the surface of the sediment. Add zinc chloride (or bromide) solution, up to a maximum of half the bottle's height, mix very well. Take care that no lumps remain. Centrifuge ten minutes at 3,000 RPM.

(9) If a blackish ring is formed on the bottle's rim, made by floating organic material, mix it well with the liquid before pouring to the next bottle. A fine painters brush, preferably with plastic hair, comes handy, but remember to clean it very well between use in different bottles, or use several such brushes, each for every sample. Pour the supernatant liquid into another bottle and discard the sediment.

(10) Fill the bottle with 10% HCl, mix very well, centrifuge 15 minutes at 3,000 RPM, decant the liquid.

Please note: if you want to save the heavy liquid, pour it into another container. If properly filtered and boiled (in a fumehood!), its specific gravity can be restored. It could also be restored by adding zinc chloride crystals, or zinc bromide solution, but in such a case you will end by having increasing quantities of heavy liquid and would not really save much. Check that after boiling the solution is still acid enough.

(11) Add 40 ml 10% HCl, stir well and transfer to 50 ml centrifuge tube. Centrifuge eight minutes at 3,000 RPM, decant.

(12) Add 10 ml water, stir well, transfer to 15 ml centrifuge tube, preferably with a conical edge. Centrifuge and decant.

(13) Wash twice with water. At this stage decision should be made whether to apply acetolysis, oxidation, or staining. Acetolysis may turn useful occasionally and does not seem to harm the pollen. However, if not indeed necessary, it is better to refrain even from this process. It is suggested not to oxidize and to avoid the use of hot bases, such as KOH or NaOH.

(14) Mount on slides.

This procedure can easily take care of samples weighing up to 500 grams. If larger samples, of several kilograms, are to be macerated, the procedure remains similar, only the ware and equipment change. The first aim in processing a large sample is to bring down the amount of sediment to manageable quantity. For this, the first stages, of swirling and flotation, need large containers. For swirling, household buckets or plastic waste bins are useful, not expensive and easy to come by. Large spoons for mixing may be more difficult to get, but any store which stocks equipment for restaurants or hotels would most probably carry them. It is possible to use either plastic or stainless steel spoons, but both types must be checked for their resistance to hydrochloric acid. Some lower quality utensils, even if made of plastic, tend to dissolve in acid, while certain stainless steels could rust.

Swirling is done in much the same manner as described above, in an appropriate container. Steps of the HCl swirling are repeated as above, but settling times have to be prolonged, up to several minutes. The sample is left overnight to settle and the liquid pumped out next morning (see above for details).

The remaining quantity of deposit and liquid would usually be too much for handling with a common centrifuge. If a large capacity centrifuge is available, the process is continued as above. If not, the mixture of sediment and liquid is stirred well, then poured into a plastic cylinder with holes drilled in it every 5 cm, vertically, with plastic taps fit to these holes. Zinc bromide solution (or zinc chloride crystals, but the former is more convenient) is added and mixed well, until the desired specific gravity of the mixture, usually 1.85, is gained. The cylinder is left to stand, covered, until all mineral matter had settled down, which may take between 24 and 48 hours. When the sample had settled, the uppermost part of the liquid is gently stirred and the uppermost tap, which would drain enough of the heavy liquid containing the floating organic matter, is opened, so that only this part is transferred into another container. Note that this latter container should hold at least twice the quantity of the heavy liquid, which has to be diluted. Another, yet simpler possibility is to use a plain cylinder or bucket and suck the palynomorphs containing upper layer with the aid of a pipette.

An equal amount of 10% HCl is added and mixed well. Concentrating the organic matter from this mixture can either be done by repeated centrifugation, or, if the container is left, covered, for 24 to 48 hours, by vacuum pumping the supernatant liquid, in which case further dilution with twice the quantity of 10%

HCl is suggested. If centrifugation is preferred, it is sometimes useful to add a carrier to the tube, to catch the organic material. This carrier is usually some silt (not clay, which is hard to get rid of), which is subsequently removed by another step of flotation. This second flotation step, incidentally, seems necessary also with the settling method, since some fine particles are almost sure to remain. Maceration then proceeds as described above. Using this procedure, samples as large as five kilograms had been processed and yielded pollen in sufficient numbers for statistically significant results.

5.4. Other substances

5.4.1. Chemical precipitates

Chemical precipitates are quite common in arid lands, some precipitated from relatively fresh water, usually carbonates such as travertines and speleothems, others deposited from warm, saline water, like rocksalt, gypsum, anhydrite or dolomite. In terms of pollen extraction prospects, the chemical precipitates are divided into three groups. The first comprises water soluble evaporites, mainly rocksalt, but also others, rarer, such as sylvite and carnalite; the second consists of precipitates which are dissolved in boiling hydrochloric acid, such as gypsum, anhydrite and dolomite; the third includes such evaporites that can usually be dissolved in boiling HCl, but do not yield well or at all to this treatment, probably due to formation of more complex compounds; this group occasionally requires special techniques, which may also be applicable to rare minerals, such as magnesite and others.

The procedure for extracting pollen from water soluble salts is as follows:

(1) Dissolve a coarsely crushed sample in water, preferably warm. Both crushing and warming the water are done to hasten the process, since usually large samples are used. An average of one kilogram is optimal, but due to the frequent very high rates of evaporite deposition, larger samples may occasionally be required. Dissolution is made in a large container, such as a household bucket. Toward the end of the dissolution, concentrated HCl is added, to dissolve carbonates and other evaporites. Heating may be helpful at this stage. If swirling is necessary, in cases when the salt contains large amounts of clastics, it is also done at this stage, following the addition of HCl.

(2) Let stand overnight, covered. When the deposit has settled down, pump out the supernatant liquid with a vacuum pump.

(3) Continue in the same manner described above for fine clastic sediments.

Calcite and aragonite precipitates are treated in much the same way as rocksalt, only instead of water, concentrated HCl is used, added until all effervescence has stopped and the sample entirely dissolved. This procedure may require very large containers for the first stage, for which plastic garbage bins were found useful. Carbonates are also soluble in a variety of other acids, such as nitric, which is not recommended, being a strong oxidant; phosphoric acid, which is difficult to handle; and organic acids such as formic, acetic and EDTA (ethylenediaminetetraacetic acid), which do the job more gently than the mineral acids. It was not proved however, that gentleness is a virtue in this particular case. Rather, it seems that the most common, useful and certainly cost effective is still hydrochloric acid (needless to say, of a technical grade).

Gypsum and anhydrite are usually dissolved, at least partially, in hot 10% hydrochloric acid:

(1) A Coarsely crushed sample is put in an appropriate container, hot 10% HCl is slowly added, while continuously stirring. It is advisable to keep the container in a boiling water bath, in a well ventilated fumehood. It is difficult to tell how long it would take for the sample to completely dissolve and the process may be lengthy. After all crystals had dissolved, the container is left, covered, overnight to settle down. If large enough amount is treated in this way, the process can be stopped any time the amount of clastics accumulated on the container's bottom seem sufficient for further treatment. Telling what this amount might be is a matter of the specific rocks treated and experience based on experimental runs. The remaining crystals are separated from the fine clastics by swirling or sieving.

(2) If lucky and all crystals had indeed been dissolved at this stage, or if stopped as described above, procedure follows in the same lines as for fine clastic sediments, including swirling, if necessary.

If the quantity of thus released clastics seems too small, it was found out that in many cases it is preferable to simply increase the initial size of the sample and use for further maceration whatever fine clastics liberated, instead of investing a lot of labor trying to achieve complete dissolution.

Attempts can be done to try and dissolve gypsum and anhydrite in other reagents. Sodium thiosulphate ("hypo" salt of photographers) is reported in Brown (1960) and in Kummel and Raup (1965) to help indeed in dissolution of these evaporites. A solution containing one part of anhydrous "hypo" to four parts of water, heated on a steam bath or gently boiled, is effective in dissolving gypsum and anhydrite. Admittedly, we did not have much success with the use of "hypo" in our laboratory. Brown further reports that these evaporites are also soluble in ammonium sulphate, or in a supersaturated solution of sodium chloride. In cases where the evaporites contain free sulphur, it can be dissolved by hot xylene, benzene, or carbon tetrachloride.

Kummel and Raup (1965) suggest trying to dissolve sulphates in 10% nitric acid. This may be hazardous, since it is a very strong oxidant, thus should be performed with care and constant monitoring. Another possibility which may be worth trying, is the use of EDTA, in 10% aqueous solution.

It may occasionally be very difficult (E. Schulz, personal communication, 1988), to obtain pollen entrapped in such evaporites as magnesite, or possibly also other rare evaporitic compounds. The only solution in these rare cases is to consult a chemist, or better still a geochemist, who might come out with an answer. The only thing to make sure is, that suggested reagents would not destroy the pollen as well. Occurrences of these rare compounds can sometimes be avoided, by sampling a different type of deposit in the same horizon.

5.4.2. Organic deposits

Organic sediments, which are usually quite rare in arid lands, can, when they do occur, serve as an important source of pollen grains. In terms of extraction possibilities, three types of such deposits occur: peat and highly organic sediments; animal midden; and coprolites or various kinds of pellets.

Peat is treated in the same manner described in many textbooks for this type of deposit (*cf.* Faegri and Iversen, 1964), with only one principal modification, which involves the use of hydrochloric acid as an initial step. The reason is that, unlike the typically acid peat formed in wet climates, peat laid down under arid conditions is in most cases basic, comprising significant amounts of carbonates.

The steps involved in extracting pollen from an arid land peat are as follows:

(1) Place the sample in a beaker, slowly add 10% HCl (concentrated HCl, if rich in carbonate), stir. Continue adding HCl until all effervescence terminates. Centrifuge and decant. If sample and amount of liquid is too much for centrifugation, let settle overnight and vacuum pump. Usually swirling is of no use with peat, but if it contains plenty of mineral matter, apply swirling following the HCl treatment, before letting settle overnight.

(2) Wash with 10% HCl.

(3) Sieve through a 210 μm screen.

(4) Wash twice in water.

(5) Add 10% KOH solution, heat over a boiling water bath (or an electric hot plate), while stirring. Centrifuge and decant.

(6) Wash several times with water, until the liquid comes clear.

(7) Float the organic material in a heavy liquid.

(8) Acetolyze.

(9) Mount on slides.

The KOH treatment would dissolve the humic acids and produce a cleaner slide. However, it may prove detrimental to the pollen, so that if humic acids are not too common, as is the case with many peats deposited under basic arid conditions, it is preferable to skip the KOH step. It may also be tempting to oxidize the sample for obtaining cleaner slides, but this process is hazardous, since pollen is not too well preserved in basic peats. If oxidation is attempted, it should be performed with continuous monitoring. It is also suggested to avoid any ultrasonic treatment, which could mechanically destroy the pollen.

Animal middens are usually cemented by their dry urine, as is the case with packrat and hyrax. An extraction technique is as follows:

(1) Place the sample in hot water, stir until cement has dissolved. Filter, through a 210 μm screen, to get rid of the coarse vegetal material. Centrifuge and decant. Some investigators place the midden in a tea strainer for this step and wash it with warm water into a container. The ensuing mixture is then filtered and centrifuged.

(2) Boil in 10% KOH. Pollen in midden are usually well preserved, so this step could be applied quite safely here.

(3) Wash with water, until liquid comes clear.

(4) Treat with 70% HF, over a boiling water bath. This step can be replaced by heavy liquid flotation.

(5) Acetolyze.

(6) Occasionally, a second 10% KOH treatment would be necessary.

Thompson (1985) and Davis and Anderson (1987) also applied a 10% HCl treatment, to get rid of carbonate particles which are trapped in the midden.

Coprolites, pellets and fecal pellets are usually not readily soluble in water and would require some mechanical disintegration, followed by 10% HCl treatment, before procedure can be continued, following the same lines brought above for the maceration of midden. In cases the coprolites are poor in organic compounds some of the above steps could be omitted.

5.4.3. Soils and subsoil crusts

These, in terms of pollen extraction, are treated similarly to corresponding sediments, namely cemented clastics. These deposits can be divided into two types: loose soils, or cemented soils and subsoil crusts.

Loose soils are treated in the same manner as clastic sediments (see above). Subsoil crusts are treated same as cemented clastic rocks, according to the type of cementation. The most common types of cements in arid lands are carbonates and gypsum, which are dissolved using methods described above for chemical precipitates.

Occasionally, subsoil crusts may be formed by recrystallization and compaction of clay minerals. In such cases a treatment by hydrofluoric acid is unavoidable. However, even then the sample should first be subject to hydrochloric acid treatment, to remove as much carbonates as possible, before HF is applied, so as to prevent formation of fluorocarbonates, which are very difficult to get rid of at a later stage.

Fish (1984) extracted pollen from soils in central Arizona, using approximately 60–70 cm^3 of soil, which was deflocculated in hydrochloric acid and swirled in water as described in Mehringer (1967; see above) in order to remove the larger particles. Sample was then subject to flotation in a zinc bromide solution, of 2.0 specific gravity, to further separate fine particles from the remaining organic matrix. In view of the low organic contents of the study area soils and in order to minimize damage to the pollen, no chemical steps for reducing organic debris had been employed by Fish. She also made comparisons in this study, of extraction techniques employing either digestion of silicates by hydrofluoric acid, or using the zinc bromide flotation method. She concluded that both methods gave similar results and the heavy liquid process seems preferable.

5.4.4. Water

Pollen grains are quite easily extracted from water, with the use of a continuous flow centrifuge. The principle of operation is, that water is put into the centrifuge tubes through a set of inlets, connected to a common inlet, into which water is entered. While the centrifuge rotates, sediment is accumulated at the bottom of the tubes and the supernatant water is extracted via another set of outlets, into a common one. This procedure requires angle heads and the process of deposition depends on both the speed of the centrifuge, which is usually quite high, up to 6,000 RPM and the velocity of the circulating water which determines the amount of water passing through the apparatus at a certain time unit. This last factor is determined by the user, usually through the use of a pump and a regulating tap at the inlet.

The only logistic problem involved with this process is the large amount of water necessary for obtaining a decent pollen spectrum, which may occasionally amount to hundreds of liters. Carrying this amount to the laboratory may be problematic, so to overcome it we usually take the centrifuge to the field instead, with an electric pump and a generator. If a simple microscope is also taken along, initial monitoring of the process can be done in the field, which could save extra work. Centrifugation equipment can thus fit in an average size car.

The pump drives water through the centrifuge at a rate that can be determined by a tap, or, if a more sophisticated pump is available, by controlling its output and measuring the amount of water with a flowmeter. A simple flowmeter can be bought in stores which stock car accessories, usually used to determine fuel consumption.

The rate of water flow in the centrifuge is determined by experience and monitoring, at least during the first runs. We found out (Horowitz, 1970) that an average rate of flow of one liter per minute, for a centrifuge with a total capacity of 300 cm^3, rotating at 5,000 RPM is adequate. In our case a sample of 100 liters yielded enough pollen for analysis. However, these figures can change with the type and speed of the centrifuge and with the pollen contents of the water, at a specific source.

The sediment collected in the centrifuge tubes is transferred into a single container, with some water, then treated in the same manner applied to unconsolidated deposits (see above). If it is anticipated that some time will pass before final steps are carried out, it is advisable to add some formaldehyde to the wet sample, to protect from bacterial activity and oxidation.

A possibility pointed out by Brown (1960), is to let the water stand in containers for a certain time, probably in the order of one to two days, until all particles have settled down, then to decant the water. This process could also be carried out in the field, where it would be helpful to have a small pump for that. Again, such pump can be bought in car accessory stores and operated from the car's battery. This process, although requiring a longer stay in the field, does not necessitate neither hauling the water to the laboratory, nor buying an expensive continuous flow centrifuge. If only a limited number of samples is to be studied, it is by far preferable. It is however necessary to make sure the containers are well covered, to prevent contamination.

Peck (1973) used sediment traps to study pollen load in water. These are costly and seem to need permanent attention, but the results were good. Traverse (1990) summarized his and others attempts at extracting pollen from water, both rivers and sea. He first used a method by which the water was evaporated after collection, then the ensuing sludge macerated in much the same way as fine clastics. More recently, Traverse filtered the water through silica screens, which were dissolved in hydrofluoric acid to extract the trapped palynomorphs. This latter method seem preferable in terms of practicality, since the amount of equipment necessary for sampling is minimal, constituting a pump, appropriate silica filters and some tubing. Traverse reported very good pollen yields by this method.

5.4.5. Exotic substances

Exotic substances are usually tied up with archeological excavations, their pollen contents can occasionally reveal interesting information concerning the composition of the substance, its provenance, use, or other related knowledge. Such substances may be honey, asphalt, all kinds of resins, amber, contents of jars, paints, embalming compounds, mortar, plaster and others, many of which would appear as small lumps of blackish material.

It is sometimes quite difficult to tell whether such a small, black lump was originally honey, asphalt, resin, or possibly some other substance. Although a chemical analysis is of much help in revealing the nature of such substances, their pollen contents may yield additional, significant information on composition and provenance.

Unless the chemical composition is known ahead, the first trial should be to dissolve the lump in boiling water. In such a way, a small black lump from a Roman site near the Dead Sea which was thought to be asphalt, was dissolved in water and yielded plenty of pollen originating from insect pollinated plants, proving that it was originally honey. Dissolution is followed by sieving, heavy liquid separation and sometimes acetolysis.

If the substance resists dissolution in hot water, the next trial should involve ethyl alcohol, which would dissolve all kinds of resins that had a variety of uses in ancient times. Resins were used for ceremonial purposes, embalming, sealing ceramic jars and preserving their contents, etc. A black lining from the inside of a Roman wine jar from Israel was dissolved in alcohol and contained pollen of pine and grape vine, indicating that Romans drank a kind of Retsina.

Girard and Maley (1987) extracted pollen from embalming resins of an Egyptian mummy and also from the dust accumulated in its thorax. The latter was treated as a mineral matter, in the same manner applied to unconsolidated sediments, involving HCl, HF and acetolysis. The resins were dissolved in two parts of acetone, mixed with one part of trichlorethylene. The organic fraction was then acetolyzed. Pollen indicating plants used for embalming and rituals were thus extracted.

Dissolution in alcohol is followed by filtration, heavy liquid separation and acetolysis. If the substance is relatively pure, the residue can be mounted right after dissolution and centrifugation.

Organic solvents, of which hot xylene is one of the best, are the next step, if dissolution by water or alcohol failed. Many asphalts would dissolve in this way and since particular asphalts frequently have a palynologic "stamp", their provenance could in many cases be traced (Horowitz and Langozky, 1965). Usually the samples had to be boiled in xylene several times, each followed by centrifugation. When the residue is comparatively clean it is washed several times with acetone (use ONLY glass tubes), followed by absolute alcohol, after which it is transferred into an aqueous solution. Further cleaning may be attempted, since pollen are quite well preserved in asphalt. This may involve application of boiling 10% KOH, followed by several steps of washing with water, heavy liquid flotation, acetolysis, or even mild oxidation, carefully monitored.

Amber proved to be one of the more difficult substances to dissolve, for which no satisfactory solution can presently be suggested. Several kinds of amber are indeed still in their resin state and can be dissolved in absolute alcohol.

Plaster and mortar should be macerated as chemical precipitates, their cement comprising either carbonate or gypsum, or occasionally both. Boiling 10% HCl should in most cases be adequate for dissolving these, while subsequent maceration is going on along the same lines as for fine clastics. Paints would in some cases dissolve in hot water, or absolute alcohol, or sometimes in acetone, depending on materials used for their manufacture. It is quite difficult to suggest a general procedure, because paints differ so much from place to place and along the ages. Chemical analysis could be of help in such cases.

Chapter 6

Identification, counting and presentation

Identification and counting of the palynomorphs, followed by presentation of the results either as raw data, or in some way or another numerically and graphically processed, are the inevitable steps taken by palynologists in order to "transfer" the grains from the microscope slides kept in the collections, into papers which could circulate the information among interested scientists. This procedure is common to all branches of palynology, thus the following chapter will only focus on limitations set on pollen analysis in arid lands due to their specific environmental constraints.

6.1. Identification

The usual problematics of palynomorph identifications, such as those grains which could only be defined to a genus, or even only to a family level with the aid of common optical equipment, also pertain to arid lands. In addition, possibilities of identifying pollen grains obtained from samples collected in arid lands are further hampered by two main aspects, namely the frequent inadequacy of reference collections on the one hand and the commonly bad state of preservation of palynomorphs on the other.

Since more often than not arid regions are prone to long distance transportation processes, a reference collection of pollen derived from the regional flora may not be sufficient for identifying the pollen spectra, in which exotic elements may take a considerable part. As an example, at the beginning of palynological studies in Israel, in the late 1950's, marine sediments from the offshore Mediterranean had been studied and yielded a considerable number of pollen grains and spores of unknown affinity. Subsequently, Rossignol (1961) found out that these are derived from tropical plants and are brought by the Nile, which led me in 1965 to Ethiopia, to collect plants for a reference collection, a distance of several thousand kilometers. This was found helpful when analyzing recent sediments from the Bay of Elat (Aqaba), which also contained the tropical visitors, this time brought by the sea currents (Horowitz, 1966).

This problem is not too acute with relatively young, late Pleistocene or Holocene pollen spectra, which could be studied by comparison with their recent counterparts. For this purpose transportation paths have to be studied at a preliminary stage and reference collections prepared in accordance with the results. It is however much more complicated with older sediments, when the provenances of palynomorphs are more difficult to ascertain. As an example when studying the late Cenozoic palynostratigraphy of Israel, a reference collection must be studied from as far as Turkey to the north, down to the equator in the south (Horowitz and Horowitz, 1985).

6.1.1. Preservation, differential destruction and identification

A detailed discussion of preservation and differential destruction of pollen grains in various types of deposits is brought at the end of Chapter 2. When identification is concerned, the reasons for a bad state of preservation are less important and the major issue which remains is, whether a significant spectrum could be obtained from a given sample, or it would better be abandoned. When studying a badly preserved pollen spectrum, it is thus crucial to count also the unidentified and unidentifiable palynomorphs and compare their numbers to those which could be identified. In this respect, the aim of the study may dictate the significance attributed to a sample. As an example, if a palynostratigraphic study based on identification of index fossils is attempted, as is frequently done, when these are located with certainty it is of less importance whether the majority of other taxa cannot be identified. In such cases a stratigraphic assignment of the analyzed sample can still be given. If, on the other hand, the paleoenvironment is of interest, the reconstruction of which is usually based on numerical relations between different types of palynomorphs, it would be essential to obtain a spectrum in which the major part of grains could be identified with certainty.

The degree of this certainty may vary and it is quite self evident that a well preserved spectrum, in which most of the taxa can be identified to the species or genus level, would bear a greater significance than a badly preserved one, in which the grains could only be assigned to families, or occasionally super families. However, even the latter case is no good excuse to completely dispense of the sample, since general conclusions, however inferior to detailed ones, may still be of use, especially in places where nothing better is available, as is frequently the case in arid regions.

The problem of differential destruction of pollen grains may be more difficult to resolve. In such cases, some of the taxa may be readily identifiable, while others could be completely obliterated from the spectrum. It is even more complicated with longer sequences, where different types of sediments may act differentially on the palynomorphs, in a way that certain taxa are destroyed in one type of sediment, while other taxa damaged by another kind of deposit. In such cases, the only solution would be to compare several sequences covering the same time span, preferably collected from different depositional basins. Only the recurrence of results would render them significance.

Hall (1981) had summarized the difficulties involved with deteriorated pollen and the interpretation of diagrams in SW USA (p. 193): "Pollen assemblages are often altered by deterioration, which results in biased grain counts and possibly erroneous paleoecologic interpretations. Distorted counts are caused by (1) progressive pollen deterioration, (2) differential pollen preservation, (3) differential recognition of poorly preserved grains (4) different kinds of deterioration in different sediment types. The best indicators that pollen assemblages have been altered are high frequencies of deteriorated grains and low total pollen concentrations". I would like to stress that only the first, or the combination of these two indicators should warn and call for attention; low total pollen concentrations alone, if the grains are quite well preserved, could result from high rates of sediment deposition, not necessarily involved with deterioration.

Some examples of the difficulties involved with identifications of pollen grains at different levels of destruction are presented by scanning electron microphotographs in plates 6.1.1 through 6.1.6. Plate 6.1.1 compares recent contaminants to well preserved fossil palynomorphs. Plates 6.1.2 through 6.1.5 bring forth microphotographs of Chenopodiaceae, conifers, Compositae and *Artemisia* pollen in various levels of preservation. It is evident that there is no problem in identifying the well preserved or slightly corroded grains, but it becomes progressively difficult with advancing corrosion. It is quite interesting to note that occasionally "ghosts" of pollen grains could be identified, even to the genus, such as *Artemisia* (Plate 6.1.5, Figs. 3–6). These "ghosts" are most probably inner fillings of the palynomorphs, made from clay, in which the general shape of the grain can still be observed although the exine had entirely deteriorated. The "ghosts" can stand HCl, but I have never seen them following HF treatment, which is why they are thought to be composed of clay. In many cases, as with badly corroded Compositae, even a single spine can be used for identification (Plate 6.1.4, Fig. 6).

It is fortunate that many of the common pollen occurring in arid lands can be identified to the family level, or sometimes even the genus, even when badly corroded. Such are conifers, Chenopodiaceae, Compositae and *Artemisia*, which may yield significant environmental information, since they usually make the bulk of pollen spectra in such regions. Others, also of importance, such as Gramineae and Cyperaceae, are more difficult to identify when corroded, which may occasionally alter some of the conclusions. The case is especially difficult with Cyperaceae, which have a rather delicate pollen grain that easily yields to destruction. Such possibilities should be taken into consideration in a region where freshwater vegetation is expected to have grown, like when marshes, lakes and spring deposits are analyzed and are found to be devoid of Cyperaceae pollen; or in sandy extremely-arid terrains, where Cyperaceae are sometimes a major constituent of the flora. Again, comparisons could be made within succeeding samples if the same phenomenon is observed along the entire analyzed sequence, but may be doubtful if Cyperaceae appear in some of the samples and not in others. A

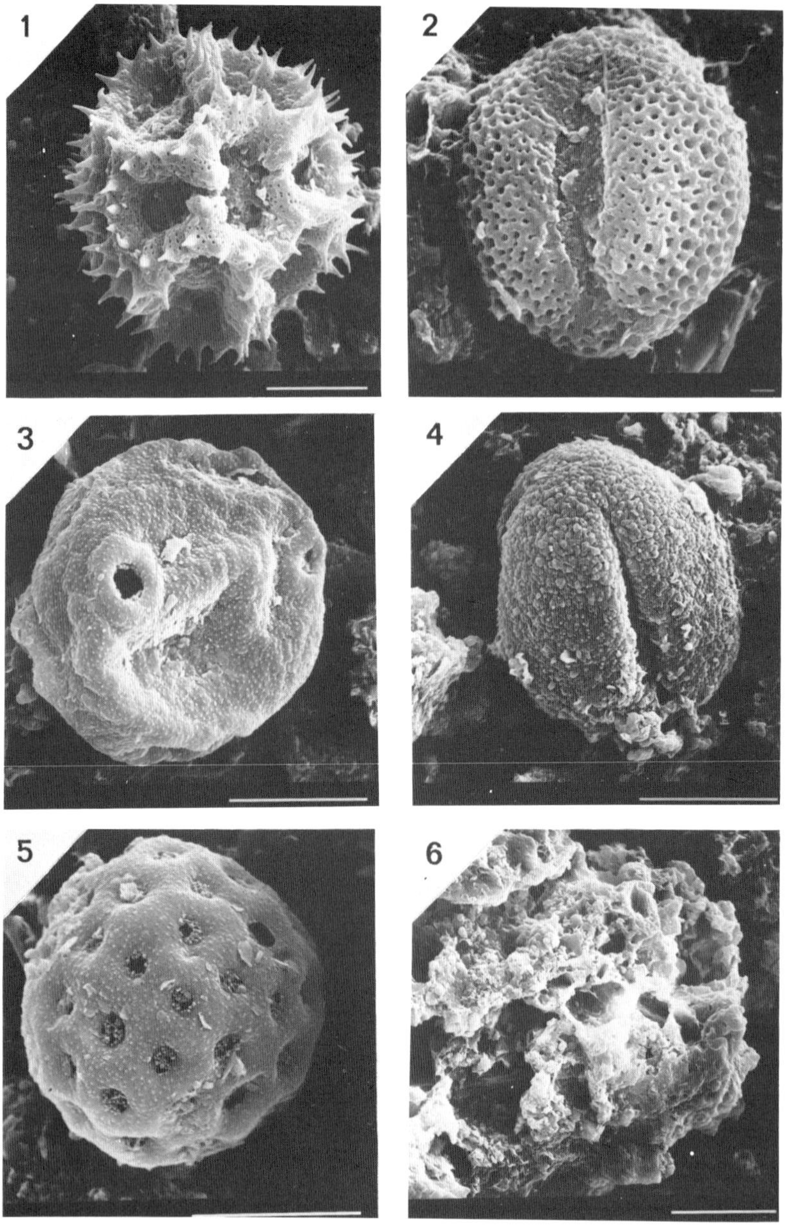

Plate 6.1.1. State of preservation, compared with recent contaminants. All samples are from En Gedi 2 borehole, drilled on the Dead Sea shore. All SEM microphotographs taken at the Geological Survey of Israel, by Mr. M. Dvorachek. Scale bar: 10 μm, except for 2, where it is 1 μm.

1. Liguliflorae pollen, still filled up with protoplasm, clearly a contaminant. 1,093–1,096 m, Palynozone Pb (late Pliocene). 2. *Tamarix* pollen, still filled up with protoplasm, clearly a contaminant. 1,093–1,096 m, Palynozone Pb (late Pliocene). 3. Well preserved *Pterocarya* (?) pollen, same sample as 1 and 2, but clearly a fossil one, since no such tree is presently growing in the region and the grain is empty. 4. Well preserved *Quercus cf.libani* pollen, from the same sample as above, most probably fossil. 5, 6. Two Chenopodiaceae pollen from a single sample, 1,153–1,156 m, Palynozone Pb (late Pliocene). 5 is probably a contaminant, since most pollen in the sample are very badly preserved, such as 6 is.

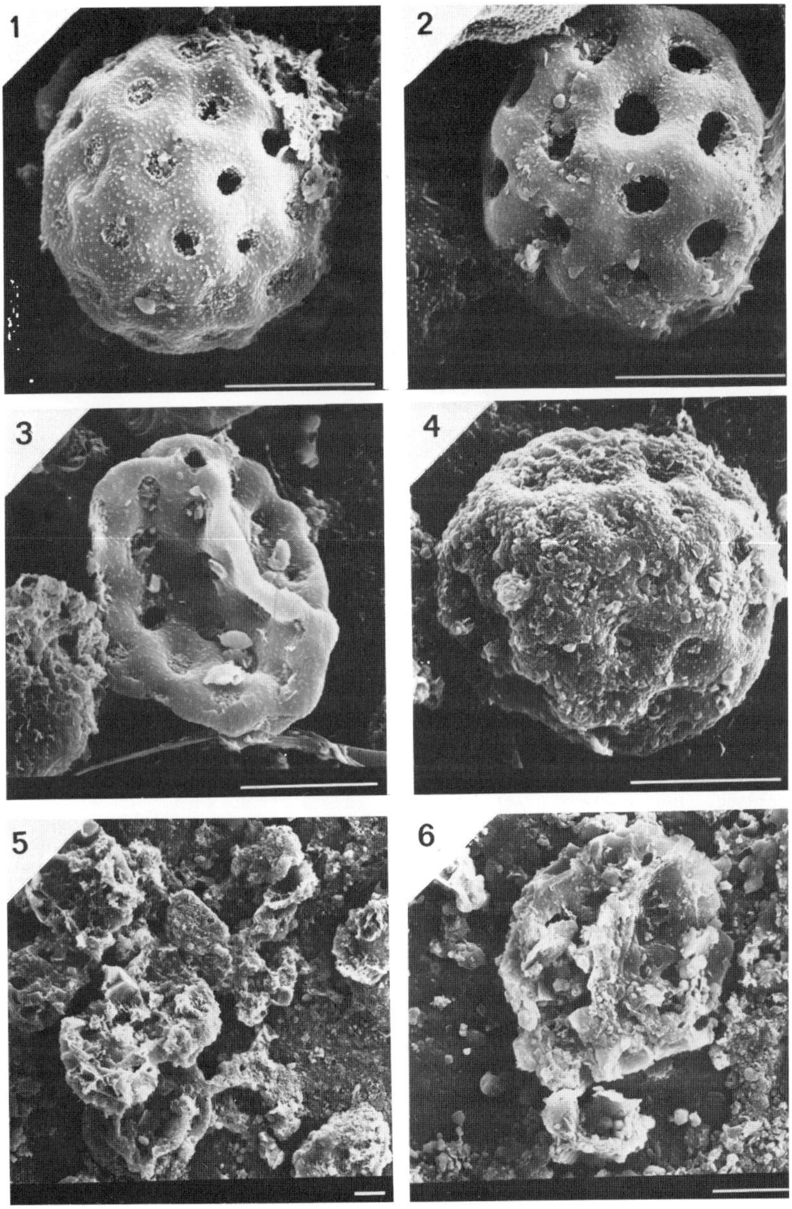

Plate 6.1.2. Different states of preservation in Chenopodiaceae pollen. All samples are from En Gedi 2 borehole, drilled on the Dead Sea shore. All SEM microphotographs taken at the Geological Survey of Israel, by Mr. M. Dvorachek. Scale bar: 10 μm.

1. Relatively well preserved, only some of the pores are slightly corroded. 1,093–1,096 m, Palynozone Pb (late Pliocene). 2. Fairly preserved, pores corroded, ornamentation almost gone. 1,093–1,096 m, Palynozone Pb (late Pliocene). 3. Fairly preserved, pores corroded, ornamentation almost gone, somewhat crumpled. 1,153–1,156 m, Palynozone Pb (late Pliocene). 4. Slightly corroded, filled up with clay, which is hard to clean, making the grain quite opaque in the light microscope. 1,093–1,096 m, Palynozone Pb (late Pliocene). 5. General view of a badly preserved assemblage. 1,153–1,156 m, Palynozone Pb (late Pliocene). 6. Badly corroded, very difficult to identify. 1,321–1,327 m, Palynozone Pa (early–middle Pliocene).

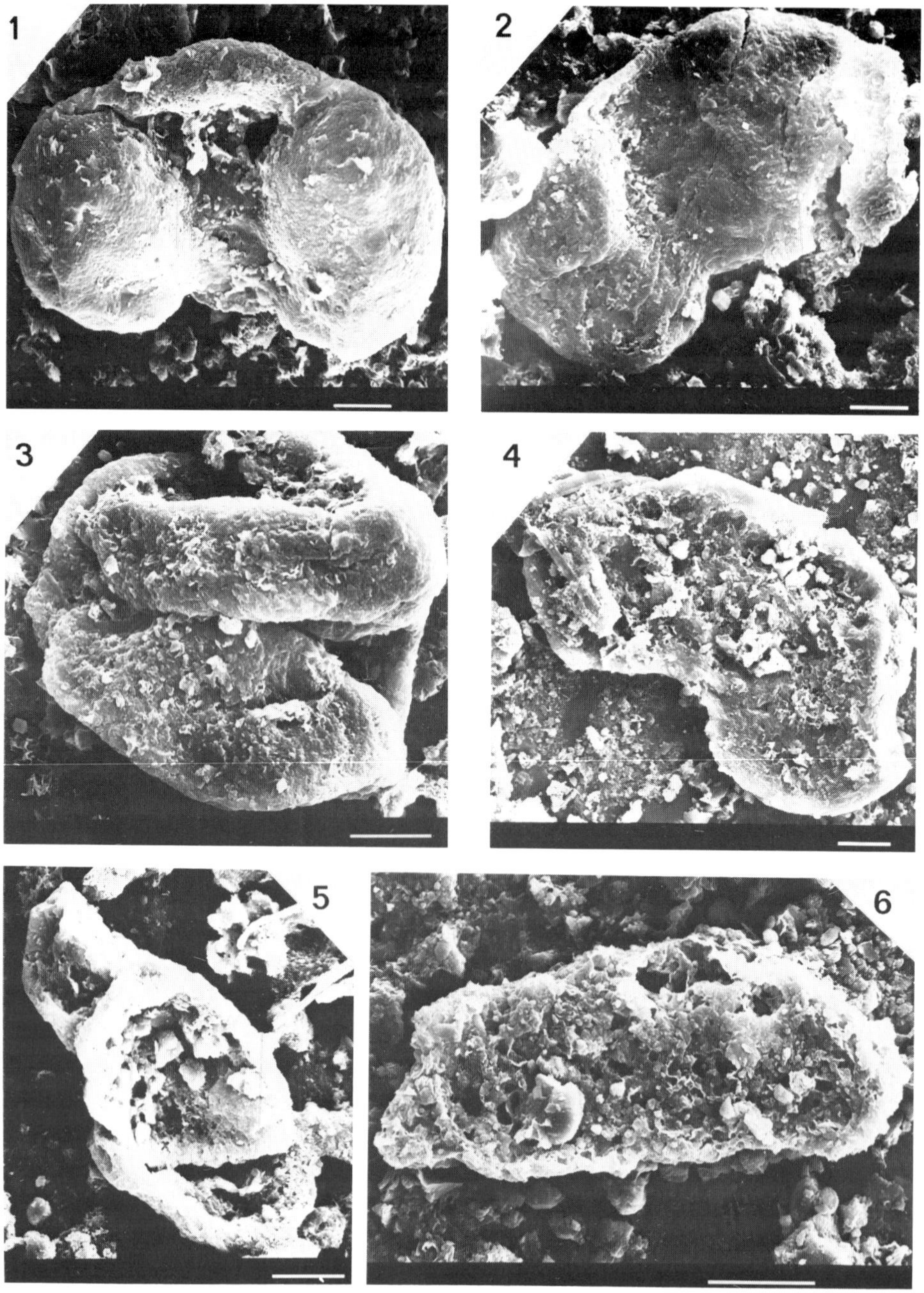

Plate 6.1.3. Different states of preservation in conifers pollen. All samples are from En Gedi 2 borehole, drilled on the Dead Sea shore. All SEM microphotographs taken at the Geological Survey of Israel, by Mr. M. Dvorachek. Scale bar: 10 μm.

1. Relatively well preserved. 1,450–1,453 m, Palynozone Pa (early–middle Pliocene). 2. Somewhat broken. 1,450–1,453 m, Palynozone Pa (early–middle Pliocene). 3. Crumpled. 1,450–1,453 m, Palynozone Pa (early–middle Pliocene). 4. Corroded. 1,153–1,156 m, Palynozone Pb (late Pliocene). 5. Crumpled and corroded, species hardly identifiable. 1,093–1,096 m, Palynozone Pb (late Pliocene). 6. Badly corroded and crushed, hardly identifiable. 1,321–1,327 m, Palynozone Pa (early–middle Pliocene)

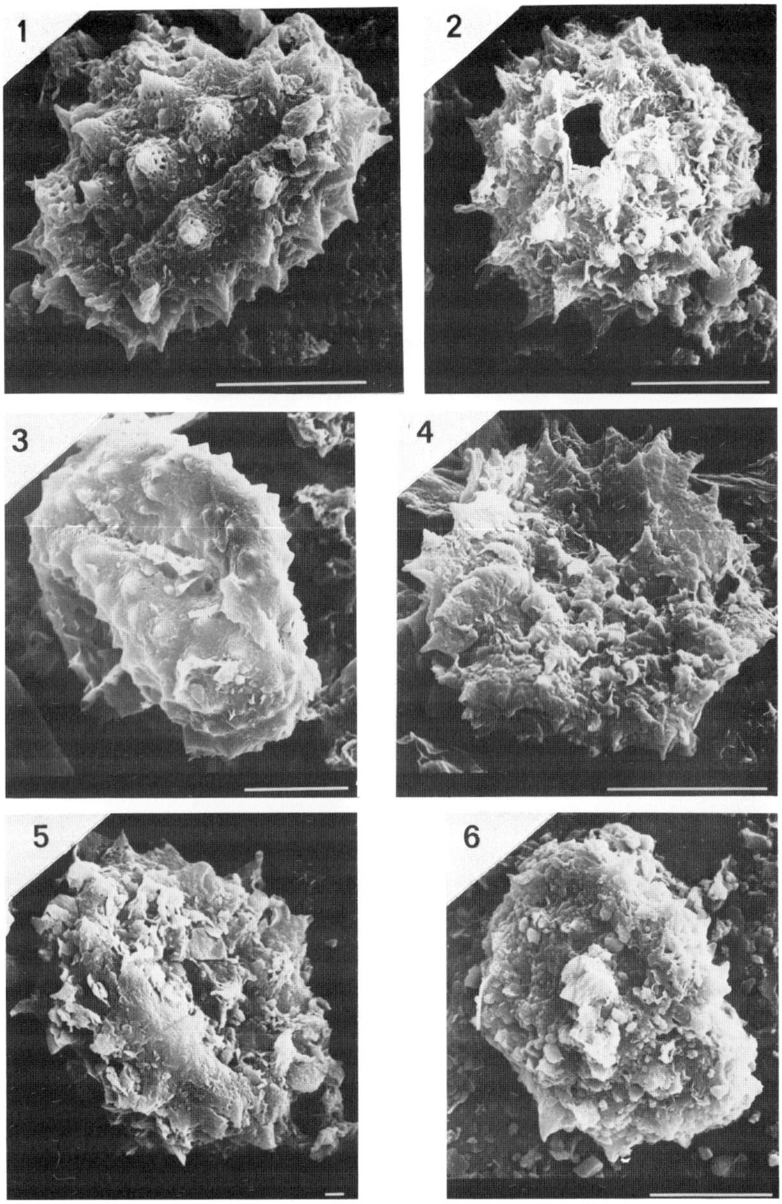

Plate 6.1.4. Different states of preservation in Compositae pollen. All samples are from En Gedi 2 borehole, drilled on the Dead Sea shore. All SEM microphotographs taken at the Geological Survey of Israel, by Mr. M. Dvorachek. Scale bar: ten micron, except for 5, where it is 1 μm.

1. Well preserved. 1,093–1,096 m, Palynozone Pb (late Pliocene). 2. Slightly corroded, mainly between the spines. 1,093–1,096 m, Palynozone Pb (late Pliocene). 3. Corroded, mainly the spines. 1,450–1,453 m, Palynozone Pa (early–middle Pliocene). 4. Crumpled and corroded. 1,450–1,453 m, Palynozone Pa (early–middle Pliocene). 5. Corroded and full of clay, some spines probably removed. 1,093–1,096 m, Palynozone Pb (late Pliocene). 6. Badly corroded and crumpled, only some of the spines left. 1,249–1,255 m, Palynozone Pb (late Pliocene).

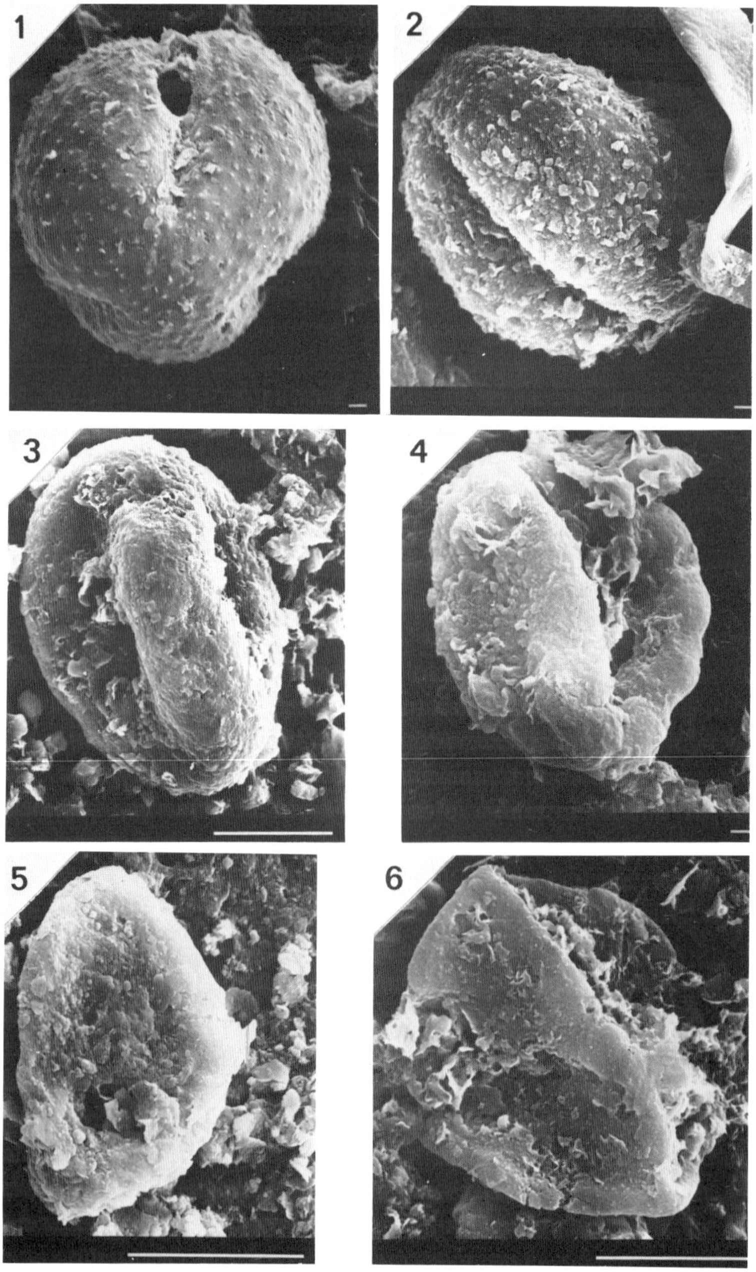

Plate 6.1.5. Different states of preservation in *Artemisia* pollen. All samples are from En Gedi 2 borehole, drilled on the Dead Sea shore. All SEM microphotographs taken at the Geological Survey of Israel, by Mr. M. Dvorachek. Scale bar: 1 μm for 1, 2, 4, 10 μm for 3, 5, 6.

1. Well preserved. 1,093–1,096 m, Palynozone Pb (late Pliocene). 2. Slightly corroded, spines hardly seen. 1,093–1,096 m, Palynozone Pb (late Pliocene). 3, 4. Ghosts, or molds of the grain. 1,450–1,453 m, Palynozone Pa (early–middle Pliocene). 5. Badly preserved ghost. 1,321–1,327 m, Palynozone Pa (early–middle Pliocene). 6. Badly preserved and corroded ghost, hardly identifiable. 1,450–1,453 m, Palynozone Pa (early–middle Pliocene).

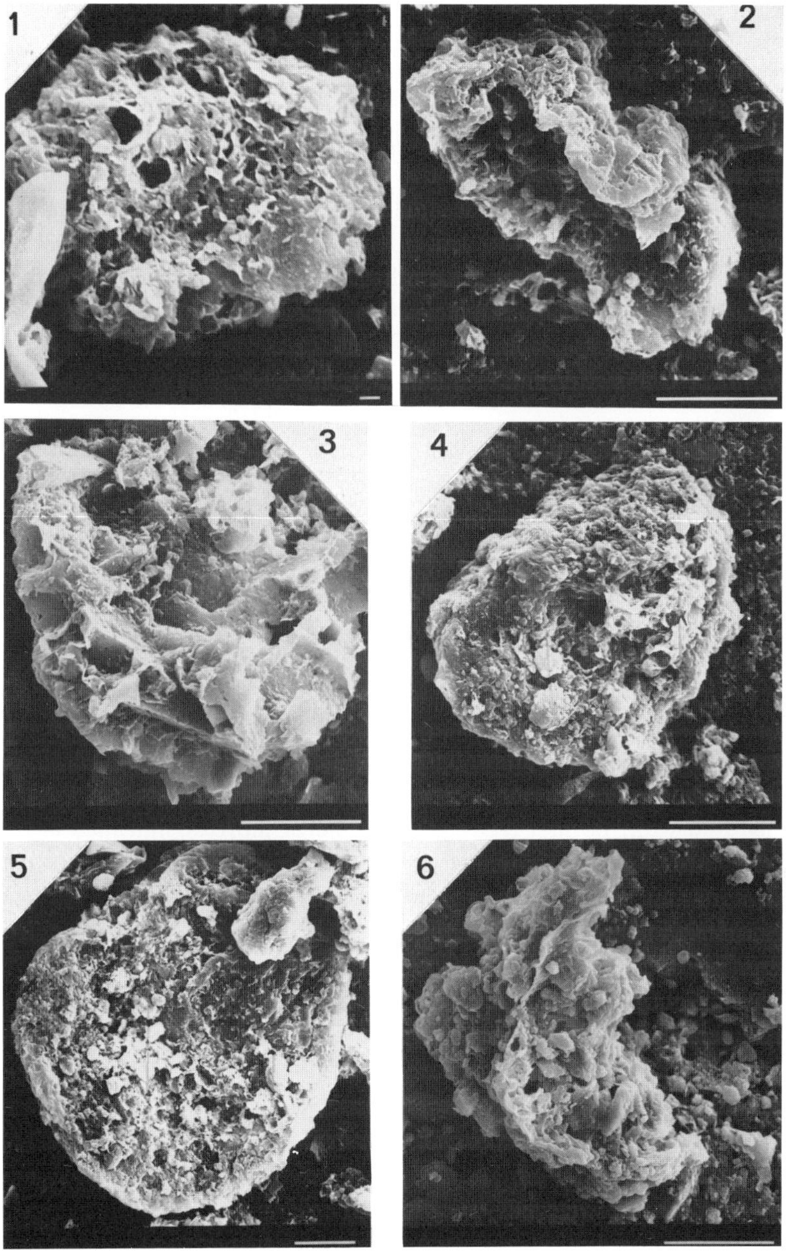

Plate 6.1.6. Unidentifiable pollen, due to severe corrosion. All samples are from En Gedi 2 borehole, drilled on the Dead Sea shore. All SEM microphotographs taken at the Geological Survey of Israel, by Mr. M. Dvorachek. Scale bar: 10 μm, except for 1, where it is 1 μm.

1. Chenopodiaceae? 1,093–1,096 m, Palynozone Pb (late Pliocene). 2. Compositae? 1,153–1,156 m, Palynozone Pb (late Pliocene). 3. Chenopodiaceae? 1,153–1,156 m, Palynozone Pb (late Pliocene). 4. Compositae? 1,153–1,156 m, Palynozone Pb (late Pliocene). 5. Conifer? 1,153–1,156 m, Palynozone Pb (late Pliocene). 6. No guess. 1,249–1,255 m, Palynozone Pb (late Pliocene).

consideration should then be made whether this is a matter of vegetation changes or of preservation properties of the sedimentary environment.

An example for differential preservation in different types of marine deposits is brought by Cheddadi (1988), who analyzed pollen from cores collected in the eastern Mediterranean, on the Nile cone. He states that pollen preservation is considerably better in the anoxic sapropels, while within the partially oxidized yellow-buff marls the grains are degraded. Cheddadi proposes that degradation acts differentially on various types of pollen, causing the better preserving sapropels to be richer in many taxa as compared with marls, whereas pollen of pines, cedars, Gramineae and fern spores are more resistant, thus dominating the marls.

6.1.2. Level of systematic identifications

Two major aspects are concerned with the possible level of systematic identifications of pollen grains and spores recovered from arid lands deposits. The first is the natural resemblance of palynomorphs produced by different parent plants, usually of the same genus or family such as Compositae or Cyperaceae (but occasionally with no apparent systematic connections whatsoever), which precludes their identification below the genus or family level. This factor is common to palynology in general and is not specifically peculiar to arid lands. The second factor, which is of much more significance in arid lands, involves the state of preservation of the palynomorphs which, in many cases is far from satisfactory in sediments collected from these regions. This may hinder identifications at a level otherwise possible in other parts of the world.

Problems involved with the first factor seem to be of more significance in the dry regions, mainly since most of the vegetation comprises plants belonging to such families as are quite difficult to identify below the family level even under the best of circumstances, such as the Gramineae, Cyperaceae, Chenopodiaceae (which also include some of the Amarathaceae) and Compositae. Only a few of the pollen grains produced by arid lands plants can be identified to the genus level, such as some of the Compositae (*Artemisia*, *Centaurea*, *Ambrosia*) some of the Liliaceae (*Asphodelus*), some of the Papilionaceae (*Astragalus*), *Ephedra* and a few others.

However, since the bulk of non-arboreal vegetation in dry lands is made of plants which cannot be identified further than the family level this may imply for a lesser significance of the results, due to a lower level of resolution. In richly vegetated regions of the globe, where the forests are the prevalent vegetation formation, most of the trees can be identified to the species level by their pollen grains. Since trees are quite rare in arid lands and unfortunately, those which do occur like thorn trees (*Acacia*), produce very little pollen grains which are preserved in sediments, one has to heavily rely on the non-arboreal pollen for drawing conclusions concerning paleoenvironments and stratigraphy (El Moslimany, 1988, 1990).

Not all, however, is bad with this situation. The inability to identify most of the non-arboreal pollen beyond the family or the genus level, combined with the rarity of trees, minimizes the number of taxa with which palynologists have to deal with while analyzing spectra recovered from arid lands deposits. Since the total number of variables in the spectra is considerably smaller as compared with those obtained from temperate regions, even much smaller as compared with the tropical, the results would evidently be less refined and resolution badly affected. However, if one is to arrive at a statistically significant pollen spectrum, the total number of grains which would have to be identified and counted could also be smaller.

Since identifications to the species level are quite difficult in arid lands even where arboreal pollen are concerned, especially those arriving from far away regions or higher elevations, compromises have to be made, for which two terms have been adopted, "confer" and "type". The term "confer" (*cf.*) denotes such palynomorph which resemble very much those produced by the plant whose name is applied to them, but were not necessarily originated in this parent plant, thus "*cf.*" only denotes morphologic resemblance. These should indeed be regarded with great care when conclusions are drawn, since their botanical affiliations are never entirely secure.

The term "type" may include two or more species under a specific heading, species for which further separation seems impossible under the circumstances. Daoud (1984) made a detailed study of recent pollen spectra in Tunisia and used several "types" to define the oak pollen transported to the analyzed sites. Thus *Quercus* type *suber* = *Q. suber* + *Q. mirbecki* + *Q. afares*; *Quercus* type *ilex* = *Q. ilex* + *Q. coccifera*; and *Quercus* type *pubescens* = *Q. pubescens* + *Q. pedunculata.* The three types differ by their sizes and the structure of their furrows. Similar subdivision into types is suggested by Daoud for various species of *Plantago*, *Acacia* and *Polygonum*. Pollen of the Gramineae are divided into four groups: the oval; the rounded; various others; and the large ones, for which the term "Cerealia type" is used. The only species of Gramineae which could be specifically identified is *Lygeum spartum*, which is distinguishable from the rest of its relatives. Grains of Compositae have been divided into the echinate and the fenestrate, while those plants which are identifiable generically — *Artemisia*, *Centaurea* and *Ambrosia* — were treated as such.

A similar treatment is given by Jacobs (1985) to pollen of different pine species occurring in sediments of the American Southwest. She divided the fossil pine pollen into four groups: *Pinus ponderosa* + *P. contorta*; *Pinus aristata*; *Pinus flexilis* + *P. astrobiformis*; and *Pinus edulis* + *P. monophylla.* Again, similarly to the oaks treated by Daoud, most differentiations are made by size analysis, aided by the shape, like the presence or absence of bladder blisters and the general outline of the grain.

Separation into types, or groups of pollen which look much the same, based only on size, should be carried out with great care, since many factors affect the size of the grain when observed under the microscope. Among these some arc natural, connected with the specific type of deposition and preservation environments, while others may depend on the maceration methods used and even on the type of mounting medium. Jacobs (1985) states that pollen grains mounted in glycerin jelly or glycerin were significantly larger than those mounted in silicone oil. Consequently, such comparisons of sizes is fairly safe when studying a series of samples collected from essentially similar deposits and prepared and mounted similarly, but may be quite hazardous when sizes are compared to data in the literature, or when the environments and maceration processes may have been essentially different than those under study.

The attribution of "confer" or "type" may serve as a useful convenience for the pollen analyst in an arid land, but it should not be neglected when conclusion are drawn, since in many cases the "type" or "*cf.*" is lost, at least as an attitude and grains are thought to represent the parent plants from which their names had only been borrowed.

The identification as "type" may vary in different parts of the world, both according to the attitude employed by a specific researcher and the occurrences of morphological types in the analyzed spectra. As opposed to the attitude taken by Daoud (1984) in the Sahara, Fish (1984), while studying pollen in the American Southwest, regarded the entire Gramineae family as a single entity, separating with certainty only the pollen produced by corn (*Zea*). Other Gramineae grains, smaller than those of corn but larger than most native grasses, were noted as possible indicators of semi-cultivated species. Traditionally, Chenopodiaceae include in the American Southwest also pollen of the genus *Amaranthus*, of the Amaranthaceae family, thus are almost always treated as "Cheno-Ams" in this part of the world. Subgrouping of the Compositae is made by Fish into the "long spined" and the "short spined" grains, which may correlate with environmental conditions, but do not seem to correspond to any systematic affiliations. On the other hand, Fish states that some of the Cactaceae, like the genus *Opuntia*, can be further subdivided into the culturally recognized subgroups of *Cylindropuntia* (chollas) and *Platyopuntia* (prickly pears).

Some of the specific identifications made by palynologists in arid lands are only based on the fact that only a single genus, or sometimes a single species, occurs in the region at the present day. Such is the case with *Larrea tridentata* (creosote bush) in the American Southwest or olive (*Olea europaea*) in the Mediterranean region. Identifications based on such single occurrences may serve well for the later periods, such as the Holocene (Bottema, 1984), however should be considered with care if sequences extend down to the Pleistocene, a period which displays major climatic and vegetation changes, thus the single occurrences could not be taken for granted any longer. A positive identification is much preferable in most cases.

Possible levels of identification drop even further when the palynomorphs are less than well preserved, as may be the case with the majority of spectra recovered from samples collected from sediments laid down in arid lands. In such cases we (Horowitz and Horowitz, 1985) found out that morphologic groups of palynomorphs which bear some resemblance have to be regarded together as a single entity, such as Gramineae and Cyperaceae. The same applies to all Compositae (except for *Artemisia* and *Centaurea*), which are only identifiable by the occurrence of spines, sometimes without even the possibility to distinguish between the echinate and fenestrate subgroups, because all that remained is a blurred ghost with several typical spines. As long as these were seen, the grain could positively be identified as Compositae (see Plate 6.1.4, Fig. 6).

It was found out that within oak pollen known from Israel and the vicinity, distinction could be made between two groups: the evergreen *Quercus calliprinos*; and the winter deciduous group of oaks, comprising *Q. ithaburensis*, *Q. libani* and *Q. boissieri*. When preservation was less favorable, all that could be done with oak grains was to assign them to *Quercus* spp. A similar problem is with the Pinaceae. When analyzing well preserved assemblages it is quite easy to discern between *Pinus halepensis*, *P. pinea* and *Cedrus libani*, all occurring at the northern part of Israel and the southern Lebanon. In damaged assemblages all these were put under the general term "conifers". Indeed, the use of such a general term is not too advisable, unless it is exactly specified what taxa are encompassed in such a definition. In the Quaternary sediments from Israel (Horowitz, 1987a; Levin and Horowitz, 1987) the vesiculate genera *Pinus* and *Cedrus* have been grouped under "conifers", while *cf. Picea* (or *Cathaya*), which is quite readily distinguishable even in badly preserved assemblages, was separately treated (see Chapter 9 for details).

Naturally, this problem becomes progressively more acute when the spectra are older. In sediments of early Quaternary and Neogene age, both the state of preservation and the lack of a reference material make the use of generic and especially specific terms questionable. In practice generic names are used even for palynomorphs recovered from Neogene sediments in most parts of the world, but some prudence should be applied with the level of significance of the ensuing paleoenvironmental conclusions.

6.1.3. Ideal vs. practical identifications

Or, in other words, how far would one go and how much effort should be put in a sample, before abandoning a pollen spectrum and declaring it useless?

There is no ideal pollen spectrum, even under the best of circumstances, due to objective reasons such as the fact that far from all plant species are represented by pollen in the fossil assemblages. Not all of those which are present, can be identified down to their species level due to morphological similarity (at least at the present state of knowledge, identification techniques and instrumentation). The considerations whether to go further with a given spectrum, or to abandon it altogether, depend therefore only on practical criteria.

It is quite clear that if a vegetation succession has to be studied in great detail, only rich, well preserved pollen spectra are to be used. These, unfortunately, are quite rare in arid lands. The ones obtained from such regions can however yield important information concerning stratigraphy and paleoenvironments of the area, which would be based on the relative distributions of larger plant communities. Fortunately arid and especially extremely-arid lands, are very sensitive to even slight changes in climates or environments and react in accord. This is readily demonstrated in extreme cases, such as the Namib Desert (Fig. 1.2.7), where a single rainstorm which occurs only once in many years, can cause development of plant communities which would be there for several years to follow. Such plants would eventually die if another rain is not coming within some time, but this period can extend for several years (Seely, 1987).

In this way, pollen profiles collected from arid regions can supply very useful information regarding regional environmental changes (Horowitz, 1989b); because of the sensitivity of such regions, they can also yield information about local events, such as the opening or drying up of springs, immediately manifested by the wealth of their typical vegetation pollen (Horowitz, 1987b); formation of salt lakes and playas is indicated by prevalence of Chenopodiaceae (Horowitz, 1989c) and so on. For such purposes, even pollen spectra which are badly preserved and can hardly be identified, usually not beyond the family level, are quite useful. Conclusions would naturally be much more meaningful if the lines of changes recur in several analyzed sequences.

It is therefore quite evident that a pollen spectrum, or a profile, should be dispensed with only when the majority of palynomorphs are so badly damaged, that no identifications could be made at all, at least for the greater part of grains. If identifications can be made at some level, some use can be found for the spectra, however the conclusions drawn could only be based on what was positively identified.

Patience plays a major role in such cases. If a sample only yields a few pollen grains, there are two possibilities: either throw it to the waste basket, or try to rerun a larger amount of sediment, until a satisfactory spectrum is at hand. The latter possibility may depend on availability of material, therefore suggestion was made earlier to collect more than seems necessary when fieldwork is carried out.

Generalizations and groupings of palynomorphs made during the identification lower the level of resolution and significance, but still, conclusions could be quite safely drawn, based on that level. If several sequences are examined from the same locality, the overall significance becomes much better established. For details and examples, please refer to the following chapters.

The level of systematic identifications in older, early Cenozoic, Mesozoic and Paleozoic sediments, depends on both the morphology of the palynomorphs and on the criteria used by a particular researcher involved with the pollen analysis, to define morphologic entities as individual taxa. Since such definitions are quite far from satisfactory, certainly not applied objectively at our present state of knowledge, it can be seen quite frequently that a certain palynologist would define many more taxa in a given spectrum, as compared to another. Splitting and grouping of taxa, both morphogenera and morphospecies, are quite common in the literature and may have a detrimental effect on conclusions drawn regarding paleoenvironments, which are based on the taxa diversity concept (for details see Chapter 11). In such cases it seems almost essential that only a single palynologist would carry on the analyses or, if a group of palynologists is involved, that coordination would be satisfactory. If the principles according to which morphospecies are defined are not entirely clear, it is sometimes better to disregard the information, rather than step directly into this pitfall.

6.1.4. Identification of polyphase processes

Polyphase processes are defined here for palynology as those by which palynomorphs from different natural provenances are transported to and included within a single sample. Broadly, polyphase processes comprise three categories: Recent contaminations; inclusion of synchronous palynomorphs, brought to the depositional environment by long range transportation; and inclusion of palynomorphs of a different age than the time of deposition of the analyzed sample, both by redeposition of older and by infiltration of younger grains.

The recognition of polyphase processes involves three criteria. The first is knowledge of the pertinent vegetations which may contribute to a given pollen spectrum. This aspect does not have to do with the possibility of identification of the grains, but rather with interpretation (see Chapter 8), provided the grains were recognized according to an appropriate reference collection. The second involves differences in the states of preservation of palynomorphs which originate from different sources, that had undergone different processes on their way to the locality of final deposition. The third concerns changes in the sporopollenin through time, which can occasionally be detected by the palynomorphs physical and chemical properties, such as differential absorption of certain pigments used for staining, or the autofluorescence characteristics of the exine. Only the second and third aspects will be dealt with here, quite shortly, because not much research was hitherto carried out on these physico-chemical aspects of palynomorphs, even less so in arid lands.

A. Identification of polyphase processes by state of preservation

Recent contaminants, which usually do not have to appear in the slides if collection and preparation of the samples are properly done, can in most cases be quite easily discerned. In many cases such recent contaminants have still some protoplasm left within them, which would immediately point them out against any fossil assemblage and preclude them from being counted. If staining is used, the recent contaminants tend to absorb the pigment more readily than the fossil ones, are usually much more brightly colored, thus clearly identified as such. The problem may however be crucial when recent pollen spectra are studied, in which case all grains, the relevant ones as well as the contaminants, look very much the same. It should be noted that the term "recent pollen spectra" may apply to three different categories: airborne spectra, those obtained from the surface and those from just under the surface. The problem of identifying recent contaminants is of greater significance with the first two categories, while if analyzing spectra from near surface deposits, which have already been subject to burial processes, the recent contaminants can easily be detected as such. Such subrecent pollen spectra are also much better representatives of the recent pollen deposition, when compared with the fossil record, since they have already undergone processes involved with differential destruction, which are quite rapid. Havinga (1984) had shown that most of the differential destruction takes place within the early years following deposition, therefore the subrecent spectra seem to be a much more reliable foundation for comparing and discussing paleoenvironments.

If recent contamination is suspected, it is advised not to acetolyze the pollen assemblage during preparation of the sample, since acetolysis may dissolve most or all of the protoplasm, rendering the recent pollen a fossil appearance.

Recent contamination can easily be identified even if the sample was acetolyzed, in such regions where introduced plants are important pollen producers. Such is the case in Israel, where various species of *Eucalyptus* have been introduced during the last decades and are quite widespread at present. When subrecent sediments are studied, the presence of *Eucalyptus* grains immediately calls attention to the possible inclusion of other recent contaminants as well (Horowitz, 1979). A similar situation is known from South Africa, by the occurrences of pine pollen, a tree which was introduced only recently to this part of the world (Scott and Bonnefille, 1986).

The question is however, what share of such known contaminants as pine or *Eucalyptus*, should preclude the use of an entire pollen spectrum. Opinions vary, as is the case with the Middle Breccia of Sterkfontein Cave (discussed below). Horowitz (1975) concluded that the lesser preserved pollen were not contaminants, while well preserved pine, Gramineae and other pollen had been precluded from the final spectrum. Scott and Bonnefille (1986) concluded that the entire spectrum should be discarded as contaminated.

Identification of polyphase processes by the different states of preservation of grains in a single assemblage is not always possible. This is mainly due to the fact that in many cases no clear distinction can be made of significantly differently preserved groups. In many cases redeposited sporomorphs, or those transported from far away localities, sometimes over thousands of kilometers, do not necessarily differ in their state of preservation from the locally deposited, autochthonous palynomorphs. Such is the case with the offshore Mediterranean sediments in Israel, where the percentage of Gramineae pollen is much higher as compared with spectra collected onshore. Only the occurrences of exotic spores and pollen, known to have been brought by the Nile from as far as tropical Africa (Rossignol, 1961), suggested (Horowitz, 1979) that some of the Gramineae pollen were also brought by the Nile, thus should in fact be regarded as allochthonous, although their state of preservation is essentially identical to the local ones.

Stanley (1966) did not observe significant differences in the state of preservation of palynomorphs recovered from Pleistocene to Recent marine sediments. He could only differentiate the autochthonous from the redeposited grains by the different absorption of Safranine-O. Van Gijzel (1971; and earlier publications) observed a similar problem in continental sediments and could only discern the autochthonous from redeposited pollen by using ultraviolet autofluorescence characteristics of the grains.

In some cases however, it is possible to identify polyphase inclusions of palynomorphs by their different state of preservation. Horowitz (1975) divided the pollen spectrum of Middle Breccia from Sterkfontein Cave into three parts, based on the states of preservation. The first was very well preserved, with some of the grains still containing some protoplasm, which included also several pine grains, a tree introduced into the region quite recently. The inclusion of these pollen in the spectrum indicates that they had been washed down and infiltrated the sample, by recent processes. It should be noted that the sample was washed well with 10% HCl prior to preparation, both by Horowitz and consequently, by Scott and Bonnefille (1986), which had not removed the recent contaminants. Although cleanliness was observed, the recent pollen were found by both investigators, which indicates that contaminants can be washed by rainwater and be included deep in the sample. The second assemblage comprised pollen grains which are quite badly damaged and hardly identified, considered fossil by Horowitz; and the third consisted of palynomorphs which have been degraded to some degree, but much less than the second group. Horowitz concluded that the second group was deposited contemporaneously with the cave sediments, while the third assemblage, of only partially degraded pollen, was included within the breccia at a later stage, by percolating waters, when the original cave sediments have turned into a hard limestone cemented breccia.

An assurance to this conclusion came from the analysis of vertebrate bones included within this breccia member (Vrba, 1975), which indicated a similar environment to the one suggested by the badly preserved and corroded pollen assemblage. It seems however that, although distinctions of this type could be sometimes made, it should be done very carefully.

B. Identification of redeposition by physico-chemical properties of the exine

I. Staining. Stanley (1966) made a thorough study of palynomorphs obtained from Quaternary ocean bottom cores and divided the spectra into three groups of grains: the ones which are certainly of Quaternary age; others that were surely reworked from older sediments; and those which resemble morphologically the Quaternary ones, but may have been reworked. Stanley applied Safranine-O staining to the palynomorphs spectra and found out that those spores and pollen which had certainly been reworked could be divided into four groups:

(1) Grains that do not accept stain, so that the exine is brown or light yellow colored. This apparent lack of stain acceptance is most common, though not restricted, to pollen and spores known to be of Paleozoic or Mesozoic age. The germinal region of the sporomorphs included within this category, may occasionally accept stain slightly.

(2) Grains in which both the endexine and the ektexine accept stain very readily. This results in a pollen grain or spore which is considerably darker than most of the others on a given slide. This type of stain acceptance was observed most frequently in grains known to be of late Mesozoic or early Tertiary age.

(3) Grains in which the endexine takes stain readily, whereas the ektexine does not accept stain, or accepts it only slightly. This type of staining characteristic was found to occur most frequently among spores and pollen known to be of Mesozoic age.

(4) Grains which accept stain very much like recent pollen and spores. Here the ektexine accepts stain more readily than the endexine, or both absorb stain about equally. This type of stain acceptance was observed only rarely among grains known to be redeposited.

Stanley indicates that some of the criteria defined above could also be applied to acritarchs and dinoflagellates.

Consequently, Stanley stated that the differential stain acceptance of known redeposited pollen and spores permits the establishment of objective criteria for the recognition of reworked grains among morphologically indistinct plant microfossils. The variations in stain acceptance observed in known reworked grains may be a reflection of exine chemistry changes which took place during the reworking process. Stanley also brings forth the possibility that other factors, such as depth of burial and metamorphism (prior to the reworking) may play an important role in bringing about changes in exine chemistry.

Great depths of burial and partial thermal metamorphism can bring about changes in the color of the sporomorphs, which become darker and darker as these processes propagate. It is however notable that thermal metamorphism can also occur in certain confined horizons in a given sequence, as a result of proximity to volcanic activity and shallow intrusions, even small scale ones. Such horizons, in which the palynomorphs are much darker than in others below and above, had been observed in a borehole drilled in the Hula basin in northeastern Israel, Notera 3 (Horowitz and Horowitz, 1985). In this region volcanic activity was quite frequent during the Quaternary and thermal events left their impression on the pollen grains in certain horizons (Bein and Feinstein, 1988).

II. Autofluorescence. Van Gijzel (1971; and earlier publications) studied the autofluorescence of pollen grains and spores, which are exited by ultraviolet light. Measurements of the resulting autofluorescence show clearly a change in color with increasing ages of the grains. Recent and Holocene grains have bright green to yellowish autofluorescence colors, which turn towards deep yellow and brown in older palynomorphs. Van Gijzel also found out a good correlation between his results, obtained by ultraviolet excitation and the results reported by Stanley (1966; see above for details), obtained by differential stain acceptance of the grains. He thus states that the changing autofluorescence colors are most probably a result of changes in the chemical composition of the sporopollenin.

This method was however never widely spread, mainly due to technical difficulties involved with the measurements of autofluorescence. The foremost difficulty is that the autofluorescence fades away quite quickly following ultraviolet excitation, sometimes in a matter of seconds, thus measurements had to be done on the spot, without possibility of rechecking. The second is the high price of the equipment necessary for exact measurements of the autofluorescence wavelengths.

Since the rather simple process of staining by Safranine-O suggested by Stanley gives similar results to those obtained by the complex autofluorescence methods, it was never found practical to use the latter technique on a wider scale.

Another drawback of Van Gijzel's methods seems more pertinent to pollen spectra obtained from arid lands deposits. Van Gijzel (1971, personal communication) stated that the use of acids during the preparation process impairs the autofluorescence characteristics of palynomorphs. It is very rare to locate, in arid regions, such sediments that could be prepared for pollen analysis without the use of acids, which practically renders the autofluorescence method useless for these parts of the world.

It still remains to find out what is the influence of acid treatment on stain acceptance characteristics of palynomorphs and whether these are not impaired, to establish whether this method could be safely applied for identifying redeposition in arid lands. Stanley (1966) does not detail his maceration procedure, but only defines it as "standard", which does not say whether or not acids had been used. However, if I recall the standard techniques used at that time, acids were not applied. Otherwise, I doubt if correlation could be found between Stanley's and Van Gijzel's results. It is interesting to note, which may be meaningful in arid lands, that Stanley did find correlation between stain acceptance and progressive oxidation applied in his laboratory.

6.2. Counting

After being identified, palynomorphs are treated in either of two ways, according to necessity. In many stratigraphic studies involving pre-Cenozoic strata, the basis for subdivision of the sequences is the occurrence/non-occurrence of the palynomorphs, which are used in this respect as any other guide or index fossils, with hardly any attention given to their botanical affinities. In this way the reconstruction of past floras, if attempted at all, only presents the taxa, with no respect to their relative shares or to the plant communities. Obviously such treatment, although very useful for stratigraphy and geology, cannot be used for reconstruction of past environments.

Quaternary palynology, in contrast, commenced by counting the various taxa and calculating their percentages, thus enabling the reconstruction of paleoenvironments. This was primarily done out of necessity, since usually no new taxa had appeared during the short periods dealt with and hardly any plants make definite last appearances. The calculations of taxa percentages allow a better understanding of the composition of plant communities, which are much more meaningful for delineating past climatic conditions. Tendency is however presently growing also among paleopalynologists to count the various taxa and give their percentage relations, which is much more helpful in any attempts at reconstructing ecological and environmental conditions.

The problem of how many palynomorphs should be counted from each sample is almost as old as palynology itself, chiefly because there seemed to be no acceptable limit. This problem is often discussed in forested regions, where samples are very rich in pollen and availability is not the restricting factor. In arid lands, where the limiting factor is very frequently the availability of pollen in a certain sample, the problem is much more acute, thus debates whether spectra should be used or altogether thrown down the drain are quite common.

The debate, as usual, is between the ideal and the practical. It is quite clear that, ideally, as many grains as possible should be counted. But surely, if one has a sediment sample in which pollen concentration could be as high as 100,000 grains per gram, it would be quite difficult to count them all and come out with a pollen diagram, during one's lifetime. On the other hand, if a sample from an arid land is studied, that only yields 100 palynomorphs, which can usually be counted within an hour or so if one is familiar with the taxa, how exactly should such a spectrum be viewed?

Statistical methods (for details see Birks and Gordon, 1985) would not define a number for counting, but rather delineate the factor of uncertainty, or the confidence interval for any of the counted taxa, in a given sum. Ninety five percent confidence intervals had been calculated by Mossiman (1965) and to facilitate their use, Maher (1972) presented them in a form of nomograms. Essentially, 95% confidence interval means that in 95% of all cases the "true" value would be found within the limits of the calculated interval. This, incidentally, still leaves 5% of cases in which the value would be outside the limits. The formula for calculating the 95% confidence interval in a binomial system is as follows:

$$P \pm 2\frac{\sqrt{P(1-P)}}{N}$$

where P is the proportion of any counted sporomorph and N is the total sum of counted grains.

Within a binomial system, the confidence interval becomes wider when the percentage gets closer to 50% and drops on both sides of this figure. Simple calculations would immediately demonstrate that the confidence interval becomes narrower with the increase of N, which is only a mathematical expression of what was said above, that the more grains counted the more accurate the picture would be. Practically, if a certain taxon takes 50% of the spectrum, its 95% confidence limits would approximately be 43–57 if the total is 200 grains, or 36–64 if only 50 sporomorphs form the total.

The question remains whether such limits are acceptable, but this consideration should always be made regarding the expected use of the spectra. Let us consider what happens if an analyzed sequence yields an average of only 50 grains per sample. The widest margins would be in the median, 50% values, which could oscillate between 36 and 64%, as said above. When constituents are in the range of 10%, they could vary between 1.5 and 18.5%, but if the range is 5% the limits would be −1 to 11%. While the range for the 50% taxa may be acceptable, however with a frown from any palynologist enjoying usually much richer samples, the ranges for the 10 and 5% taxa would hardly be considered acceptable by anybody.

But let us analyze the practical information: if stratigraphy is concerned, based on relations between two, three or even four palynomorph groups, as is usually the case in this type of studies in the arid lands, the prevalent group would always have sufficient significance for palynostratigraphic assignment. Further, since palynozones are defined by a series of samples, significance is even better. In this respect, it is worthwhile stressing again that, especially in geologically younger sediments where zonation involves rather short periods, it is indispensable to analyze sequences rather than single samples. A palynostratigraphic assignment based on analysis of a single sample in which only 50 pollen grains had been counted is entirely dubious; while if a sequence was analyzed, results are much more meaningful.

Similar considerations can be applied where the composition of past vegetations is discussed. When a limited number of plant associations is involved, such as relations between Compositae, Gramineae and Cheno-Ams, if several samples are available for each period a count of 50 grains per sample could be sufficient, while a single sample is almost useless.

Cases in which only 50 grains are used for drawing conclusions are quite rare, since in most arid lands deposits more pollen can usually be counted. The discussion however indicates that even under such unfavorable conditions pollen analysis is not entirely useless, as long as evident restrictions on the significance of results are understood and appropriately dealt with.

A practical example from a desert fringe area, in which AP/NAP ratios are taken as a basis for palynostratigraphy, as is the case in Israel (see Chapter 9), would show that if wet climate palynozones are characterized by AP values of 50% and up, separated by dry ones, in which AP drops to less than 10%, a total count of 50 grains per sample is sufficient for defining such palynozones. These considerations may be very helpful especially where palynostratigraphy is analyzed from commercial boreholes material, where additional samples are impossible to get under normal drilling circumstances.

40% relative 95% confidence intervals are calculated in Faegri et al. (1989) and show that if, within a binomial system, the required limits would be 40–60 for a taxon that takes 50% of the total, a total count of 100 grains is sufficient. When the percentage of this constituent drops, the total number required to maintain the 40% relative interval considerably increases, so that if a taxon takes only 1%, a total count of 9,900 grains is needed to keep the limits, 0.8–1.2%. The immediate application of this calculation is that, within ordinary pollen spectra, the shares of minor elements do not have any statistical significance, but should be regarded as occurrences only. Non occurrences of such minor elements do not mean, statistically, that they are not there, if a sufficient number of grains was not counted to prove their absence.

An example pertaining to arid lands palynology is brought in Suc (1978), who could only distinguish between cold and warm *Artemisia* steppe by the occurrences of minor elements, such as *Phlomis*. Only their positive occurrences are of significance.

Groupings, both of taxa and of spectra, naturally increase their statistical meaning, thus should be considered when the number of pollen from a single sample is not sufficient for sound conclusions. Possibilities for such groupings are discussed in Chapter 7.4.

Another solution which may serve to overcome some of the problems of low pollen concentrations and countings in samples collected from arid lands is the analysis of sequences, rather than single samples. A paleoenvironmental inference based on a single sample from which only 50–100 grains are counted would be quite dubious. However, if a sequence of some 5–10 samples in which such numbers are counted, bears similar characteristics, the conclusions are much more sound and could be relied upon. The same could be said for analyzing several chronologically parallel sequences, discussed above.

The aim of the study should be considered when the number of palynomorphs to be counted is decided upon. If all one wants is a general palynostratigraphy, say for the Quaternary or the late Cenozoic, in cases where samples are not too widely vertically spaced, moderate counts (50–100 grains per horizon) would suffice. If a detailed reconstruction of the vegetation is attempted over a short period, several hundred grains have at least to be counted, more if possible. Practically, when one is concerned with arid lands, there is only one advice: identify and count whatever you can under the circumstances and, referring to the results, try and find out what conclusions could be drawn with relative safety. Do not throw any spectrum before you are definitely sure it is useless. You will find out that in most cases no spectra have to be dispensed with, rather each one would have its significance, however limited. You may find out that rather poor pollen spectra recur once and again, thus gaining more and more significance as further samples are analyzed.

An example, again from Israel: pollen diagrams have been constructed for the later Holocene from the Hula Basin, in the north of the country (Horowitz, 1969) and from the Haifa Bay (Horowitz, 1979), based on some 100–200 grains. A pollen diagram made by Baruch (1986) from Lake Kinneret for the same period, based on 1,000–1,500 grains in each horizon, displays essentially similar trends for the natural vegetation, when the anthropogenic influence is removed from the calculations (Fig. 7.4.3).

6.3. Presentation of the results

Once the palynomorphs are identified and counted, the results have to be presented, either in the form of tables or graphically, since a text description usually misses the point when too many data are concerned. Except for the simplest way, in which the actual countings are presented in a table, all other ways of presentation involve numerical and statistic treatments of some kind or another. The numerical treatment of pollen spectra is discussed in great detail in Birks and Gordon (1985) and does not seem to differ when arid regions are concerned, as compared to the rest of the world. The only difference may be the lower total number of palynomorphs counted from a certain horizon, but the treatments suggested by Birks and Gordon are also adequate and applicable in such cases.

The major difference between arid and forested regions lies with the level of confidence attributed to the results. This level of confidence is an important statistic factor in pollen analysis, which should not be overlooked or neglected while calculating the data for presentation. See, for example, levels of confidence (or probability margins) calculated and presented in the pollen diagram drawn by Oller et al. (1984–85). Admittedly, very few authors include the levels of confidence in their pollen diagrams and it is suggested that these figures are always added. Preferably, the level of confidence should be as high as possible. In regions where no limitations are on the total number of counted grains, in terms of both sample yield and manpower for identifying and counting the palynomorphs, a level of confidence of 95% may be practical.

However, due to the scarcity of grains in most types of sediments deposited in arid lands, it is advised not to expect such high levels of confidence, but rather to make the best use of whatever is at hand, as long as the level of confidence is known and preferably also displayed with the results in some way, when discussing the conclusions.

Presentation of the results has two aspects, the factual and the selective. The factual presentation supplies the reader with all the information gathered from the sample, either in the way of countings or in the form of percentages, without any further treatment like grouping of taxa or omission of certain others due to conceptual considerations, except for the primary grouping which depends on the possible level of identification.

The selective presentation of data is always concerned with subjective considerations of the particular researcher involved, usually geared at displaying, proving or strengthening the point under discussion, or the line of interpretation adopted toward the desired conclusions.

Such ways of presentation, both factual and selective, are discussed in many previous books and publications, thus will be treated here only superficially. The interested reader is referred to Birks and Gordon (1985), Faegri et al. (1989), Moore et al. (1991) and others, for detailed discussions.

6.3.1. Factual presentation

Simplest of all, naturally, is if the raw data are presented, in the same way as these emerge out of the microscope slide, namely a list of taxa identified and how many of each had been found in a given sample. This way of presentation usually takes the form of a table, which hardly ever appears as such in the palynological literature, mainly due to the fact that these tables take a lot of printing space, are quite hard to follow and therefore far from practical (Birks and Gordon, 1985). The merits of this way of presentation are quite clear: first, this seems to be the most objective way of presenting data, especially if the taxa are accompanied by descriptions, photographs and drawings; second, each reader can subject the data to any desired kind of statistical, numerical or graphic treatment which may be favored. This could be indispensable for clearing, proving, or occasionally disproving a certain case, not necessarily the same topic dealt with by the original researcher.

A step further in this way of presentation is by displaying the pollen spectra as percentages of the total, or of a partial number of the counted grains. If these percentages are presented together with the numerical basis for their calculations, it is quite straightforward to reconstruct the raw, original data for any further use. Such a set of percentages is also usually presented by way of a table and similarly to the raw data, takes a lot of printing space and is hardly seen in the literature, except for cases when only a limited number of samples was analyzed.

Occasionally, when printing space is of no problem, such data are available to the reader. This however is usually only the case with monographs, theses and reports, which in most instances have a very limited circulation due to the restricted number of printed copies, thus are usually very hard to get.

Further modes of presentation, which are so devised that a considerable number of data could fit into a printed page, by way of tables or diagrams, seem to depend on graphic methods, developed through the years, all of which bear some difficulties when attempting at reconstructions of the raw data.

6.3.2. Graphic considerations

Graphic considerations are not peculiar to palynology of arid lands, but rather general in nature, thus will be dealt with only briefly, since over the years these have been discussed time and again in many publications, recently summarized in Traverse (1988), Faegri et al. (1989) and Moore et al. (1991). The most common graphic treatment is geared to present the changes in percentages of the constituents of pollen spectra, either in the spatial plane or along the temporal axis. Hardly ever both sets of data are presented together, in a single figure.

The most important warning given by many authors concerns the popular tendency to overcrowd the figures, thus seemingly presenting more data, but in actual fact these become difficult to conceive, until in many cases the picture is lost.

Spatial distribution of palynomorphs, of selected taxa or groups, is done by either a set of "pie charts", which represent several taxa or groups, superimposed on a map or a transect of the studied region (Fig. 2.2.2), or by sets of circles of varying diameter, which is proportional to the percentage, drawn for each taxon or group on a separate map (Fig. 7.4.2). A set or bar graphs, or histograms (Fig. 11.3.1) can similarly be used, while in some cases isopollen maps serve to present the data. Some other graphic symbols may also occasionally be in use. Choosing between the graphic possibilities is mostly by personal preference and it is not rare to find several options used in a single publication, each describes best a particular discussed case.

Temporal variations in the relative distribution of palynomorphs are traditionally expressed by a pollen diagram, ever since Von Post (1916) suggested this method in his pioneering work. From the graphic point of view, three main types of diagrams are presently in use, the cumulative, composite and resolved (Faegri et al., 1989). These can be used either for displaying percentage calculations or for absolute values like pollen influx and pollen concentration within the sediments.

The cumulative diagram is chiefly used for percentages which, for each sample level, add up to 100%. The distribution, or percentage curves are therefore added to each other, so that areas between the curves correspond to the relative share of the constituents. In most cases, different shading is used, to clear the picture, so that results are more easily observed. Personally, I found out that such a cumulative diagram is very useful when two or three variables are presented (see Fig. 8.5.8, right column). If more are displayed in this way, the picture comes out blurred and the data cannot readily be observed. This type of diagram is therefore convenient for showing AP/NAP ratios, the relations between local and regional pollen, or other such cases.

The composite, or overlapping diagram, is so constructed that all taxa or pollen groups have a common base line, or occasionally two or even more base lines, from which the percentages are plotted (see for example Fig. 8.5.8, left column). When two base lines are used, one on the right hand side and the other of the left, it is common that the left is reserved for arboreal pollen and the right for non-arboreal, so that the dividing line between the two is indeed a cumulative line, indicating the AP/NAP ratios. The composite diagram readily displays the relations between percentages of various constituents, thus seems the preferable way to present up to a maximum of six variables. When it is attempted to plot more than this number of curves, the ensuing picture is less and less clear, making the following of data quite difficult (Traverse, 1988). Symbols are usually used for each of the variables, which are occasionally replaced by lines of different thicknesses, or various types of broken lines. These seem easier to draw, especially with a computer driven plotter, but are certainly more difficult to follow.

The resolved diagram shows each of the taxa on a separate curve, with its own base line. This type of diagram is very clear, thus seems presently preferable to more and more researchers. There are however two problems concerned with the graphics of the resolved diagrams. The first involves attempts at showing distribution curves for too many species in a given sequence, which makes the width of such a diagram out of proportion, especially when the section itself is not too long. Traverse (1988) rightly states that foldouts are quite difficult to comprehend, thus the main points are lost in plenty of paper. Attempts to overcome the graphic problems of proportions lead to another problem. These attempts are in two directions: either to cut the diagram and put its parts one below the other, or to use different scales for some of the constituents, so that all would fit in the proportions of a printed page. In both cases it is quite difficult to make comparisons, even when the different scales are clearly marked.

The solution to all these graphic problems, suggested by Traverse, is to present graphically only the most significant data necessary to prove the discussed case, which are used as basis for conclusions, giving the rest of the data in the form of a table or a separate diagram. Such a separate diagram could be used for the minor elements, or other data of no immediate importance. This suggestion, incidentally, would equally apply to all three types of curves.

There are two main tendencies in the way of presenting pollen distribution, or percentage curves. One is showing values for each of the levels by bar graphs or histograms, based on the concept that each of the levels is analyzed as a single entity and should be so presented, without attempting to connect the horizons artificially. This way is especially useful and seems even mandatory, in cases where samples are located quite far apart, either temporally or spatially. The definition of "far apart", incidentally, is always subjective and depends on both personal considerations and demands placed for a particular study. The other way, which seems presently more common, is to connect the values for each of the taxa, with ensuing sawtooth curves. This way is based on the concept of evolutionary trends in plant communities, due both to gradual genetic and environmental changes through time, thus viewing the ensuing "continuous" curves as being composed of a certain number of dots.

Graphically, the sawtooth is much clearer to the observer as compared with the bar graph diagram, thus preferred by most researchers and presently widely used.

To conclude, it does not really matter which kind of graphics is used for presenting the results, as long as these can clearly be observed and are not overcrowded with unnecessary details.

6.3.3. Selective presentation

Selective presentation is usually done by either of three ways, which are frequently also combined in a single diagram, or table: grouping of several taxa into larger numerical units of specific significance, such as environmental, depositional, etc.; calculations of percentages based on a total which is only part of the total number of counted grains; and omission of certain palynomorphs from the calculations of particular pollen sums. Straightforward, rather general examples of selective presentation are common in almost any pollen diagram from anywhere in the world. Such are the separate grouping of arboreal and non-arboreal pollen, the calculations of relative shares of arboreal pollen based on the total sum of this group, or the omission of pollen grains produced by submerged water plants such as *Myriophyllum* and others from the total pollen sum.

Selective presentation becomes however of prime importance when dealing with pollen spectra obtained from samples collected in arid lands. The relatively small number of pollen produced by the regional xerophytic vegetation is easily masked when no selective presentation is applied, for several reasons: grains arriving from sources other than the regional dry land flora, by winds, rivers and other transporting agents; palynomorphs produced by a well developed vegetation which may surround the basin from which samples had been collected, such as a spring, small lake, or an archeological site; inclusion of redeposited or infiltrated pollen, etc. A well developed local vegetation may also act as a filter to pollen produced by the regional, thus diminishing the latter representation in the sample. Whenever the regional vegetation is of interest, the "foreign" elements must be separately presented, apart from the regional pollen and vice versa. This becomes particularly important when palynostratigraphy is attempted, as the local pollen may vary from one locality to the other, while the regional is a much more sound basis for palynozonations and correlations. For details and use of this approach please refer to Chapter 9.

A. Grouping

The most common way of grouping practiced by palynologists in arid lands is according to environments in which the pollen producing plants are growing, or are supposed to have grown in times past, when fossil material is dealt with. Eventually, the compositions of taxa involved in this way of grouping change from one region to another, according to the relevant plant communities, thus no general rules can be suggested. Grouping according to provenance environments could be of a general nature, or may become quite detailed, depending on practical considerations, regional characteristics, the number of palynomorphs identified and counted, or the state of preservation.

General grouping is used for the Quaternary palynostratigraphy of Israel (see Chapter 9 for details), where the pollen produced by regional vegetation — a grouping by itself — are divided into three groups, representing the Mediterranean forest or maquis type vegetation, the Irano-Turanian steppe and the Saharan desert environments. Such grouping evidently involves a considerable margin of overlapping, which should always be clear when results are discussed. Plants which grow in each of the three major environmental domains may also occur in the other two, however comprising smaller shares of the flora. Such are trees, which can also be found in the steppe and occasionally in the desert environment; Compositae, considered typical for the deserts, which occur in both steppes and Mediterranean maquis and forests; and typical steppe plants such as *Artemisia* and *Ephedra*, which are also found in the other environmental domains.

The conceptual basis for subdivision of these three groups is the prevalence of certain plants in a particular environment. Thus, the boundaries between the three environments, as expressed by distribution curves in the diagram, should not be regarded as sharp, exact numerical entities, but rather, the always existing range of overlapping must never be overlooked. From the graphic angle, this may be difficult, since usually the eye takes definite lines as sound data. It is also quite difficult to exactly calculate the range of overlapping, which may add to the uncertainties discussed above.

The grouping of pollen grains into those originating from local *vs.* regional provenances also has quite a wide margin of overlapping. Gramineae, as an example, usually regarded as of local origin, also comprise some part of the undergrowth of forests and may even have been transported from far away, such as those brought by the Nile to the eastern Mediterranean, thus regarded as allochthonous (Horowitz, 1979).

The problem of overlapping is incidentally typical for any kind of grouping and seems unavoidable, except in rare cases — for instance, when redeposited palynomorphs considerably differ in age from the studied deposits. The range of overlapping naturally increases with decreasing levels of systematic identifications. The increased overlapping is particularly relevant in arid lands, due to both the general poor state of preservation of palynomorphs in the deposits and the prevalence of certain plant families in which generic, not to mention specific identifications are almost impossible, even in the better cases (Gramineae, Cyperaceae, Compositae, Chenopodiaceae — to mention a few).

A more detailed grouping of pollen into environmentally or ecologically oriented provenances evidently depends on a better state of preservation, good pollen yield of the samples and eventually, the number of environmental provenances represented by the pollen spectra. The latter factor may change from one locality to another due to topography, adaptation and the potential of transporting agents. Scott (1982), followed by A. Cadman (1988, personal communication) could distinguish as many as eight ecological provenances for pollen spectra from Transvaal, South Africa:

(A) Forest and forest-margin taxa.
(B) Bushveld trees of rocky and relatively moist habitats.
(C) Bushveld trees with relatively wide range of habitats.
(D) Plains bushveld trees of relatively dry habitats.
(E) Small shrubs and herbs.
(F) Herbs of dry habitats.
(G) Sedges.
(H) Grasses.

On top of these, two other groups are added, not ecologically related, the known contaminants and the unidentified palynomorphs.

When large numbers of pollen are identified and counted and transporting agents play an important role in bringing grains from far away localities, as is the case in a study of recent and Holocene samples from the Sahara, Cour and Duzer (1976) could identify as many as 22 different groups, based on both geographic provenances and different local environments.

Grouping of the pollen comprising a spectrum according to provenance biotopes is very common in the palynologic literature, especially when arid lands are concerned, where it is quite common that more than a single environment is represented. Grouping could be however done according to any necessary factor which serves at the time to prove a particular case. Such are groupings into autochthonous and allochthonous palynomorphs (Rossignol, 1961, 1962, 1964, 1969b), which depend on the distance travelled by the grains before deposition. This distance is not generally defined, but rather by the conditions prevailing in any particular studied area, which should therefore be well stated in each case. Groupings into the natural *vs.* anthropogenic vegetations (Dimbleby, 1985; Behre, 1986), synchronous and redeposited (Horowitz, 1970), terrestrial and aquatic (Fall, 1985, 1987), transport relations of river *vs.* winds (Buchet, 1982), etc.

No strict rules or clear definitions exist as to which plant belongs to each group, of any type of grouping which may be exercised. It seems therefore of prime importance not only to state the groups by general terms, but to give, as part of the presentation, a description of the composition of the various groups, preferably accompanied by discussion of the rationale used as a foundation for the specific way of grouping adopted in each case.

B. Partial presentation

Partial presentation usually accompanies the process of grouping palynomorph taxa, discussed above and refers to any numerical or statistical treatment, carried out on a certain group of taxa, which comprises only a part of the total number of identified and counted palynomorphs in a particular spectrum or set of spectra. In such cases, the more usual basis for calculation is the total number of palynomorphs belonging to the chosen group. However, other bases for calculation are also used, such as the components of the chosen group compared to another group taken as 100%, or a variety of other possibilities.

The most frequent partial presentation, in almost any palynological publication, is the way by which arboreal pollen are treated, the various taxa calculated from the total sum of components of this group. Relations between the shares of various arboreal pollen are usually used for concluding about the forest composition. This is however slightly different in arid lands, where most arboreal pollen are allochthonous, thus changes in the shares of constituents do not necessarily reflect the remote forest composition, but may represent changing wind directions or other factors. Thus the arboreal pollen diagram, which frequently also appears in studies dealing with arid lands, may bear different significance as compared with forested regions. This may sometimes call for a different way of presentation, that would better display the share of transportation in the spectra.

Other presentations, which are characteristic of spectra obtained from arid lands, display relations between percentages of pollen produced by hydrophil against halophil vegetations in an aquatic depositional environment, which reflect the hydrological regime in the area. Hydrology may depend on the general climate, but frequently only in part, when a much stronger influence may be exerted by local conditions, such as faulting (Horowitz, 1987b) which may have a considerable influence in tectonically active basins. Another partial presentation used in arid regions displays the relations between pollen produced by the local and regional vegetations. The local would usually comprise wadis, playas, springs, etc., while the regional may either refer to the dry land vegetation alone, or combined with some or all allochthonous elements. The latter group may also deserve a separate presentation at times. Examples for such partial presentations are given in Chapter 9.

Occasionally, percentages are computed for selected groups, as was done by Horowitz et al. (1978), who calculated relations between Compositae and Gramineae and between Chenopodiaceae, Compositae, Gramineae and Cyperaceae, for reconstructing local paleoenvironments in a salt-pan near Kimberly, South Africa. Cour and Duzer (1978), who had studied a transect of recent sediments in the Sahara, presented numeric and percentage relations between various pollen groups, such as Gramineae *vs.* Cyperaceae, Gramineae *vs.* Compositae, Gramineae *vs.* Chenopodiaceae, Compositae *vs.* Cruciferae, *Ephedra vs.* Umbelliferae, *Plantago vs. Tribulus*, echinate *vs.* fenestrate Compositae and spherical *vs.* oval Gramineae. Cour and Duzer found out that calculations of these relations could be successfully used in defining climatic, edaphic and sedimentologic domains, rather than merely using the shares of the various taxa, which seem to bear a lesser significance. Similar approach was also adopted by El Moslimany (1990) in Arabia.

Again, it does not seem practical to suggest any definite rules in the way of partial presentation, or which group should be treated in what way. It also seems that no way of standardization is advised for spectra or sections, especially if collected from far away localities. It is though advisable to consider such ways of partial presentation and possibly others, not discussed here or never thought of before, to try and see whether the use of different bases for calculations would not eventually make for a better display of the issue under discussion.

C. Omissions

Omissions are usually a necessary outcome of grouping and of calculations based on partial sums of palynomorphs in a given spectrum. Omissions can be classified into two principal categories: taxa which are omitted from a group to which they seem to belong and treated within another, like hydrophil trees which may be grouped with other, non-arboreal vegetation of this biotope, rather than the arboreal pollen; and taxa which seem unnecessary for a certain discussion, thus altogether dropped, such as introduced plants, contaminants, etc.

Some straightforward omissions had for a long time been practiced by palynologists, such as the almost general disregard of pollen and spores produced by submerged water plants; omission of various kinds of spores, the rationale of which is critically doubted at present (Faegri et al., 1989); omission of pollen produced by cultivated and introduced plants from the recent pollen spectra, when used as a basis for reconstructing past natural environments; calculations of arboreal pollen spectra without the hydrophil and river bank trees such as *Salix* and *Populus*.

Shares of such omitted taxa, when at all represented in the pollen diagrams, are usually calculated on a basis of the total sum of counted grains, excluding the omitted ones, which thus may frequently display values exceeding 100%, especially if pollen produced by submerged plants are concerned. It is also a tendency in such cases, usually when percentages are quite high, to display the rejected ones on a different scale, which may occasionally confuse the reader.

The omitted taxa may take a considerable share of pollen spectra, especially when obtained from arid lands, particularly if the regional dry land vegetation is of interest, or if regional palynostratigraphy is studied. In such cases (see Chapter 9 for details) the entire group of pollen representing local vegetation may be altogether omitted, since usually the local environments, especially in taphrogenic basins, do not have to do anything with regional changes. The latter are in most cases used as basis for palynostratigraphy, particularly in the younger geologic periods, the Neogene and Quaternary. In other cases allochthonous sporomorphs may be omitted (Rossignol, 1969b), so that an accurate picture of the close environment could be obtained, regardless of the transporting agents. Again, in some such cases the omitted groups are separately represented in some way or another, while occasionally they are completely dropped from the presentation.

It is always the question whether a group of palynomorphs which are not used throughout a certain article should be at all presented by way of a table, diagram or otherwise. It seems advisable (see also Traverse, 1988) to present the entire set of data and not to dismiss any as unnecessary, even if not presently used. This may be found indispensable at some future opportunity, for further research or for use whenever different approaches or ideas are raised by another researcher, who would not be able to acquire the data for some reason. It is frequently found out that such data cannot be retrieved, particularly if the original publication had been printed a long time ago, thus are practically lost. The savings in small amount of extra labor and printing space are not worth facing such risks.

Another type of omissions is concerned with the possible levels of specific identifications of palynomorph taxa. This is frequently carried out without even paying attention to the fact that it is done. Such is usually the case with Compositae, which are presented in most cases without the inclusion of such genera belonging to this family which can be readily identified, like *Artemisia*, *Ambrosia* and possibly others. However, quite frequently it is not even mentioned in the presentation which taxa are excluded from the Compositae calculations. Sometimes this is self evident, when these curves follow each other and are clearly marked, but in other cases it may only be mentioned in the text, or not at all. It is therefore suggested that such omissions will be clearly mentioned in the presentation.

Cour and Duzer (1976) and Buchet (1982), while studying recent pollen spectra from a rather long transect in northern Mali and a section obtained from a sebkha for the Holocene, have tried omitting progressively several taxa from the spectra, in an attempt to obtain a better, or more objective representation. Buchet commenced by omitting pollen produced by river plants, then Cyperaceae, followed by pollen produced by local plants and finally Gramineae, until the curve obtained for arboreal pollen seemed a better representation of the actual distribution of this group. Similar experiments had been carried out by Cour and Duzer, to obtain a better representation of the paleoenvironments.

D. Combination of adjacent samples

Combination of adjacent samples, usually between two and four, into a single spectrum displayed on a diagram or a transect, is done for several reasons. The first, which concerns very much spectra obtained from arid lands, is due to poverty in palynomorphs in each of the individual samples which, when combined together, allow for a significant spectrum. The combined spectrum will thus have a wider stratigraphic or geographic range (equals to lower resolution), but a higher level of confidence for the paleoenvironment. Another reason, which equally applies to all territories, is to smooth the curves and rid of minor "noises" in the diagram, transect, map, or any other possible way of presentation. The smoothing should evidently follow relevant statistical rules. Such smoothing has the tendency to loose minor details (Faegri et al., 1989), but on the other hand the regional picture is invariably much clearer, thus palynostratigraphic correlations can be made more easily.

An example can be seen from two boreholes drilled in the Dead Sea Rift (Fig. 9.3.2), Notera 3 in northern Israel, where Mediterranean environment prevails and Amazyahu 1, some 250 km southward, in the Dead Sea region which is surrounded by a desert. It can clearly be seen that both smoothing and higher credibility, resulting from combining adjacent samples, make a sound foundation for the palynostratigraphic correlation of the two boreholes, even though drilled in different environments.

Such combination of samples could be done in many ways and applied to different variables, but the one precaution to be remembered is that in every combination attempted, the combined samples must have common palynological characteristics. These characteristics should not necessarily apply to the entire spectrum, but may be arbitrarily chosen for a particular group, or groups of palynomorphs, such as, for example those representing the regional vegetation. If the differences between combined spectra exceed acceptable limits, which naturally should be well defined since no standard exists, the average would only have a limited meaning and in most cases be found useless, or even erroneous for environmental or stratigraphic applications. The definition used in every particular case for palynologic resemblance should be clearly stated, to avoid confusion, preferably in a numerical way (see Birks and Gordon, 1985).

Such combinations can also be used for the planning of future work on transects and sequences obtained from a certain region and may give a very good clue as to what should be the geographic or stratigraphic spacing of samples in forthcoming studies. Such procedure would, on one hand, ensure that the desired level of significance is attained, while on the other, labor and time could be saved. Subsampling should also be considered while planning a palynological study, detailed in Chapter 4.

E. Correction factors

Corrections of raw data prior to presentation are not attempted too frequently, because of two principal reasons: first, correcting factors are not always easy to establish, because the processes involved are more often than not difficult to transform into mathematical equations, since they are not entirely understood; second, usually conclusions are drawn by comparison with recent (or better still, subrecent) pollen spectra, which represent the present day vegetation and environment and had already gone through all kinds of differential processes; when these relations are directly applied to the fossil ones, the need is lessened for intervening correction factors.

The correcting factors cannot be readily calculated. Even when such are calculated and seem pertinent to given conditions, they may change without notice while going down the section or moving to another territory. The more common corrections are based on the fact that certain taxa produce different amounts of pollen per plant, calculated as numeric factors. Frequently such corrections involve factors set to each identified taxon, by relating pollen production of this particular plant (or group of plants) to a certain predetermined value. This value is in most cases the share of a known and common taxon, arbitrarily chosen for reference, which may be estimated to be understood in terms of pollen production and differential preservation.

Several authors, based on observations concerned with differential pollen production (Birks and Birks, 1980; Bradley, 1985), had suggested correction factors for various taxa and have presented pollen diagrams calculated according to these factors.

In most cases however, differential preservation is ignored, otherwise for each type of depositional environment and every kind of sediment one would have to use a different set of factors, as shown by Havinga (1984). Cheddadi (1988), who analyzed sediments of the Nile delta, maintained that the state of preservation is different in the two types of deposits comprising his cores, the somewhat oxidized yellowish clays and the highly organic black sapropels. Cheddadi consequently suggested correction factors to account for the differential destruction of palynomorphs in these two depositional environments, in the ensuing pollen diagrams. As was carefully done by Cheddadi, the concept on which correction is suggested should be clearly stated in the presentation. Preferably, the raw data should also be presented in some way.

The situation is somewhat different in arid lands, where both differential preservation and long distance transportation play an important role in defining the composition of pollen spectra. The differential processes involved in preservation and destruction of palynomorphs in these regions, discussed in Chapter 6.1.1, evidently require correction factors, however these are quite difficult to define in a general sense, thus it seems that each particular case should be treated at its researcher's discretion.

Correction factors involved with long distance transportation seem to need some clarification. The factor for a particular taxon may change over its travelled distance and depends also on the type of transporting agents involved. For example, vesiculate conifer pollen travel rather long distances by winds, but also float well.

The correction factor, when these are compared with oak pollen which do not travel so well, will change with the distance from the source area which, as it grows longer, conifers would become progressively over-represented and oak under-represented. As an example, both oaks and cedars grow in the Lebanon and their pollen travel south by northern winds, along the Dead Sea Rift. However, the share of cedar pollen in the Hula Valley, to the north, is considerably lower than in the Dead Sea, some 250 km to the south, while in the Bay of Elat (Aqaba), some 450 km to the south, cedar dominates as compared with the share of oak in the recent sediments (Horowitz, 1966, 1971, 1979).

Another example comes from pteridophyte spores, which are quite bad air travellers, usually deposited close to their growing habitat. This is however different when water transportation is involved, whereby spores could travel thousands of kilometers. Such are spores carried by the Nile to the eastern Mediterranean offshore (Rossignol, 1969b) from equatorial Africa which, as compared with other sporomorphs of tropical provenance and even with the vesiculate *Podocarpus*, increase their relative shares along the Nile, since all others do not travel as well. The correction factor which would be applied to fern spores thus considerably changes along their way via the Nile (or the Red Sea currents, see Horowitz, 1966; and a discussion in Chapter 2.2.3).

Thus differential pollen production alone could not be used for calculating correction factors in dry lands, without taking into consideration processes of transportation and destruction, which are also of a differential nature and of prime importance in these parts of the world. Moreover, it may sometimes be possible to calculate correction factors for the present day, when conditions are fairly well known, but application of these to the past can prove extremely hazardous.

Chapter 7

Principles of interpretation

Once the palynomorphs are put on paper, in the form of tables, diagrams, or whichever way had been chosen for their presentation, it is time for the more interesting part of the game. Truly, thinking and concepts have been put into the palynological work much before, ever since sampling strategy was decided, through choosing the way by which the samples should be prepared and pollen extracted, via considerations made for the identification and presentation of the results. But the present stage of palynology is even more intriguing, mainly because of the variety of possibilities for conceptual diversifications, which may occasionally end with the situation in which two palynologists would arrive at entirely antagonistic conclusions based on the same set of data. This had already happened in the past and is sure to happen again, only manifesting the beauty of scientific thought, which lets everyone wander in individual conceptual paths with complete freedom.

For this reason, it is quite difficult to sum up the topic of interpretation. There seem to be no strict rules by which one should behave, neither a standard line of thought by which one is bound. Conclusions should be eventually based on the data, but this seems to be the only limitation which pertains throughout the process of interpretation. One is reminded that almost all breakthroughs in science had been achieved by those refusing to think in the traditional way, or along already existing lines and instructions.

It would therefore be advised to regard the present chapter and the others that follow, which are however being written, with a mixed attitude. On the one hand, examples are brought from palynological studies hitherto carried out in arid lands, to show the various possibilities and ways of thinking, which resulted in interpretations being thought of and conclusions being drawn; on the other hand, a warning and hope, that future students of these regions would not restrict themselves to what has already been done, but will try for new, possibly wider horizons, which would make palynology of arid lands even more meaningful.

A plea, concerning previous studies: these are not necessarily obsolete, while new ideas are not always the better ones. Great care (and possibly some prudence) should be taken before attempting to dispose of older interpretations and conclusions, or turn the entire picture upside down. Usually, two possibilities exist while attempting a new interpretation: either conclusions are based entirely on a novel understanding of the previous data, or a new set of data is acquired, that would justify a different interpretation which also includes the older data. In both cases, it is advised to refer to previous studies and present a detailed discussion, which could base the new approach beyond doubt. Disregarding or ignoring is by no means an appropriate approach and should better not be adopted in such cases.

This chapter, incidentally, is crammed with warnings against possible pitfalls in which palynologists have occasionally found themselves, usually without any prior notice. This rather long list is not brought here to discourage researchers, but only to indicate some of the possibilities of arriving at hasty interpretations and conclusions, or initially disregarding such facts and processes to which more attention had to be primarily given. Not all such warnings are shared by everyone and opinions vary. Some interpretations or ways by which the data are treated are accepted by certain palynologists, while entirely rejected by others. It is not always possible to take a definite attitude in such cases, which should be left to individual discretion and criticism.

7.1. Recent pollen spectra — a tool for interpreting the past

Recent pollen spectra are undoubtedly the best possible reference for interpreting the past (Faegri et al., 1989). This concept is however based to a great extent on negative grounds, since it seems that no better reference or tool, or any other method at all for that matter, is presently available to us.

The basic principle of geology, uniformitarianism, which claims that the present is the key to the past, was also adopted as a principal foundation for palynology and paleopalynology. Three aspects of the present day vegetation and its ecology have to be taken into consideration while attempting to apply this principle for interpretation of past vegetations and paleoenvironments. Two of these aspects are natural, while the third is human induced, either in an active, intentional way or passively, unintentionally, but however is there to consider.

The first aspect concerns the degree of dependability of the vegetation on specific natural environmental characteristics, seemingly necessary to maintain a certain plant community in a given locality. The principal question raised here is whether a given environmental setting would have resulted in the past in a development of a vegetation formation similar to the one which occupies such an environment at the present day, or could also have invoked a different floral pattern. The second aspect concerns the question of relationships of a certain vegetation formation, or plant community, with its representative pollen spectrum. The third comprises problems raising from changes made in the natural vegetation by human activities. Such changes can be of two principal types: the first is caused by direct activities of man, involving processes such as deforestation and agriculture, which result in expansion of cultivated and introduced plants, consequently diminishing areas occupied by natural vegetation; the second is indirect and results in development of ruderal vegetation, comprising a host of opportunistic taxa which are much better developed close to human settlements. To these are added activities which change pollen transportation paths.

7.1.1. Vegetation–environment relationships

Relationships between vegetation formations or plant communities and their environmental demands seem to be quite well established for the present day conditions. These are described in great detail in many botany and ecology books and publications, for the different parts of the world. It was indeed by the distribution of major vegetation formations that the climatic zones of Earth had been established (see Chapter 1). However, the main problem which bears on application of these relationships for interpreting past environments is yet unsolved. Although the present day pattern of the connections between plant communities and their environments is well established, it is not quite clear if these could safely be applied also to the past, mainly because no entirely undoubted assumption can be made that such exactly have the connections been in the past. In other words, is it sure that a certain given environment would have had always resulted in the development of a vegetation similar to the type growing in such an environment at the present day? The limits of doubt further expand when the flora is interpreted by its pollen, since not all plant taxa are represented in the spectra and many that are, cannot be identified to the species.

One of the principal problems with the present day plant communities concerns the climax vegetation maintained by a certain environment. There are many approaches among botanists, some opposing, toward the concept of a "climax" vegetation. Certain researchers even entirely deny its applicability, or the existence or such possibility as an ultimate development toward a plant community that would not change anymore if environmental conditions remain as they are. These view a certain plant community in a given locality as a temporary stage in an ever developing process of its composition. This problem becomes more acute if one remembers that climatic changes, even though minor, are quite frequent throughout time. Indeed, within the last 10,000 years most places on Earth had undergone through such changes, which have had directly inflicted on the environments.

The question is therefore, if a vegetation typical for a certain environment is not indeed a climax association but merely a stage in the development, could the environment be precisely defined by its present day plant community? Such definition is naturally possible in the wider sense, but when one goes down to details, pictures may be confusing. It seems that no definitely positive answer could be given to this question at our present state of knowledge, so that scientists would have to be content with a more general interpretation of the past. The reasons for uncertainty are many, but the more significant concern ecological constraints placed on growth and expansion of a certain plant taxon by its individual physiology, or similarly on a plant community; and the adaptation possibility displayed by plants when their environment changes, such as could easily be demonstrated when they are artificially removed from their typical environment into a different one, in which they may even better develop. These factors are always relative, since competition among various plants also plays a role here, thus it is not enough that one species is well adapted — it should be better off than others at the same biotope to succeed.

The problem is that not much is known about constraints applied to a certain plant taxon and these of course vary from plant to plant, even within a well defined plant community. Naturally, when an entire plant community is treated, the range of its constraints is considerably narrower than for each of its individual constituents. Thus it would be much more significant to rely on plant communities, when interpreting past environments and refrain from dealing with individual taxa. This can easily be done in forested regions, where the composition of forests could be used as a significant indicator of paleoenvironmental conditions. There are indeed limitations to the use of entire plant communities even in the temperate and tropical domains, particularly because the pollen spectra do not represent all plants comprising the community. However, in arid lands the situation is much worse, mainly due to the limited variety of pollen in the spectra which results from poor vegetation and pollen production, the fact that many plants are insect pollinated and the considerable differential destruction involved with the transportation and deposition processes. On top of that, many pollen types characteristic for arid lands cannot be identified further than the family or genus level even when well preserved, as opposed to the situation with arboreal pollen in most humid domains.

An example from the Zagros Mountains, in Iran, can readily show the problematic of relying on a limited number of taxa for interpreting paleoenvironments. Van Zeist (1967; and further publications) found out that during the later part of the Pleistocene the nowadays typical oak forest of the Zagros, represented by abundance of its pollen in the recent and Holocene spectra, was preceded by an *Artemisia* steppe, whose grains dominate the assemblages recovered from the older deposits, where oak is hardly occurring. Van Zeist concluded that aridity prevailed in the Zagros Mountains during the late Pleistocene, expressed by the development of a typical steppe. In contrast, but based exactly on the same data, El Moslimany (1986) claims that the more significant constraint on the development of oaks in the Zagros is the winter temperature which, when too low, would not permit the growth of young oak seedlings which cannot survive their first year and freeze to death. El Moslimany therefore concludes that the Zagros was colder in late Pleistocene times, but not necessarily drier, as Van Zeist postulated. Rather, the higher winter snowfall and the lower temperatures hindered the development of oaks, thus the region was readily invaded by the opportunistic *Artemisia*, which could easily spread and cover considerable areas. There seems to be no way, at the present state of knowledge, to safely decide which of these two opposing interpretations would better describe the late Pleistocene Zagros paleoenvironment.

The problem of unknown critical constraints applicable to development of plants is further complicated by the usual lack of plant stocks in arid lands, which could have readily replaced a certain type of plant community when environmental conditions are changing. The richly vegetated regions maintain almost always such plants that would spread and occupy a niche which may have been evacuated by others, due to changing ecological or environmental conditions, thus signifying such new characteristics. The usual non availability of suitable vegetation stocks in arid lands precludes such rapid replacements by plants typical to the new scene and the void is then occupied by opportunistic plants. These have much wider limits and better physiological adaptability to constraints, which enables them to occupy and develop well in a wider variety of environments. A refined interpretation of the paleoenvironment is therefore much more difficult, or sometimes impossible.

Adaptation and tolerance shown by various plant taxa to different environmental conditions can easily be seen when plants are introduced from one region to another, where some of them can develop and spread even better than in their original environment, although (or because) conditions are different. Such, for example is *Eucalyptus*, typical for the semi-arid regions of Australia, which, when introduced to other countries had developed very well in environments typified by high groundwater tables, as was the case in Israel, where this tree was planted to help dry and reclaim marshes. Examples of such plant adaptations are indeed numerous.

Adaptation of plants to changing environments can also be observed in nature. Such is the case in the Negev Desert, southern Israel, where a great variety of typically Mediterranean environment plants are presently growing, supposedly as relics from a former period (Danin, 1972), during which the region enjoyed this kind of climate. Within these plants, incidentally, are also included trees, such as *Pistacia atlantica*. The problem with such relics may become quite confusing in arid lands, especially when collecting samples from a site which is intimately connected with a high groundwater table in a wadi, or a spring, where relics usually find refuge and may flourish, thus could even obscure the otherwise dry regional picture. Such may be the case with travertines analyzed from the extremely-arid Arava Valley in southern Israel (Weinstein-Evron, 1987a), where oak and other arboreal pollen were found in considerable quantity, as opposed to other synchronous lake deposits, where the pollen spectra indicate extreme aridity (Horowitz, 1987b).

7.1.2. Anthropogenic effects

Human activity can induce changes in the natural conditions, which would be represented in the recent pollen spectra used for reference. These changes are not peculiar only to the present, but had commenced ever since human settlement began on Earth, especially when agriculture is practiced. The influence on pollen spectra is however much more pronounced in the recent sediments, when this activity has gone beyond any natural boundary and expressed almost everywhere, practically in any type of the present day environments. The human influence on the recent pollen spectra may be either due to intentionally made changes of the vegetation by deforestation, cultivation, domestication and introduction of plants from other regions; or indirectly by the development of ruderal (otherwise termed "weed"), opportunistic plant communities around habitations, cultivated areas and even along roads.

Other changes made in the landscapes by man, either intentionally or unintentionally, may have caused changes of transportation paths and energies, directly affecting the transport of sediments and pollen. Such activities would be reflected in the recent pollen spectra and may considerably change their compositions, so that the natural environmental conditions would not be accurately represented, occasionally deviating to a great extent.

Arid lands are more prone to effects exerted by human activities on the recent pollen spectra, than forested regions, mainly due to the low masking factor of the dry areas local pollen components. The long history of human habitation in these regions may also be reflected in the subrecent pollen spectra; while even in the densely populated regions of western Europe and North America there are still many localities where recent pollen spectra can be obtained and studied from relics of natural forests, in which the intervention of man would be minimal, or can easily be detected and subtracted. Also, even when forests are cleared, the internal composition of their arboreal components does not alter much within their relics, as long as the environment does not change. Since environments are generally interpreted from the composition of arboreal pollen spectra, these remain a dependable basis as long as relics are still in the region, which seems rather the rule than the exception. In the arid lands, where interpretations are chiefly based on the non-arboreal taxa, it may sometimes be quite difficult to locate sampling sites in which man's activity could be neglected or easily separated from the natural components of the spectra, since hardly any herbaceous vegetation can withstand activities such as overgrazing (Horowitz et al., 1981).

The agricultural activities of man and the introduction of exotic plants from various different regions are probably the easiest to recognize in the recent pollen spectra. Most cultivated plants, including those directly derived from ancestors still growing in the same region, are recognizable by their pollen grains. The same is true for most introduced plants, which are not natural to the studied region. This part of the recent pollen spectra can easily be subtracted from the total and studied separately. This could also be readily done in arid lands, since most of the cultivated and introduced plants are foreign to these regions, thus the occurrences of their pollen in the recent and subrecent spectra places very little difficulties. A difficulty may however arise, since it is not known which natural plants had been replaced by the introduced ones, and to what extent.

The problem with ruderal weed associations is much more difficult to tackle in arid lands, mainly since most of these are plants that at any rate grow in the dry environment, only better developed and much more spread close to human habitations. Moreover, most of the weed species belong to such families as Chenopodiaceae, Compositae and Plantaginaceae, members of which consist also a major part of the natural vegetation. Since in most cases the pollen cannot be identified further than the family, or rarely the genus level, there is very little indication which pollen is derived from the natural part of the vegetation and which one from the ruderal weeds. The phenomenon of weeds is evidently also known from the richly vegetated regions, but there, as stressed above, it is the composition of the forest which is generally used for interpreting environments. In the arid lands, in contrast, it is the non-arboreal pollen on which paleoenvironmental conclusions are usually based, thus weeds may considerably distort the picture, when not positively identified.

The third group of human activities, both intentional and unintentional, comprises processes which may affect the recent pollen spectra by changes made in the transportation agents. Intentional changes are usually those connected with irrigation and other hydrological operations, building and military activities, while unintentional changes may be induced in the composition of fauna, its migration routes and its feeding habits.

Irrigation and hydrological operations are of two main types: construction of dams and diversion of river paths. Dams, which are presently constructed across many rivers, especially in arid lands, considerably reduce or sometimes entirely stop water flow in such dammed rivers. Even when water flow is not entirely stopped, the dammed reservoir formed upstream acts as a sediment trap, which causes at least partial trapping of palynomorphs as well. Such partial trapping is sometimes more problematic than total deposition of the palynomorphs, since it may cause differential sedimentation. This may change the composition of both spectra recovered downstream and those obtained from the reservoir deposits. Dams are also quite frequently built across wadis, where they usually completely stop sediment and pollen transport, except during stronger than usual floods. In both cases, either when all sediment is deposited or when a strong flood carries it further downstream, sorting and differential deposition do not seem to affect the finer fractions due to the higher energies involved. Therefore, all kinds of wadi sediments could be considered a more reliable source for recent pollen spectra.

Rivers are sometimes diverted from their original courses and water is being pumped, used for irrigation and other purposes. Pumping can be either done from a dammed reservoir, from an intermediate lake or from the river itself. All such activities may affect the fluviatile, both perennial and intermittent transportation paths of sediments and eventually palynomorphs.

Dams are also frequently constructed in richly vegetated regions, but usually differ in their influence on the recent pollen spectra as compared with the arid lands. This difference is caused by the essentially stronger and perennial water flow in rivers of the fertile countries which, following the initial filling up of the dammed reservoir continue on their way in an almost unchanged or minimally changed manner. Since most of the water flow in such reservoirs is on the surface, the overflow capacity of transporting palynomorphs remains practically similar to what was the situation before damming. This situation is entirely different in arid lands, where the dams usually considerably affect the flow and overflows may be quite rare, or restricted to a certain season.

When a large river crossing an arid land, whose water sources are far away in a humid environment is considered, such as the Nile or the Tigris which, although being dammed are perennially flowing, such dams constructed quite high upstream would usually diminish the contribution of pollen brought from the humid provenances to spectra recovered from arid land sediments downstream. Consequently, a comparison of recent sediments with older ones from the same locality in an arid land downstream, may reveal seemingly environmental changes, because in the older deposits pollen transport had not been affected by the dam (or dams, as may happen) and brought more grains indicating a humid environment, which came from the intake area and do not have any bearing concerning the situation in the sampled site.

Pumping of groundwater, extensively done in arid lands, can also cause similar effects on pollen transported by this agent. Such pumping, when extensive, can cause a complete drying up of a perennial river or spring of local origin, thus may considerably change the recent pollen spectra by elimination of one of the palynomorph sources. A special case of pumping which had changed the recent pollen spectra of Lake Kinneret, Israel, happened when saline springs which surround the lake have been artificially diverted to lower the lake's water salinity. The saline springs contain palynomorphs from both their intake area in the forested hills to the west and from a salt body of Pliocene age dissolved at depth of some 1,800–2,000 meters. The thus redeposited palynomorphs formed an integral part of the subrecent spectra in Lake Kinneret, but do not occur anymore in the recent spectra.

Intentional human activities which may enhance redeposition and be reflected by inclusions of older palynomorphs in recent sediments are usually those connected with the active exposure to the surface of older strata, or by removing their protective vegetation cover, thus exposing the deposits to more extensive erosion by both winds and water. Such activities comprise bulldozing of soil and rock for foundations and roads, the constant use of dirt roads in arid lands which raises considerable amounts of dust to the atmosphere, or agricultural activities such as overgrazing, typical to many arid regions, that may remove most of the natural vegetation cover. This process, besides causing considerable changes in the plant communities, exposes the soil to wind deflation (Fig. 1.2.12).

Military activities, which are unfortunately quite common in arid lands, may raise huge amounts of dust to the atmosphere, thus causing considerable redeposition of previously laid sediments and palynomorphs. For understandable reasons, arid lands are a perfect terrain for drill activities of heavy military equipment such as tanks (this paragraph is incidentally written right in the middle of "Desert Storm" war between the allies and Iraq, which turns upside down much of the surface and older deposits in the Arabian desert regions).

Thus, even though human activities do not seem to have the capacity of causing significant changes in wind directions and energies, they can and do open new sources of sediments to the activity of winds and water. These sources are readily eroded and transported, supplying sediment and pollen which had not been previously available for redepositional processes, that would reflect in the recent pollen spectra.

Intentional human activities also include animal domestication and husbandry, which result in widespread populations of these animals attached close to human habitations. Such animals may cause, by the pollen contents of their feces, changes in pollen spectra recovered from the close vicinity of settlements. This seems of no problem for interpreting recent pollen spectra, since such sites are well defined and easily recognizable, thus the influence of domesticated animals can readily be detected and treated accordingly. It may however be more problematic in far away grazing terrains, as it is the habit with many pastoral communities to wander into remote regions during certain seasons.

Unintentional human activities concerned with fauna are reflected in processes of pollen transportation by various animals. These are altered by man's activity by both causing changes in the composition of natural fauna, or its feeding habits. Some animal species are more opportunistic than others and seem to take advantage of human activities and neighborhood, thus they multiply and spread much beyond their previous natural shares and territories. Birds such as the barn owl, which transport pollen in their pellets and excrements, the latter being sometimes an important source for recent pollen spectra especially in places like caves and rockshelters, are much more abundant presently in Israel than before the last 50 years of extensive settlement of the country. But not only barn owls are more abundant, their feeding habits and diets had considerably changed throughout the time of human activity. Nowadays they feed mainly on rodents that are themselves opportunistic, feeding and multiplying in cultivated fields, such as *Microtus*, which are not feeding anymore on natural vegetation. This change of their diets is evident in recent pollen spectra recovered from the owls pellets. A similar process can be seen with all kinds of birds of prey, both diurnal and nocturnal, which feed on rodents that had changed their diets following agricultural activities. Similar phenomena are also observed in pollen spectra recovered from feces of mammals, both herbivores which may feed on cultivated crops and weeds, rather than the previous natural grasses and predators, which in turn feed on such herbivores.

Unintentional human induced changes have also been observed in the migration paths of certain birds, especially birds of prey, which may also cause changes in recent pollen spectra recovered from their pellets and excrements. The changes in these pollen spectra are a combined result of both changes in birds migration routes, which thus transport the pollen to localities different than the previous natural ones, or bring the pollen from different provenances; and from the change in diets of the animals on which those birds of prey feed.

Bats seem to be rarer presently than before and their feces, which used to comprise a considerable part of cave deposits in so many localities, are very much affected now by the human induced vegetation, both agricultural and ruderal. Most bats feed on insects, but these mainly eat agricultural crops at present; bats which feed on fruit (flying foxes) turned very much to eating cultivated ones, rather than the natural, as they had previously done.

Such changes in animal populations which follow human activity are by no means restricted to the higher phylla, but can also be seen with insects, worms and others. Thus for example, pollen spectra recovered from food reservoirs of burrowing bees and wasps are very much dominated by cultivation and introduction of plants by man. Honey and associated bee products which used to be natural in the good old days are presently mainly based on cultivated plants, which is clearly reflected by the pollen spectra recovered from these substances. In fact these are so much affected by human activity and by the types of crops grown in the vicinity of beehives, that recent honeys and wax cannot anymore be used to define natural palynological processes, but are restricted to the identification of plants visited by the bees, which are only useful for the apiculture and honey production industry.

7.1.3. Recent pollen spectra as vegetation representatives

In general, the so termed "recent" pollen spectra could be subdivided into three principal groups: assemblages obtained directly from transporting agents, of which the most frequently studied are spectra of airborne palynomorphs; spectra recovered from flowing water, either rivers or groundwater, are also included within this group. The second group comprises pollen spectra obtained from recent sediments, here the term "recent" applying literally and strictly to sediments which are presently under the process of deposition and accumulation. The third group includes pollen spectra recovered from subrecent sediments, which are nevertheless usually referred to as "recent".

The first group, of airborne and waterborne spectra is easily sampled and studied, representing the bulk of palynomorphs liberated to the transporting agents, prior to the major influences of depositional processes, most of which would have detrimental effect in terms of differential destruction of at least part of the grains. The second group comprises the part of palynomorphs which had actually been deposited from the transporting agents, but had not gone through processes of burial and possible alterations connected with this process, such as further differential destruction and preservation. Sampling is somewhat more tricky as compared with the first group, particularly since care should be taken to collect those parts of a sediment that are in the process of deposition and had not yet gone through any further stage, a task which however sounds simple may at times turn quite complicated. The third group, of subrecent sediments, could present technical and conceptual problems in terms of sampling, which may affect its significance as a foundation for possible interpretation. Primarily, rates of deposition have to be established and ascertained and it should be well defined how long, in terms of time, the "sub" is. On the one hand, the longer it is the better representation of preservation and differential destruction processes would be available; on the other hand, one must be quite sure that the deeper part of the sediment was accumulated under environmental conditions similar to the present day, so that the spectra are indeed representative of the "recent" processes. Assumptions made here should be well founded and justified beyond reasonable doubts.

The number of taxa that can be identified within each of the three groups considerably diminishes from the first through the third. The first group would usually produce highly diversified spectra, rich in species, with very well preserved grains, however many of these would never become properly fossilized under normal conditions. The second group is not often as varied in its number of taxa, but most palynomorphs are still quite well preserved, since the delicate ones which are almost readily oxidized, or otherwise deteriorated while still being carried by the transporting agents, usually entirely disappear at this stage, hardly ever at all deposited. The third group usually consists of less taxa than the second, since some are destroyed through their burial. Differential destruction may also considerably change the numeric relations of the various constituents, as compared with the first group.

The third group, of subrecent spectra, despite difficulties involved with its sampling procedures, indeed represents environmental conditions in the best way pertaining to interpretations of the past. This is primarily because the palynomorphs had gone through at least part of the differential destruction processes, of which the preliminary stages are quite rapid, in the order of tens of years (Havinga, 1984). A distinction must be made between the first group, which represents the vegetation and environment in the most accurate way and the third group which, although much less accurate and more general in nature due to exclusion of many components which could not stand fossilization, is however a much better reference to the past. This, since analyzing paleoenvironments is always based on palynomorphs which had already gone through burial and differential destruction processes.

A thorough study of the recent pollen should better include the three groups, each of which could provide additional information concerning processes involved in the formation of pollen spectra and the exact way in which they represent vegetation. Understanding of the ways by which the vegetation is displayed by its pollen spectra, through the processes of transportation, deposition and differential destruction, could significantly broaden the basis for interpreting the past. Thus pollen recovered from the air and other transporting agents can yield valuable numerical information as to various contributions made to the spectra at a certain locality from other provenances. Only by separately analyzing pollen loads of various transporting agents an exact picture of the various constituents of a given pollen spectrum can be assessed with an acceptable degree of certainty. By "separately" it is meant that winds from different directions would be individually analyzed, continuously throughout at least a whole year, a method which may also turn interesting when analyzing pollen loads of water, animals and so on.

Analysis of the second group would provide valuable information as to which types of grains are practically deposited at a certain site. Some of these may not withstand burial and destruction processes characteristic for a particular depositional environment, but may appear in sediments of another, either at a different locality in the same region, or even occasionally down the same sequence, where sedimentary conditions had been more favorable for its preservation.

It would naturally be interesting to find out why some of the palynomorphs occurring in the first group do not ever find their way to the sediments, but this seems to be of lesser significance to the interpretation of past environments. It would at any rate be quite risky to jump to conclusions based on comparison of the first and second groups, such as that when similar spectra are obtained from different localities for the second group, other plants whose pollen occur in the first group at one place should also have been present in the other.

Analysis of the third group, which had at least gone through part of the differential destruction and preservation processes, would indicate more precisely than the former groups the actual expected relations of fossil pollen spectra with any given environmental conditions. This group would thus be a better basis for any interpretations attempted for the past. However, care should also be exercised in this case, since some of the differential destruction does not take place before a much longer time had passed, as compared with what is represented even by the subrecent spectra. Such are for instance destruction caused by cementation and crystal growth which usually do not affect as yet the subrecent sediments (except maybe for evaporites), chemical or thermal alterations and so on.

Three fundamental differences exist between pollen spectra of the arid regions and those obtained from temperate and to some extent tropical environments. The first concerns the considerably greater rates of pollen destruction in sediments of arid lands; second is the lower levels of possible specific identifications; and the third is the lower masking factor of local dry regions pollen, which necessitates a much more thorough study of transportation paths and the import of exotic pollen from other, remote provenances, grains which are incorporated within deposits laid down in the arid lands.

Differential destruction of palynomorphs is evident also in the temperate regions, but usually most plants growing in these areas are represented in the subrecent pollen spectra, although in somewhat different proportions to their share in the actual composition of the vegetation. This is, in addition to differential preservation, a result of different rates of pollen production by each of the various plant taxa, different susceptibility to transportation and so on. Following these processes, which had been long studied in temperate regions, correction factors could be introduced to account for them, so that the subrecent spectra could give an accurate representation of the flora, in many cases even down to the species level. This is evidently also a result of the fact that many plants of the temperate regions can be identified by their pollen down to species and are truly wind pollinated.

The situation is entirely different in arid lands, due to both the considerable rates of differential destruction and the problematics of systematic identifications. Not only differential destruction is usually more damaging to the palynomorphs, it is also the much greater variety of depositional environments in arid lands, as compared with the temperate, that makes general predictions almost impossible while attempting comparisons of the fossil to the subrecent spectra. One has only to remember, in this respect, that almost 98% of the Holocene pollen diagrams from Europe are obtained from the analysis of a single type of deposit — peat.

The problem of systematic identifications in arid lands, even with well preserved grains, would make for the subrecent spectra to represent families, or in the best cases genera. Thus the spectra are far from including all plant taxa growing in these regions, or if they do, no specific differentiation can practically be done. Combination of the two factors mentioned above would make pollen spectra obtained from arid lands of a much more general nature, thus a refinement in vegetation reconstruction in the way one may be used to in temperate regions, is practically impossible.

The low masking factor of regional arid lands vegetation pollen is more pronounced in extremely-arid and arid environments, as compared with the semi-arid where pollen production is higher. This calls for much more attention paid to contributions made by aerially restricted local vegetations to the pollen spectra. Such local vegetations, which may be maintained by a spring, an oasis, a salt-pan and so on, could contribute a considerable share of the pollen spectra, but would usually only represent the immediate vicinity, with no bearing on the regional environmental conditions, but rather the local hydrology. The local influence may be further complicated by palynomorphs brought to the site by the groundwater, from rather far away catchment area. In this case, an analysis of pollen loads of the water may give the necessary clue to factors affecting the composition of pollen spectra.

The solution for problems connected with local vegetation is rather simple in temperate and tropical regions, where usually only the arboreal pollen are used for interpreting the regional environment, in most cases after omission of these trees which grow in direct connection with open water, such as willows and poplars. It is much more complicated in arid regions, since both the local and regional vegetations mainly comprise non-arboreal plants, and occasionally representatives of the same family can be found in both environments. Although frequently it is not the same genera which occur in the regional and local environments, distinction is very difficult due to problems of the possible levels of specific identifications.

As an outcome of all factors mentioned above, it would usually be advisable to subdivide the "recent" pollen spectra in arid lands into three principal groups: taxa truly representing the regional, arid land vegetation; palynomorphs representing the capabilities and provenances of transporting agents; and those produced by restricted occurrences of better developed local vegetations. Such subdivision would eventually involve overlappings, which seems inevitable under the arid lands circumstances, but however can give a good idea of processes responsible for the composition of pollen spectra and their relations with the environment. Occasionally, a fourth group may be necessary, comprising these palynomorphs which may occur in all three groups in approximately even shares (emphasis is here on "even"), thus having no particular significance.

Another eventual difficulty with such subdivision is with the total number of sporomorphs which have to be counted for an acceptable level of confidence. When each of the subgroups is individually treated, the grand total is meaningless, thus care should be taken to provide for sufficient number of grains that will make the subdivisions internally significant, regardless of the others. Eventually, it will be found that in some cases it is not necessary to provide enough palynomorphs for all such subgroups, but only for the one (or more) which serves for a certain particular study.

To sum up, it seems that it would be advisable to study subrecent sediments as reference to the past, rather than deposits which are indeed laid down at present. However, to obtain a reasonable understanding of the subrecent spectra, so that these could serve as a sound basis for interpretation, a study should also be made of pollen loads of the various relevant transporting agents. These should include winds of all different directions and seasons, groundwater, flowing water (also in all different seasons) and floods in wadis. Occasionally, when the case is appropriate, also contributions made by other agents such as various animals and man. Attention should be paid to processes involving redeposition in this context. All these would contribute to the better understanding of polyphase inclusion of palynomorphs in sediments, a process so typical for palynology of arid lands.

7.1.4. Relationships of exact sample location and recent pollen spectra

The various aspects discussed above which affect the composition of recent pollen spectra had also been active in the past, except for human intervention which diminishes quite rapidly as progressively older strata are dealt with. The influence of these processes changes from locality to locality, mainly depending on directions and energies of various transporting agents. Thus, even if the regional pollen rain is considered more or less uniform over large areas, which is a principle agreed upon by most investigators both in arid and fertile lands, the influence of transporting agents may considerably distort the basic composition of pollen spectra obtained from different localities within a given region.

As discussed above, such distortions are subordinate and can readily be dealt with in forested regions, in which the arboreal pollen spectra are taken as reference, but seem to pose considerable problems in arid regions, where most interpretations are based on the composition of the non-arboreal group, which is much more affected by these differentiation processes. When trying to interpret past environments in arid lands it is of prime importance to exactly define the location of sampling sites for recent spectra taken as reference, in relation to transporting agents. These transporting agents can occasionally bring to the sampled site considerable amounts of exotic palynomorphs, that can sometimes even entirely obscure the regional, arid land assemblages. Transportation could also cause marked preferences in the representation of local plants. Examples for such processes are indeed numerous and only some are brought here, to demonstrate possibilities and limitations.

The Jordan–Dead Sea Rift Valley which runs from north to south across Israel (Fig. 9.3.3) transects a steep environmental gradient, from Mediterranean domain to the north through extremely-arid terrain to the south, a distance of some 450 kilometers. This valley is subject nowadays almost continuously throughout the year to a daily northern wind, originating north of Israel, which sweeps across its environmental gradient. This wind carries along pollen grains originating from the Mediterranean vegetation to the north and deposits them on its way south. In addition, both eastern and western winds arrive at the Rift from higher regions in which vegetation is quite rich. Thus, although the Dead Sea is surrounded by arid to extremely-arid terrain, the pollen spectra obtained from its recent sediments (Rossignol, 1969a) do not considerably differ from those obtained from the Hula Valley to the north (Horowitz, 1979), lying in the middle of the Mediterranean region, with only subordinate representation of the local xerophytic flora (Table 2.2.5).

Just outside the path of this wind, only several kilometers to the west of the Dead Sea, the recent pollen spectra are entirely different, representing the local arid flora rather than Mediterranean vegetation to the north (Alon, 1976; Darmon, 1988, 1989). When the northern wind continues its way southward it sweeps over the 200 kilometers stretch of the Arava Valley and arrives at the Bay of Elat, where again pollen of Mediterranean origin are deposited and are quite common in recent pollen spectra of the sea bottom sediments (Horowitz, 1966).

Another factor seems to affect pollen deposition from the northern wind and cause considerable differences between various types of deposits. It seems that the windborne pollen are much better trapped in water bodies, as compared with terrestrial and wadi sediments, which are almost devoid of Mediterranean elements, although lying in the path of the northern wind. The lesser ability to trap pollen from the wind may also be due to differences in the periods of flood activity and deposition in the wadis, during the winter, as compared with the flowering season of most Mediterranean plants, in spring time. Thus wadis serve as pollen traps only during the rainy winter, while most blooming takes place in the desert. The muds deposited by wadis dry up very quickly, usually in a matter of days, thus during the rest of the year the wadi beds do not accumulate sediments and eventually pollen, but are rather subject to wind deflation. This seasonality may explain the differences observed in recent pollen spectra obtained from wadi deposits and water laid sediments, which only emphasizes the need to sample both localities before any interpretation is attempted.

This phenomenon, of open water acting as a better trap for airborne pollen, was also noted by Leroi-Gourhan (1982) in Syria, who found out that recent sediments collected from lakes and marshes yield pollen spectra representative of the regional vegetation, while those obtained from soils only reflect local conditions, in a rather restricted way.

Fluviatile transportation can also cause considerable differences in pollen spectra obtained from recent sediments laid down where the river enters a depositional basin, as compared with areas of this basin which are remote from the river's influence. This was observed in Lake Kinneret where, in its northern end the Jordan River enters and causes the pollen spectra to be much richer in grains originating from Mediterranean and bank vegetations, as compared with the south of the lake, into which no fluviatile contribution is made (Horowitz, 1969). It is also probable that some of the Mediterranean palynomorphs observed in recent sediments of the Dead Sea are brought to the basin by the Jordan River, for which this basin is the terminal base level. The richness in Mediterranean elements in the Bay of Elat, into which no rivers flow at all, seems however to indicate that wind transportation is the dominant agent by which these Mediterranean pollen grains are carried, while water transport is subordinate. An almost similar influence of a river inlet on pollen deposition is also seen in Lake Tchad (Maley, 1972), discussed in some detail in Chapter 2.2.2.

Differences in susceptibility of pollen grains to wind transportation can also cause distortions in the recent spectra in connection with the sample location. It can be seen that across Lake Kinneret (Fig. 7.4.1) AP/NAP ratios are higher at the center of the lake as compared with localities closer to its shores. This seems to result from combination of the rather uniform regional airborne pollen rain over the entire lake's area and contributions made by the non-arboreal shore vegetation. Pollen grains derived from the latter are not carried far away, since they are not lifted high enough by the winds to be mixed and become part of the regional airborne pollen, that rather originates from plants growing on the higher surrounding hills.

Recent pollen spectra obtained from rockshelters and caves are subject to at least two aspects concerning the exact location of sampling, which are the direction of opening and differential deposition within the enclosed space. The effects of the direction to which the cave (or rockshelter) opening faces, its relations to prevailing winds directions and local distribution of plant communities are shown by Weinstein (1981), discussed in detail in Chapter 2.2.1. Caves facing different directions in the same area, such as those situated on two opposite wadi slopes in a hilly terrain, may yield completely different recent pollen spectra, which has a direct bearing on interpretations made for the past. Thus a recent pollen spectrum for reference must be collected, as far as possible, under similar conditions in which a sequence was deposited, preferably from exactly the same site. Any implications drawn from a regional pollen spectrum taken as a basis for interpretation, attempted at a sequence collected from a cave or a rockshelter would usually be misleading. The worst recent reference for such a sequence may be the one collected from another, however nearby, cave whose opening may be facing a different direction. A clear prevalence of pollen derived from coastal plain vegetation can be seen in sediments from Tabun Cave, Mount Carmel, Israel, which is situated on a sea facing cliff (see Fig. 8.4.3 and discussion in Chapter 8.4).

Pollen deposition is not uniform within a cave, where spectra differ when collected from the center or closer to the cave's walls. This is especially apparent when pollen is brought to the cave by more than a single transporting agent, the two most common being winds and animals of various sorts and behavioral patterns (Davis, 1990). Davis suggests that the most representative would be pollen samples collected near the center of the cave's chamber, where eolian deposition is most rapid. The differences in rates of deposition in various parts of a cave may cause, accordingly, differential pollen sedimentation. It is evident that sequences which are studied should also be collected from the same locality, or at any rate comparisons of recent and fossil materials should be very carefully done, to make sure both had been collected in localities subject to similar processes of pollen deposition. Incidentally, this may be found quite difficult at times.

Consideration should also be given to occurrences of well developed local flora, such as those supported by springs, saline playas and salt-pans, or by human habitations, which are usually very restricted in area but maintain a lush vegetation as compared with the poor regional arid land flora. Such point occurrences may almost entirely mask the regional part of the spectra by their extensive pollen production, deposited very close to the vegetation stands. The main problem in arid lands is that most of the plants supported by different local conditions such as mentioned above belong to families whose plants are also an integral part of the regional vegetation. Since specific identifications are usually impossible, the differentiation between local and regional pollen is difficult, or occasionally impossible.

Two problems are involved with occurrences of rich local floras in arid lands, which should be considered while samples collected from or close to such sites, are taken as reference for interpreting past environments. The first concerns human habitations, which could be located in a certain place and deserted for causes which do not have to do anything with natural environments, such as wars or economic reasons. The second concerns hydrological conditions necessary to maintain any water supported vegetation, which may alter following structural, rather than environmental changes. Lakes in the Dead Sea region had shrank following tectonic movements with no change in climate (Horowitz, 1979), springs appeared in the Arava Valley following exposure of freshwater aquifers to the surface by faulting (Horowitz, 1987b) and so on.

To account for all above mentioned aspects, which may considerably affect recent pollen spectra taken for reference to the past, it is advised to sample for present day assemblages as many as possible different localities, that may give a better idea concerning the way by which the past should be regarded. In this way the basis for interpretation would be much wider and conclusions concerning paleoenvironments more sound. This suggestion is pertinent even when a single sequence is interpreted and was indeed carried out in many arid lands (see, for example, Davis, 1983/84; Schulz, 1990; and others).

7.2. Arboreal pollen as a basis for interpretation

Arboreal pollen found in spectra obtained from arid lands deposits can be of two main provenances, either produced by indigenous trees, or brought to the site of deposition by various transporting agents from other environments, that may quite frequently be far away from the analyzed location. Another group, which is only relevant to sediments of Holocene age, comprises arboreal pollen produced by trees that only occupy the dry regions following human activity. Each of these groups would naturally be indicative of different environmental factors, thus could be used for interpretation of its relevant sources. Whenever pollen from two or more groups are present, it may be found necessary to treat them separately. This is not always a straightforward procedure due to various degrees of overlapping of at least some of the components, which have to be safely established before interpretation is attempted.

7.2.1. Pollen of indigenous trees

There are two main subenvironments which could maintain indigenous trees in arid lands. The regional, which comprises elevated terrains, or the "local" that consists of such places where secondary enrichment of water is available. The first group would evidently better represent the regional environment, while the second is more dependent on local hydrologic regimes. These do have some connection with the regional environment, but the degree of connection may considerably depend on factors such as faulting, availability and paths of groundwater, occurrences of aquifer or aquiclude layers and so on. In some extreme cases there may be no connection at all between the regional and local environments, such as where a large river crosses an arid land.

It is however not always possible to distinguish between the two subenvironments, since many tree species may grow in both, thus such overlappings are hard to differentiate in the pollen spectra. Frequently the same species which occupies elevated terrains only grows to a bigger tree in wadis. A typical example is *Acacia*, but others do the same. In such cases it may occasionally be possible to consider the minor elements of pollen spectra, both of arboreal and non-arboreal origin, as a source of information that would clear the actual situation.

A third group of trees also occupies local niches in arid lands, those introduced by man or otherwise affected by his activity. This group is only important in very young sediments but should not be overlooked, since these are frequently analyzed for their pollen and also, since recent pollen spectra are used as a key to the past.

A group of arboreal pollen which could be considered either indigenous or representing an environment different than the arid comprises grains produced by trees growing on higher mountains within the arid region itself, where conditions are humid enough due to elevation, to even maintain forests. Such is the case in the Basin and Range country in Southwestern USA, where pine forests cover the higher places and their pollen found in great numbers (Martin, 1963) within the arid lowlands sediments. Opinions as to which attitude should be adopted vary, since on the one hand these trees grow within the geographical limits of the otherwise arid land, but on the other represent an altogether different environment.

It seems however that the intimate connections between these two different environments justify inclusion of such arboreal pollen in the indigenous group. It seems that any change in environmental conditions on the mountains, which would affect tree growth up there, would also be felt in some way or another in the arid lowlands. The trees may at times act as a more sensitive indicator to such changes in temperature and precipitation.

A. Regional indigenous trees

It is essential to differentiate between two principal types of arid regions in terms of their indigenous regional arboreal components. The first involves arid lands in which environmental changes had been subordinate during a long period, in geological terms, or which have had direct geographical connections with arid environments even when their climate ameliorated. These environments could, over a long time, serve as sources or reservoirs for xerophytic vegetation, which could move to the newly formed desert almost immediately following inception of a dry climate. The second comprises geologically young dry regions, a matter of no more than several thousand years, in which the environment was more humid not long ago, that havc not had a continuous source of available xerophytic stock.

The main difference between the two types is the degree of adaptation through which the plants (and eventually, also animals) had undergone. In places where the dry environment had persisted for a rather long period, at least several hundred thousand years, such as the Namib (which is serving as xerophytic stock for the neighbouring Kalahari for long periods of time) or Atacama deserts, adaptations had led to the development of trees or tree-like plants, from otherwise typical non-arboreal families, such as the Compositae, from which *Tarchonanthus camphoratus* (Fig. 7.2.1) had developed into a tree covering considerable areas in the drier parts of southern Africa, various Euphorbiaceae (Fig. 7.2.2) in the Old World, Liliaceae and Cactaceae in the Americas (Fig. 7.2.3). These plants had developed their adaptation, thus in addition to genuine trees typical for arid regions, such as acacias (Fig. 1.2.5) comprise the arboreal constituents of desert vegetations.

Evidently, one should not expect desert trees and tree-like plants to bear appearance similar to trees of forested regions (Fig. 7.2.4) but however, in terms of physiology, life cycle, ecology and occasionally even internal structure these are practically trees. The principal difference between the latter trees and those of the forested, especially the temperate, lies however with their pollination. Most forest trees are wind pollinated, thus serve as a prime source of environmental information, while the majority of xerophytic trees are either insect pollinated, or their pollen is not adequately represented in spectra obtained from arid lands, for various reasons.

Many of these trees or tree-like plants may be misidentified in the pollen spectra as non-arboreal components, especially in such cases where preservation is not best. As an example, *Tarchonanthus* can be identified to the genus (Scott, 1982) in well preserved pollen assemblages, but otherwise may be included within the rest of the Compositae, or may occasionally be mistaken for *Artemisia*, with whose pollen it bears some resemblance. The same holds true for pollen of other tree-like plants, such as those belonging to Cacatceae, Liliaceae or Euphorbiaceae.

A subgroup of deserts such as Southwestern USA or the Kalahari, had gone through periods of alternating dry and humid climates throughout the Quaternary (Jacobs et. al., 1985; Scott, 1989). But these had always maintained close geographic connections with neighboring arid environments, where xerophytic vegetation was continuously adapting to drought. This vegetation served as a stock, ready to migrate to whichever close environment that became dry enough. Thus, those dry lands which had shifted longitudinally following climatic changes maintain typical adapted xerophytic floras, although a given locality may not have been dry for any long period.

Semi-arid regions which maintained their environmental conditions for long enough periods, or had an available stock of adapted trees, are usually characterized by development of a bush, comprising low trees which may frequently achieve a considerable coverage, forming a very dense vegetation (Fig. 1.2.6).

In contrast with arid lands which had enjoyed a long history of drought or have

Fig. 7.2.1. *Tarchonanthus camphoratus*, a composite that had developed into a tree, Kalahari. Courtesy of L. Scott.

Fig. 7.2.2. Tree-like *Euphorbia damarana*, Euphorbiaceae, Namib Desert.

Fig. 7.2.3. Tree-like *Opuntia*, Cactaceae, Galapagos Island.

had available stocks of xerophytic plants, stands the group of relatively new dry regions, such as the Sahara (Williams and Faure, 1980) or the Negev (Horowitz, 1979). In these the flora comprises almost solely non-arboreal plants, faster to migrate as compared with trees. Such trees as acacias, which are well adapted to arid environments and their seeds easily spread, even from far away stocks, usually by birds, do grow also in some of the newly formed deserts.

On the other hand, these rather "young" arid lands frequently maintain trees which are relics of the former humid environment, usually in such specific refuge microenvironments as moister wadi beds, rain facing slopes, or the vicinity of

Fig. 7.2.4. Tree-like *Aloe pillansii* (Kokerboom), Liliaceae, Namib Desert.

springs and streams. Such occurrences are typical for both arid and semi-arid environments, less for extremely-arid.

Unfortunately the most typical arid lands trees — acacias — are hardly represented in the pollen spectra due to reasons unknown (maybe their relatively large size?). Truly, these trees are insect pollinated, but nevertheless produce considerable quantities of pollen grains liberated to the air. These grains, for some reason or other, only travel very short distances and have never been found to comprise any considerable shares of pollen spectra recovered from arid lands deposits, even when samples are collected from wadis along which acacias are very common. As an example, Van Zinderen Bakker and Müller (1987) reported only 2.3% *Acacia* pollen from a recent sample collected at Sossusvlei in the Namib Desert, a place in which these trees are indeed abundant (Fig. 3.3.2.1). The scarcity of pollen derived from *Acacia* and other xerophytic trees or tree-like plants and the frequent impossibility to identify grains of trees belonging to otherwise non-arboreal families, leave the scene almost exclusively to non-arboreal pollen. The latter are thus an indispensable source of information for interpreting regional environmental conditions in arid lands.

B. Trees of water enriched environments

The usual occurrences of trees in arid regions are restricted, or better developed in such biotopes where secondary enrichment of water, either fresh or somewhat saline is available. Enrichment can either occur around open water bodies such as springs, perennial streams, water filled salt-pans, marshes and lakes, or at places where groundwater tables are not too deep, such as oases, wadis and other low lying depressions. Some of the trees growing in such environments may be peculiar to this specific biotope, while others may occur elsewhere, but are frequently more abundant closer to localities with secondary enrichment of water. Trees characteristic for the first group are various types of palms, such as the date palm (*Phoenix*), while the other group is typified by thorn trees and closely related genera, *Eucalyptus* and many others. This second group thus inevitably involves considerable overlapping of regional and local elements.

Overlapping also exists in trees occupying both saline and freshwater maintained environments. Although some taxa would exclusively grow only in one of the two biotopes, others, such as *Tamarix*, may flourish in both saline environments and in wadis or close to springs. These overlappings may pose difficulties in assessing significance of arboreal pollen derived from trees growing in water enriched environments, especially those typical to both wadi beds and elevated terrains. This difficulty also results from the frequent impossibility to identify these pollen further than the genus or even family level. Some help may be sought for in the minor elements of the pollen spectra, both AP and NAP, which would frequently require further counting of the slides.

The source of problems involved with interpretation of hydrophil (in this term are included here all trees growing in freshwater enriched environments) arboreal pollen is that usually samples are collected in arid lands from the low lying localities. The reason for this preference is because these are very frequently the only places where sediments accumulate, but this is also where these trees always grow. The stress here is on "always", since occurrences of pollen produced by such trees in wadi sediments would not indicate whether the trees only grew in the wadi bed, or indeed covered the entire area, which makes quite a difference in reconstructions of past environments. The picture is further complicated by the fact that most of these trees are under-represented in pollen spectra, while within wadi sediments they may at times be over-represented, due to the short distance between the trees and locality of deposition. Such over-representation can evidently be used for defining the environment of deposition.

Another problem is that frequently trees considered typical hydrophil, such as *Tamarix* and *Populus*, may also grow outside water enriched localities (Fig. 1.2.10). On the other hand, characteristic regional trees such as cypress or even oaks may be attached, especially in semi-arid regions, to the vicinity of springs or other freshwater bodies.

The question whether to use hydrophil trees pollen as regional or local indicators could in the younger deposits, of late Pleistocene–Holocene age, be considered by comparison with the present day vegetation and recent pollen spectra, but such considerations become more and more complicated as progressively older deposits are dealt with. The degree of adaptation to arid conditions shown by various arboreal plant taxa in different regions is another factor which should be carefully studied. Differences in this degree of adaptation may cause some taxa to be better representatives of certain environments, such as *Salix* and *populus* indicating open water in Israel, thus inflicting on interpretation of the entire hydrophil group. Again, minor elements may turn to be of importance, thus extended countings can become indispensable.

C. Trees of anthropogenic environments

Arboreal pollen produced by trees growing in anthropogenic environments seem, at a first glance, to bear significance during only the last several thousand years. But indirectly they may prove quite important while any palynological study conducted in arid lands is based on recent pollen spectra, as reference for interpretation. The reason for this is that human settlements have a twofold possible influence on arboreal pollen composition and shares: a direct way, by which trees foreign to the dry region are introduced and cultivated, whose pollen could be readily identified as such; and an indirect way, by which changes caused by human activities result in expansion or reduction of the natural arboreal components. The first type of influence is only significant during habitation times, but the second may last until the present day, occasionally thousands of years after the settlements had been deserted, changing the percentages of the seemingly natural AP components in the recent pollen spectra. The two ways of anthropogenic influence on the arboreal pollen spectra depend chiefly on methods used for agriculture, but also on the necessity for building materials and for shade.

When discussing anthropogenic influence on trees in arid lands, initial differentiation should be made between periods during which a settlement is inhabited and other times when it is deserted. Pollen spectra recovered from layers deposited concurrently with a habitation period, either in the site itself or its close vicinity, especially downstream or downwind, may yield arboreal components produced by cultivated and introduced trees. The pollen of introduced trees are in most cases easy to identify, since they do not naturally grow in the region. Some trees may however had been introduced from close environments and could possibly grow in former days in the nowadays arid land, if it had enjoyed somewhat different climatic conditions at the times of settlement. Another problem may be encountered with trees that had been introduced such a long time ago, that nobody knows now whether they are introduced or native to the region. Examples for such trees in Israel include pines, cypress, olive, fig tree and pomegranates. Truly, pollen of the first three had been encountered in older than habitation deposits (the latter two hardly represented by pollen anyway), but the impossibility to positively identify species in less than well preserved grains makes the issue questionable.

To check for such trees it would be necessary to compare pollen spectra from the site with those obtained from synchronous sediments collected upstream or upwind. Upstream sediments are usually quite easy to obtain, however their age must be safely established; upwind deposits are more problematic, since if there is suspicion that a different climate existed in the period of settlement, it may have had also affected wind directions, so that discretion should be applied in such cases. If settlements are not too numerous it would probably be best to compare the spectra in question with those obtained from a locality separated from the site by some topographic barrier, such as a parallel wadi.

Cultivated trees could also be of two provenances. They may have either been introduced from other regions, such as almonds grown in archeologic sites in the Negev, Israel, or bred from stock available in the arid land itself, like date palms. In both cases, problems of discrimination and definition of anthropogenic influence are similar to those discussed above for the introduced trees. Pollen of such cultivated and introduced trees are easily carried from the site by winds and floods, deposited downwind or downstream, thus may be incorporated in synchronous sediments found at quite a distance from the archeologic remains.

Depending on methods used for agriculture, the human influence could also be extended to the natural arboreal vegetation long after the site itself was abandoned. The introduced and cultivated trees would usually not grow much longer after human care is stopped, when the settlement is deserted, since they are not too well adapted to drought. Local trees, such as date palms may persist much longer. Hydrophil wadi trees can take the place of the cultivated and introduced ones, especially where dams and artificial terraces had been constructed across wadis, a method quite frequently exercised in arid lands agriculture throughout most archeological periods (Fig. 7.2.5). Such dams and terraces can hold water for long periods and serve as fine grain sediments traps, which also include pollen. These thus form a preferable biotope for hydrophil trees, which would grow better in such places as long as the dams or terraces are not ruined. Such constructions are known from Israel, dating up to more than 3,000 years ago, still holding richer vegetation at the present day. Thus an increase in the wadi vegetation, which could affect

Fig. 7.2.5. Nabatean dam (100 BC–100 AD) holding better developed vegetation even at the present day. Central Negev, Israel.

recent pollen spectra, may be an outcome of agriculture which had long ago been deserted, rather than a consequence of natural conditions.

Another facet of anthropogenic activity which could considerably affect arboreal pollen spectra is the collection of wood for fire and building. Such activity could help desertification and reduce arboreal pollen shares. It is quite difficult to assess the time needed for regeneration of the natural vegetation after desertion of a site, since it depends on adaptation of the trees, their rates of reproduction and the environmental conditions.

7.2.2. Imported arboreal pollen

Imported arboreal pollen comprise the part of spectra obtained from arid lands deposits, which was produced by trees growing in a different, more humid, usually remote environment. This straightforward definition is self evident when recent sediments are studied and their pollen spectra can be compared with the present day distribution of vegetation. The only problem here seems to define the use of the term "remote". Clearly, when arboreal pollen are carried over distances of several hundred or more kilometers, this would satisfy almost everybody as remote enough. Second thoughts occur frequently when distances are shorter, as is the case when the boundary of an arid region and its fertile neighboring land is quite sharp. Such cases could happen if mountains enjoying higher rainfall are situated within the arid land, discussed in Chapter 7.2.1, or when the boundary is orographic, as happens between the Mediterranean and desert domains in some parts of Israel.

However, when dealing with fossil pollen assemblages the use of the term "imported" and the grouping of imported arboreal pollen as such may need proof, which could occasionally be quite difficult to obtain. More often than not such proof would be interpreted from the non-arboreal part of the spectra or, if these do occur, from pollen of indigenous trees. Usually proof is obtained after excluding the arboreal pollen for which suspicion is raised as being transported from another provenance. If the rest of the spectra indicate an arid environment, of a type which cannot maintain such trees that had produced the suspected grains, this can be used as a guide.

Quite frequently the necessary proof is hard to achieve due to possible ambiguity of pollen comprising the rest of the spectra. An example of such ambiguity is demonstrated in El Moslimany (1990) who analyzed recent pollen spectra across the Arabian Peninsula, presented in Tables 7.3.1 and 7.3.2. Compositae pollen are dominant in the arid environments, but increase their shares again when the transect gets to the mesic biotopes, where arboreal pollen naturally occur. In this case, if the arboreal pollen are excluded for suspicion of being imported (this would never happen in recent spectra where the situation is clear, but may occur with older, where vegetation is unknown and sequences not continuous), the rest of the spectrum would seemingly indicate an arid environment, by the prevalence of Compositae. The only possible solution to the problem may occasionally be found with the minor non-arboreal elements of the pollen spectra, which could be used to define the environment and help decide whether the arboreal part comes from indigenous trees, or had been imported. The other solution, of trying to identify the Compositae further than the family level, is usually impractical in fossil spectra, except if they are indeed very well preserved, which unfortunately is rare.

Another possibility of uncertainty exists when no arboreal pollen occur in recent sediments of a certain region, as is the case in the central Negev, Israel. In the same region, older strata recovered from Chalcolithic sites dating some 5,000–5,500 years ago yielded several percents of oak pollen (Horowitz, 1979). It is very difficult to decide whether oaks indeed grew close to the sites during Chalcolithic settlement, or their pollen brought by winds from the north, where these trees presently grow. However, the oak pollen can serve as indirect indicators to some difference in climatic conditions between the Chalcolithic and the present day. This is based on the assumption that even if oaks had not grown there at that time, presuming their territory was similar to the present day, thus their pollen only brought to the site by winds, this indicates an increase in wind intensity. Such increase would naturally bring more rains to the Chalcolithic sites, since precipitation is dependent on northwestern winds in this region. The possibility that the southern limit of oaks was, in Chalcolithic times, further south than it is today, cannot be ruled out.

It is even plausible that all three possibilities are true, namely that some oaks grew close to the sites, the northern winds had been stronger and oak stands indeed expanded further south. It is impossible to decide at this stage which possibility, or combination, is more plausible and whether the indication of oak grains is direct or indirect regarding their actual growth locality. At any rate this hints at some climatic change from the Chalcolithic to the present day, although it remains uncertain whether this difference could actually bring oaks to grow in the nowadays arid environment.

Sediments correlative with the Chalcolithic site of the Negev mentioned above, in which some oak pollen were found, contain in northern Israel considerable amounts of pollen produced by winter deciduous oaks, mainly *Quercus ithaburensis*, which outnumber those produced by evergreen oaks, *Q. calliprinos* (Horowitz and Gat, 1984; Baruch, 1986). This is in contrast with the present day situation, when pollen of winter deciduous oaks are subordinate in comparison with evergreen in the recent spectra. Based on both the prevalence of pollen of winter deciduous oaks and on isotopic composition of water dated to the same age, Horowitz and Gat concluded that during Chalcolithic times at least some of the rain fell in the Levant in summer time. Since such summer rains could only have arrived in Israel during periods when the summer high barometric pressure over the Sahara was not too well developed, it would be sensible to conclude that the environment must have also changed in the desert regions of Israel in some way. Thus, while pollen spectra from the arid environment have only shown a slight difference from the present day, by the occurrence of minor amounts of oak grains, the analysis of arboreal composition in the neighboring Mediterranean region yielded much more conclusive results.

The central Negev is quite far from the main path of northern winds in Israel, which are more active along the Jordan–Dead Sea Rift Valley. Indeed, the arboreal portion of spectra recovered from the southern Dead Sea for Chalcolithic times display the prevalence of winter deciduous oaks, by their pollen which had been carried there by the winds (Horowitz, 1987a). The difference between pollen spectra obtained from Chalcolithic layers in the central Negev, where AP is subordinate and those recovered from the Dead Sea, where AP is well represented and clearly displays the internal composition of this group, emphasizes the importance of the transporting agents. It is thus possible, by analyzing concurrent pollen spectra, to interpret also, at least to some extent, transporting agents in a given period and their relative activity.

Changes in the total shares of imported arboreal pollen in spectra obtained from arid lands may frequently indicate some kind of environmental change. This may have happened as expansion or reduction of the forest territory at its provenance area, change in wind regime, or actual occurrence of trees on the site. On the other hand, in some cases such changes proved to be only local, depending on hydrologic factors which could cause spreading or disappearance of some hydrophil or halophil non-arboreals, thus affecting AP percentages, if calculated from the total. In such cases the change in AP percentage would only be apparent, not real. These considerations are further discussed below, in Chapter 7.4.

Once a local origin for the arboreal pollen is excluded, they can serve as a good basis for interpretations pertaining to both environmental conditions in their provenance area and the potential of transporting agents.

The internal composition of imported arboreal pollen spectra, when their total number is the basis for calculations, could serve for interpreting environmental conditions of their provenance region. If the provenance region is too far away, for example tropical forests which contribute *Podocarpus* and Combretaceae pollen to offshore sediments in Israel, it would be quite risky to interpret any connections between the tropical forest composition and arid environments of the northern Sahara, thus indirect implications are very hard to prove in such cases. Moreover, it should be remembered that if arboreal pollen arrive from considerably distant source, transportation could be differential, in which case changes in the internal composition of the arboreal pollen group may have only resulted from variations in transport, not in provenance environments.

On the other hand, arboreal pollen derived from a relatively close provenance, such as Mediterranean and arid environments in Israel, or tropical and subtropical regions and the southern Sahara, the internal composition of arboreal spectra may yield very valuable, although indirect information on conditions of both the humid provenances and the neighboring arid regions. Pollen spectra from arid lands may occasionally be of lesser sensitivity to minor environmental changes due to difficulties involved with specific identifications, thus their resolution may be inferior to arboreal pollen from a neighboring region. As an example, the humid/arid boundary oscillates in Israel following minor climatic changes, with no apparent influence on the composition of the xerophytic pollen flora, but a drier phase causes prevalence of pine over oak in the Mediterranean flora, which is readily recorded in the pollen spectra obtained from the arid region, into which the AP are brought by the northern winds (Horowitz, 1987a).

Such pine/oak pollen relations are apparently further emphasized by their different susceptibility to wind transportation. Pine pollen are better carried by the wind over longer distances. This could be proved when analyzing recent sediments from the Bay of Elat, where pine and cedar pollen predominate over oak, both growing some 300–400 kilometers to the north of the locality of deposition (Horowitz, 1966); this is in contrast to these ratios in the Hula basin, which is situated within the Mediterranean domain (Fig. 9.3.3), where oak grains considerably dominate those of conifers (Horowitz, 1979; Weinstein-Evron, 1983).

Pine/oak pollen ratios seem to stress environmental changes by their different susceptibility to wind transportation. When climate becomes drier and the

Mediterranean environment boundary retreats, pine would spread better on the hills (Zohary, 1959), being better adapted to drought than oak. This spreading is accentuated by the differential wind transportation in samples collected from arid regions, where pine is anyway over-represented in comparison to oak. The picture would naturally be reversed when humid conditions are set. It goes without saying that the specific climatic and palynological relations between the neighboring humid and arid environments should be individually established for any particular region.

Pollen spectra obtained for the entire Quaternary sequence in Israel (Horowitz, 1987a; Levin and Horowitz, 1987) show quite clearly that arboreal pollen imported from the Mediterranean environments, southward to the Dead Sea, comprise different relations of components when comparison is made between periods corresponding to the Alpine glacials (even oxygen isotope stages) and those correlative with the interglacials (odd oxygen isotope stages). The glacials are typified by pollen derived from winter deciduous oaks, while interglacials are characterized by prevalence of evergreen oaks and pine. Winter deciduous oaks are more dependent on availability of water as compared with evergreen and pines, which are better adapted to drought. Thus the composition of imported arboreal pollen spectra is indicative of environmental conditions in the Mediterranean region of the country, although the analyzes have been carried out on sediments laid down in an extremely-arid desert region. It should be noted that the differences between humid and dry periods are further accentuated in the arid land basin due to preferential transportation.

If more than a single provenance is defined for the imported AP, both environmental conditions and relative potentials of transporting agents can be interpreted for the different source regions. Such situations of multiple provenances are reported from various parts of the Sahara (Beucher, 1975; Van Campo, M., 1975; Maley, 1981; Hooghiemstra et al., 1986; and see also Tables 2.2.4 and 2.2.7).

The relations of arboreal pollen groups from different provenances can be used for defining past wind transportation (Hooghiemstra et al., 1987), or to follow migrations of certain vegetation belts, those for which interpretations based on AP are more sensitive than when relying on non-arboreal spectra (Hooghiemstra and Agwu, 1986, 1988; Hooghiemstra, 1988b).

7.3. Non-arboreal pollen as a basis for interpretation

Non-arboreal pollen comprise the most common group in spectra recovered from arid lands deposits, usually making up the majority of the total number of grains in samples. This is common to arid lands all around the world, except for such localities where considerable relief can maintain trees on the higher elevations, whose pollen may be over-represented in sediments accumulated in the low lying basins, as is the case in the Basin and Range country of Southwestern USA (Martin, 1963; Mehringer, 1967; and many others); or in places where large quantities of allochthonous palynomorphs are supplied by potent enough transporting agents from far away provenances, in which trees are common, as is the case along the Dead Sea Rift in Israel (Horowitz, 1979; and others).

At any rate, since the majority of pollen contributing plants growing in arid lands is non-arboreal, while the few arboreal taxa do not contribute much to the pollen spectra, it is the analysis of this group which can provide information concerning environmental conditions in these regions. In contrast, arboreal pollen recovered from arid lands deposits can yield valuable information about topography, transportation processes and the AP provenance environments, data which could strengthen conclusions drawn from the non-arboreal group.

There are three principal approaches to the treatment of non-arboreal pollen recovered from arid lands sediments, both recent and fossil. The first deals with the entire pollen spectrum as a basis for calculations and interpretations (Mehringer, 1967; D'Antoni and Markgraf, 1977; Van Zinderen Bakker, 1984; Brun, 1985; Boyd, 1990; El Moslimany, 1990; Singh et al., 1990; and many others). A second group of researchers (Cour and Duzer, 1980; Horowitz and Horowitz, 1985; and others) attempts at distinguishing between pollen produced by two groups of plants: these that cover the entire dry land area, such as Compositae in certain regions, which represent regional environmental conditions; or allochthonous palynomorphs, together with grains produced by plants growing in restricted biotopes such as wadi beds, where secondary enrichment of water controls growth, as the Chenopodiaceae. Hydrological conditions, such as groundwater levels in wadis, are not singularly dependent on regional regime. They also result from such local characteristics as the depth of an aquiclude layer below the wadi gravel, or water supply to wadis which can come from rains in far away regions. These cause floods, supplying water that can be stored in the gravel for a long period. Local hydrology can depend on springs, whose aquifers may have a distant intake area.

Since the local conditions in wadis, springs or salt-pans and playas do not necessarily or directly reflect the regional environment, it seems preferable to deal separately with pollen produced by plants growing in either of these two principal biotopes of the arid lands. Needless to mention that there is some overlapping of taxa from the two environments, especially since specific identifications are impossible, which enables only a general nature of interpretation. Such generalization is unavoidable however with any treatment of non-arboreal pollen, regardless of whether the entire spectrum is dealt with, or separated into subgroups. It seems that there is no objective way for eliminating such overlappings.

The third approach (Van Zinderen Bakker, 1957; Coetzee, 1967; Cour and Duzer, 1978; Horowitz et al., 1978; El Moslimany, 1990; and others) experiments with internal numeric relations between selected non-arboreal pollen groups as possible environmental and paleoenvironmental parameters.

7.3.1. The entire non-arboreal spectrum

The first approach toward interpreting the non-arboreal pollen as representative of climatic and environmental conditions in arid lands takes into account all taxa occurring in the spectra. Since the plant groups represented by non-arboreal pollen occupy different habitats in the dry region, they are differentially laid down in various depositional environments within the area. Their significance depends to a great extent on the exact sampling locality for any particular palynological study. Thus samples collected from higher terrains would be chiefly representative of the regional pollen flora, while those collected from lowlands, in which water is more abundant, such as wadis, salt-pans, etc., could in most cases be dominated by pollen produced by locally better developed plant associations, which may occasionally almost entirely mask the regional picture. Since most analyzed sequences are usually collected from the latter type of localities, in which the occurrence of water and vegetation in a low lying place helps trap the sediments and pollen grains, such sections would display quite considerable shares of local pollen in the obtained spectra.

This does not seem to pose particular problems when only a single sequence is analyzed, especially when local conditions may have more meaning than the regional. This is usually the case with studies geared toward archeology, where the immediate environment controls possibilities of life and agriculture far more than regional conditions. However, such sections in which the local pollen dominate may turn quite problematic when correlations are attempted between several sites located within different microenvironments, which may differ in water salinity, groundwater table and so on. This problem can readily be seen in studies such as by D'Antoni and Markgraf (1977), who analyzed several sequences in the eastern arid part of Argentina, which are clearly different from one another in their non-arboreal pollen assemblages, although correlation was attempted and seems possible.

The problem of intercorrelating sequences is not too grave in late Quaternary sediments, where much help could be sought in radiocarbon datings, which seem to solve difficulties with most analyzed archeological sites. However, with older deposits such correlations are very difficult to perform wherever local elements are predominant. Late Quaternary correlations had been locally suggested by many investigators in various desert regions around the world, such as the Sahara (Brun, 1985; Lézine, 1989), India (Vishnu-Mittre, 1976; Singh et al., 1990), Australia (Boyd, 1990), The Arabian Peninsula (El Moslimany, 1990) and others.

The main problem discussed by many students of arid lands concerns the question, which pollen taxa or associations are characteristic of a certain dry environment and to what extent could be used as relatively "safe" indicators. In other words, when turning to interpret past pollen sequences, increase or decrease in the share of which pollen type or assemblage would be indicative of changes in the degree of aridity, or in which direction these changes go. Such characterization would naturally differ from one region to another, depending on the composition of respective plant associations.

Surprisingly, the number and variability of pollen taxa involved with the definitions of extremely-arid, arid and semi-arid environments is rather limited even on a global scale. This results from a combination of two aspects pertaining to arid lands palynology: the impossibility to identify species or frequently genera in pollen produced by most plant families occupying such environments; and the scarcity of pollen produced by many typical desert plants in the sediments, such as acacias, Cactaceae and Euphorbiaceae, where they only occur sporadically and cannot thus be safely used for interpretation. A third possible aspect may concern the global development of deserts, in late Cenozoic times, when most plants which adapted to dry conditions had developed. These adaptations take a rather long time, which was probably only available during the Miocene, when climatic oscillations had a much lower frequency as compared with following periods (Horowitz, 1990). Thus, the adapted xerophytes had time enough for an almost global spreading, most probably aided by migratory birds, during the last 12–15 million years, which may explain the great worldwide similarity in dry lands vegetations.

Consequently, the palynological scene is left to a very limited number of taxa, which in most cases include Gramineae (Poaceae), Cyperaceae, Chenopodiaceae, Compositae, *Artemisia*, *Ephedra*, *Calligonum* and not too many other non-arboreal types, the latter usually only occurring in limited quantities. this worldwide similarity may facilitate pollen based global correlations of past arid environments, provided possible migrations of paleoclimatic belts are taken into consideration. A discussion of problems involved with this way of correlation is brought in Chapter 11.1.

The characteristics of substrate have a marked effect on vegetation communities which can develop in arid and especially in extremely-arid environments. The two major regional types (besides local, restricted occurrences in localities richer in water) are rocky surfaces and sand dunes. Numerous crevices in rocks may hold primordial soil-like deposits or accumulate eolian dust, forming a suitable substrate for plants, protected from wind deflation. Such crevices may also maintain some humidity, which is protected from evaporation by the surrounding rock and facilitates plant growth (Danin, 1972). Sand dunes, especially abundant in extremely-arid regions, are particularly hard on plants, since sand rarely maintains any humidity, thus plants growing in this habitat require specializations which are quite rare, shared by only a limited number of taxa such as *Cyperus conglomeratus* or some species of *Calligonum* (El Moslimany, 1990). The main reason for the limited number of substrate types in arid and especially in extremely-arid lands, is that soil formation processes are not active in these regions due to lack of water. Thus, the landscape is subject to either wind deflation which exposes rocky surfaces, or eolian deposition usually limited to sand dunes, while finer particles are carried away to the semi-arid and more fertile regions, where they are finally laid down.

Rocky substrates in extremely-arid areas are usually typified by Gramineae (Fig. 1.2.7), which seem to grow all over the region following occasional rare rains. There is hardly any growth in these areas in periods when rain does not fall, which may even take years. Thus pollen spectra obtained from the higher terrains in extremely-arid regions in the Middle East and southern Africa are characterized by an almost total prevalence of Gramineae pollen (Horowitz, 1977a; Van Zinderen Bakker and Müller, 1987), taking 90–98% of the total number of the grains in the spectra.

Sand dunes of the Arabian Peninsula are typified by an association of *Cyperus conglomeratus* and *Calligonum* pollen (El Moslimany, 1990). The same, or possibly some other species of Cyperaceae is typical for such environments in India (Vishnu-Mittre, 1976), while *Calligonum* is also frequent in the extremely-arid Sahara (Cour and Duzer, 1980) and southern Negev.

The arid environments of Kuwait and southern Iraq, with a rainfall of 50–150 mm/year are dominated, according to El Moslimany (1990) by pollen of Chenopodiaceae, *Plantago*, Compositae and minor shares of various insect pollinated herbaceous species. Arid environments in many localities around the globe seem to be characterized by a considerable share of Compositae, Gramineae and Chenopodiaceae in the pollen spectra. This was pointed out in Southwestern United States, (Meyer, 1977), in the Near East (Leroi-Gourhan, 1982; Horowitz and Horowitz, 1985), South America (D'Antoni and Markgraf, 1977), the Sahara (Cour and Duzer, 1978), India (Vishnu-Mittre, 1976), eastern and southern Africa (Van Zinderen Bakker, 1957; Coetzee, 1967; Horowitz et al., 1978), the Arabian Peninsula (El Moslimany, 1982, 1990) and many other localities. The characteristic pollen assemblages for an environmental gradient across the Arabian Peninsula, from the extremely-arid Rub' al-Khali to the south through the mesic Zagros to the north, are shown in Table 7.3.1.

Both cold and warm semi-arid regions are very frequently characterized, almost all around the world by considerable occurrences of *Artemisia* pollen (Van Zeist, 1967; Suc, 1978; Brun, 1985; Cheddadi and Van Campo, E., 1985; Horowitz and Horowitz, 1985; El Moslimany, 1990; and many others). An interesting problem concerned with high occurrences of *Artemisia* pollen, which results from inability to identify its grains further than the genus level, is whether such abundance represents a cold or warm steppe, both typified by the prevalence of this taxon. It is therefore only by occurrences of minor elements that distinction between cold and warm *Artemisia* steppes can be made. Suc (1978), while analyzing a late Quaternary sequence in southern France, had observed that an *Artemisia* rich horizon was also accompanied by however small percentages of *Phlomis* pollen, indicating that this horizon indeed represents a warm steppe. The main problem with such minor elements in pollen spectra rich in *Artemisia* is that usually the latter is so abundant that extended countings of the spectrum in question are frequently necessary, to

TABLE 7.3.1

Recent pollen percentages in assemblages collected across the Arabian Peninsula (total counted = 100%)

Environment	Extremely-arid	Arid	Semi- arid	Submesic	Mesic
Locality	Rub' al-Khali	Kuwait–S Iraq E Arabia An Nafud	Iranian Plateau Zagros and Syrian steppes	Syrian forest- steppe	Zagros forest
Number of samples	8	40	26	12	43
Pollen of					
Cyperus congl. + *Calligonum*	56	6	–	–	–
Chenopodiaceae	4	38	31	15	6
Compositae + *Plantago*	–	31	18	17	27
Artemisia	–	2	17	17	10
Gramineae	–	6	12	18	13
Arboreal pollen	–	2	5	21	18
Others	40	17	17	12	26
Chenopodiaceae/ *Artemisia* ratio	n.a.	19.0	2.6	0.8	1.1

Reprinted with permission from El Moslimany (1990).

obtain at least some representation of minor components. Notably though, is that such countings of minor elements do not have to be statistically significant and their occurrence alone would usually be a good enough indicator. Their absence, clearly, would not make any contribution to the interpretation.

Markgraf (1985) had also shown that accompanying elements could be used to distinguish between a cold and a warmer, arid steppe in southern Patagonia. Another indicator used by Markgraf is the species diversity, which is limited in samples interpreted as representing a cold steppe, but increasing in layers concluded to have been deposited under a warmer, though drier environment.

7.3.2. Grouping and partial non-arboreal spectra

The second approach to non-arboreal spectra involves their subdivision into groups of pollen produced by plants growing in specific environments, that show similarities in their response to ecological constraints. This approach is based on the concept that pollen grains produced by plants which have rather ambiguous significance should be excluded from any discussion concerning the general, regional environmental conditions. The significance of such ambiguous grains should only be reserved to such environments which they undoubtedly characterize, usually of a local rather than regional nature. In most cases the excluded palynomorphs comprise those produced by hydrophil and halophil vegetations, which are considerably more abundant in such environments as wadis, springs, salt-pans and so on, not considered as representing regional conditions.

As with any kind of such grouping, especially with non-arboreal pollen, that cannot be identified further than the family or occasionally genus level, overlappings of the included and excluded taxa evidently exist and should be considered while interpretations are attempted. Typical hydrophil vegetations are dominated by pollen of Gramineae (Poaceae), Cyperaceae and *Typha*, while the halophil mainly comprise Chenopodiaceae. It seems that such general grouping may be relevant to almost all arid lands around the globe. Exceptions to this general grouping must be especially made in extremely-arid regions, where pollen spectra may be dominated by Cyperaceae, indicating psammophilous vegetation (El Moslimany, 1990), or by Gramineae (Van Zinderen Bakker and Muller, 1987), growing on a rocky substrate.

When pollen spectra collected from sequences in an arid land are dominated by either Cyperaceae or Gramineae, with hardly any other component occurring in significant percentages, the problem is whether to interpret the paleoenvironment as hydrophil, possibly of a restricted local nature, or as a regional, extremely-arid domain. The answer may frequently be found with occurrences of minor elements of the pollen spectra, which may indicate either way. Freshwater occurrences are almost always accompanied by the growth of plants such as *Typha*, which never appears on elevated terrains in arid lands. Thus *Typha* pollen observed in an otherwise Gramineae or Cyperaceae dominated spectrum would indicate that the latter had most probably originated from a hydrophil vegetation; in contrast, pollen derived from plants which may accompany Gramineae or Cyperaceae on the higher terrains, such as *Calligonum*, could be used to indicate an extremely-arid regional environment.

The need to rely on minor components for defining more precisely the paleoenvironments, would naturally involve counting or searching the slides until such are found. When minor elements are missing for any reason, the more common being a poor state of preservation, interpretation could still be made, but with a lower degree of safety. Interpretation in such cases could be based on the variety in number of taxa present in a given pollen spectrum. Usually, the lesser the variety the better chances are that the spectra represent an extremely-arid environment. Such interpretation should only be made when other spectra are available for comparison, since the species diversity concept is based on relative numbers, rather than absolute (see also Chapter 11).

Horowitz and Horowitz (1985) attempted grouping of the non-arboreal pollen in order to systematically follow paleoenvironments throughout the late Cenozoic of Israel, in sequences obtained from boreholes. The pollen spectra have been primarily divided into regional and local elements. Of the regional, two groups pertaining to the present discussion are the taxa representing arid environments and those signifying semi-arid steppes. The local elements either represent hydrophil or halophil vegetations, which also include typical wadi vegetation. It is believed that such subdivision is useful to differentiate between the regional and local environments, the latter signifying hydrologic developments.

Arid environments are indicated by prevalence of pollen produced by Compositae, with some Umbelliferae and occurrences of *Plantago* and *Asphodelus* and rarely some other minor elements; the semi-arid steppe vegetation is chiefly represented by pollen of *Artemisia* and *Ephedra*, occasionally accompanied by *Scabiosa*, *Echinops*, *Astragalus*, *Centaurea*, Cruciferae, Papilionaceae, Liliaceae, Malvaceae and other minor constituents.

Freshwater supported environments are typified by dominant Gramineae with some *Typha* and occurrences of *Rubus*, *Polygonum*, Labiatae and others. To these had been also added several arboreal pollen produced by trees growing in these biotopes, such as *Tamarix*, *Acacia*, *Salvadora*, *Phoenix* and a few others. Saline conditions are characterized almost solely by pollen of Chenopodiaceae.

The use of this subdivision had enabled correlations of sequences collected from various localities along a stretch of some 250 kilometers, from the north of the Jordan Valley down to the Dead Sea and south of it, in locally different environments. These correlations could not have been done if pollen produced by hydrophil and halophil vegetations were also taken into account; indeed, attempts at correlating spectra based on the total number of counted grains simply failed, due to pronounced dominance of local elements, which masked the regional picture (Fig. 7.4.4). For more details, see Chapter 9.

To compare this way of treatment, the table based on totals (El Moslimany, 1990) shown above (Table 7.3.1) is also recalculated according to the suggestions made in Horowitz and Horowitz (1985). The results (Table 7.3.2) clearly show the prevalence, more than 88%, of Compositae and *Plantago* in more than 40 recent samples collected from the arid regions of the Arabian Peninsula, as compared with approximately equal shares of these arid elements and *Artemisia* in 26 samples from the semi-arid regions, where arboreal pollen are somewhat more abundant as compared with the arid. These relations of arid and semi-arid components are also maintained in samples collected from submesic environments, but there the arboreal pollen shares are considerably higher.

It is quite interesting to follow this transect into the mesic environment, where AP maintain approximately similar values to the submesic, *Artemisia* drops, but Compositae and *Plantago* rise again. This is due to development of the forest undergrowth, which maintains many Compositae in this region. Since specific identifications are impossible, the differentiation between an arid and a mesic environment would have to be based on occurrences of arboreal pollen and accompanying minor elements. These minor constituents may turn very important in localities where considerable amounts of arboreal pollen are brought by winds or water to a depositional basin situated in an arid land. Some help may be sought in the ratios of Chenopodiaceae to *Artemisia*, discussed below.

TABLE 7.3.2

Relations of arid, semi-arid and arboreal components in recent pollen assemblages collected across the Arabian Peninsula (total counted = 100%)

Environment	Arid	Semi-arid	Submesic	Mesic
Locality	Kuwait–S Iraq E Arabia An Nafud	Iranian Plateau Zagros and Syrian steppes	Syrian forest-steppe	Zagros forest
Number of samples	40	26	12	43
Pollen of				
Compositae + *Plantago*	88.6	45	30.9	49.1
Artemisia	5.7	42.5	30.9	18.2
Arboreal pollen	5.7	12.5	38.2	32.7

Recalculated and reprinted with permission from El Moslimany (1990).

7.3.3. Ratios of pairs of taxa

A third approach for interpreting paleoenvironments was attempted by several investigators, by calculating the ratios of pairs of non-arboreal pollen taxa. Cour and Duzer (1978) point the applicability of such calculations, which they have tried with all kinds of combinations, in a transect made across the Sahara, analyzing recent sediments.

El Moslimany (1990) presents Chenopodiaceae to *Artemisia* ratios (Table 7.3.1) for a transect made from the southern, extremely-arid part of the Arabian Peninsula up north to the mesic Zagros Mountains in Iran. She shows quite conclusively that these ratios progressively drop with increasing humidity. It is however noteworthy that this ratio is not pertinent to the extremely-arid Rub' al-Khali desert, where these plants hardly occur. Pollen spectra obtained from other extremely-arid regions in Israel (Horowitz, 1977a) or in the Namib Desert (Van Zinderen Bakker and Müller, 1987) are entirely dominated by Gramineae and practically devoid of either *Artemisia* or Chenopodiaceae. It is therefore quite clear that the ratios suggested by El Moslimany could only be used for arid through mesic environments. These calculations should be tried when arboreal pollen are transported into an arid region (see above), but cannot be used to differentiate between arid and mesic conditions, in such cases where Compositae are common in the spectra.

Van Zinderen Bakker (1957), Coetzee (1967) and Horowitz et al. (1978) had tried the ratios of Compositae to Gramineae as a possible indicator for aridity in southern Africa. They suggested that the higher this ratio is, the more arid the paleoenvironment seems to have been.

It is quite interesting to follow the conceptual rationale for calculations of ratios of selected pairs of taxa. It is evident that environmentally significant ratios indeed follow similar principles as discussed above for the subdivision into ecologically oriented groups. Both more useful ratios, Chenopodiaceae to *Artemisia* and Compositae to Gramineae, have in common the comparison of a regional element (Compositae, *Artemisia*) to a local one. Since environmental changes are usually better felt in the more sensitive regional flora, which does not have the backup of groundwater, while the local vegetations are more stable, the ratios indeed represent regional conditions. The local flora is therefore only used as a rather constant reference, while the true picture is obtained from the regional.

7.3.4. Pollen morphology

An altogether different approach to the environmental significance of some non-arboreal pollen grains is presented by several French researchers, who indicate that the grain's morphology, or in some cases exine ultrastructure in certain taxa, are different when the plants grow in arid environments, as compared to close relatives, of the same genera, which enjoyed a more humid biotope. Thus Lobreau-Callen (1980) showed that morphology of pollen grains of *Thesium* (Santalaceae) differs according to geographical distribution of the species — the pollen become heteropolar under environmental stress, in arid or in polar regions. Cerceau-Larrival and Roland-Heydacker (1980) discovered morphological differences in the ultrastructure of exine in several genera of Umbelliferae, which seem to correlate with the environments where the plants grow. Similar bioclimatic adaptations are also reported by Cerceau-Larrival and Carbonnier-Jarreau (1980).

Of lesser importance to arid lands, but of a similar significance are variations in the size of birch (*Betula*) and possibly hazelnut (*Corylus*) pollen observed in a transect from southern to northern Europe, reported by Usinger (1980).

It should however be noted that such variations and differences in size, shape or ultrastructure between pollen grains of a certain genus, observed for different environments, can only be used for interpretation of the past when two practically rather difficult prerequisites are met. First, a transect must be made across a variety of environments, since these variations are relative and could only be defined by comparative studies. Second, which makes such data even less applicable for interpreting past environments in arid lands, is that the state of preservation must indeed be excellent, to clearly observe the minute variations, especially those connected with the exine ultrastructure.

7.4. Grouping in pollen spectra and diagrams

Pollen spectra can be subdivided into groups of palynomorphs each comprising several taxa, or in some cases even only a single one. Subdivision could be initiated in various instances: where each group bears a specific environmental or ecological affinity; is subject to a different transportation agent; is of anthropogenic origin;

has a peculiar adaptational or evolutionary significance; and so on, as the specific case may be. It is occasionally even found beneficial to present and discuss the results of pollen analyses in more than one way of subdivision and grouping, as bases for interpretations of different problems. Such subdivisions are thus based on common characteristics of the constituents of each group. Although the more common are environmental affinities, others could be used as well in appropriate opportunities, not necessarily involving botanical aspects such as close systematic relations, arboreal or non-arboreal origin and so on.

The individual groups are usually treated in two ways. First, by regarding the entire spectrum as a single entity, defining numeric relations among the various groups and their temporal or spatial changes, which would imply on the relative distribution of environments represented by taxa involved within each of the groups. The second views spectra as consisting of several composite entities and deals with each of the groups separately; thus, relations among each of the groups' constituents would be used for interpreting changes a specific biotope or environment in space and time, with no indications of its relative areal distribution as compared with others.

The way of subdivision and method of grouping applied should probably be peculiar to each given region, thus no strict rules or suggestions can be set. Examples are brought for various ways of dealing with subdivision and grouping in Israel, in the hope that these will indicate possibilities involved with this way of interpretation. By no means should subdivisions be done in other regions in exactly similar ways, where environments are different and pollen spectra are affected by other factors.

7.4.1. AP/NAP ratios

The most common of these subdivisions is into groups of arboreal and non-arboreal pollen, always practiced in forested regions, where it indeed has a direct environmental significance. AP/NAP relations are used there to interpret the relative extension and coverage of forests; the internal composition of the arboreal pollen group indicates regional environments; while the variety of NAP are used for understanding local conditions. This ideal, rather straightforward approach is self evident especially where sequences are collected for pollen analysis from very small depositional basins such as peat bogs, which is a frequent procedure in these parts of the world. In such localities the pollen rain is quite uniform and AP/NAP ratios would readily reflect the forest coverage.

AP/NAP subdivision and relations may frequently have different bearings in arid lands, where non-arboreal pollen are used as general environmental indicators, while arboreal would frequently only yield information concerning transportation. AP can provide information about composition of their provenance forest, if indeed all its constituents are represented in pollen spectra obtained from the distant arid region. Pollen produced by trees which grow in arid environments are very frequently highly under-represented in the spectra, so that most arboreal pollen found in samples are at least to some degree allochthonous; trees thus only grow some distance away, in a neighboring more humid environment or on mountain tops, in which case the potential of transporting agents and differential susceptibility of each type of grain would play an important role in the obtained pollen spectra; seasonal strong winds may cause preferences in imported arboreal pollen distribution, both in relative quantity and in internal composition, in basins located at different positions in relation to wind directions and flowering seasons of respective trees.

Another concern is that quite frequently long sequences can only be obtained from relatively large depositional basins, which maintain well developed non-arboreal shore vegetation all around, especially in bordering marshes or playas. If the arboreal pollen is only supplied by winds, the pollen rain could be considered rather uniform all over the depositional area. In this case the shore vegetation pollen would be over-represented near the shore as opposed to the center of the basin, in relation to pollen produced by the arboreal. The reason is apparently that pollen produced by the low shore vegetation are not lifted high enough by winds, to become properly mixed in the overall pollen rain, thus are usually deposited in greater numbers close to their sources, where they readily sink. Some preferences may also be seen in basins fed by a river arriving from a more humid provenance, carrying arboreal pollen which are deposited where the river's transporting energy fades away.

Two cross sections, north–south and east–west, of recent sediments from Lake Kinneret, Israel (Horowitz, 1979) are presented here (Fig. 7.4.1). It is clearly seen that in both cross sections arboreal pollen shares increase in the middle of the lake as compared with near shore spectra. The Jordan River enters the lake from the north, but the sample collected near its entrance does not show higher AP, although the river originates where trees are common. This seems to be a result of the river's water energy, that does not allow the grains to be deposited immediately, thus they are carried deeper into the lake.

Pollen derived from common non-arboreal taxa growing close to the lake, such as Compositae and Cruciferae are considerably more abundant along the shores (Fig. 7.4.2). In contrast, grains of *Artemisia*, although non-arboreal, are very frequent in the center of the lake, somewhat less along its eastern shore. This seems to be a consequence of the greater susceptibility of these pollen to wind transportation and the distribution pattern of *Artemisia* in the region — the plants are considerably more abundant on the highlands to the northeast than around Lake Kinneret.

The marked differences in AP/NAP ratios between the center and near shore samples may create a false impression of climatic changes when single sequences are analyzed from large enough basins, whose configurations had been changing through the studied periods for reasons other than climatic, such as tectonic or volcanic. An example can be brought from the late Pleistocene Lake Lisan, which covered a stretch of over 250 kilometers along the Dead Sea Rift in Würmian

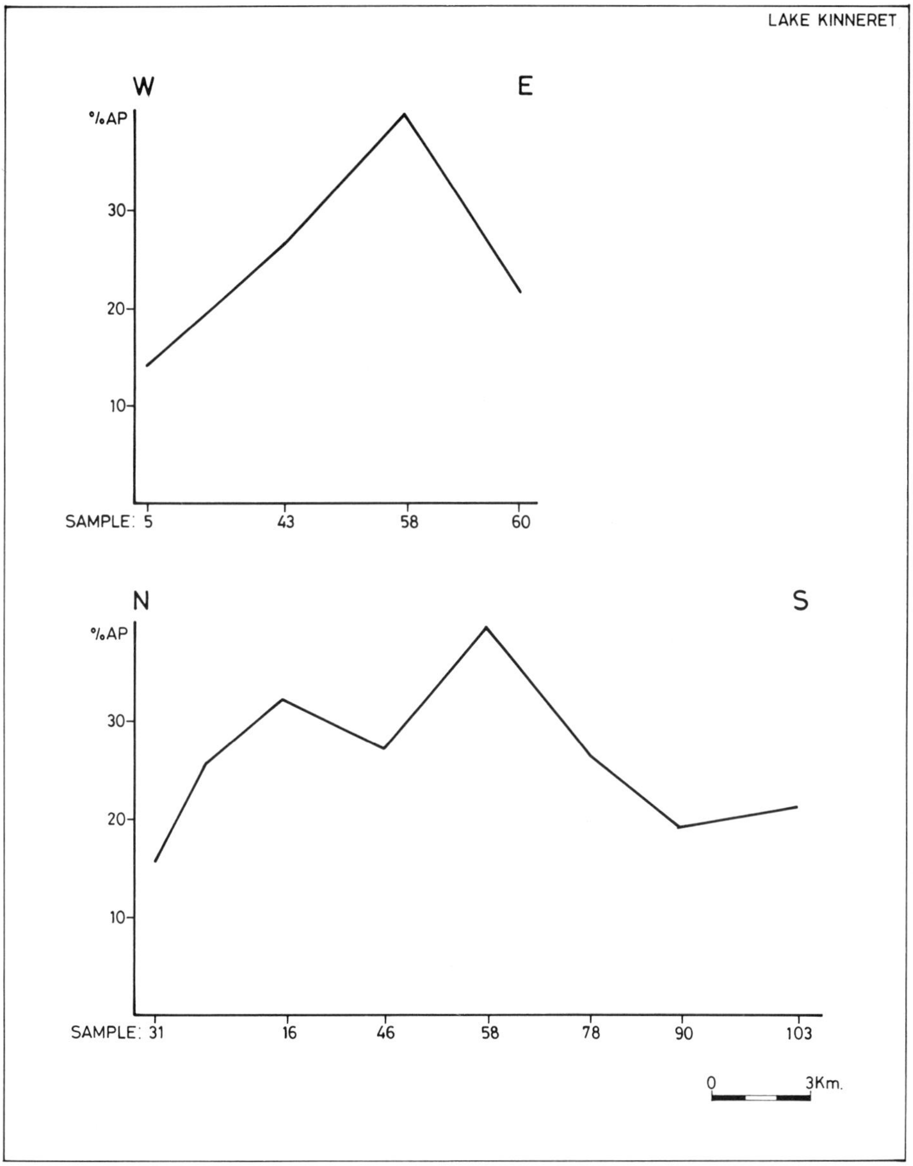

Fig. 7.4.1. West–east and north–south sections across Lake Kinneret, Israel, showing arboreal pollen ratios calculated from the total number of counted grains, from subrecent sediments. Cultivated and hydrophile trees are omitted. Higher AP frequencies are recorded in the middle of the lake, where the shore vegetation influence is minimal. Data from Horowitz (1969).

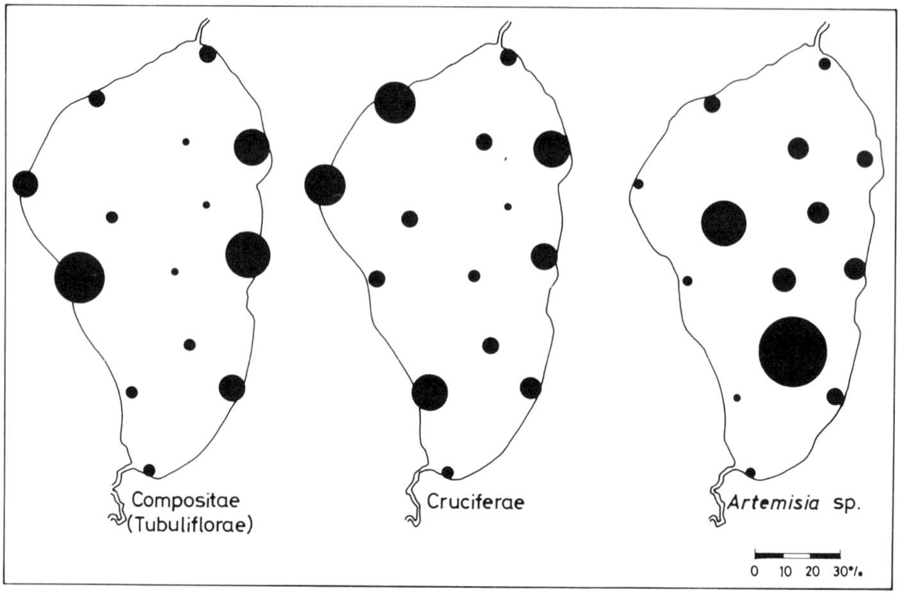

Fig. 7.4.2. Relative frequencies of Compositae, Cruciferae and *Artemisia* pollen, calculated from the total number of counted grains, from subrecent sediments in Lake Kinneret, Israel. Reprinted with permission from Horowitz (1969).

times. This lake had shrunk to less than one third of its area during the late Würmian, following a tectonic disturbance which resulted in formation of the Dead Sea and Lake Kinneret (Fig. 9.2.13), still at the height of pluvial conditions at that time, when the regional climate was quite humid (Horowitz, 1979).

Similar changes can be seen when springs develop due to faulting (Horowitz, 1987b), causing widespread expansion of freshwater vegetation which alters considerably AP/NAP ratios (Fig. 8.5.4) in pollen spectra from the Dead Sea, with no climatic significance. For further details see Chapter 8.5.

When shoreline changes are indeed caused by the climate, AP/NAP ratios in a given sequence would usually display it in an emphasized, somewhat amplified way. If consideration is made of a section through the middle of a basin such as lake or marsh, climatic deterioration would result in diminution of its area, thus bringing the shores closer to the site. This alone would lower the relative AP shares, in addition to lesser amounts of pollen produced by the reduced arboreal vegetation, caused by the drought. Such reduction is frequently accompanied in arid regions by a geographic shift of vegetation belts (Weinstein-Evron, 1990). The picture would naturally be opposite with climatic amelioration, but again amplified. In such cases the amplified AP/NAP fluctuations are going in unison with the regional climatic trends, further amplified by their doubled sensitivity.

The only way in which correct interpretations and clear understanding could be achieved in such cases, whether the AP/NAP ratios fluctuations are caused by climatic, tectonic or other processes, is by analyzing several sequences from different basins, for comparison. Whenever such sequences show similar trends in several basins, the regional climatic factor is the more prominent in affecting arboreal/non-arboreal ratios.

Another problem concerned with distance from the shore where a single sequence is collected and studied, is the initial AP/NAP ratio on which interpretations are based. These ratios are compared in Fig. 7.4.3 for several sequences collected from various depositional basins in Israel. One of the sequences from the Hula, UP 15, consists chiefly of peat deposited in marshes bordering the lake; UP 6 was drilled in a locality where the lake existed up to about 500 years ago, presently covered by marshes following the shrinking lake; sequences are also presented from lacustrine sediments of Lake Kinneret, the Dead Sea and offshore Mediterranean deposits. The initial AP shares are considerably higher in lacustrine and marine sediments as compared with peat. This difference is seen in UP 6 at the transition

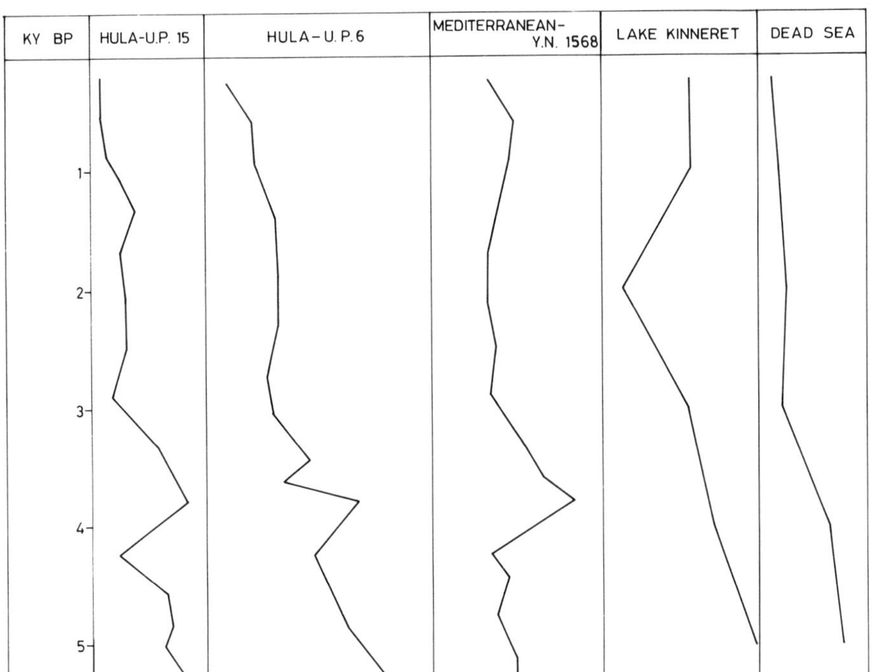

Fig. 7.4.3. Arboreal pollen ratios, calculated from the total number of counted grains, from five sequences in Israel spanning the last 5,000 years. Pollen of cultivated trees are omitted. Data from Horowitz (1979, 1987a), Baruch (1986).

from chalk to peat, at depth of one meter, when the neighboring UP 15 does not display such a change, which is therefore concluded to be of a very restricted local significance.

Regardless which of the sequences is dealt with, the shapes of the curves bear similar trends in all sections, even if the absolute arboreal pollen percentages differ, except for minor variations that should probably be attributed to local peculiarities. This similarity is maintained in the sequences collected from Mediterranean as well as arid environments, indicating that regional climate is responsible for the changes in tree coverage throughout this period. It is therefore the shape of the curves which is taken for interpretation, rather than the absolute values of arboreal pollen percentages.

Incidentally, shrinking of the Hula Lake at this time is climatically controlled, as can be seen from the parallelism in AP/NAP ratios and reconstructions of the lake's shoreline based on sedimentary analysis (Horowitz, 1973a), thus structural disturbances could be ruled out. Both Lake Kinneret and the Dead Sea show some shrinking during the later Holocene, however the shoreline had not moved much, since these lakes are deeper than the shallow Hula and their flanks are steeper.

Another factor which may change the starting point, or initial percentages taken as basis for interpretation when AP/NAP ratios are discussed, is the frequent over-representation of imported arboreal pollen in arid lands deposits. This results from combination of several factors: the local arid land arboreal vegetation, even when well adapted and developed, does not contribute much to the pollen spectra for various reasons; on the other hand, quite frequently the first trees to comprise considerable stands at the boundary between arid and humid environments, or submesic biotopes, are conifers adapted to lesser water availability, which produce considerable amounts of windborne pollen, easily spreading to the neighboring arid lands (see Martin, 1963; Jacobs et al., 1985; and others for over-representation of pine pollen in Southwestern USA; and Leroi-Gourhan, 1982, for the Middle East).

Initial over-representation of arboreal pollen is often common along wind paths, as discussed in Chapter 7.2. This may create a false impression that trees grew much closer to the sampled basins or sites, especially if no continuous sequences are analyzed and no control of the present day is available or relevant, as happens with older deposits (such as Neogene formations). If continuous sequences going up to the present day are analyzed, the initial reference point could be set by the present day conditions, preferably by analysis of subrecent spectra to minimize anthropogenic influence. As discussed above, also in this case it is the shape of the arboreal pollen curve that counts, rather than absolute percentage values.

7.4.2. Grouping by environments

Different ways of subdividing and grouping pollen spectra are by creating several groups, each bearing a specific ecological or environmental significance. A detailed example of such subdivision and interpretation of various palynomorph groups and their internal composition is brought in Chapter 9, thus only outlined here in a general way. Such subdivision could be treated by regarding numerical relations between the main groups, by analyzing proportions of taxa constituting each group, or both.

Comparisons of relative frequencies of the main groups would create a sound basis for interpreting the relations and distributions of cardinal environment types. As an example, pollen spectra from late Cenozoic Dead Sea Rift deposits, in several basins, have initially been subdivided into taxa derived from regional vegetation on the one hand and those produced by local flora, intimately connected with the hydrology of each specific basin, on the other. The relations between these two groups usually reflect fluctuations in local conditions, such as development or drying up of springs, marshes, playas, changes of lakes' shorelines and so on. This is based on the concept, which however occasionally needs proof, that the regional pollen production could be thought of as more or less uniform along the sequence, in the same manner discussed above for arboreal pollen distribution over a large depositional basin. Variations in local pollen production would thus affect more considerably the relations between the two groups, rather than the regional changes. Again, only comparisons among sequences collected from a variety of basins could confirm the validity of this concept.

The two groups had been further subdivided, the regional taxa into those produced by plants growing in either desert, steppe or Mediterranean environments; the local into pollen originating in hydrophil or halophil vegetations. It is quite obvious that some degree of overlapping exists between the various constituents, since subdivision was done according to predominance of a certain taxon in a given biotope. It seems that due to overlapping interpretation would be less refined, so that one is advised not to pay too much attention to minor fluctuations in the curves.

Oscillations of the regional components may be verified if they recur in several sequences collected from different basins. Confirmation of fluctuations in the shares of local constituents is usually hard to get, because even if several sections are analyzed from the same basin, their local taxa percentages may vary due to distance from springs or other supply points for the local pollen flora. It is therefore suggested that the regional setup be first interpreted and may be used for inter-basin correlations, while the local hydrologic conditions subsequently determined for each of the basins separately.

It is however quite clear, when looking at the pollen diagrams brought in Chapter 9, that regional conditions are to a great degree persistent in the analyzed sequences. When stratigraphy and regional environments are concerned, a much more significant general picture could be obtained if the so-considered local elements are taken out of the calculations. The total used for regional interpretations is evidently the sum of pollen produced outside the local, restricted biotopes.

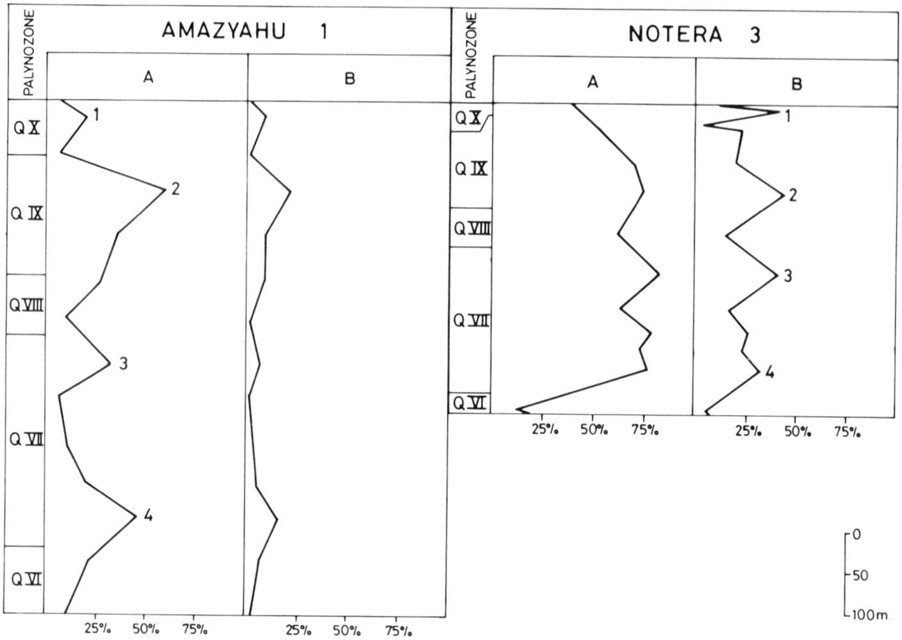

Fig. 7.4.4. Arboreal pollen ratios, calculated from (A): the total number of pollen produced by the regional flora; (B): the total number of counted grains. Sequences are from the Hula basin (Notera 3) in northeastern Israel, located in Mediterranean environment, while the southern Dead Sea basin (Amazyahu 1) is in the midst of an extremely-arid region. Peaks 1–4 are clearly seen in the arid sequence only when AP is calculated from the regional components, due to the masking effect of local ones; in contrast, the Mediterranean section displays these peaks clearly only when calculations are made from the total number of counted grains. Each of the peaks in these curves is an average of four to six grouped samples, to minimize noise. Data from Horowitz (1987a). For location see Fig. 9.2.1.

Fig. 7.4.4 shows two sequences from the Dead Sea Rift, comparing AP/NAP ratios calculated (A) as part of the regional components of the pollen spectra or (B) from the total number of counted grains. It can readily be seen that no clear correlation is observed when either the total spectra or the regional ones are compared. If the local elements are subtracted from the curve obtained from Amazyahu 1, drilled in the extremely-arid southern Dead Sea where local, especially halophil vegetation is well developed, regional trends can be readily correlated. However, if the same taxa are omitted from the sequence of Notera 3, in the Hula Valley where Mediterranean climate predominates, correlation is again impossible. This results from the fact that plants of the same families constituting hydrophil and halophil vegetations of the Dead Sea region are part of the regional flora up north, where "local" vegetation is quite subordinate as compared with regional.

It is thus quite clear that if a more objective picture is to be interpreted from pollen diagrams based on samples collected from arid lands deposits, it would be preferable to consider them without reference to the local elements, which should only be used for interpretations of hydrologic conditions pertaining to a specific sedimentary basin. However, the case may occasionally be that correlation is needed between the arid region and its neighboring submesic or mesic environments. If the latter are for some reason analyzed from a smaller basin surrounded by trees it would be preferable to attempt both ways of calculations (Fig. 7.4.4), based on total and regional components, to find out which one gives better correlation with the arid region's diagrams.

The subdivision into groups of taxa from arid and semi-arid environments, *vs.* local elements may need a thorough consideration where extremely-arid conditions may have had prevailed at certain periods in particular regions, depending upon the

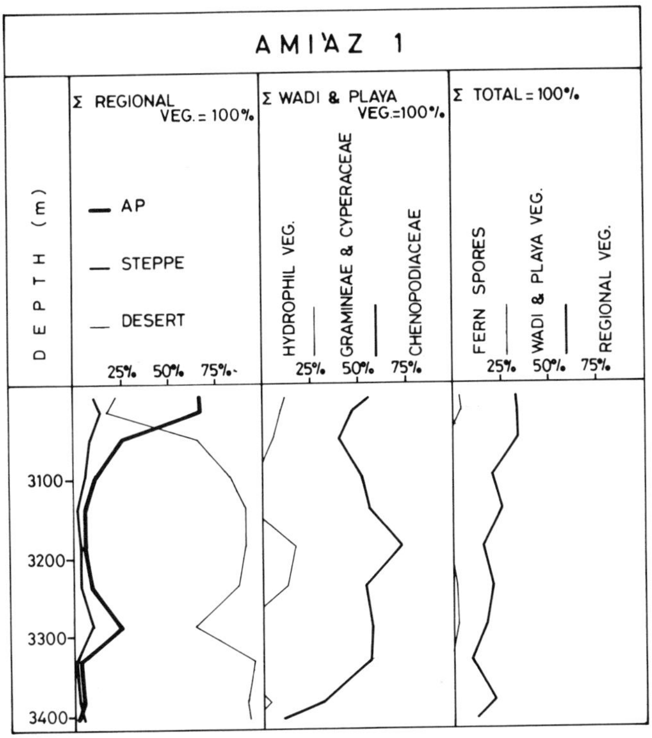

Fig. 7.4.5. Extreme aridity as seen in a sector of Ami'az 1 borehole from the southern Dead Sea region, in early Miocene (Palynozone Ma) times, by prevalence of desert elements in the regional flora, Gramineae and Cyperaceae in the local. Regional elements are calculated out of the total, indicating very poor local vegetation and probable allochthonous origin for most recovered pollen grains. A peak of the hydrophile pollen around 3,200 m may indicate a temporary spring, possibly resulting from faulting. Please note that curves in the left column are overlapping, while the center and right are cumulative. Data from Horowitz (1990). For location see Fig. 9.2.1.

characteristics of xerophytic vegetations. The best example is probably portrayed by Gramineae and Cyperaceae, which are frequently thought of as representing local freshwater environments in the Near East (Horowitz and Horowitz, 1985). The problem is that members of these two families are almost the sole constituents of regional floras in extremely-arid biotopes in this same region, thus the question may arise whether a peak displayed by either Gramineae or Cyperaceae should be interpreted as representing a spring or marsh development, or a regional extreme aridity.

Since species identifications are practically out of the question, other factors should be considered. Such could be occurrences of pollen produced by minor spectral elements, which would differentiate between extreme aridity, like *Calligonum* in the Middle East (El Moslimany, 1990), or the existence of freshwater, as *Rubus*, Labiatae and others.

A possible example for representation of extremely-arid conditions during the late Miocene in the Dead Sea region is brought in Fig. 7.4.5. The local elements are subordinate as compared to regional, dominated by Compositae, as opposed to pollen produced by trees and steppe plants, which hardly occur at all. The combination of prevailing arid elements and Gramineae/Cyperaceae possibly indicates extremely-arid conditions. Naturally, such considerations must be thought of individually for each region and depositional domain, taking into account the respective vegetations, transportation agents and sedimentary processes.

Specific implications of numerical relations among and within various groups of palynomorphs, for interpretation and reconstruction of paleoenvironments, are brought in Chapter 8. An example for reconstruction of paleoenvironments during the late Cenozoic in Israel, based on these principles is presented in Chapter 9.

Chapter 8

Reconstruction of past environments

Reconstruction of past environments is usually the ultimate goal of palynological studies. In this respect arid lands are no different than forested regions, however the principal difference lies with the lesser accuracy of regional vegetation indicators, for reasons discussed above. To this is added the greater importance of markers denoting various microenvironments in arid lands such as springs, salt-pans and others, or exotic sporomorphs arriving by long distance transportation. These may on the one hand mask to a certain degree the regional indicators, but on the other could provide indispensable information on local habitats, winds, animals and a variety of other factors which affect pollen spectra obtained from arid lands deposits.

A good stratigraphic control is the basic foundation for reconstructing past environments. This could be achieved by various methods, either independent of the palynological research, such as radiometric datings or any way of assigning ages to the studied sample or sequence, or be a product of the pollen analysis, in the way of establishing sufficiently accurate palynostratigraphy. The stratigraphic principles and methods applied to arid lands palynology are essentially similar to those used in other regions (see Birks and Gordon, 1985, for numerical analysis of zonation), thus do not necessitate repetition in the present context.

Palynostratigraphy is applied in much the same way as elsewhere, except for one point which should be observed, but however seems to be frequently overlooked. It is quite often tempting to use all counted pollen for establishing palynostratigraphy, so that results would be statistically significant when based on a larger number of sporomorphs. However, it is much more useful (and evidently logical, too) to use for regional palynostratigraphy only these palynomorphs produced by regional vegetation (Horowitz, 1987a; Levin and Horowitz, 1987), excluding those derived from local, restricted plant occurrences, or the ones brought by long distance transportation only to a certain locality, but not to others within the same region.

The cardinal problem in arid lands is however that it is not always easy to tell which pollen come from regional and which from local vegetation. An example is the case of Gramineae and Cyperaceae, which grow in arid lands in intimate connection with freshwater sources such as springs, but in extremely-arid regions may comprise the regional vegetation, in addition to freshwater occurrences, however rare. It is therefore suggested that several ways of grouping and reference would be applied before palynostratigraphic principles are decided upon, but only those which consistently recur in several sequences be used as markers.

All the above discussion pertains to the late Cenozoic, where stratigraphy is quite detailed, usually based on climatostratigraphy and environmental changes rather than on evolutionary trends. In older sediments, where time stratigraphic units are of a longer duration, it is frequently useful to establish palynostratigraphy on evolutionary principles of first appearances and extinctions. This is also because there is no way to decide which of the pollen spectra components is of a local, or regional significance.

Once the stratigraphy is established, reconstruction of environments goes on along lines described in the former chapter, or hopefully involving new concepts which will come up. The following is a series of examples of what was done in this respect, by various investigators.

8.1. Vegetation

Reconstructions of late Cenozoic arid lands vegetations are usually based on the recent pollen spectra, as discussed above in Chapter 7.1, while evidently taking into consideration all processes which may have affected the assemblages, such as transportation, anthropogenic influence and so on.

Reconstructions are evidently easier when dealing with geologically younger deposits, of late Pleistocene and Holocene age, but may become more and more difficult when the sediments get progressively older. The problem here is that occasionally no recent counterparts are known for the fossil pollen spectra. Such for example is the case with early Quaternary assemblages in Israel, dominated by up to 95% pollen of *Picea orientalis* (see Chapter 9 for details). In these cases presumptions have to be made and reconstructions based more on conceptual considerations than on factual evidence.

One of the cardinal problems involved with reconstructions of dry region floras is that almost invariably the palynologically analyzed sequences, hence the pollen spectra and diagrams that serve as foundation for reconstructions, are collected from low lying basins of deposition. This is an understandable result of the fact that higher terrains are almost always subject to erosion, especially in extremely-arid and arid zones, except for rare occurrences of sediments accumulated in caves and rockshelters. In semi-arid regions where loess is prevalent, it could yield uplands sequences directly representing the regional conditions.

Three factors are involved with pollen spectra obtained from low lying basins: these are very frequently surrounded by a local vegetation that reflects hydrological conditions rather than regional environments; a supply of sediments and palynomorphs could occasionally arrive from considerable distances by rivers and winds; and, especially in arid regions people have been inhabiting the lower basins for long periods in history and prehistory, being the only localities where subsistence is possible in these environments. These factors have affected both the older and recent deposits, but the extent of particularly the last factor is very difficult to assess for the past. It is therefore of prime importance to distinguish between the regional, allochthonous and local components while reconstructing past vegetations, with attention paid to the possibility of anthropogenic influence on the latter. The main problem with such distinction is however, the considerable overlappings of constituents, discussed in Chapter 7. A discussion of anthropogenic induced vegetations follows in Chapter 10, while reconstructions of pre-late Cenozoic floras of arid lands, based on different principles and approaches, are discussed in Chapter 11.

Reconstructions of past vegetations could be presented in a verbal form, in way of discussion (see for example Mehringer, 1967; Ritchie and Haynes, 1987; Ritchie, 1987; Lézine, 1989; and many others), may be summarized in a table (see Chapter 9), or accompanied by a graphic presentation of the interpretation. Reconstructions may concentrate on a certain aspect of the vegetation, such as fluctuations of vegetation belts boundaries over the studied period, fluctuations that can be expressed as spatial or altitudinal; can be presented as cross sections or block diagrams, usually of localized phenomena; or, when information seems sufficient, displayed as paleovegetation maps.

Spaulding (1990) presented changes in altitudes of vegetation belts in the Sheep Range, southern Nevada (Fig. 8.1.1), Mojave Desert, based on a synthesis of previously published pollen diagrams for Tule Springs site (Mehringer, 1967) in this region. The high mountains of the Mojave Desert display marked vegetation zonation, where the modern subalpine zone is a woodland dominated by bristlecone pine (*Pinus longaeva*). Exposed ridges and peaks above 2,700 meters maintain stands of black sagebrush (*Artemisia nova*). Below the subalpine zone occurs a forest of ponderosa pine (*Pinus ponderosa*) and white fir (*Abies concolor*), while below that a woodland grows, dominated by single needle pinyon (*Pinus monophylla*) and juniper (*Juniperus osteosperma*). Sagebrush (*Artemisia tridentata*) can be codominant in this woodland and is also locally dominant in the mesic biotope of the underlying desert-scrub. The present day desert-scrub at 700 m elevation at the bottom of Las Vegas Valley is characterized by creosote bush (*Larrea tridentata*), white bursage (*Ambrosia dumosa*) and saltbushes (*Atriplex* spp.).

Pollen spectra used for reconstruction of the late Wisconsin vegetation belts have been dated from ca. 8,000 to 14,500 yr BP and from 22,600 to 31,300 yr BP, obtained from the valley floor. Mehringer interpreted the abundant (35–85%) pine pollen and the consistent presence of fir in late Wisconsin sediments as indicating lowering of the vegetation zones by at least 1,000 meters. Depression of lower limit of the woodland by at least 700 meters was further supported by high percentages of *Artemisia* in certain horizons. Relatively high percentages (15–20%) of *Juniperus* also point out that these plants were abundant on the valley floor. Using recent

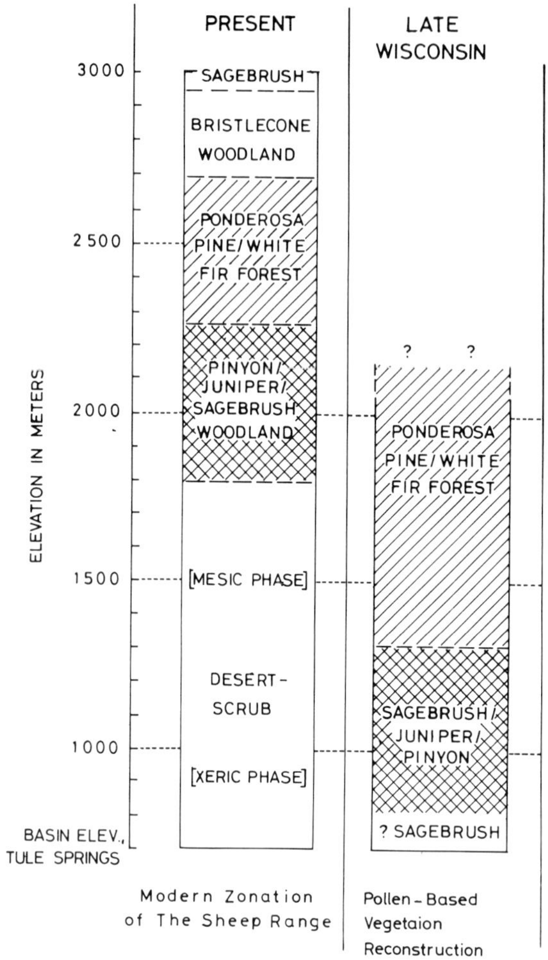

Fig. 8.1.1. Comparison of modern vegetation zonation in the Sheep Range, southern Nevada, with models of late Wisconsin zonation based on the Tule Springs pollen data. Reprinted with permission from Spaulding (1990).

surface samples for comparison, Mehringer (1967) concluded that there was a lowering of the vegetation zones by at least 1,220 meters during the peak of the Wisconsin pluvial conditions. Thus juniper–pinyon woodland was mapped as occurring between 800 and 1,300 m elevation, with ponderosa pine and white fir forest above. Mehringer also notes that there would be extensions of continuous stands of sagebrush in the deep soiled valleys, below 800 m and that species which normally occur above ponderosa pine would have colonized the higher peaks, above ca. 2,000 m in the northern Mojave Desert. The diagram comparing present day altitudinal belts with the late Wisconsin ones is brought here (Fig. 8.1.1).

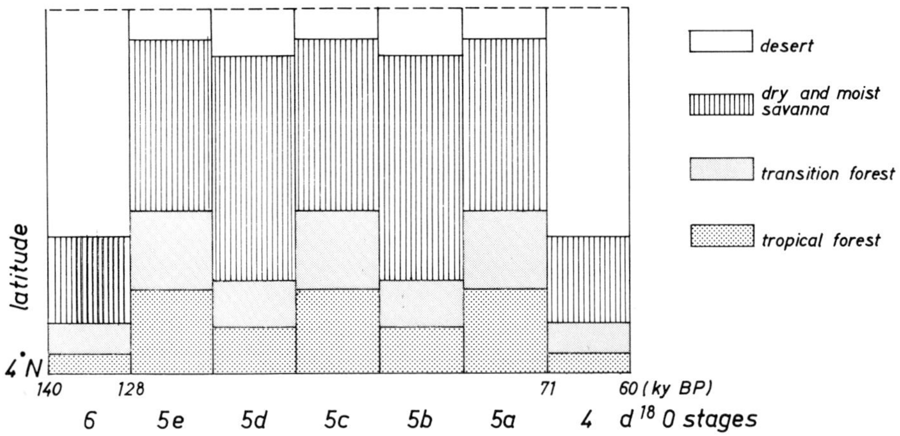

Fig. 8.1.2. Idealized diagrammatic of latitudinally shifting vegetation zones south of the Sahara during the interval 140–60 ky BP. Reprinted with permission from Hooghiemstra and Agw· '1988).

Hooghiemstra and Agwu (1988) had analyzed palynologically two cores obtained from offshore sediments of the Atlantic Ocean west of the Sahara, where palynomorphs arrive by winds from all over the neighboring part of Africa, for the time interval 140–70 ky BP. Based on these they reconstructed diagrammatically the spatial and temporal migrations of vegetation belts across the Sahara, summarized in Fig. 8.1.2. It seems that tropical forests had maximal expansion at ca. 144 ky (oxygen isotope stage 6), ca. 98 ky (transition of 5.c to 5.b) and around 70 ky (first part of stage 4). An increasing delay of the response of tropical forest to global warming is apparent during the last interglacial. This may be a result of the tropical forest need of continuous humidity, thus this forest monitors rainfall rather than high temperatures south of the Sahara. During the last interglacial the southern boundary of the Saharan vegetation belt shifted only slightly. Expansions and contractions of the tropical forest area are correlated with contra-oscillations of the grass dominated savannah zone.

On the other hand, considerable latitudinal shifts of the desert/savannah boundary to the north had occurred during the penultimate glacial/interglacial transition (ca. 128 ky) and during the last interglacial/glacial transition (ca. 65 ky) to the south.

Dupont and Hooghiemstra (1989), again based on pollen analyses of sequences collected from the Atlantic Ocean west of the Sahara (ODP 658), had extended the previous study and reconstructed migrations through time of the Saharan/Sahelian boundary along the last 700 ky, presenting the results in a way of a generalized map (Fig. 8.1.3).

Schulz and Whitney (1986, 1987) made a more realistic presentation while reconstructing the late Pleistocene–Holocene vegetation in the An Nafud Desert,

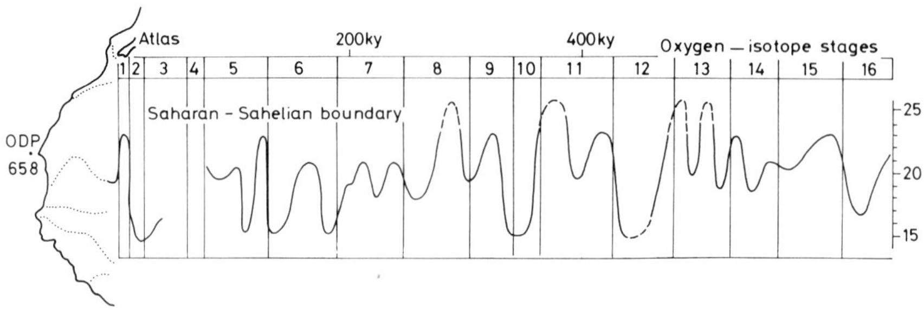

Fig. 8.1.3. Summary diagram on a linear time scale of the estimated latitudinal shifts of the Saharan–Sahelian boundary, based on pollen data of ODP Site 658. Reprinted with permission from Dupont and Hooghiemstra (1989).

northern Saudi Arabia, from ancient lake bed sediments accumulated in shallow water bodies at these times. This region is part of the large belt of semi-deserts, bordering the Mediterranean Sea on its southern end. Schulz and Whitney present detailed pollen diagrams for this region and mention the general characteristics of the pollen spectra, referring to three periods: the present day; the Holocene ca. 8,400–5,400 yr BP, corresponding to at least a part of the Atlantic stage; and the late

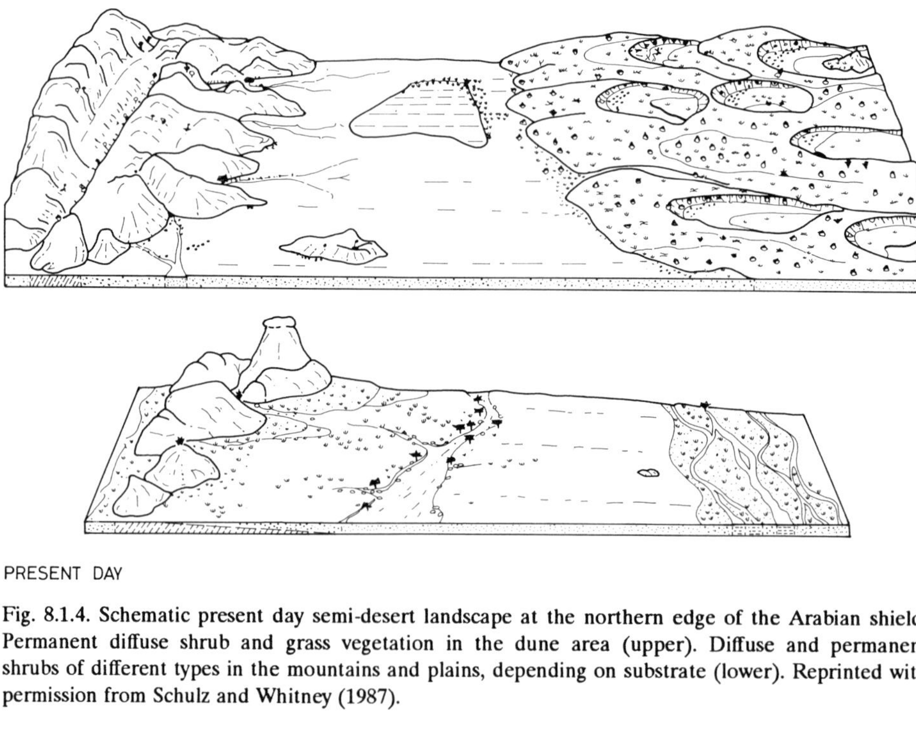

PRESENT DAY

Fig. 8.1.4. Schematic present day semi-desert landscape at the northern edge of the Arabian shield. Permanent diffuse shrub and grass vegetation in the dune area (upper). Diffuse and permanent shrubs of different types in the mountains and plains, depending on substrate (lower). Reprinted with permission from Schulz and Whitney (1987).

Pleistocene, between 34 and 24 ky, a time span for which global glacial conditions had not been fully developed. The recent vegetation comprises communities of *Acacia* and *Panicum* in the large wadis, *Anabasis*, *Salsola* and *Rantherium* on the plains and *Artemisia*, *Stipagrostis* and *Cyperus* on the dunes (Fig. 8.1.4). The different types of vegetation mainly depend on geomorphology and substrate types, according to their water holding capacities. The present day pollen rain indicates that most vegetation types are represented in its composition, thus serving as a suitable tool to reconstruct the former types of plant covers.

The Holocene pollen spectra are characterized by Gramineae and Cyperaceae, shrub-like *Convolvulus*, *Cornulaca* and herbs such as *Moltkiopsis* and *Polycarpaea*; Compositae and *Plantago* are common. Trees are only represented by pollen of *Acacia*, *Maerua*, *Balanites* and *Hyphaene*. It seems that the portion taken by allochthonous palynomorphs is negligible. These spectra indicate a semi-desert vegetation (Fig. 8.1.5) quite similar to the present day, but in some areas with a denser herb cover and *Typha–Phragmites* stands around the shallow water bodies.

The late Pleistocene pollen spectra are dominated by Gramineae, Cyperaceae and shrub-like *Calligonum*, *Fagonia* or *Salvadora*, while *Acacia*, *Maerua* and *Balanites* are hardly represented. Common pollen of herbs are derived from *Moltkiopsis*, *Plantago*, *Polycarpaea* and others. These spectra led Schulz and

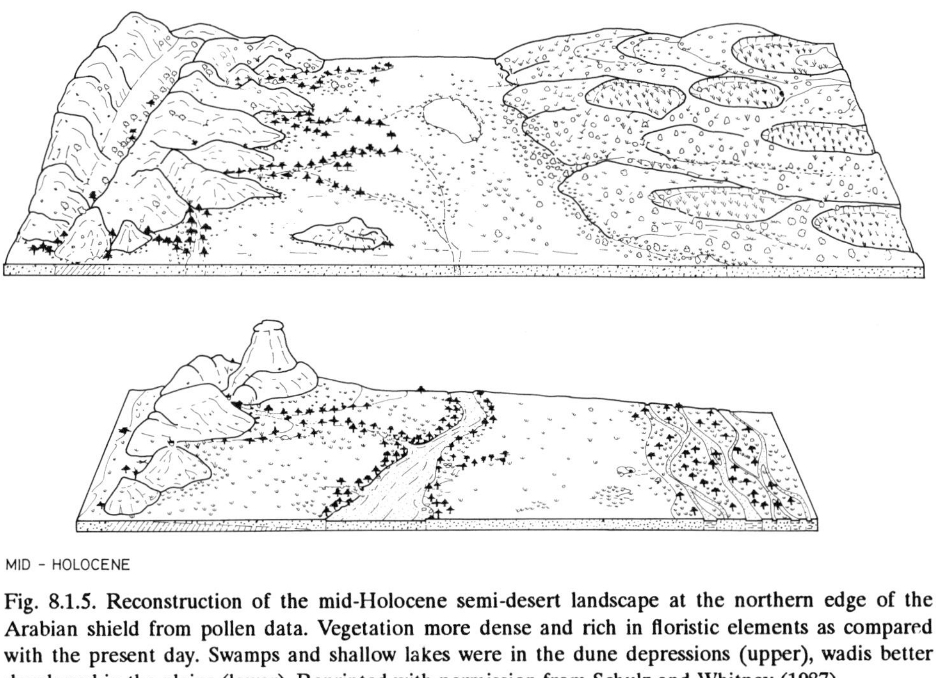

MID - HOLOCENE

Fig. 8.1.5. Reconstruction of the mid-Holocene semi-desert landscape at the northern edge of the Arabian shield from pollen data. Vegetation more dense and rich in floristic elements as compared with the present day. Swamps and shallow lakes were in the dune depressions (upper), wadis better developed in the plains (lower). Reprinted with permission from Schulz and Whitney (1987).

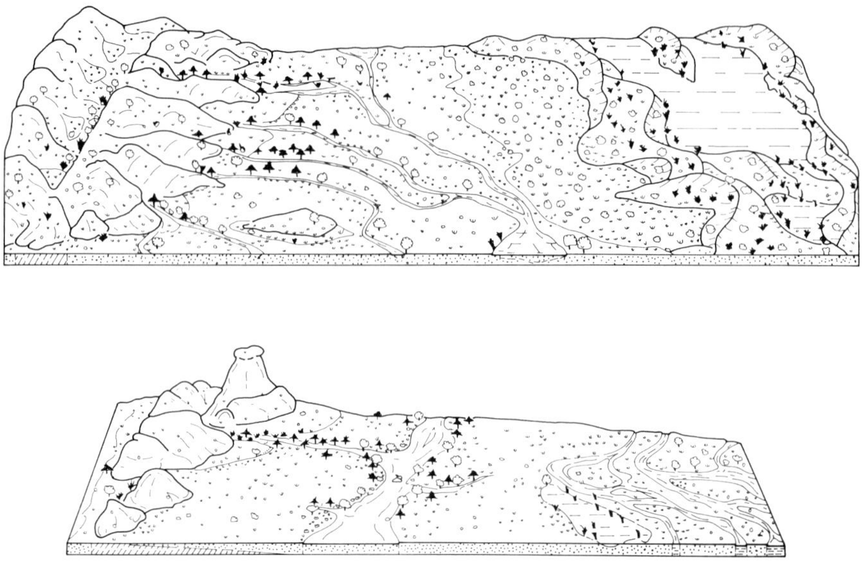

Fig. 8.1.6. Reconstruction of the late Pleistocene landscape at the northern edge of the Arabian shield from pollen data. Dominance of longitudinal dunes with large lakes in interdunal depressions (upper). Extended wadis and shallow lakes in the plains (lower). Rich, dense vegetation as compared with the present day. Reprinted with permission from Schulz and Whitney (1987).

Whitney to conclude (Fig. 8.1.6) that the region was a semi-desert with stabilized dunes, with herb cover more extended than today. The water bodies (lakes or marshes) had been surrounded by *Typha* and *Phragmites* belts, but there is no sign for a denser tree cover. *Acacia*, *Maerua* and *Balanites* had possibly grown on the nearby mountainous areas.

Schulz (1987a, b) presents several pollen diagrams obtained from sediments accumulated in ancient "lake" beds and marshes in the central Sahara during the middle Holocene, on which he based a reconstruction of the vegetation in the form of a paleovegetation map, which is compared to the present day plant distribution in the region (Figs. 8.1.7, 8.1.8). The desert vegetation of the Sahara is bordered to the north by Mediterranean semi-deserts and to the south by Sahelian savannahs and consists chiefly of *Acacia–Panicum* or *Tamarix–Stipagrostis* communities. The transition to the north is mainly by a more dense vegetation, forming a steppe. To the south, the transition to the Sahelian savannah is seen in the pollen spectra by the high percentages of Gramineae and Cyperaceae in the dune areas, or Gramineae alone in the rocky and mountainous regions of the Sahara.

The pollen diagrams presented by Schulz indicate that during the middle Holocene the limits of the Sahelian savannahs have been displaced at least some

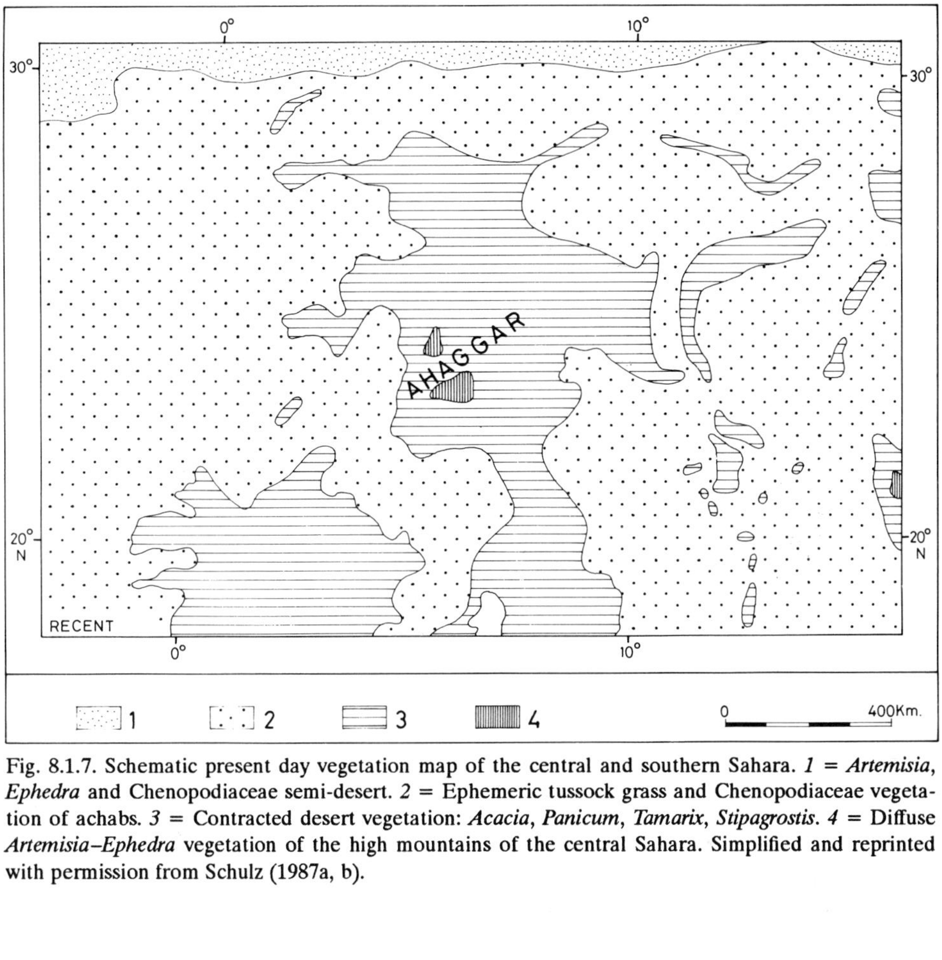

Fig. 8.1.7. Schematic present day vegetation map of the central and southern Sahara. *1* = *Artemisia, Ephedra* and Chenopodiaceae semi-desert. *2* = Ephemeric tussock grass and Chenopodiaceae vegetation of achabs. *3* = Contracted desert vegetation: *Acacia, Panicum, Tamarix, Stipagrostis*. *4* = Diffuse *Artemisia–Ephedra* vegetation of the high mountains of the central Sahara. Simplified and reprinted with permission from Schulz (1987a, b).

400 kilometers to the north of their present day position. In most localities in the Sahara the lakes had dried out some 6,000 years ago, while only a single site in the south yielded deposits which gave information concerning the younger part of the Holocene. Between 6,000 and 4,000 years ago the Sahelian savannah was progressively replaced by Saharan plant communities, while from this last date and until the present day this transition was completed and vegetation degraded to its present day composition, dominated by Chenopodiaceae–Amaranthaceae.

These few examples are brought here to indicate the possibilities, but had been only randomly chosen. The reconstruction of past vegetations in arid lands, when safely tied up with chronology, can serve as a basis for almost any other, progressive interpretations.

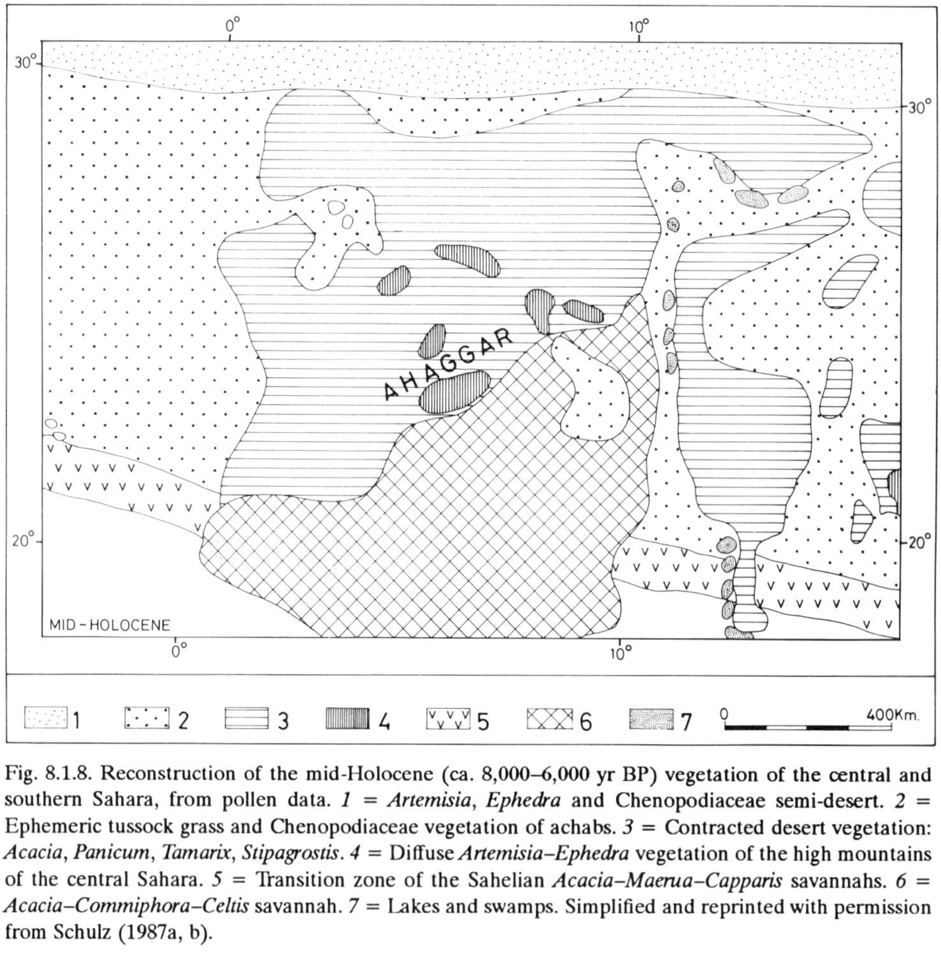

Fig. 8.1.8. Reconstruction of the mid-Holocene (ca. 8,000–6,000 yr BP) vegetation of the central and southern Sahara, from pollen data. *1* = *Artemisia*, *Ephedra* and Chenopodiaceae semi-desert. *2* = Ephemeric tussock grass and Chenopodiaceae vegetation of achabs. *3* = Contracted desert vegetation: *Acacia*, *Panicum*, *Tamarix*, *Stipagrostis*. *4* = Diffuse *Artemisia–Ephedra* vegetation of the high mountains of the central Sahara. *5* = Transition zone of the Sahelian *Acacia–Maerua–Capparis* savannahs. *6* = *Acacia–Commiphora–Celtis* savannah. *7* = Lakes and swamps. Simplified and reprinted with permission from Schulz (1987a, b).

8.2. Climate

Reconstruction of paleoclimates is fundamentally based on considerations related to past vegetation communities, or plant distribution patterns, as interpreted from the pollen spectra. Consequently, similarly to the situation with reconstructing past vegetations, a clear distinction must be made when discussing late Cenozoic climates or dealing with earlier periods. The late Cenozoic climates could usually be reconstructed based on understanding of the flora, since very frequently recent counterparts could be found and used as a reliable basis for interpretations. This is hardly so in periods predating the late Cenozoic, for which reconstructions should imply different concepts, discussed in Chapter 11.

Very frequently however the stage of reconstructing a vegetation is skipped while interpreting paleoclimates, as it is a common habit among investigators to refer, for example, to arboreal pollen as indicating a moist or humid climate, while dealing with arid lands spectra, with no direct regard of the actual floras. It is however understandable that in the background, it is the composition of vegetation which plays the role and skipping is usually done for the sake of ease and simplicity of presentation.

Two stages in the reconstruction of past climates are common: the first involves defining the characteristics of an interpreted paleoclimate, such as being humid, dry, warm, cool and so on; the second implies attempts at modelling such climate, trying to reason out the type of atmospheric circulation which may have been responsible for an ensuing climate and vegetation at a given period. An example of this working process, from palynology through vegetation, climate and finally proposed paleoclimatic models, is detailed in Chapter 9 for the late Cenozoic of Israel.

Occasionally, only a certain particular component of a past climatic system is interpreted or emphasized in the discussion, which seems of cardinal importance to the solution of a given problem, or when the palynological data are insufficient for more general conclusions. Such could be a change of humidity without reference to its sources, extent or type, that may in cases be either of a regional, climatic significance or as a result of restricted local conditions such as development of a fault spring (Horowitz, 1987b); wind directions (Van Campo, E. et al., 1982, Hooghiemstra, 1988b), discussed in more detail in Chapter 8.3; past temperatures (El Moslimany, 1988), or other possible factors. Such partial interpretations of past climates usually concentrate on the more important factor pertaining to the discussed period and environment, as defined and conceived by individual investigators (see for example the preferential importance given to paleotemperatures over humidity off the southeastern African coast by Van Campo, E. et al., 1990). Another aspect of past climates is the style of rain, which may at times have a greater effect on plantlife than the total average annual amount.

Evidently, precautions brought above for interpreting past vegetations equally apply also to paleoclimates, thus should preferably be observed. The most significant of these precautions in arid lands are the partial independence of local vegetation from the general climate, being controlled by hydrology; long distance transportation, where pollen brought by winds do reflect the climatic regime, while those grains carried by rivers may only bear on conditions in their intake areas, which may be located quite far away. Paleoclimates of the source regions for long rivers such as the Nile (Rossignol, 1969b) could be by themselves of importance, but do not necessarily have connection with, or influence upon a site analyzed from an arid region. A third important factor is the anthropogenic influence on recent spectra taken for reference. The significance of these factors is considerably more influential in arid lands as compared with forested, where the masking factor of pollen derived from regional xerophytic flora is low.

It is quite interesting to note that in some cases a paleoclimatic regime can be interpreted in more detail for the provenance regions of long distance transported palynomorphs, rather than for the desert environment in which these had been finally deposited. This may especially happen when the dry region is extremely-arid, while the depositional basin is surrounded by well developed local vegetation, that masks the regional pattern. Such is for instance the case in Israel, where late Cenozoic sequences recovered from the desert surrounded Dead Sea basin (for details see Chapter 9) yielded a more detailed information on composition of Mediterranean vegetations 200–300 kilometers to the north, as well as on hilly flora to the west and east. The patterns and possible changes in the arid flora could be referred to merely as of a "desert" character. It is however the possibility to discern paleoclimates from two regions in a single sequence, that makes pollen analysis so significant in arid lands.

Reconstructions of paleoclimates can, same as for past vegetations, be a verbal discussion (*cf.* Cour and Duzer, 1976; Davis et al., 1985a); summarized in a table (*cf.* Markgraf, 1985; and see also Chapter 9); or presented in some graphic form. Such graphic representations of past climates could be curves delineating interpreted developments through time of commonly reconstructed factors as temperature, moisture or both, in a single site (*cf.* Weinstein-Evron, 1983, 1990) or several sequences (*cf.* Caratini et. al., 1976, 1978; Maley, 1989). Probably the ultimate way of presenting paleoclimatic evolution is through a series of maps, covering as close as possible successive time increments. However, since most paleoclimatic researches in arid lands are presently only in their initial stages, this way of presentation is as yet rarely used.

The components of pollen spectra thought significant for interpreting various paleoclimatic factors are evidently different from one locality to the other, but may also differ from one researcher to another in the same region. Such differences in opinions among various investigators are quite common and may lead to opposing views, based on similar data.

Some examples of differences in opinions relating to the paleoclimatic significance of certain pollen taxa have already been discussed above, such as the controversy between Van Zeist (1967) who attributed disappearance of trees from the Zagros Mountains to drought, opposed by El Moslimany (1988) who considered this phenomenon a result of lowering winter temperatures and heavy snowfall.

Disappearance of mangrove pollen from offshore sediments off southwestern India is described by Van Campo, E. (1986), who attributed this trend to further aridization of the region. This interpretation is also supported by the composition of pollen derived from land vegetation at that time, which seemingly corroborates this postulation. It seems, however, that since at present mangroves grow even in extremely-arid regions such as the Sinai coast, this should be viewed as a

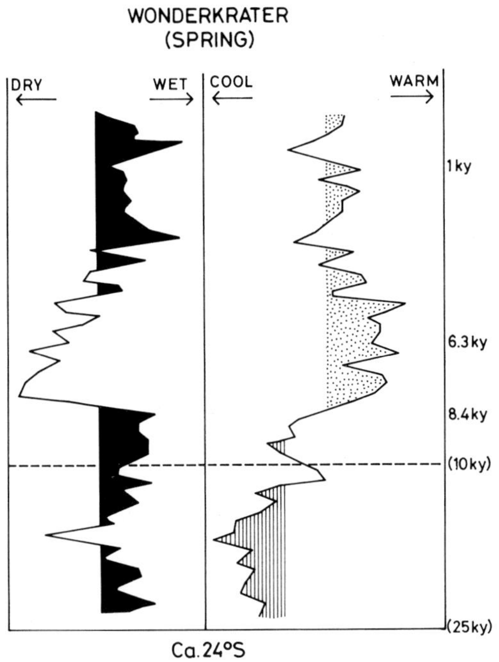

Fig. 8.2.1. Comparison of factor scores identified with temperature and moisture for Wonderkrater Spring, Transvaal, South Africa. Different shadings aid viewing of consistent pattern. Reprinted with permission from Scott (1989).

coincidence, connected with the general lowering of the global sea level at the peak of the last glacial, which may have had deprived these trees from their suitable biotopes. This was a dry period in the monsoon belt, for global reasons concerned with lowered solar insolation input, that were responsible also for accumulation of ice on the poles, which in turn caused lowering of sea level.

It is mandatory to state exactly on which components each part of the interpretations is based, to avoid future misunderstanding. As an example, Scott (1989) brings a detailed list of taxa for various localities in semi-arid southern Africa (Table 8.2.1) for founding his interpretations, representing either cool or warm temperatures, moist or dry habitats. Based on this list and its proposed paleoclimatic implications, Scott brings forth curves for both temperature and moisture developments, covering the late Pleistocene and Holocene in several sites from southern Africa. One of these, for Wonderkrater spring deposits, is brought here as an example (Fig. 8.2.1).

Caratini et al. (1976) had analyzed several cores of offshore sediments from the Atlantic Ocean west of the Sahara coast and presented pollen diagrams, based on which they drew an aridity curve for this region, covering the last 150 ky, for which they also show a good correlation with oxygen isotopes curve obtained

TABLE 8.2.1

Pollen taxa from Wonderkrater sequence, Transvaal, South Africa, showing strong loadings in factors identified with moisture (wet *vs.* dry) and temperature (cool *vs.* warm) conditions; arrangement of taxa in each group is according to descending relative values of their factor loadings

Moisture factors		Temperature factors	
Dry	Wet	Cool	Warm
Tarchonantheae	*Podocarpus*	*Stoebe* type	Combretaceae
Capparaceae	*Typha*	*Tulbaghia*	Capparaceae
Stoebe type	*Olea*	Ericaceae	Tarchonantheae
Tulbaghia	*Anthospermum*	*Passerina*	Mimosoideae
Ericaceae	*Burkea*	*Artemisia*	*Burkea*

Reprinted with permission from Scott (1989).

TABLE 8.2.2

Taxa indicating moisture conditions in the western Sahara, from offshore sediments of the Atlantic Ocean

Aridity	Humidity	Of no significance
Chenopodiaceae	Pteridophyte spores	Echinate Compositae
Amaranthaceae	Combretaceae	*Euphorbia*
Caryophyllaceae	Cyperaceae	Gramineae
Ruellia patula	*Elaeis guineensis*	Papilionaceae
Boraginaceae	*Rhizophora mangle*	Allochthonous
Tribulus	*R. racemosa*	
Fenestrate Compositae	*Mytragyna inermis*	
Cruciferae	*Celtis*	
Boerhavia	*Salix*	
Acacia	*Dodonaea viscosa*	
Prosopis	*Typha*	
Phoenix	Sapotaceae	
Hyphaene thebaica		
Plantago		
Calligonum		

Data from Caratini et al. (1976).

for the same cores (Fig. 8.2.2). The aridity curve is based on a subdivision of the pollen taxa observed in the spectra into two groups, representing either arid or humid conditions (Table 8.2.2). Although some inaccuracies seem inevitable, mainly due to overlappings resulting from the insufficient levels of identifications, it seems that the recurrence of results in several sequences and the correlation with oxygen isotopes curves justify the use of these taxa as reliable foundations for the paleoclimate reconstructions proposed in Caratini et al.

A similar study had been conducted by Caratini et al. (1978) in the Arabian Sea, off the coast of Oman, also resulting in a definition of aridity index which depends

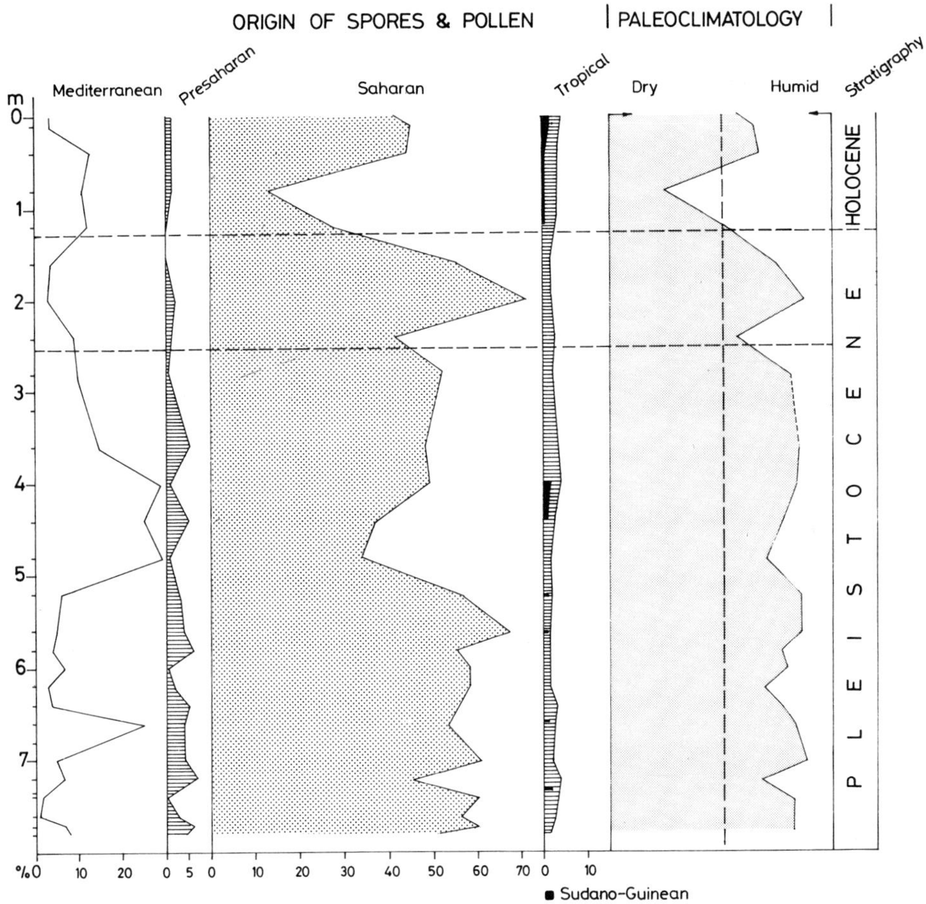

Fig. 8.2.2. Provenances of major pollen groups (left) and aridity index curve (right) for Core KL 15, collected offshore Cap Blanc (Mauritania). Reprinted with permission from Caratini et al. (1976).

on relations between pollen grains derived from plants adapted to drought such as Chenopodiaceae, Amaranthaceae, several Caryophyllaceae, *Tribulus* and *Ephedra*, against those derived from plants needing higher amounts of moisture, such as Cyperaceae, *Typha* and spores of pteridophytes and bryophytes. These permitted reconstruction of the aridity index and its fluctuations throughout the last 30–35 ky (Fig. 8.2.3). Good correlations have been obtained among the cores collected from the Arabian Sea, thus the recurrence of shapes of aridity index curves seem to justify their use for stratigraphic and paleoclimatic implications.

It is noteworthy, considering differences in opinions concerning use of certain taxa as paleoclimatic indicators that Caratini et al. (1978) regarded Cyperaceae pollen as moisture indicators, while in the same region El Moslimany (1990) states

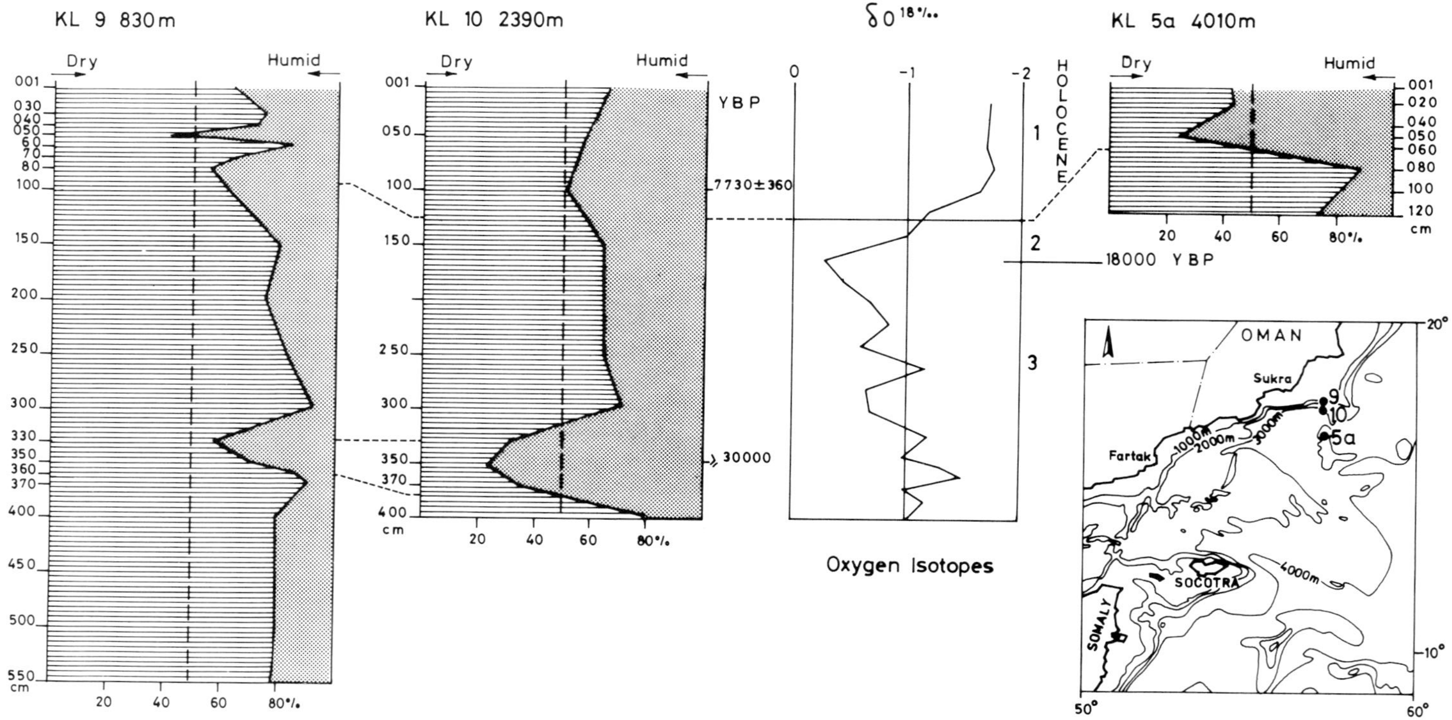

Fig. 8.2.3. Aridity index curves for three cores collected offshore Oman, from pollen data, compared with the oxygen isotope record. Reprinted with permission from Caratini et al. (1978).

TABLE 8.2.3

Comparisons between shares of evergreen and winter deciduous oaks pollen in three basins along the Dead Sea Rift, Israel, for the present day, mid-Holocene and the late Würmian pluvial phase

	Hula Valley			Lake Kinneret			Dead Sea		
	A	B	C	A	B	C	A	B	C
Present day	11	77.2	22.8	14.6	62.7	37.3	12.1	87.2	12.8
Mid-Holocene	26.3	27.9	72.1	48.5	35.5	64.5	19.5	46.4	53.6
Late Würmian	33.4	11.1	88.9	no data —			54.5	10.5	89.5

A = Share of oak pollen from the total spectra; B = share of evergreen oak pollen from the total number of counted oak grains; C = share of deciduous oak pollen from the total number of counted oak grains. Reprinted with permission from Horowitz and Gat (1984).

that the same grains are characterizing extremely-arid environments, such as the Rub' Al-Khali Desert.

The possibility of summer rains in Israel during the so termed "wetter" periods, characterized by considerably higher arboreal pollen shares in respective sediments, is discussed by Horowitz and Gat (1984), as opposed to the present day situation in which rains in Israel are confined to the winter time. The relative distribution of different species of oaks pollen is taken as basis for interpreting the rain style. At the present day evergreen oaks dominate the Mediterranean environments and recent pollen spectra. In the mid-Holocene Atlantic stage and late Würmian wet pluvial stage, general shares of oak pollen are considerably higher than at present, but the upper hand is significantly taken by pollen of winter deciduous ones, rather than the present day evergreen. The respective percentages are presented in Table 8.2.3.

The prevalence of winter deciduous oaks pollen in periods for which a "wetter" climate was concluded from the greater abundance of AP (Horowitz, 1979) is considered a cause for interpreting at least some summer rains for the above mentioned periods, which enabled a better distribution and areal spreading of such winter deciduous trees. Incidentally, some other winter deciduous trees had been more abundant in these periods, but are represented by only a minor part of the pollen spectra, most of them being insect pollinated. The relics of these tree species are still abundant in Israel, but in greatly reduced numbers. It is noteworthy that these winter deciduous relics make two thirds of the arboreal taxa today, while in areal coverage the evergreen trees are much more widespread, probably since they are better adapted to the present day dry summers, by maintaining at least some growth during the wet winter season. The assumption of summer rains is also supported by results obtained from the composition of stable hydrogen and oxygen isotope in paleowaters, dating to these wetter periods. For more details see Chapter 9.

8.2.1. Pollen–climate transfer functions

Transfer functions have been used by several investigators to directly tie up, in a mathematical way, the pollen percentages with various paleoclimatic factors. It should be primarily stressed that all kinds of mathematical computations are based on present day models, when recent pollen assemblages are taken as reference. This implies that all limitations affecting interpretations of fossil pollen spectra based on recent ones, are not overcome by applying mathematical computations, anymore than when using other presentation techniques. Although dealing with numbers and formulas, this method does not seem more accurate, neither more objective, than other ways of interpreting paleoclimates. This warning is directed at students who seem to be under the impression that mathematical applications would render a higher degree of precision or objectivity to interpretations. As in other cases, it is the basic concepts and ideas which would result in interpretations making sense or not, rather than the way of presentation. The initial checks of results obtained by mathematical treatments are also of the usual nature, namely whether a different method would corroborate or contradict the paleoclimatic interpretations.

The main problem, as with other ways of interpreting past climate and ancient processes, is that very frequently the only available control is the model constructed on the present day data. The problem involved with not having a more objective control for the past are discussed in Chapter 7.1.

Roeland et al. (1988) proposed pollen–climate transfer functions for East Africa by applying statistical methods such as multiple regression to modern quantitative pollen data, based on more than 300 samples and 695 taxa. Their regression equation is built between climatic parameters and percentages of 70 taxa from 227 pollen spectra. The results displayed good correlation between estimated and measured values of mean annual temperature and rainfall, thus can apparently be quite safely used for further studies, of fossil assemblages from this region. An attempt to reconstruct paleoclimates for the last 40,000 years in tropical Africa is presented by Bonnefille et al. (1990).

Swain et al. (1983) have made estimates of Holocene precipitation for the Rajasthan Desert, India, based on both pollen and lake level data. Taking into account the present day pollen spectra and climatic data relating to the "winter" (November through May) and "summer" (June through October) monsoons and correcting for locally over-represented pollen taxa. Another correction was the omission of cereal pollen, evidently being of anthropogenic origin, but without regard to other possible human influences on the recent pollen spectra, such as discussed in chapters 7.1.2 and 10.

Based on studies of recent pollen spectra by Singh et al. (1973) correlation coefficients with the summer and winter precipitations had been calculated for 12 taxa, presented in Table 8.2.4. It was found out that some of these taxa bear positive correlation coefficients, while other display negative ones, with either the summer

TABLE 8.2.4

List of pollen taxa used in the pollen sum on which estimates of Holocene precipitation in Rajasthan are based, together with their correlation coefficients with the "summer" and "winter" precipitation

Pollen of	Correlation coefficient	
	Summer	Winter
Calligonum [a]	−0.76	0.46
Capparis	−0.03	−0.25
Pinus [a]	0.39	0.79
Gramineae	0.08	−0.23
Cheno-Ams [a]	0.55	0.18
Aerva [a]	−0.86	−0.35
Artemisia	0.47	0.66
Oldenlandia	0.46	0.48
Prosopis	0.05	−0.32
Ephedra	0.43	0.02
Mimosa	0.31	−0.19
Syzygium [a]	0.46	0.83

[a] Taxa included in the final calibration function. Reprinted with permission from Swain et al. (1983).

or winter precipitation. Of these 12 taxa five were found to have more significant correlation coefficients, namely *Calligonum*, *Aerva*, *Chenopodium* type, *Syzygium* and *Pinus*, thus their percentages have been calculated from the sum of the 12 pollen taxa listed in Table 8.2.4.

The pollen diagram thus obtained (Fig. 8.2.4) served as a basis for calculations of the estimates pertaining to summer and winter precipitations, which seem to coincide quite well with lake levels in this region, considered to be directly influenced by the quantities of rains throughout the Holocene. The pollen diagram is modified from Singh et al. (1974), who presented a detailed sequence for this region. The calculations made by Swain et al. also coincide pretty well with the paleoclimatic interpretations of Singh (1971), based on reconstructions of Holocene vegetations and their connections with amounts of precipitation in Rajasthan.

It should be noted that Vishnu-Mittre (1974) stated that Singh (1971) had incorrectly assessed the botanical remains and proposed a different paleoclimatic scheme. The two conflicting viewpoint are partially based on a different attribution of the climatic significance of *Artemisia* and Cyperaceae, whether they indicate moisture or drought. Singh considered both to indicate higher humidity, while Vishnu-Mittre regards these as indicators for increased aridity. Needless to say, both researchers base their opposing views on the recent pollen spectra.

Swain et al. presented a third opinion, when concluding that they believe Cyperaceae should not at all be used as paleoclimatic indicators, since they are extremely variable in distribution, preferably responding to local moisture conditions rather than to the regional climate.

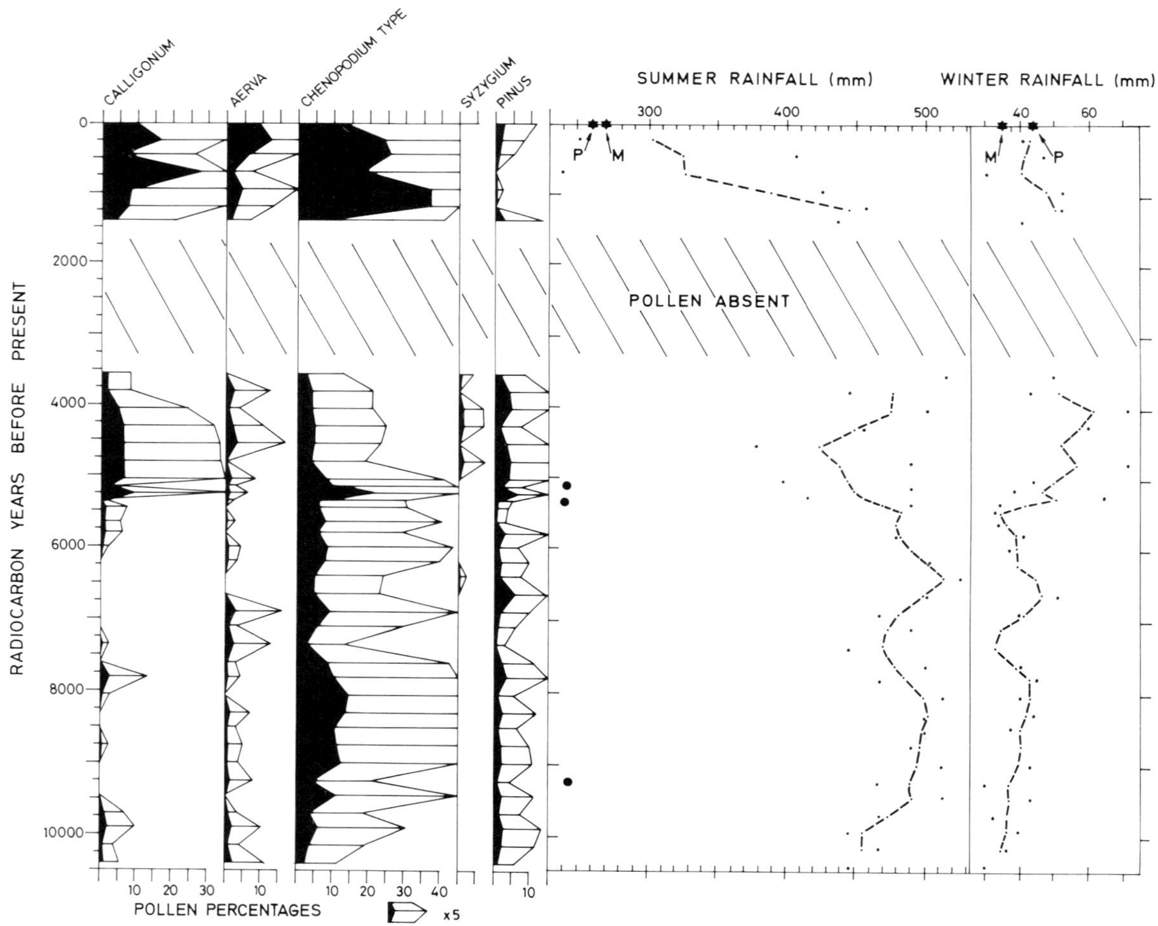

Fig. 8.2.4. Pollen profiles and estimates of summer and winter precipitation derived from pollen data from Lunkaransar saltwater lake, Rajasthan, India. The pollen percentages are based on a sum of 12 taxa listed in Table 8.2.4. Radiocarbon dates are indicated by black circles. *M* represents modern summer and winter precipitation based on 1930–1960 meteorological data from Bikaner; *P* represents reconstructed estimates based on modern pollen data from the same vegetation zone as Lunkaransar. Individual precipitation estimates for each pollen sample level are indicated with dots; the dot-dashed line is a 1-2-1 smoothing of the individual estimates. Reprinted with permission from Swain et al. (1983).

The estimates published by Bryson and Swain (1981) for the same region somewhat differ from those of Swain et al. (1983) by using a different pollen–climate calibration factor. They also included several other pollen types: *Capparis*, Gramineae, *Artemisia*, *Oldenlandia*, *Prosopis* and *Ephedra*, which seem to have a greater range of variations at several points along the sequence, as compared with the recent, surface samples. It can clearly be seen from a comparison of the two studies that the interpretations arrived at depend very much on the basic assumptions employed by either group of students for calculating the precipitation correlation coefficient factors, as well as the pollen taxa included or excluded from these computations. Views on the latter issue evidently differ among various investigators.

Contrary to the attitude taken by Bryson and Swain (1981) and by Swain et al. (1983), which was based on application of the knowledge of climatic limits of a small number of taxa, Kershaw and Nix (1988) have based their quantitative paleoclimatic estimates from a sequence obtained in the region of Cairns, northeastern Queensland, Australia, on a considerably greater number of taxa. Kershaw and Nix examined an alternative approach to the one discussed above, utilizing both range distributions and abundance of a relatively large number of taxa to arrive at paleoclimatic estimates. Pollen deposited in the study area have had two main provenances discussed by Kershaw and Nix, a rain forest in which the annual precipitation may reach up to 4,000 mm and a xeric environment, where yearly rainfall is less than 300 mm. By calculating the percentages and coefficients of many pollen taxa they could show possible distributions of mean annual temperatures and precipitation for both the driest and wettest environments along the last 10,000 years.

Van Campo, E. et al. (1990) used recent pollen spectra for calibration of transfer functions, to be used for interpretation of paleotemperatures east of the African coast, in a core collected somewhat north of the Zambezi River outlet to the Indian Ocean. Van Campo, E. et al. maintain that temperature is the prime environmental factor in this subtropical region. They focused on Combretaceae and Ericaceae, for establishment of their functions, since these are well represented in the pollen spectra and their general distribution patterns in East Africa follow contrasting altitudinal and thermal affinities.

By using transfer functions the authors show that results displayed by the pollen records coincide quite well with those shown from both the analysis of oxygen isotopes and plankton, carried out in the same offshore sequences. Van Campo, E. et al. stress that the three paleotemperature curves thus obtained are based on independent estimates, each with quite different data, sensitivities and assumptions. Since there seems no other connection between the land originated pollen, composition of plankton and oxygen isotopes but temperature, the authors expect correlation between estimates only for large scale trends of climate changes, not necessarily for the minor deviations, which certainly reflect different environmental settings and processes.

The discussion concerning some inconsistencies between terrestrial and marine paleotemperature reconstructions for the last glacial maximum is not yet closed, according to Van Campo, E. et al. (1990). However, they maintain that despite the different processes involved, both lowland terrestrial and marine estimates of the surface paleotemperatures at this site seem to accurately record the average value and range of temperature changes over major glacial/interglacial cycles. Their study does not reveal any noticeable discrepancies between land and sea trends in paleotemperatures, the latter measured from oxygen isotopes.

It seems that the interpretations by Swain et al. (1983) and Van Campo, E. et al. (1990) are better founded than those relying on pollen alone. The interdisciplinary approach, which presents independent corroboration of interpretations based on palynology, may provide a sounder basis for future development of pollen–climate transfer functions.

8.3. Transport

Transporting agents have a marked effect on pollen spectra obtained from arid lands, due to combination of both the frequently long distances involved and the low masking factor of local pollen production. This may occasionally seem a drawback, while trying to reconstruct regional environments, but could be rather useful while attempting to establish energies and paths of transporting agents in the past. The two most important long distance pollen transporters, winds and rivers, have been studied in several localities in arid lands and their contributions to spectra taken as a basis for interpreting the paths and energies of these agents in the past. The past activity of some other transporting agents, like the sea and ocean currents, were quite rarely studied in dry regions, while contributions of minor pollen carriers such as animals and groundwater are only occasionally mentioned in the literature.

Naturally, the influence of fluviatile transportation is more prominent in marine or endoreic continental basins of deposition, fed by relatively large rivers arriving from far away provenances, which are usually also affected by winds to some degree. Pollen spectra recovered from marine sediments bordering an arid land could also serve in certain opportunities to reconstruct sea currents activities, as expressed by pollen deposition in the studied locality.

Marine sediments are generally the best source for reconstructing transporting agents activity, primarily because they display relatively low percentages of pollen produced by aerially restricted plant communities, such as those maintained by local hydrologic peculiarities. This makes it much easier to separate and group the palynomorphs taxa into associations arriving by either transporting agents, which also facilitates defining the probable provenance for each of the groups.

The main problem involved with reconstructions of past transporting agents activity is that any change which might have had affected these agents, would possibly have also resulted in changes of locations of vegetation belts and provenances for the transported palynomorphs. It is therefore necessary to interpret the results very carefully, taking into consideration both possibilities. In this respect too, marine sediments seem to make a better source for information, since usually it is quite easy to discern between river and wind transported assemblages. Comparison of changes in the relative percentages of palynomorphs contributed by either source can facilitate interpretations of paths and relative influence of each of the transporting agents. Pollen spectra recovered from marine sediments deposited in such localities which are not fed by any river, as the Bay of Elat (Aqaba), display a more pronounced influence of sea currents and winds.

Pollen spectra obtained from continental sediments are usually dominantly affected by wind transportation, but whenever an endoreic lake such as the Dead Sea (Fig. 2.2.7, Table 2.2.5) occurs in an arid region, fluviatile contribution should be accounted for.

8.3.1. Winds

The validity of recent pollen spectra as indicators of wind directions through their share of allochthonous pollen was demonstrated by various authors in different arid regions (Horowitz, 1966; Horowitz et al., 1975; Van Campo, M., 1975; Nieto and D'Antoni, 1984; Cheddadi and Van Campo, E., 1985; Hooghiemstra and Agwu, 1986; Caratini et al., 1988; and many others).

Studies by Hooghiemstra and Agwu (1986), Hooghiemstra et al. (1987) and Hooghiemstra (1988a, b, 1989) deal with pollen transportation and deposition processes in the Atlantic Ocean, offshore western Sahara. The construction of a present day model enabled interpretations of paleowind directions and intensities in this region throughout the last 140 ky. Hooghiemstra and his collaborators used different ways of grouping palynomorph taxa observed in the spectra for either reconstructing the paleovegetation belts discussed above (see Chapter 8.1), or the paleowind data, detailed in Hooghiemstra (1988a, b).

Hooghiemstra maintained that intensified trade winds would transport pollen over a greater distance, or import a larger amount of grains from far away provenance, as compared with periods in which these winds are of lesser intensity. He expected therefore that the present day distribution patterns of trade transported palynomorph taxa would shift southward when these winds became stronger. *Pinus* best comes up to the requirements of an index taxon for that purpose, since it grows in distant localities and its pollen production is high, well suited for wind transport. The main provenances for *Ephedra*, *Artemisia* and Compositae are situated north of the Sahara (Van Campo, M., 1975; Cour and Duzer, 1976), making also these taxa suitable indicators for trade winds activity. The extension of their source areas and consequently their pollen contribution, possibly changed to a greater extent during the past, following climatic changes in the region, as compared with *Pinus*, which grows farther away from the desert. Pollen of Chenopodiaceae, Amaranthaceae and Gramineae seem to be indicative of trade winds activity in transportation as well, as far as pollen from northern provenances are concerned.

Pollen grains brought to the deposition site from the east can reach the location only while west blowing winds are active, thus they record the African Easterly Jet transport.

Based on these principles Hooghiemstra (1988b) subdivided pollen taxa in spectra obtained from the offshore cores into seven groups:

(1) Forest trees and shrubs of western and southern Europe and northern African provenances (*Alnus*, *Betula*, *Corylus*, *Picea*, *Pinus* and *Tilia*).

(2) Mediterranean trees and shrubs, including deciduous oaks (*Erica*, *Olea*, *Quercus* spp., *Rhus*, *Thymelea* and others).

(3) Plants of steppe and desert communities (*Balanites*, *Ephedra*, *Calligonum*, *Capparis*, *Tamarix*, *Zygophyllum* and others).

(4) Taxa with many representatives in steppe and desert environments, pollen of which could only be identified to family or genus level (*Artemisia*, *Paronichia*, Chenopodiaceae/Amaranthaceae and Gramineae).

(5) Taxa which have their center of distribution within the tropical forests, but reach into savannahs including the Sahel (*Acacia*, *Celtis*, *Elaeis*, *Cassia*, *Rhizophora*, Combretaceae and others).

(6) Taxa often characteristic of moist or wet biotopes (Cyperaceae, *Isoëtes*, *Sparganium*, *Sphagnum*, *Typha* and *Utricularia*).

(7) Other taxa not included in the previous groups, such as Compositae, Cruciferae, *Plantago*, *Polygonum*, pteridophyte spores and so on.

Groups 1–5 represent distinct latitudinal vegetation belts, while 6 and 7 are less associated with such zonation. Thus two types of pollen diagrams have been calculated, based either on the total of groups 1–5, indicating vegetation belts shifts throughout the analyzed cores, or including also groups 6 and 7 which have taxa significant for wind activity, as Compositae and Cruciferae. The most characteristic trade wind indicators are found within pollen groups 1, 3, 4 and 7.

Based on the relative distribution of these pollen groups along the cores, Hooghiemstra (1988b) had constructed a diagram of trade winds intensity during the period from 20–5 ky BP. Taking into account other indicators as well, he could also detect shifts in the savannah belt, changes in discharge of the Senegal River, or contributions made by Mediterranean vegetation (Fig. 8.3.1). Hooghiemstra (1989) had further extended these records to cover the last 140 ky.

Van Campo, E. et al. (1982), Prell and Van Campo, E. (1986) and Van Campo, E. (1986) have shown, using recent pollen deposition processes as a model, variations in monsoon intensity over the last 150 ky, in pollen diagrams obtained from deep sea cores. Their results seem to display rather good correlation with oxygen isotope curves for the same sequences, offshore India (Van Campo, E., 1986) and the

Arabian Sea (Van Campo, E. et al., 1982; Prell and Van Campo, E., 1986), which had demonstrated direct connection between the degree of activity and intensity of the monsoons, with paleotemperatures records obtained from foraminifera.

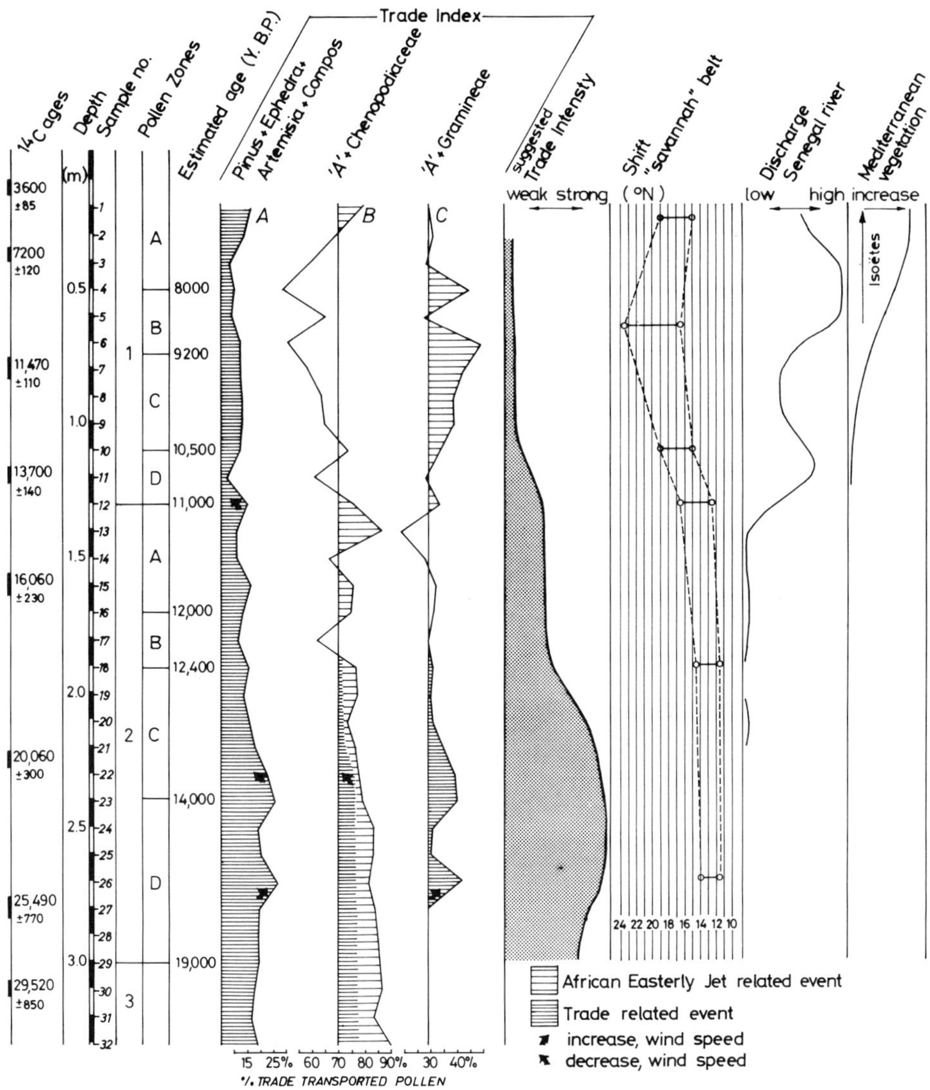

Fig. 8.3.1. Summary of changes in vegetational and climatic variables in northwest Africa around Termination I, based on pollen data. Estimated relative intensity of the trade winds based on pollen influx of trade-transported pollen. Also shown are estimated latitudinal shifts of the grass-rich "savannah" vegetation zone, or southern border of the Sahara, discharge of the Senegal River based on the *Rhizophora* record and abundance of Mediterranean type vegetation. The *Isoëtes* record is indicative of seasonal climatic humidity north of the Sahara. Reprinted with permission from Hooghiemstra (1988b).

8.3.2. Rivers

Studies concerned with past river activities are quite rare in arid lands, due to the obvious fact that these transporting paths are not too common in such regions, usually playing important roles only in very specific environments. Studies of recent pollen deposition processes from several localities around the world show the potential importance, involved with analysis of pollen transported from far away provenances, in determining past fluviatile history (Rossignol, 1969b; Maley, 1972; Hooghiemstra, 1988a), but however systematic reconstructions of paleorivers through time are quite rare.

An example brought above from Hooghiemstra (1988b; Fig 8.3.1) for discharge of the Senegal River, which crosses the southwestern part of the Sahara and pours its waters into the Atlantic Ocean, seems to be one of the few exceptions to the present state poverty of research in this respect. Although amounts of water carried by the Senegal River evidently reflect climatic conditions in its tropical provenance, quite far from the Sahara to the south, some influence on the arid environments have most probably been exerted, at least in biotopes connected with the river. Amounts of water in such a large river and others of its kind crossing deserts, directly affect groundwater table in their floodplains, which may extend quite a distance from the river itself. Thus the large rivers determine vegetations and possibilities of life and human settlement in extended areas, as is the situation in Egypt, located in an otherwise Saharan environment, where most of the population occupies the Nile Valley and Delta.

Changes and variations in arid lands biotopes are usually not directly connected with conditions in the provenance area of the rivers, but pollen spectra obtained from arid localities affected by the rivers, such as deltaic and offshore marine environments, may be significantly influenced by the amounts of water and the ensuing bank vegetation. River activity has thus two aspects in an arid region, displayed in its contribution to the pollen spectra. The first is of grains brought along from its provenance region, which can usually be easily detected; while the other concerns palynomorphs produced by bank vegetation, which are difficult to discern from pollen produced by the xerophytic flora, such as most frequently are Gramineae and Cyperaceae, which may even mask the regional spectra.

Horowitz (1979) showed that a clear definition of river activity, based on these palynomorphs which are allochthonous beyond doubt, such as pteridophyte spores and pollen of *Podocarpus* and Combretaceae brought to the Israeli offshore by the Nile (Rossignol, 1969b), could be used to estimate the amounts of pollen produced by the river's bank vegetation, which comprises taxa similar to those arriving at the sea by the eastern winds from Israel. In this way a more precise reconstruction of the regional flora could be obtained, by subtracting this part of the spectra contributed by the Nile River.

Some mentions of past changes in river activities in arid regions, based on pollen analyses are, among others, by Van Campo, E. (1986), who observed that the western Ghats rivers of southwest India had reduced runoff during an aridity phase between 22 and 18 ky BP; Sneh et al. (1986) and Cheddadi (1988) described fluctuations of the Nile River during the late Pleistocene and Holocene; D'Antoni et al. (1984) bring some notes from semi-arid Argentina.

It seems that the potential lying in this way of treatment, of analyzing contributions made by rivers to past pollen spectra and subsequently reconstructing the fluviatile history, are far from fully utilized at the present day in arid lands and can be further explored.

8.3.3. Other transporting agents

Sea currents have proved to be significant contributors of allochthonous palynomorphs (Horowitz, 1966) to recent sediments, but it seems however that no reconstruction of their past activity in arid lands had ever been thoroughly discussed. Similarly, potential possibilities of reconstructing past groundwater activity were hardly explored in arid lands, except for initial considerations regarding the past activity of Quaternary springs in the Dead Sea Rift, identified by their redeposited Pliocene sporomorphs (Horowitz and Horowitz, 1990).

Transportation by animals was studied for the past especially in the American Southwest, considering packrat (*Neotoma*) midden (Betancourt et al., 1990) and in southern Africa (Scott, 1987; Scott and Bousman, 1990), but these transportation paths are very short, confined by the ranges of activity of involved animals. This way of transportation seems however responsible for part of the composition of pollen spectra from midden, coprolites and the like. Long distance transportation by animals such as migratory birds was, to the best of my knowledge, never studied for the past.

8.4. Environments of deposition

Almost any environment of deposition from which sequences are collected for pollen analysis is typified by a characteristic local vegetation, which is represented, or occasionally even over-represented in the ensuing pollen spectra. The importance of reconstructing past environments of deposition seems to increase progressively when older strata are studied. This may become especially significant while analyzing boreholes, particularly the deep ones penetrating into sediments deposited in unknown environments, which may occasionally be difficult to interpret from the rock samples brought up by the drilling rig. Quite often such layers do not have any counterparts cropping out, as is for instance the case with many tectonic basins, thus borehole samples are the only source for information pertaining to environments of deposition (see Chapter 9).

Reconstructions of past depositional environments, besides the obvious scientific interest, may have quite important bearings on economic issues, such as the search for possible traps of oil and gas (Horowitz and Horowitz, 1990), or defining the subsurface extension of prospective aquifers.

Reconstructions of past environments of deposition are carried out by referring to both the internal composition of pollen taxa derived from the local flora and to numeric relations between this particular group and other palynomorphs produced by the regional vegetation. In this respect it should be noted that these latter relations do not solely depend on the shape and size of the depositional basin, but also on general climatic conditions. Thus, in order to correctly interpret the depositional environments, as far as possible, it is mandatory to primarily establish the general climatic conditions, as interpreted from the regional components of the pollen spectra; define possible contributions made by transporting agents, again by considering percentages and composition of the allochthonous part of the spectra; and only then proceed to discuss local factors directly affected by the environment of deposition, which could serve as indicators for reconstruction, only after exclusion of all possible external influences on the pollen spectra.

A detailed example of such separation of the regional from local components, the use of the former to conclude about general climatic conditions and application of the latter for interpreting local environments of deposition, is brought in Chapter 9 for the Dead Sea Rift in Israel. A detailed discussion of anthropogenic environments is given in Chapter 10.

8.4.1. Marine

Conditions which had prevailed in past marine environments are usually interpreted from sequences collected from the sea or ocean bottom cores, a very common technique at present, but which unfortunately only spans younger geologic periods, hardly ever reaching beyond the Neogene/Quaternary boundary. Older marine rocks may be exposed or drilled on land, or penetrated by offshore boreholes, mainly those connected with search for oil and gas. Although it is impossible to directly define changes in seawater depth by palynomorph assemblages, coastline fluctuations can be studied from both spectra obtained from marine sediments and those accumulated on the continental side.

A clear differentiation is made while studying marine sequences, between palynomorphs of terrestrial provenances, arriving by winds and rivers and those of marine origin, such as organic walled dinocysts and related forms. The distribution patterns of both groups in the marine sediments is controlled to some degree also by sea currents. Dinocysts are usually living in the uppermost layer of seawater, where light is plentiful. They are rather sensitive indicators of salinity and temperature in their specific biotope, information which is gathered from the internal composition and variability of taxa within this group of palynomorphs.

Since some of these microfossils are very sensitive to salinity changes, they could occasionally be used as indirect indicators of aridity (Morzadec-Kerfourn, 1980, 1988). It is however of prime importance to make sure that such interpreted salinity changes had indeed resulted from changing climates, not a consequence of changes in oceanic circulation or freshwater discharge by rivers into the sea. Truly, these two factors may depend on general climate, but may also be a result of structural changes, especially in seas such as the Mediterranean or the Red Sea which have a very narrow connection with the oceans. It is also possible that an increase of salinity caused by a global change of climate and oceanography at a certain locality, would be accompanied by a more humid climate on land, not necessarily by aridity, which is too often connected to salinity. This connection needs a positive proof, rather than the usual feelings.

A second method of treating organic walled microfossils for establishing past oceanographic conditions is by comparing, along the sequence, their numeric relations with the land originated palynomorphs. The ensuing curve is used for interpreting possible changes in the distance of sampled locality from the coastline, or in other words, interpreting past sea level fluctuations. There are however several pitfalls in applying this method, which should be clearly established beforehand. Changes in the primary productivity of the sea, which determine the amounts of organic walled microfossils, are not necessarily resulting from, or accompanying sea level changes; variations in terrestrial palynomorphs influx into marine sediments could be initiated by several factors other than the elevation of coastline; the combination of these two factors calls for great attention while interpreting sea level changes from dino/pollen ratios.

An example of these variables can be seen (Fig. 8.4.1) in a study made by Caratini et al. (1978) on bottom sediments of the Arabian Sea off the Oman coast. Two antagonizing tendencies affect the dino/pollen ratios: during the late Pleistocene sea level was low and pollen production on land was also low, due to the arid climate (compared with index of aridity for the same cores, Fig. 8.2.3). These trends have been inverted in the Holocene, when sea level rose, but so did pollen production, due to the more humid climate and better developed vegetation. When observing the dino/pollen ratios in the cores it is evident that in KL9, close to the coast, pollen take the upper hand in the Holocene; farther away, at KL10, the dino/pollen ratios remain almost constant over the Pleistocene/Holocene boundary; while in KL5, still farther seaward, dinos are much more abundant in the Holocene, reflecting sea level changes. It is therefore evident that even the location of sampling may determine the respective representations, thus only an analysis of several sequences could clear the picture and lead to a more meaningful interpretation in such cases.

Sea level changes can also be reconstructed from the relative percentages and absolute concentrations of typical coastal vegetation pollen in the sediments, for which mangroves are one of the best indicators in the hot arid regions. The virtue

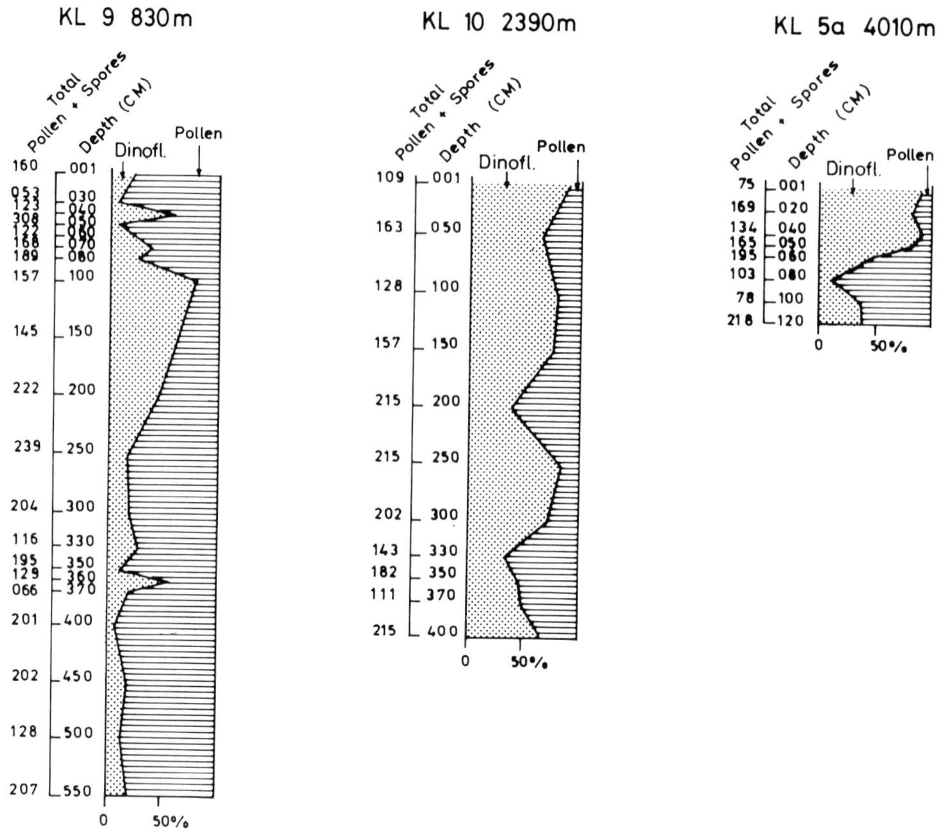

Fig. 8.4.1. Ratios of dinoflagellates and terrestrial palynomorphs in three cores offshore Oman. For location of cores see Fig. 8.2.3. Reprinted with permission from Caratini et al. (1978).

of mangroves as a coastal indicator lies in the fact that these trees grow in no other environment but coastal. This is in contrast to other typical coastal plants such as Chenopodiaceae, Gramineae or Cyperaceae, which are also found along rivers and in other environments, thus bringing into consideration the factor of overlapping, which would make the ensuing interpretation less refined.

Although unintentionally, Van Campo, E. (1986) had demonstrated (Fig. 8.4.2) that both pollen percentages and concentrations in pollen per gram sediment of Rhizophoraceae (mangroves) can be used to follow sea level changes during the last 20 ky, in a sequence collected off the southwest Indian coast. Van Campo, E. had attributed the changes in distribution of mangrove pollen to the general climate on land, thus regarding these grains as aridity indicators. It seems however that they should be considered as indicators of sea level changes rather than the general climate, as also maintained by Lézine (1988, p. 141): "Mangrove development in the Lake Guiers area follows the general known pattern for the sea level fluctuations".

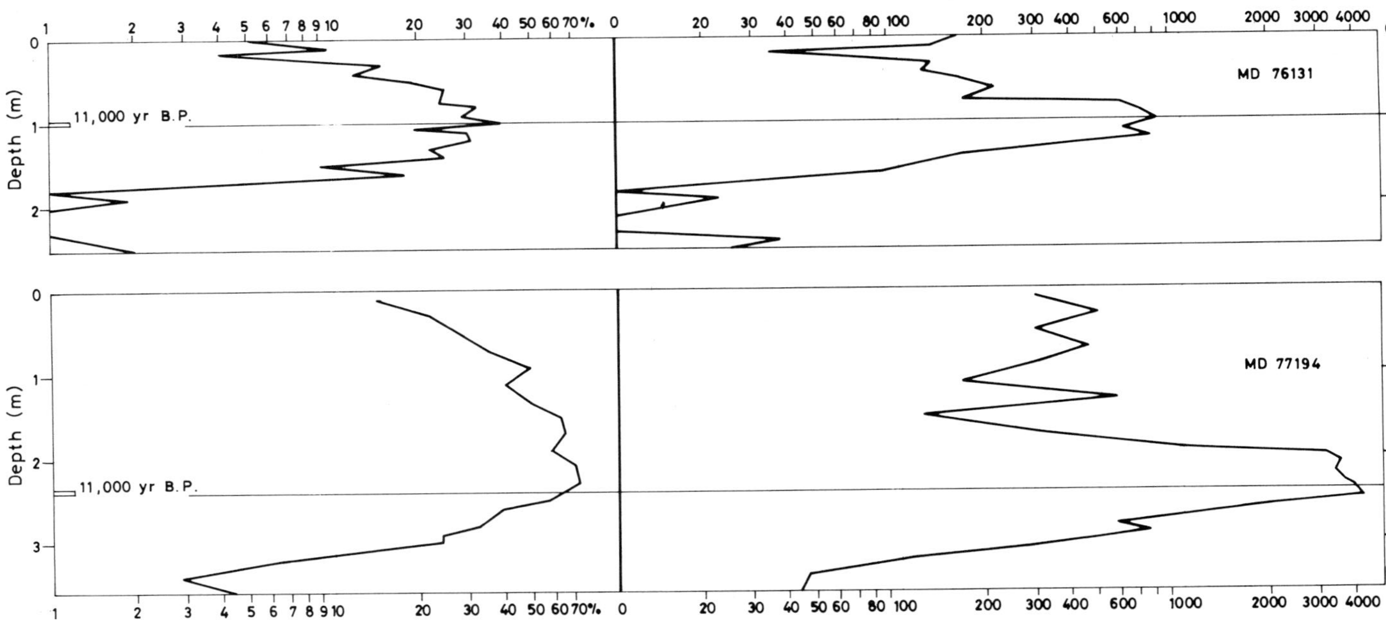

Fig. 8.4.2. Rhizophoraceae (mangroves) pollen percentages on the left and concentrations per gram sediment on the right, in two sediment cores offshore southwestern India. Reprinted with permission from Van Campo, E. (1986).

Naturally, there is an almost direct relationship between sea level changes and the glacial/interglacial cycles during the Quaternary, which makes a feasible connection between the two factors, sea level and aridity, thought to affect mangrove development and distribution.

Cheddadi (1988) had shown that some aspects of the seawater circulation could be detected by the state of preservation of palynomorphs recovered from deep sea cores. Distinct differences in both pollen concentrations and their state of preservation, could be used for indication whether bottom sediments in which the palynomorphs had been embedded contained oxygen, or comprised a reducing environment. In the former case palynomorphs were rather poorly and differentially preserved, while in the latter pollen concentration was up to three times higher and a significantly larger number of taxa have been observed.

This directly bears on possible reconstructions of seawater circulation, whether the sea was stratified, as happened in the Mediterranean numerous times throughout the Quaternary (Rossignol-Strick, 1983), or if circulation had affected the entire depth of the water body, bringing oxygen down to the bottom. Incidentally, defining the Mediterranean stratification helped in constructing paleoclimatic models for the region (see Chapter 9).

The relative frequencies and pollen concentrations per gram sediment (pollen influx) depend to a great extent on transporting agents, bringing terrestrial palynomorphs to the sea, aided by the activity of sea currents in redistribution of the grains. Needless to say however, that the initial factor controlling pollen influx is pollen production on land and to some degree rates of sedimentation in the studied basin. The distribution pattern of pollen and spores was studied in detail for recent Mediterranean sediments off the Israeli coast by Rossignol (1961, 1962, 1964, 1969b), who found out that palynomorph concentrations vary, depending on both distance from the shore and lithology (see also Chapter 2.2 and Fig. 2.2.10).

The amount of freshwater discharge by rivers pouring into the sea could be defined for the past by comparing relative frequencies of bank vegetation pollen with the bulk of other terrestrial palynomorphs. Such a study was carried out by Hooghiemstra (1988b) for the Senegal River outlet, in sediment cores collected off the Sahara coast, in the Atlantic Ocean (see Fig. 8.3.1) and is also mentioned in Van Campo, E. (1986) for the Ghats rivers in southwestern India.

A pollen diagram obtained for Tabun Cave, situated on a cliff facing the Carmel coast in Israel (Horowitz, 1979) demonstrates the possibility of studying sea level changes from the continental side. The diagram shows two distinct parts: the lower (Fig. 8.4.3), for which a dry climate is known for Israel (see also Chapter 9), manifested by very low arboreal pollen ratios and the wealth in Compositae grains; while the upper part of the diagram is typified by considerably higher shares of AP and an almost total absence of Compositae, indicating a humid climate.

Changes in percentages of pollen produced by the local vegetation, Gramineae and Cyperaceae growing in connection with freshwater environments, or Chenopo-

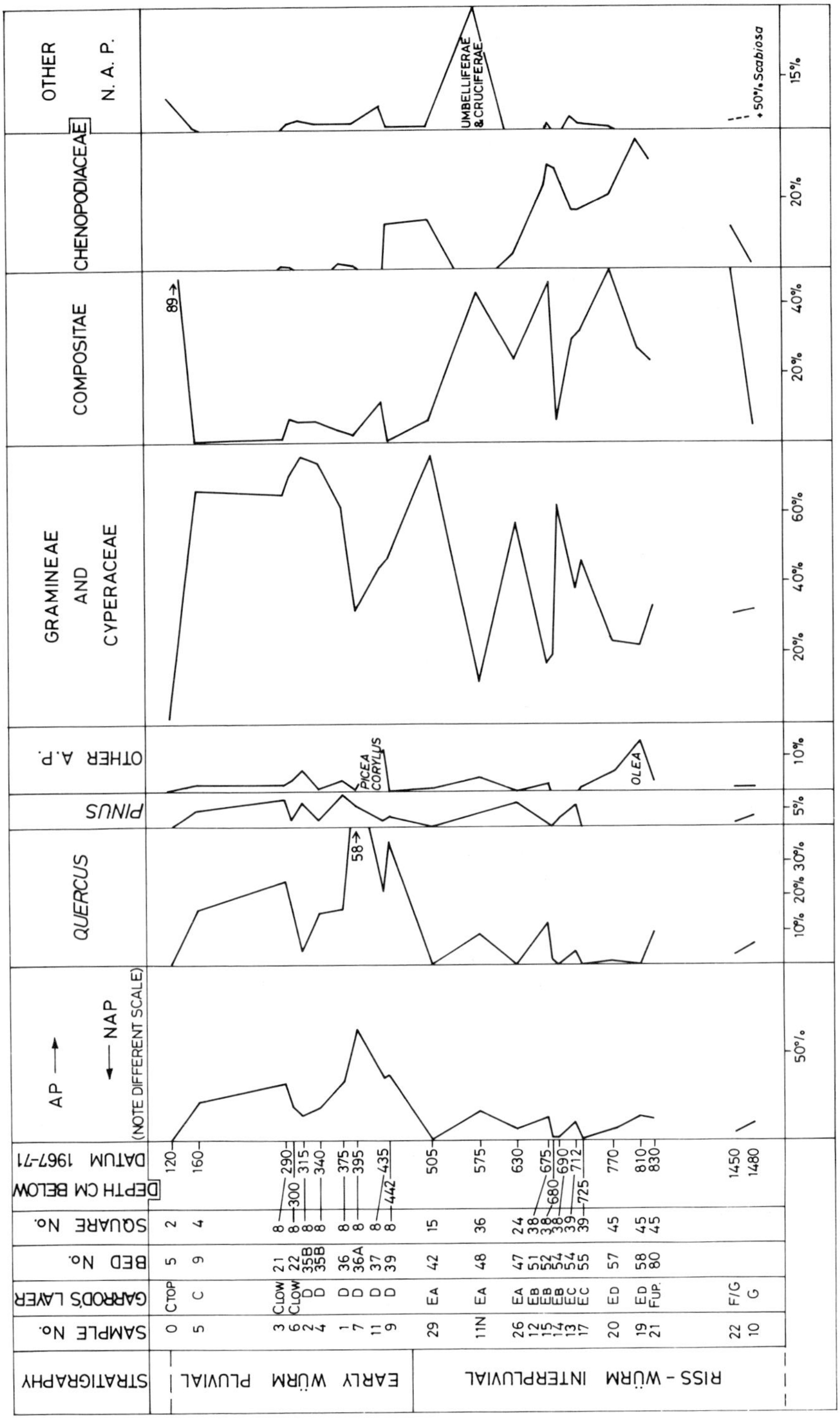

Fig. 8.4.3. Pollen diagram from Tabun Cave, Mount Carmel, Israel. Percentages based on the total number of counted grains. For location see Fig. 10.3.1b. Reprinted with permission from Horowitz (1979).

diaceae maintained by saline biotopes, had been used to reconstruct the nearby coastal plain hydrological conditions, which depend to a great extent on the relations of the freshwater coastal aquifer and sea level (see also Jelinek et al., 1973). The lowermost sample displays very low percentages of Gramineae and Cyperaceae, an almost total absence of Chenopodiaceae, but very high, more than 50% of pollen produced by *Scabiosa prolifera*. This signifies that there is hardly any coastal plain which would have maintained marshes or playas and the non-arboreal components are only contributed by plants growing on the hills, where *Scabiosa* is abundant in semi-arid climates (Weinstein, 1981). The poverty in other local NAP may have been a result of the coastal plain being covered by the transgressive sea, whose sediments are known from the Carmel foothills (Horowitz, 1979).

Further up the sequence (although it is not clear whether it is continuous at this sector), within its lower half, Gramineae and Cyperaceae oscillate around the values they show at the bottom, while Compositae and Chenopodiaceae considerably increase. These trends probably indicate regression of the sea, followed by inception of higher humidity in the region, causing development of saline coastal marshes and playas typified by these non-arboreal plants. The wealth of Chenopodiaceae in the lower part indicates that these marshes had been more saline at the beginning of the regression. Salinity decreased with the retreat of the sea, as evidenced by the very low shares of Chenopodiaceae in the upper half of the sequence, in which the local vegetation contribution is almost entirely dominated by Gramineae and Cyperaceae, indicating the formation of marshes, by inflow of freshwater resulting from the combination of a retreating sea and general amelioration in climate, which is evidenced by higher percentages of arboreal pollen in this part of the section.

8.4.2. Continental

Several factors involved with continental depositional basins can be determined from analysis of the local component of the pollen spectra obtained from their respective sequences. The size and shape of a basin and their fluctuations, can be interpreted from numeric relations between the local and regional components of pollen spectra, as discussed in detail in Chapter 7.4. In this way, possible causes for such fluctuations and shapes could also be determined, whether they are controlled by the general climate, structural disturbances and so on.

In a rather similar manner, the occurrence and water discharge of springs surrounding a depositional basin could be determined by the composition of the local pollen flora. The appearance and disappearance of springs, or changes in their freshwater output, which also cause changes of salinity in the depositional basin, would have an immediate reflection on composition of the surrounding vegetation. Such changes in spring activity may either be caused by a general climate change accompanied by increase or decrease in precipitation, or due to structural activity, when faulting may cause exposure or blocking of aquifers. A more detailed discussion of structural considerations is given in Chapter 8.5.

Relative salinities and changes in salinity can be reconstructed for the past in inland depositional basins by relations between pollen of the halophil Chenopodiaceae, against the hydrophil Gramineae and Cyperaceae, combined with pollen derived from other hydrophil plants. These relations seem to be almost global indicators for salinity, but however in certain regions different marker taxa could be found of better use for this purpose. By comparisons of the relative salinities interpreted from the halophil/hydrophil pollen ratios with the regional components, it is sometimes possible to say whether a freshwater influx to an otherwise saline basin, such as a terminal lake, a salt-pan and so on, had resulted from the general amelioration of climate or from other causes.

Fig. 8.4.4 compares two sectors from Admon 1 borehole, drilled in the southern Dead Sea basin. In one sector (Fig. 8.4.4a) the trends of hydrophil/halophil pollen ratios are mostly parallel to the curve displaying shares of arboreal pollen out of the regional vegetation components, while in the other sector (Fig. 8.4.4b) the two curves sometimes antagonize. It is quite evident that in the first sector changes in salinity were caused by the general climatic conditions affecting precipitation, whereas in the latter by faulting, exposing aquifers and creating springs at the fault scarps facing the subsiding rift valley floor, where the sampled site is located (Horowitz and Horowitz, 1990).

The composition of the local pollen flora and its comparison with the regional can occasionally yield information concerning the type of a depositional basin in arid lands, whether it forms a terminal base level or acts as an intermediate lake en route a river system. Since the formation of terminal depositional basins is frequently connected with faulting and rifting processes, dealt with in Chapter 8.5, this way of interpretation is further detailed there.

Rates of deposition in various basins could sometimes be estimated from absolute frequencies of palynomorphs in the sediments (pollen influx). If the regional pollen production is taken as a measure, if estimates can be made for its stability or amount of change and if no differential destruction is suspected (or can be accounted for), greater abundance of pollen per gram sediment would signify a lesser rate of deposition. A diagram presenting absolute pollen frequencies for a sequence collected from Lake Bogoria, Kenya, is brought in Vincens (1986) and reproduced below (Fig. 8.4.5). The sequence, which is not continuous, represents in its lower part a humid environment, followed by an arid in the upper, interpreted from the composition of the pollen spectra. The lower part, in which pollen production was quite steady under the humid conditions, shows correlation between the pollen influx and type of sediments, where in general the coarser deposits contain less pollen, most probably due to higher rates of accumulation as compared with the finer fractions.

The upper part, which was deposited in times of poor pollen production due to

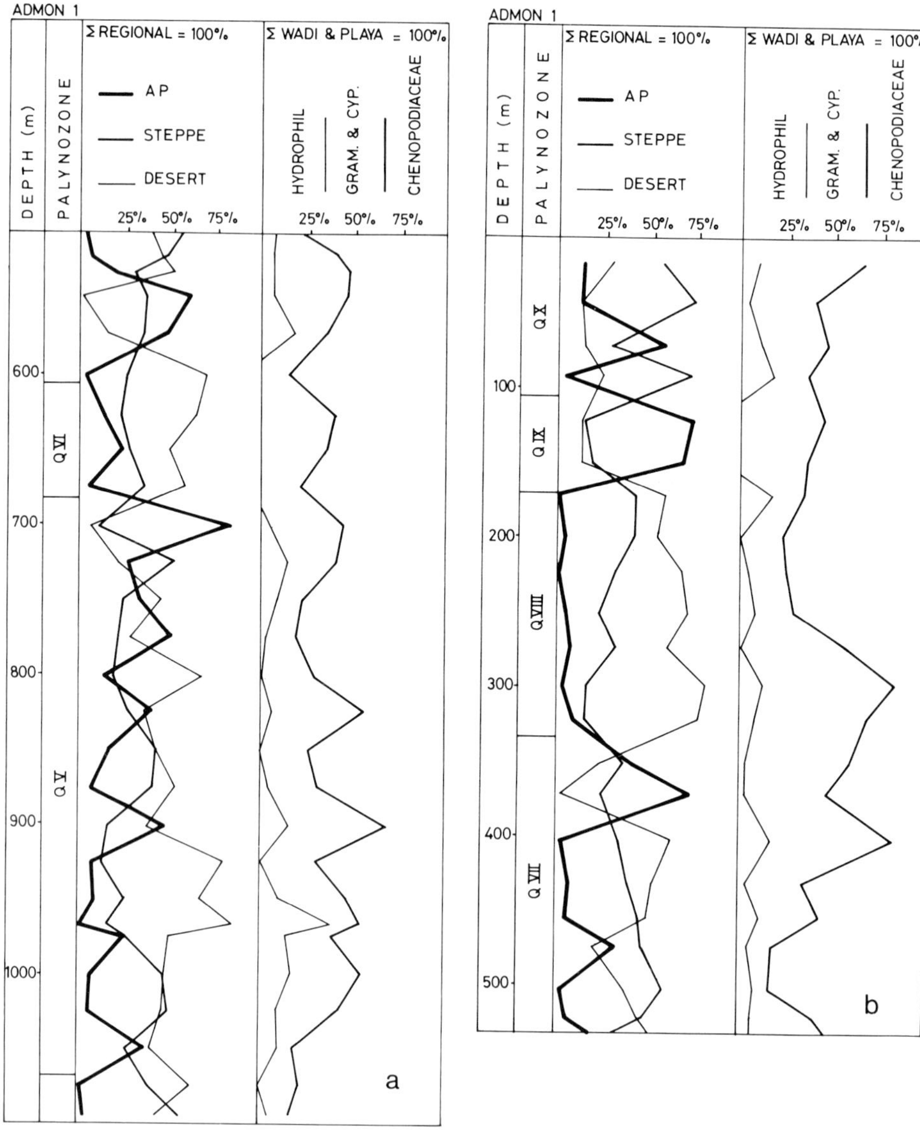

Fig. 8.4.4. a, b. Two sectors from Admon 1 borehole, southern Dead Sea basin. For explanation see text. For location see Fig. 9.2.1. Reprinted with permission from Horowitz and Horowitz (1990).

the arid climate, does not show any significant correlation. This indicates that when pollen production varies considerably over short periods, as is the usual case in arid environments, no estimates of rates of deposition could be based on the pollen influx.

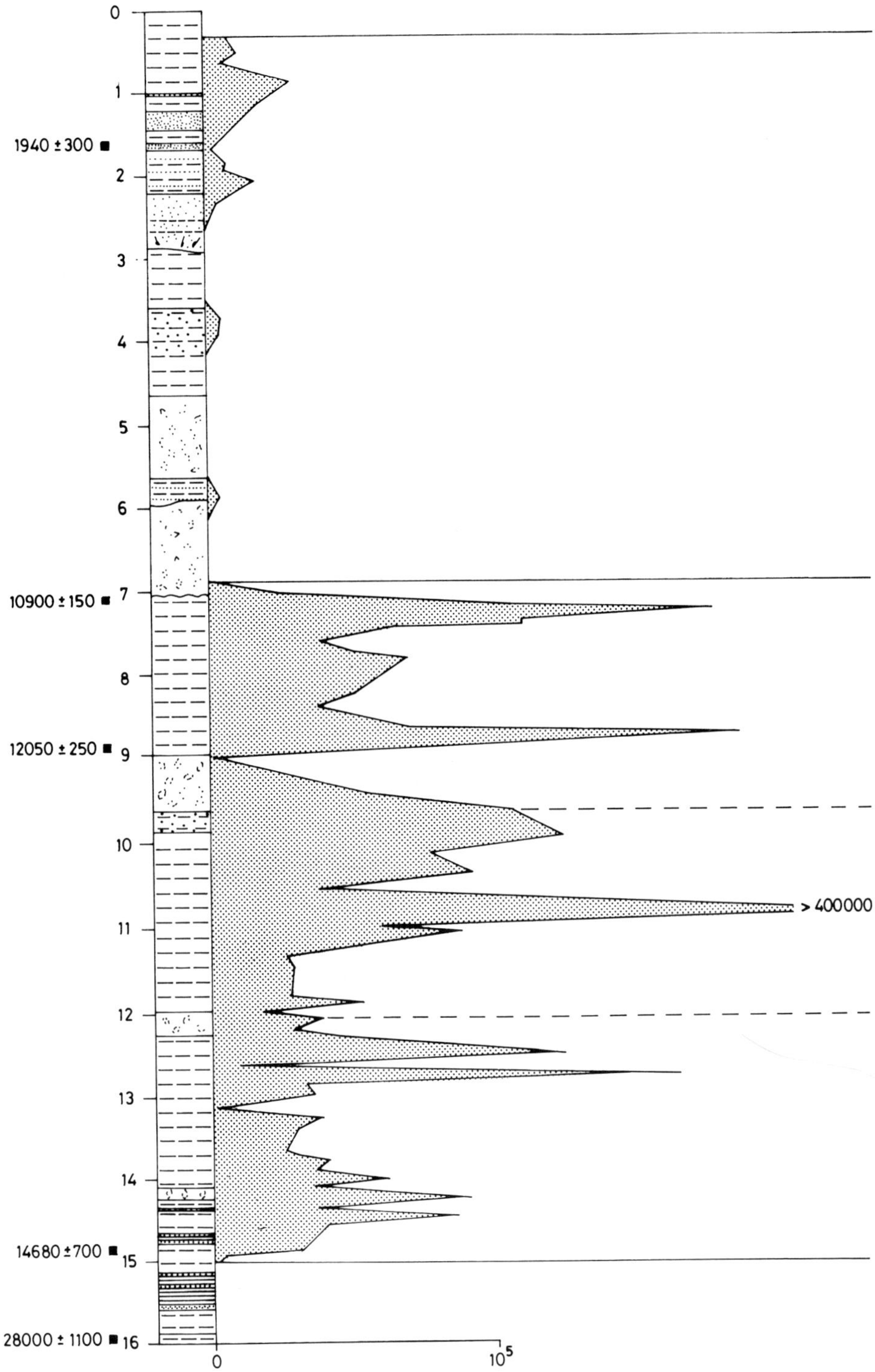

Fig. 8.4.5. The connections between sediment type, its rate of accumulation and the absolute pollen frequencies (number of pollen per gram sediment), in a core from Lake Bogoria, Kenya. Core depth given in meters; black rectangles indicate radiocarbon dates. Reprinted with permission from Vincens (1986).

Evidently, pollen analysis could be used for estimates of rates of deposition through its palynostratigraphic implication. Although this is not particular to arid environments, it has importance due to the fact that in many dry regions palynomorphs comprise the only fossil group which is represented in the sedimentary column, thus making palynology the sole tool which could be used for stratigraphic studies, particularly those concerned with landlocked basins. A detailed discussion of these possibilities is given in Chapter 9.

An example of the use of palynostratigraphy for determining rates of accumulations of sedimentary formations within the Dead Sea Rift in Israel and the ensuing structural considerations and interpretations, is detailed in Chapter 8.5.

8.5. Structure and landscape

Palynology has several merits in arid lands for reconstructing and dating structural disturbances such as faulting, uplift, rift formation and its deepening phases and establishment of ensuing landscapes. The most common way of carrying out such reconstructions through time is by assessing stratigraphy of the layers, followed by comparisons of their relative rates of accumulation. Although this is an accepted method in all regions, it is usually in the arid that no other fossils but palynomorphs are found in the deposits, thus serving as a sole foundation for stratigraphy. This tool becomes of utmost significance while studying subsurface stratigraphy of rocks laid down under continental conditions, where virtually no other microfossils occur and larger ones are impractical, since they are very rare even in outcrops. Certainly no continuous records of any other fossils occur in such sediments.

Truly, some other microfossils may occasionally occur in continental borehole sequences, such as diatoms or ostracods in some lacustrine sediments, but their presence depends on the availability of proper living environments in the depositional basin, a prerequisite which is not often met. Palynomorphs, in contrast, are produced outside the basin and are thus found even in sediments deposited in hypersaline lakes, such as rocksalt and gypsum, which are devoid of any other life remains. Thus palynomorphs can also be used for regional correlations, where the application of such may shed light on structural development of inland basins such as the Dead Sea Rift (Horowitz, 1987c, 1989c).

Structural changes can occasionally be detected by the appearance of springs. Again, although this may also be the case in richly vegetated region, it is only in the arid ones that a sudden appearance of a spring would leave an immediate, considerable mark on pollen spectra collected from any nearby site (Horowitz, 1987b).

Other such indirect interpretations could be applied to the reconstruction and dating of structural movements. The sudden development of a local hydrophil vegetation along a sequence, accompanied by unusual thicknesses of the succeeding chronostratigraphic unit, may indicate rift formation at a certain locality. Usually such tectonic movements can also be defined from other evidence, but it is palynology that would mark the date and sometimes, in subsurface sequences, even call initial attention to such a possibility.

Examples for structural reconstructions are brought below, from the Dead Sea Rift in Israel, based on many years of pollen analyses on both outcrops and boreholes.

8.5.1. Regional palynostratigraphy and structure

The late Cenozoic sequence was subdivided in Israel into 15 palynozones (Horowitz and Horowitz, 1985), presented and discussed in Chapter 9 (see Tables 9.2.1 and 9.3.1 for details; Fig. 9.2.1 for locations of boreholes; Figs. 9.3.1, 9.3.2, 9.4.1 for outline of dating). Rates of accumulations of the various palynozones are calculated from the pollen diagrams, obtained for the various boreholes. Outcrops never present complete continuous sequences due to rapid subsidence of the rift valley floor and consequent burial of the strata, while those remaining outside centers of subsidence have been subject to erosion.

Rates of subsidence of the various subbasins of the Dead Sea Rift are displayed in Fig. 8.5.1 for the Neogene palynozones, in boreholes where these have been encountered, namely Arava 1 and Ami'az 1 in the southern Dead Sea, En Gedi 2 on the border of the southern and the northern Dead Sea basins, Zemah 1 in the central Jordan Valley and Notera 3 in the Hula Valley. Rates of accumulation for the late Pliocene and Quaternary palynozones are very detailed for Notera 3 and Amazyahu 1 boreholes, where calculations could be made in 100 ky increments thanks to detailed sampling (Horowitz, 1989c); the latter is shown in Fig. 8.5.2. Rates of accumulations are given for entire palynozones in other boreholes (Fig. 8.5.3). Figure 8.5.4 summarizes rates of accumulations of palynozones deposited during the last 3.5 million years in the Hula, central Jordan Valley, northern and southern Dead Sea subbasins.

Palynozone Ma, of latest Oligocene through early Miocene age, was only positively encountered in a single borehole in the southern Dead Sea region, Ami'az 1, questionably at the bottom of Sdom 1 (the different spelling for "Sdom" and "Sedom", for the same locality, are traditional, thus each appears in the way it is mentioned in the relevant literature). In Ami'az 1, the palynozone attains almost 400 meters in thickness, overlying Senonian chalk. Correlative sediments in the vicinity are much thinner than the Ma sequence in the subsurface, never attaining more than several tens of meters at maximum. It is, therefore, quite clear that in the early Miocene the southern Dead Sea region was already subsiding as compared to surrounding regions. However, it most probably did not subside in the form of a deep trough, which would have captured the drainage.

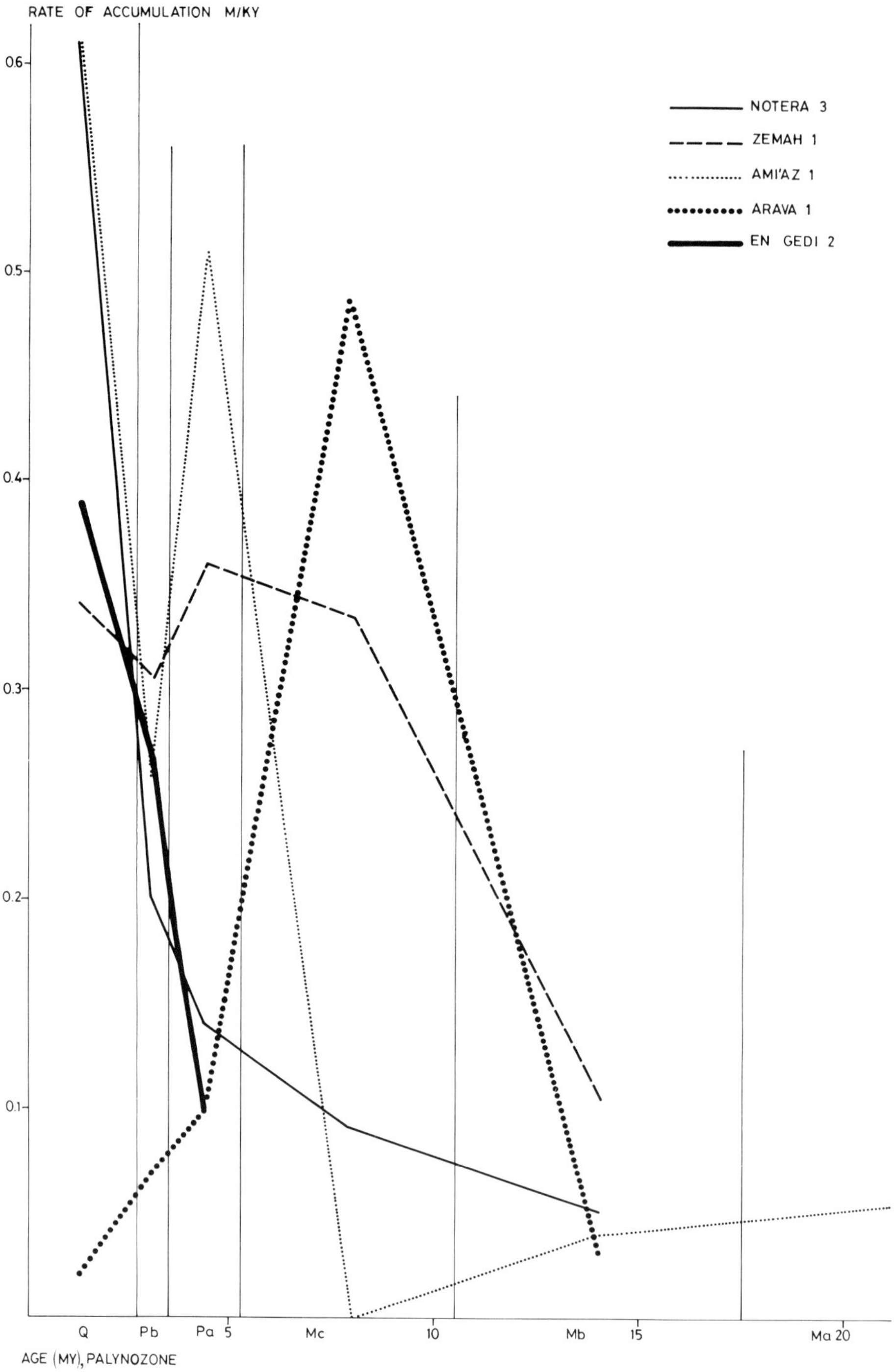

Fig. 8.5.1. Neogene rates of subsidence and accumulation in various sub-basins of the Dead Sea Rift, calculated from palynostratigraphy. For locations see Fig. 9.2.1.

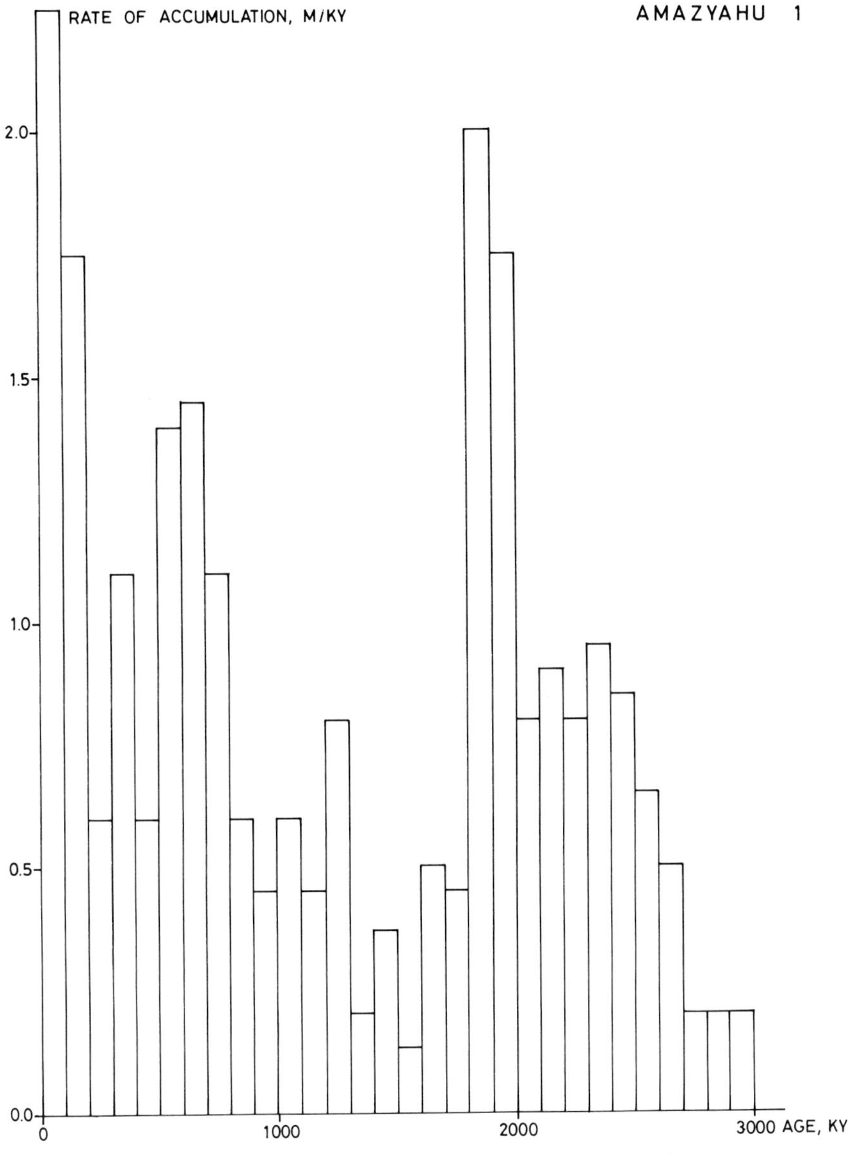

Fig. 8.5.2. Rates of subsidence and accumulation in Amazyahu 1 borehole, southern Dead Sea basin, in 100 ky increments. For location see Fig. 9.2.1.

Mb palynozone, of latest early through middle Miocene age, was encountered in several boreholes in the region. However, a seemingly complete sequence is only known from Ami'az 1, where somewhat more than 300 meters have been penetrated. Since the overlying, Mc palynozone is missing from Ami'az 1 and the

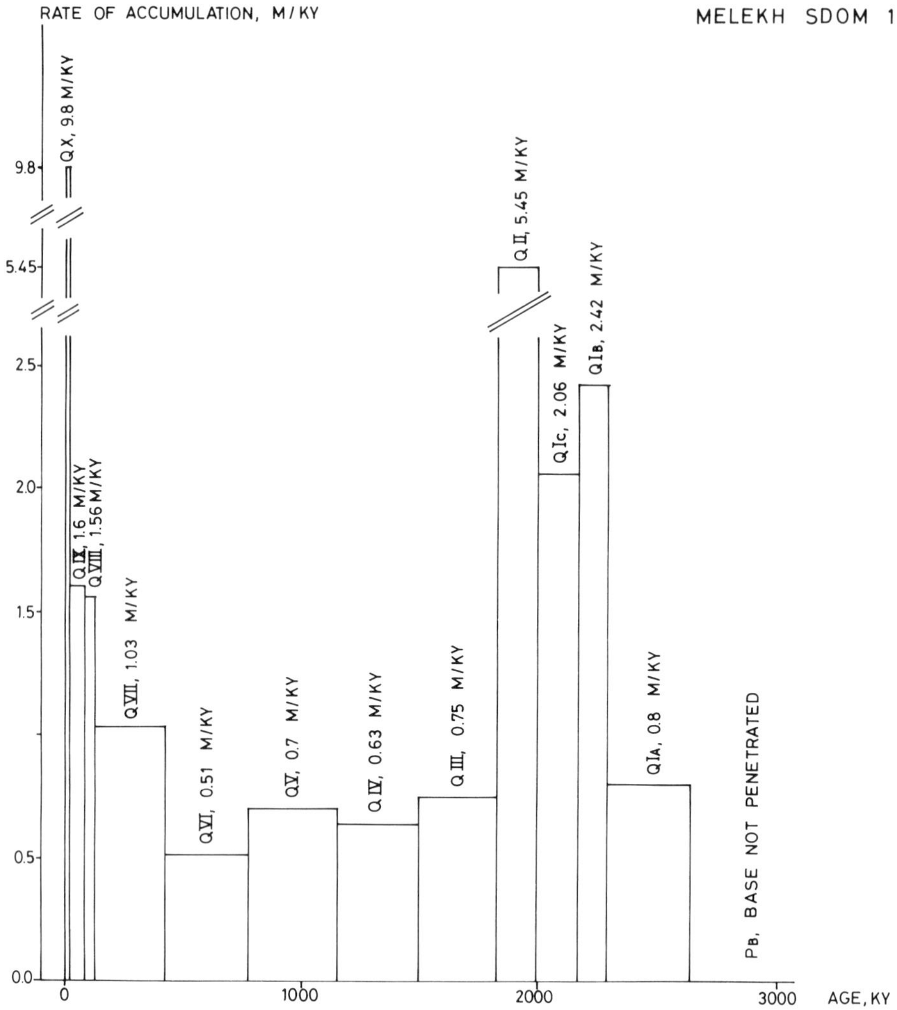

Fig. 8.5.3. Rates of subsidence and accumulation of the Quaternary palynozones in Melekh Sdom 1 borehole, southern Dead Sea basin. For location see Fig. 9.2.1.

Pliocene Pa sediments directly overlie Mb, it may well be that the uppermost part of Mb is missing, either due to erosion, which is most probable, toward the trough formed later at a close location (see below), or due to non-deposition, if this area became somewhat elevated in Mc times.

Another borehole in which Mb was encountered, south of the Dead Sea, is Arava 1 where almost 400 meters of this palynozone have been drilled, however the base was not reached at the total depth of 2,704.5 meters. The palynozone was also encountered in the central Jordan Valley where 750 meters have been

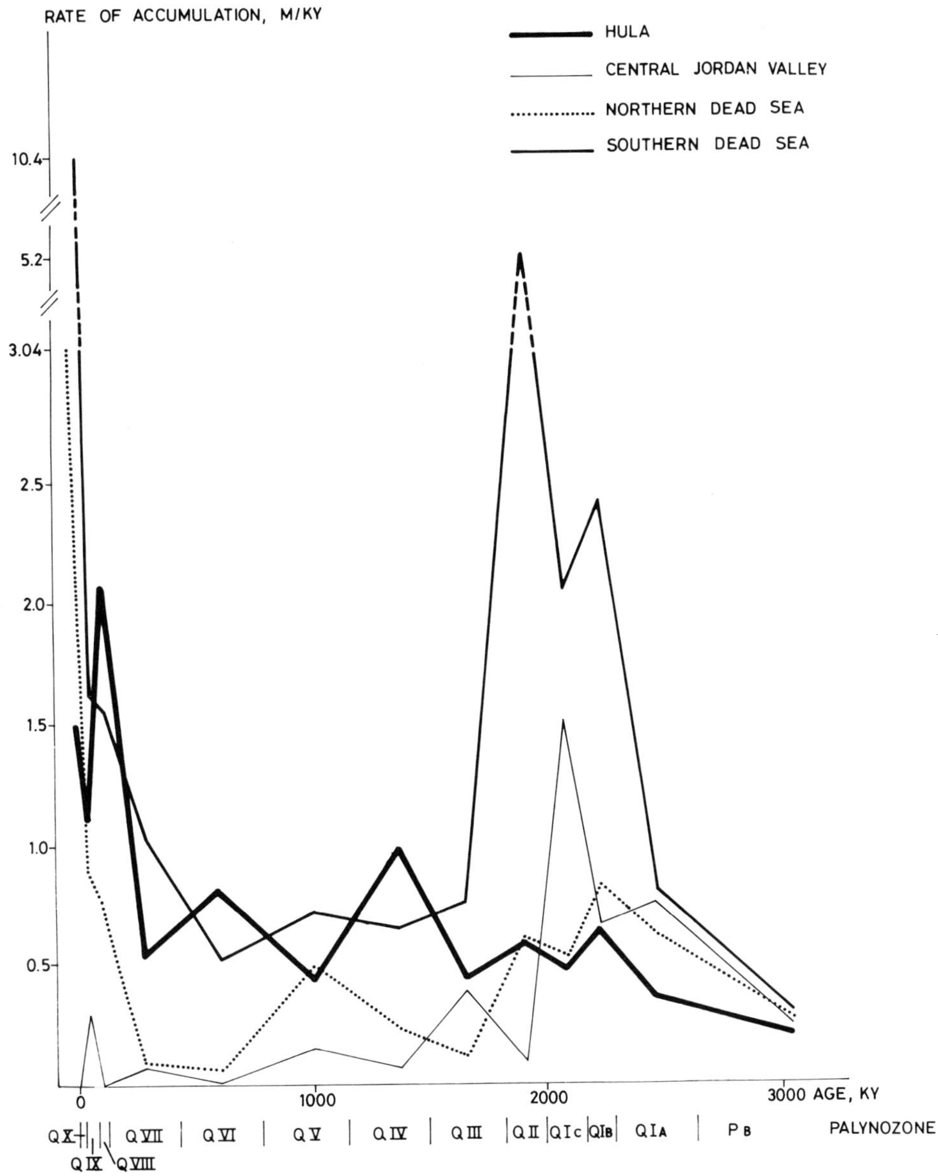

Fig. 8.5.4. Summary of the Quaternary rates of subsidence and accumulation in various sub-basins of the Dead Sea Rift.

penetrated and in the Hula Valley, where 270 meters were drilled. In both latter boreholes, the base of Mb was not reached. The sequence in Zemah 1 is most probably considerably tilted, so that the true thickness is not more than 60–70% of the drilled interval.

Mc palynozone, of late Miocene age, was encountered in a considerable thickness, some 2,000 meters, at both Arava 1 borehole, south of the southern Dead Sea Basin and in the central Jordan Valley, at Zemah 1. Correlative sediments are not known from outcrops at the vicinity of the Dead Sea or the central Jordan Valley, indicating, together with the great Mc thickness at Arava 1 and Zemah 1 boreholes, capturing of the previous river systems. These had been diverted into troughs which had opened south of the present day Dead Sea, where Arava 1 was drilled and in the central Jordan Valley, where Zemah 1 is located. Incidentally, the existence of this tectonic phase was entirely unsuspected until palynological studies have been initiated in this arid region.

Pa palynozone, of early through middle Pliocene age, was encountered in considerable thicknesses at Sdom 1, almost 2,400 meters, while at Ami'az 1 approximately 900 meters, both drilled in the vicinity of Mount Sedom, in which almost 2,000 meters of this palynozone crop out. The lithology at Mount Sedom and its close vicinity is dominated by rocksalt deposited in lagoons connected with the Mediterranean, via a narrow strait which ran northward, occupying regions of the present day Jordan and Yizre'el valleys. The lithology of Pa at Arava 1, where about 180 meters of this palynozone have been penetrated, is dominated by chalk of lagoonal or shallow marine origin.

Most outcrops of Pa palynozone correlative formations are in the order of 100–250 meters thick, thus it seems that a trough was present in the Mount Sedom region at that time. This trough differed from the previous Mc trough, both in its different location and by the fact that it served as an intermediate basin of the early middle Pliocene system, rather than the terminal basin of the late Miocene. This is evidenced from the local pollen which display higher shares of freshwater components for Pa times as compared with Mc, even when recovered from rocksalt. In the central Jordan Valley, Zemah 1 borehole shows some subsidence during Pa times, with accumulation of evaporites of no considerable extent. The palynozone was also encountered by a sequence of some 200 meters of lagoonal-lacustrine sediments in Notera 3 borehole, in the Hula Valley to the north.

Pb palynozone, of late Pliocene age, is usually no thicker than 400 meters where fully penetrated, but in some places attains much lesser thicknesses. Several boreholes which were drilled in the central parts of the southern Dead Sea Basin, such as Melekh Sdom 1 and Amazyahu 1, only penetrated the top of Pb.

Pb sediments, in those boreholes where the bottom was not reached, are similar in nature to other, fully penetrated sequences, not indicating a possibility of trough existence in these places which were later, during QI times, affected by considerable subsidence (see below). It seems that in Pb times the previous Pa paleogeography still persisted, however, with a lesser influence of the Mediterranean due to a generally lower sea level (Horowitz, 1979), with no considerable rates of subsidence.

QI palynozone, of Preglacial Pleistocene age, was penetrated by almost all boreholes drilled in the Dead Sea Rift, except for Arava 1 and Sdom 1, in which drilling commenced in older strata (Levin and Horowitz, 1987). The drillings penetrated the entire sequence of the palynozone, which is always in the order of several hundred meters thick. The best developed sequence is recorded from Melekh Sdom 1 borehole, where some 870 meters are encountered, while the other three boreholes in the vicinity — Amazyahu 1, Sdom 2 and Ami'az 1 — only recorded some 500–550 meters of QI sediments. The same order of thickness was also penetrated by En Gedi 2 borehole. The sediments usually comprise fine to medium, rarely coarse clastics, laid down in a river system.

Thicknesses at the outcrops vary, from some 400 meters within the rift borders and several tens of meters in outcrops of channel deposits on the surrounding hills.

The nature and thicknesses of the sediments constituting QI palynozone indicate that they had accumulated in a river system draining from the east to the Mediterranean, for which the southern Dead Sea region only served as an intermediate basin. This intermediate basin could have become quite deep at times, as is evidenced by the thicknesses accumulated at Melekh Sdom 1. Further discussion of the nature of pollen spectra typical for QI and the transition to the overlying QII, is brought below, in Chapter 8.5.2.

A major change in paleogeography occurred at the beginning of QII times, when the southern Dead Sea basin became terminal base level for the Jordan–Dead Sea Rift internal drainage system. QII attains some 360 meters at Amazyahu 1 and 870 meters at Melekh Sdom 1 boreholes, while in En Gedi 2 the sequence is only a little more than a hundred meters. This indicates a considerable subsidence of the southern Dead Sea Basin at that time, with rates of accumulation attaining up to more than five meters per thousand years. No outcrops of this palynozone are known from the region. It is quite interesting to note that these considerable rates of subsidence are only typical at that time for the southern Dead Sea basin, while to the north much lesser thicknesses of QII have been accumulated, in the order of several tens of meters in the central Jordan Valley and the Hula basin.

The Glacial Pleistocene–Holocene sequence of the Dead Sea, comprising palynozones QIII through QX, is some 1,500 meters thick in the southern Dead Sea basin, at Melekh Sdom 1 and Amazyahu 1, somewhat thinner — around 1,000 meters to the west, at Ami'az 1 and Sdom 2. To the north, it comprises some 1,250 meters in the Hula basin, but only 240 meters in the central Jordan Valley, at Zemah 1.

Fig. 8.5.4 summarizes average rates of accumulation for the Quaternary palynozones in four subbasins of the Dead Sea Rift, the Hula, central Jordan Valley, the northern and southern Dead Sea basins. The last phase of major subsidence, which had created the northern Dead Sea and Lake Kinneret basins is not expressed in this figure, since it is chiefly manifested by troughs which are filled up with water, not yet with sediments (Horowitz 1987c, 1989c) and are most probably still subsiding.

A look at Fig. 8.5.4 reveals general common trends in structural evolution of the subbasins, manifested by the rates of accumulation of sediments. Increase in rate of subsidence at the beginning of the Quaternary, as compared with the underlying Pliocene; decrease and a rather steady rate from some 1.5 million years ago; and a considerable increase in the rate of subsidence toward the present day, especially during the last 15–20,000 years.

However, when details are analyzed the differences between the four subbasins become apparent. The subsidence commenced at approximately the same time, during QIa. The peak of subsidence at the southern Dead Sea basin was already reached in QIb times and lasted for quite a while. However, such a peak is not seen at all in the northern Dead Sea. In the central Jordan Valley, the peak is during QIc and much shorter in duration, while the peak in the Hula was also quite short, only reached during QIV times. At this time, QIV, subsidence in the central Jordan Valley practically stopped, not renewed until the present day. At the same time, during most of the Glacial Pleistocene, both the Hula and southern Dead Sea basins continued to subside at a rather steady rate, while the northern Dead Sea basin subsided at a lesser rate, except for a peak during QV times.

Notably, subsidence of the four subbasins was not always in phase. It seems that during periods when the Hula subsided faster than the southern Dead Sea it may have temporarily become a terminal erosion base level of its own, disconnected from the south, which was most probably helped by inactiveness of the central Jordan Valley sector. Rate of subsidence again increased toward the present day, in QVIII times, in the southern and northern Dead Sea, as well as in the Hula, but again not in unison. This time, subsidence commenced at the north, its rate decreasing toward the present day.

The peak in the southern Dead Sea, although starting at about the same time as in the Hula, had only reached a considerable value toward the present day. A similar situation can be seen in the northern Dead Sea basin, where the peak of subsidence is even much greater, with the formation of the northern Dead Sea, in which some 400 meters of water are filling the recent trough. It is, of course, impossible to say whether this really is the peak, or an increase in the rate of subsidence should be expected at the future. The southern Dead Sea basin is the only one in which more than one deep well was sunk, so the data are not confined to a single point, thus showing that subsidence was quite complex, with the subsiding block or series of blocks alternately tilting in different directions.

8.5.2. Local vegetation, structure and landscape

Composition of the local vegetation pollen part of the spectra and relations of this group to the regional, have been used (Horowitz and Horowitz, 1990) to reconstruct local environments and landscape (see also Chapter 8.4.2) in the southern Dead Sea basin (see Fig. 9.2.1 for locations of boreholes). High percentages of Gramineae, Cyperaceae and hydrophil, freshwater vegetation pollen may represent two possibilities. Either an amelioration in climate, resulting in more rains and a reduced salinity of the better leached soils, as well as lesser possibility for the formation of saline playas; or, the other reason for prevalence of freshwater pollen may be due to a better developed spring vegetation, when the spring formation is connected with faulting processes, which have nothing to do with the general climate (see Fig. 8.4.4).

Fig. 8.5.5 is an excerpt from the pollen diagram of Amazyahu 1 borehole covering palynozones QI and QII, in which formation of the rift valley in its present day configuration, at the beginning of QII times, can be deduced from the non-arboreal pollen.

An example for the climatic controlled spreading of freshwater vegetation can be clearly seen in Amazyahu 1 pollen diagram for QI palynozone, between depths of approximately 1,800–2,400 meters (Fig. 8.5.5, sector A). The freshwater vegetation pollen curve fluctuates almost parallel to the arboreal pollen curve represented in the first column. Since the arboreal pollen curve quite clearly displays regional climatic conditions, it is obvious that in this case the freshwater vegetation pollen production, or the relationship between freshwater and saline vegetations, is also controlled by the regional climate. This phenomenon is also accompanied by a lower share of local vegetation pollen, as compared to the regional vegetation pollen seen in the right column, probably due to a better development of arboreal vegetation.

The lesser shares of wadi and playa vegetation pollen, combined with parallel curves for the arboreal and freshwater vegetation pollen, also indicate that pollen spectra at QI interval represent a rather flat, riverine environment, with no development of escarpments surrounding the locality of deposition. This conclusion, in turn, shows that during QI times subsidence was quite mild, with no considerable faulting, thus the rift valley was not yet developed as such.

When going up the section in Amazyahu 1 to QII palynozone (Fig. 8.5.5, sector B), no connection can be seen between the trends of arboreal and freshwater vegetation pollen curves. This indicates that an environment has developed, in which the hydrologic regime depends much more on local conditions, less on the regional climate, which implies faulting and trough formation. The fact that during this interval local pollen only attained some 50% of the pollen spectra, same as during QI, may indicate that the initial trough was quite steep walled, with only limited areas which would have been suitable biotopes for hydrophil and halophil vegetations, apparently quite restricted by the structure and shape of the newly formed basin.

Further up the sequence local vegetation takes the upper hand in the pollen spectra and fluctuates between 60 and 80% of the total (Fig. 8.5.5, sector C), indicating a well developed local environment in the form of the present day rift valley. A drop in this curve during QV times may indicate that the QV lake was

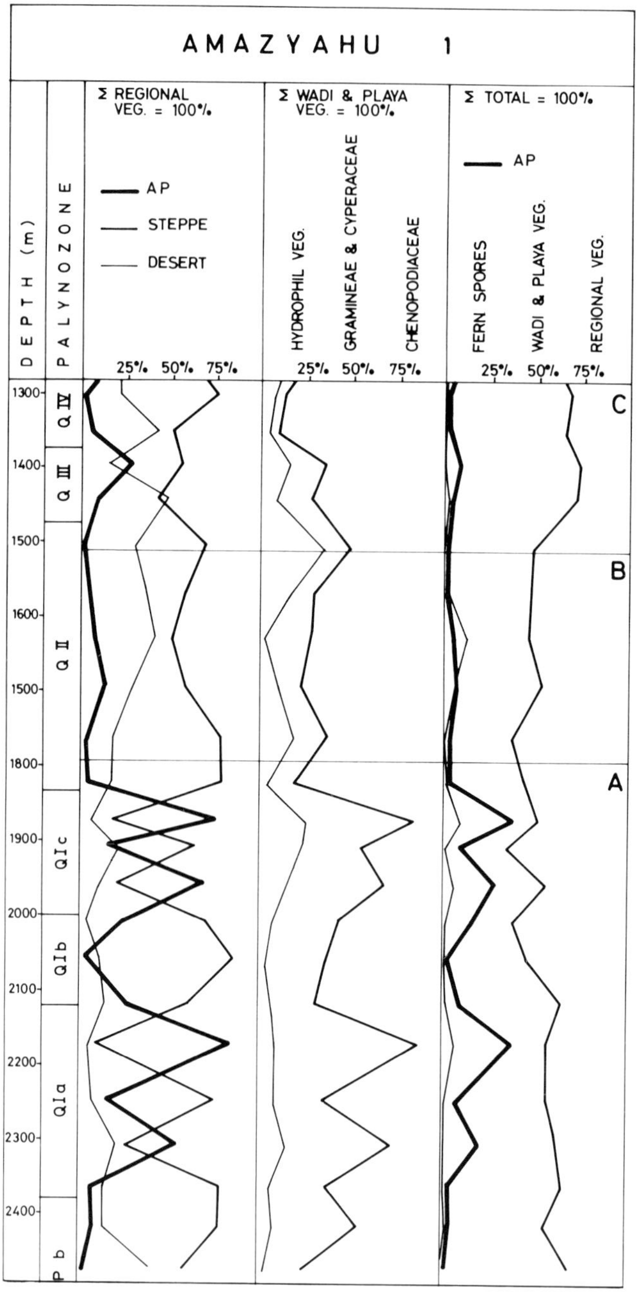

Fig. 8.5.5. Pollen diagram of Amazyahu 1 borehole, palynozones QI through QIV, showing formation of the Dead Sea endoreic system. For details see text. For location see Fig. 9.2.1.

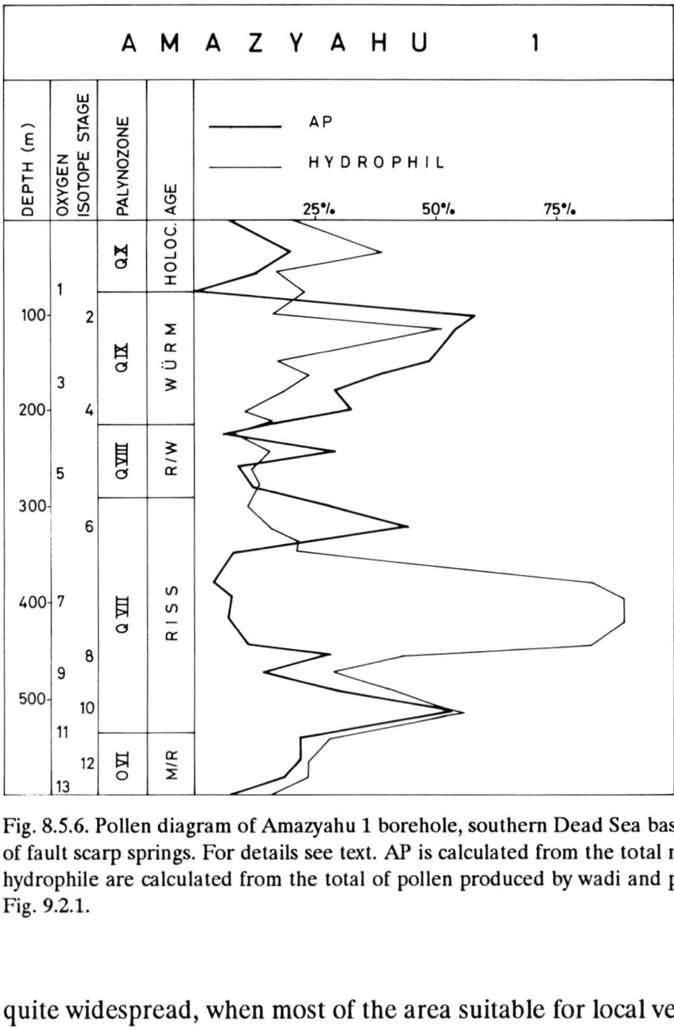

Fig. 8.5.6. Pollen diagram of Amazyahu 1 borehole, southern Dead Sea basin, showing the development of fault scarp springs. For details see text. AP is calculated from the total number of regional elements; hydrophile are calculated from the total of pollen produced by wadi and playa plants. For location see Fig. 9.2.1.

quite widespread, when most of the area suitable for local vegetation was inundated by its waters (Fig. 9.2.15).

Another reason for high percentages of freshwater vegetation pollen may be connected with faulting and structural disturbances, which caused opening of active aquifers to the surface, thus producing springs which maintain typical vegetation, comprising hydrophil plants, mainly Gramineae and Cyperaceae.

Fossil travertines, which are quite widespread in the northern Arava south of the Dead Sea, (Horowitz, 1979; Livnat and Kronfeld, 1985), indicate the existence of such springs, while their vegetation is very well represented by considerable peaks in the freshwater and hydrophil vegetation pollen of Amazyahu 1 borehole, at depths of around 400 and around 700 meters (Figs. 8.5.6, 9.2.15).

Fig. 8.5.6, an excerpt from the upper part of Amazyahu 1 pollen diagram, shows relations of hydrophil to halophil pollen on the one hand, compared with the

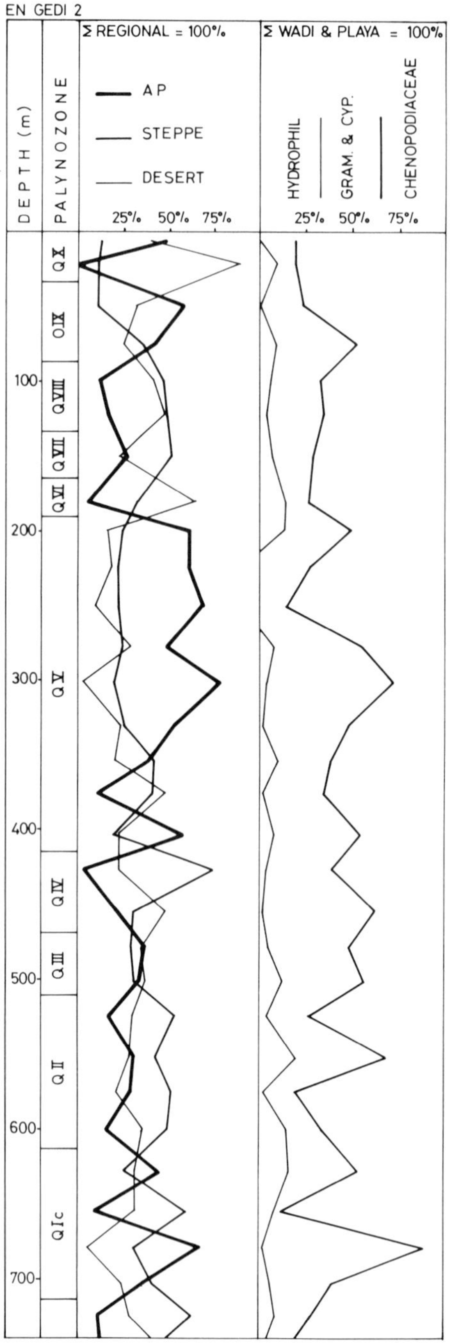

Fig. 8.5.7. Pollen diagram of En Gedi 2 borehole, Dead Sea, showing the activity of springs along the Glacial Pleistocene. For details see text. For location see Fig. 9.2.1. Reprinted with permission from Horowitz and Horowitz (1990).

share of arboreal pollen from the regional vegetation on the other. It can clearly be seen that in some cases trends of these curves coincide, while in other cases, especially during QVII palynozone times, at depth of around 400 m, these two curves antagonize, indicating spring formation by faulting, which caused deposition of extensive travertines. An almost parallel picture can be seen in the pollen diagram obtained from Melekh Sdom 1 borehole (Horowitz, 1987a) or Admon 1 (Fig. 8.4.4).

En Gedi 2 borehole, situated at the present Dead Sea shore, displays a somewhat different picture for the Glacial Pleistocene palynozones QII through QX (Fig. 8.5.7). While the non-arboreal pollen curves behave similarly to other boreholes, the rift formation in QII times caused opening of aquifers feeding the En Gedi springs, which seem to be active ever since QII times. The activity of these aquifers and their freshwater discharge, depend very much on the climate since they are replenished by rainwater. This is the reason why the freshwater vegetation curve is approximately parallel to the arboreal pollen one in this particular borehole, which differs in this respect from other boreholes in the southern Dead Sea basin.

It is also quite interesting to follow, for an older period such as the Miocene, the hydrophil vegetation, Gramineae and Cyperaceae pollen curves, as compared with the arboreal pollen. During middle Miocene (Mb) times the freshwater pollen curve is almost parallel to the AP (Fig. 9.2.3), when regional vegetation pollen take the upper hand, around 50–70% of the total. This indicates that the landscape was rather flat, with no trough morphology occurring at that time, that could have maintained a well developed non-arboreal vegetation.

At the beginning of the late Miocene Mc (Fig. 9.2.6), several considerable peaks, up to 90%, of the hydrophil vegetation pollen, Gramineae and Cyperaceae, are clear. These peaks seem to indicate faulting, which most probably caused formation of a trough in the locality where Arava 1 was drilled (see also Chapter 8.5.1). Following these peaks, wadi and playa vegetation pollen increase at depths of around 1,700–1,300 m, which is possibly due to formation of a structural trough and development of typical rift valley habitats and vegetations. This trough was apparently slowly filled up with sediments up to about 850 m, when a new faulting phase is indicated by the hydrophil, Gramineae and Cyperaceae curves, followed by an increase of the wadi and playa vegetation pollen on account of the regional. Thus two stages of faulting can be seen in Arava 1 borehole from the non-arboreal curve, one at the beginning of Mc time, the other one toward its end.

Summing up the indications of non-arboreal pollen curves, distinction can be made between periods when morphology was rather flat, dominated by a river system which crossed the present day area occupied by the rift valley, such as Mb, Pb and QI times; and periods in which troughs had been formed, surrounded by rather steep morphology, developing local habitats in which mainly playas have been formed, together with springs, such as Mc and QII through QX times, when these troughs acted as terminal base levels for the regional drainage systems.

Chapter 9

Regional reconstruction of past arid environments: the late Cenozoic of Israel

Israel is brought here as a case study, to show the possibilities of palynologically based regional reconstructions of arid and neighboring environments, throughout a rather long stretch of time — the entire late Cenozoic (Miocene, Pliocene and Quaternary). The reason for choosing Israel as an example is, because this is almost the only arid region, to date, for which continuous sequences had been drilled and palynologically analyzed, spanning this entire period. Evidently, the other is a personal excuse, since I am living in this land and most of the studies had been carried out by my students, colleagues, my wife and myself, which seems to justify my interest.

Structurally, Israel is being located throughout the late Cenozoic between two subsiding basins in which sediments are being accumulated at a considerable rate, the Jordan–Dead Sea Rift to the east and the Mediterranean to the west. Both basins had been drilled in search of natural gas or oil and many of the penetrated sections have been subject to pollen analysis. Initially these studies had been carried out to obtain a palynostratigraphic scheme of the region, which would enable correlations of the marine and continental sequences; consequently, paleoenvironments could be reconstructed on a regional scale, continuously for the entire period.

To realize the data basis, the Mediterranean offshore Neogene sequences are in the order of 2–4 kilometers thick (Gvirtzman, 1970) and the Quaternary attain up to 700 meters (Derin and Gerry, 1971). Subsidence had been continuous throughout the period, which ensures a complete, almost gap-free reconstructions. The Jordan–Dead Sea Rift is also accumulating sediments ever since the latest Oligocene, but in a somewhat different manner. Several subbasins had developed along the Rift during the late Cenozoic, in different parts, each of which accumulated great thicknesses of sediments during the sector of time of its accelerated subsidence (Horowitz, 1979). When subsidence of such a subbasin slowed down, there was always another in which sediments had already been deposited at a rather fast rate. This way, of new subbasins opening every once in a while, resulted in a cumulative sequence up to ten kilometers thick for the late Cenozoic, which by inter-correlations of the various sectors and with the marine sections, yielded a continuous evidence for the entire period.

In addition, volcanic activity along the Rift and on its shoulders resulted in many basalt flows interfingering with the late Cenozoic sediments, which made possible absolute datings (Horowitz et al, 1973; Siedner and Horowitz, 1974; Heimann and Steinitz, 1989; Heimann, 1990). In the Mediterranean offshore, detailed studies of the subsurface Neogene foraminifera biostratigraphy (Derin and Reiss, 1973) resulted in exact definitions of stages and substages for this period. Correlation of continental sequences from the Dead Sea Rift and the marine Mediterranean deposits was possible by studying both palynostratigraphy and foraminifera biostratigraphy in the same sequence (Horowitz and Derin, 1987). This combination of marine biostratigraphy, palynostratigraphy and radiogenic datings seems a rather firm basis for the chronostratigraphic framework of the late Cenozoic in Israel (Horowitz and Horowitz, 1985; Horowitz and Derin, 1987; Horowitz, 1987a, c, 1989c, 1990)

The sequences on which the presented reconstruction is based are from boreholes drilled by oil companies, therefore comprising only well-cuttings as material available for palynological studies. The sampling intervals for the well-cuttings had been in the order of 10–25 meters. The uppermost part, approximately representing the last 150 ky, is from continuous cores studied in more detail (Rossignol, 1969b; Horowitz, 1971, 1979; Weinstein, 1976; Weinstein-Evron, 1983, 1990; Baruch, 1986; Cheddadi, 1988) and better dated, thus serving as a sound foundation for the presented paleoclimatic models. These detailed studies enabled correlation of environmental changes in Israel and the Near East with oxygen isotope curves for the oceans (Weinstein-Evron, 1983, 1990; Cheddadi, 1988). They also served as foundation for correlation of the last 3.5 my (Horowitz, 1989b) with Hole 552A in the Atlantic Ocean (Shackleton and Hall, 1984) and with eastern Asia, for the Brunhes chron (Fuji and Horowitz, 1989). Incidentally, the correlations with deep sea sequences also helped further establishment of chronology for the last 3.5 my in Israel.

The combination of well-cuttings and seemingly long sampling intervals would render, on a first glance, some inaccuracy to the obtained results and reconstructions. These however seem to be at least partially overcome thanks to the combination of considerable rates of subsidence and accumulation, which makes the sampling intervals rather close in terms of time. Intervals are usually no more than 10,000 years for the older part of the Quaternary, less for the younger, but up to 100 ky for the Miocene. The analysis of some 15 such sequences, with all of them showing recurrence of main trends in the pollen diagrams (Horowitz, 1990), seem to considerably raise the level of confidence in the results. To these should be added the fact that the Dead Sea subbasins had frequently been terminal base levels, thus ensuring the accumulation of complete, continuous sequences.

9.1. Terminology

Much confusion still exists in the use of stratigraphic terminology for of the late Cenozoic, in both general and local planes. Terms used in the stratigraphy of marine deposits are better defined than those applied to continental sequences, but even with the marine late Cenozoic stratigraphy there is hardly any agreement among various groups of investigators. The main reason for that lies in the fact that most late Cenozoic sequences have been defined and recognized as local sequences. Due to frequent facies changes no global, or even wide regional distribution of the units could be ascertained, until sequences had been studied from deep sea sediments. Even then, correlation of deep sea sediments with the nearshore or continental sections, is quite doubtful in many cases, further so due to the frequent absence of reliable datings. Disagreements become even more bitter when the Quaternary is discussed, even on basic issues such as the boundary between the Tertiary and Quaternary. Regarding the various opinions, we had adopted a stratigraphic terminology, which seems to suit quite well the rock units of Israel (Horowitz, 1979). However, our approach is contradicted undoubtedly by many, maybe even for good reasons, thus it remains to see how long it still holds.

The term "late Cenozoic" encompasses the Miocene, Pliocene and Quaternary epochs. The Miocene and Pliocene conform and correlate with foraminifera zones defined by Blow, as discussed in Horowitz and Derin (1987).

The Miocene is divided into three palynozones. Ma, which had commenced already in latest Oligocene, but mainly comprises the early Miocene; Mb, of middle Miocene age and the late Miocene Mc palynozone. At the present state of knowledge, palynostratigraphy does not allow further subdivision of the Miocene, thus until more data are acquired this subdivision is retained.

A synthesis of late Cenozoic stratigraphic terminology for the Negev, southern Israel, is presented in Goldsmith et al. (1988). According to this study, which also includes correlation with the palynostratigraphy obtained from a few boreholes in the northern Arava and southern Dead Sea, Ma palynozone, of latest Oligocene through early Miocene age, corresponds to the Aquitanian and Burdigalian, from approximately 25 million years up to some 16 million years ago, encompassing foraminifera zones N4 through N7 (according to Derin and Reiss, 1973). Mb palynozone, of middle Miocene age, from some 16 my up to approximately 10 my ago, corresponds to the Langhian and Serravalian, foraminifera zones N8 through N14. The late Miocene Mc palynozone corresponds to the Tortonian and Messinian, from some 7 my to about 5.3 my ago, foraminifera zones N15 through N17.

The Pliocene is divided into two distinct palynozones, each correlative with a transgression. Pa correlates with the Tabianian, Pb with the Piacencian rise in sea level. Thus Pa comprises the early and middle Pliocene, while Pb is of late Pliocene age. The boundary between the Pliocene and Quaternary, as adopted here, is at approximately 2.65 my ago and corresponds to the maximum retreat of the terminal Pliocene regression.

Pa palynozone corresponds to the Zanclean, or Tabianian, which are apparently parallel terms for the same transgressive period, that commenced some 5.3 my ago

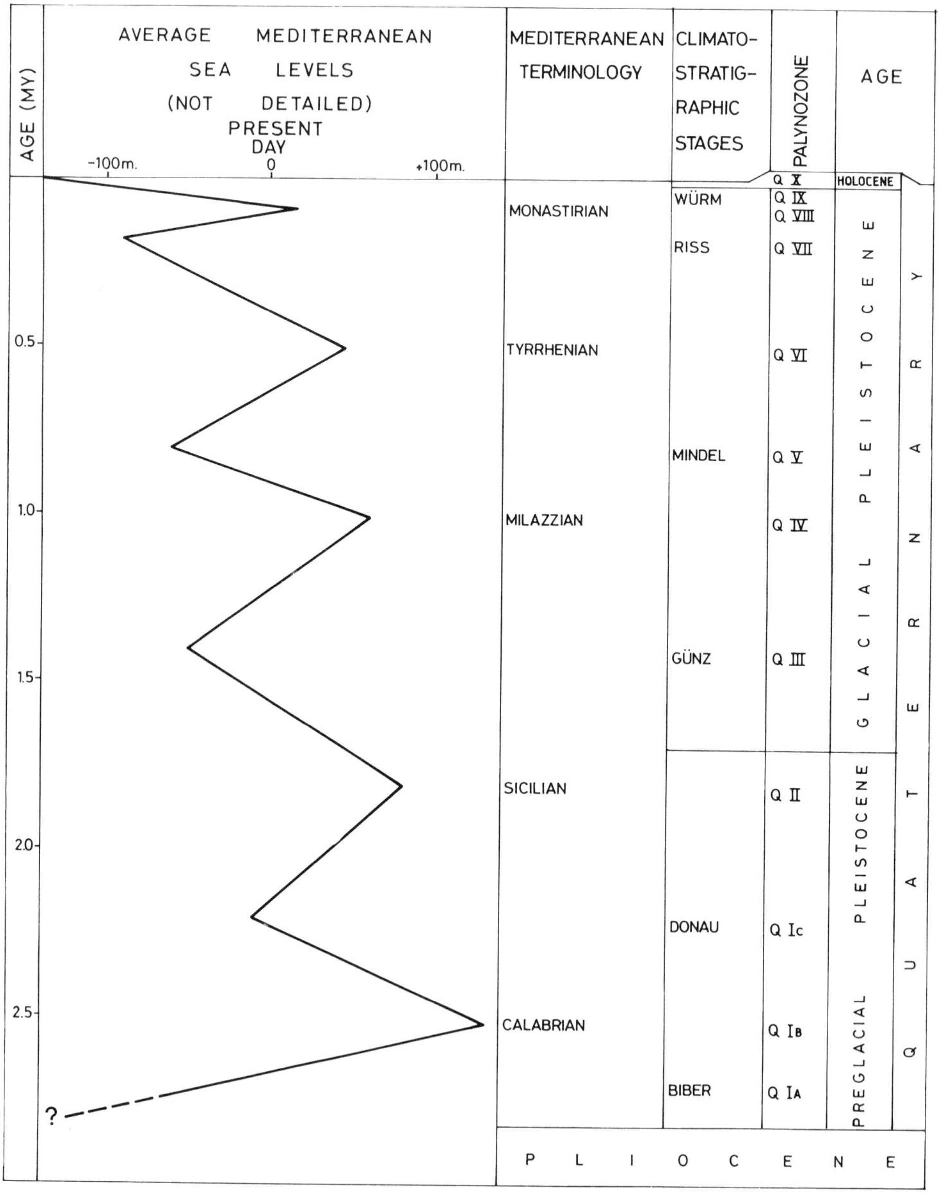

Fig. 9.1.1. Eustatic stages of the Mediterranean, climatostratigraphy and the Quaternary stratigraphic framework of Israel. Reprinted with permission from Horowitz (1979).

and ended at about 3.5 my, corresponding to foraminifera zones N18, N19 and the lower part of N20. Pb palynozone, corresponding to the Piacenzian, late Pliocene, from 3.5 my ago up to 2.65 my. The last figure is not in agreement with Goldsmith et al., who placed the Pliocene/Pleistocene boundary at 1.8 million years ago, corresponding to the upper part of foraminifera zone N20 and most of N21. It is however my opinion (see Horowitz, 1979) that the Pliocene/Quaternary boundary should be at 2.65 my.

Eustatic ingressions of the Quaternary Mediterranean (Fig. 9.1.1) are taken as the basis for subdivision of this epoch and correlation of the Israeli Quaternary sequence with the European and North African ones. Consequently, the marine ingression stratigraphy is used primarily in the coastal plain, while the European climatostratigraphy is used throughout Israel. The possibility of correlating the Israeli climatostratigraphic Quaternary sequence with the Alpine glacial/interglacial alternations is discussed in detail in Horowitz (1979; and other publications) and seems to provide a sound stratigraphic basis for the palynostratigraphic subdivisions. The Quaternary is subdivided into ten palynozones, QI through QX, of which the oldest one is further divided into three subpalynozones, QIa through QIc.

The Preglacial Pleistocene palynozones, QI and QII, from approximately 2.65 my ago up to approximately 1.8 my, correspond to foraminifera zone N22, while the Glacial Pleistocene and Holocene palynozones, QIII through QX, correspond to foraminifera zone N23, from approximately 1.8 my ago until the present day.

Sequences covering the last 3.5 million years, encompassing the late Pliocene, Preglacial Pleistocene, Glacial Pleistocene and Holocene, have been correlated on the basis of inferred climatostratigraphy with oxygen isotope curves obtained from deep sea cores from the northern Atlantic Ocean (Horowitz, 1989b). These correlations further support the chronostratigraphy adopted here for this time span (Figs. 9.3.1, 9.3.2) and help securing the paleoenvironmental reconstructions within the internationally accepted oxygen isotope curves framework. Please note that contrary to all the pollen diagrams presented in this chapter, the arboreal pollen curves in Figs. 9.3.1 and 9.3.2 commence from the right, to clearly visualize correlations with the oxygen isotope curve.

9.2. Pollen diagrams for the late Cenozoic of Israel

Pollen diagrams had been obtained for the late Cenozoic of Israel from a dozen deep boreholes drilled in the Dead Sea Rift, which penetrated the graben fill to depths ranging between 2,000 and 4,500 meters (Horowitz and Horowitz, 1985; Horowitz, 1987a; Levin and Horowitz, 1987). Additional five boreholes reaching similar orders of depths have been analyzed in the last years (Horowitz and Horowitz, 1990), but the results are not yet publicly available. A deep borehole from the Mediterranean offshore was also analyzed (Horowitz and Derin, 1987), to correlate marine and continental formations. Locations of the boreholes are shown in Fig. 9.2.1.

Sampling intervals vary from well to well according to availability of material, drilling techniques and necessity for palynostratigraphic details during the drilling operations, indicated on the pollen diagrams. All samples comprised well-cuttings and had been prepared for pollen analysis applying methods discussed in detail in Chapter 5. These involved dissolution of carbonates and sulphates by hydrochloric acid, followed by gravity separation between mineral and organic substances, with the aid of zinc chloride solution of a specific gravity 1.85. The slides are deposited

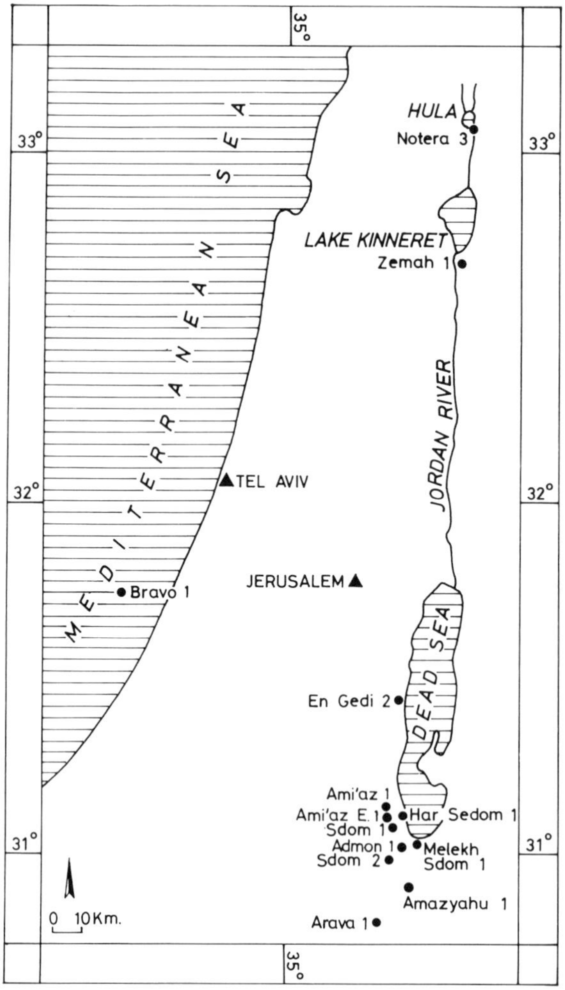

Fig. 9.2.1. Location map of deep boreholes palynologically analyzed in Israel.

at the Palynological Laboratory, Institute of Archaeology, Tel Aviv University and at the Geological Survey of Israel, Jerusalem.

9.2.1. Presentation

The pollen diagrams are divided into four columns. The left shows relations within the arboreal pollen group, the total AP taken as 100%. The curves in this column are of the "overlapping" type, thus each measures independently from zero value, on the left.

The second from the left displays relative shares of grains originating from three vegetation groups which comprise the regional flora, omitting pollen such as might have been derived from hydrophil and halophil vegetation of wadis, or the Rift Valley lakes, playas and springs. One group comprises pollen from arboreal origin, the second from steppe vegetation, while the third includes grains originating in desert plants, the total of pollen derived from regional vegetation taken as 100%. The curves in this column are of the "overlapping" type (see above).

The third column presents composition of the local, mostly hydrophil and halophil pollen flora, derived from plants attached to water bodies close to the locality of deposition, or growing in places where secondary enrichment in water is available, such as wadi beds. The total of pollen from the local vegetation is 100%. The curves are cumulative, thus percentage values are expressed by the areas between two neighboring curves.

The fourth column is based on the total number of counted pollen taken as 100% and gives the ratios of local to regional elements; also, arboreal pollen and fern spores are shown as part of the entire spectrum. The curves are cumulative for the fern spores, local and regional vegetations, but overlapping for AP, to facilitate comparisons of the latter with the curve in the second column.

The arboreal pollen presented in the left column are divided into four groups, representing different environments: *Quercetalia*, comprising mainly oaks, typical of the Mediterranean garrigues and maquis; Conifers, such as pine, cypress, cedar and unidentified Pinaceae (mainly in the Miocene), which seem to grow in dryer domains; *Picea orientalis* which, although also a conifer, is known today only from some secluded localities in western Turkey, where it grows in an euxine forest, thus is probably indicative of a rather wet, temperate environment. This is also evidenced by other fossils occurring together at the same horizons (Horowitz, 1979; Horowitz and Horowitz, 1985), such as Betulaceae and Juglandaceae, of wet environments affinity.

The second column from the left displays relationships among three pollen groups: arboreal, discussed above, generally representing moister environments; steppe, of the Irano-Turanian domain; and desert, indicative of the Saharan regime. The steppe vegetation is represented by pollen of *Artemisia* and *Ephedra* as main constituents, accompanied sometimes by *Scabiosa*, *Echinops*, *Astragalus*,

Centaurea, Cruciferae, Liliaceae, Papilionaceae and Malvaceae. The most common types of pollen indicating the desert environments originate from Compositae and Umbelliferae, accompanied by occurrences of *Calligonum*, *Plantago* and *Asphodelus*.

The third column shows composition of pollen derived from the local, hydrophil and halophil vegetations, which are shown cumulatively. The left curve is for hydrophil, both trees such as *Tamarix* which is the principal component, accompanied by *Acacia*, *Moringa*, *Zyziphus*, *Phoenix*, *Populus*, *Acer*, *Salvadora* and herbaceous plants such as *Typha*, *Rubus*, *Polygonum* and Labiatae. All these grow in intimate connection with perennial water, either springs or high groundwater tables. Some of these plants can withstand certain degrees of salinity, but in general the group seems to indicate freshwater occurrences near the locality of deposition.

The following curve is for Gramineae and Cyperaceae, which seem to bear environmental characteristics similar to the former, thus the two left curves are in most cases regarded together as indicative of freshwater occurrences in the region (except for the rare cases when Gramineae or Cyperaceae entirely dominate the spectra, indicating an extremely-arid environment). The rest of the area on this column represents the share of Chenopodiaceae in the local spectra, considered as representing saline conditions.

The last column to the right is based on the total number of counted pollen, primarily used to show relations of local and regional elements within the pollen flora. The arboreal pollen curve given in this column is less conclusive, as compared with the curve displaying AP as part of the regional vegetation and is only given here for comparison. The fern spores are also shown in this column and seem to follow the hydrophil vegetation curve trends.

In some boreholes, especially these which were sampled in more detail, or those where analyzed sequences are very long, such as Notera 3, Zemah 1, Amazyahu 1 and Arava 1, pollen spectra have been grouped together for the sake of clarity, to avoid congestion and to rid of minor noises in a manner previously found useful for the diagram obtained from Notera 3 in the Hula basin (Horowitz and Horowitz, 1985). In these cases each 2–4 samples which bear similar palynological characteristics are combined, thus every horizon on the diagram is representative of at least some 300–500 microfossils. All thus grouped sample depths are indicated on the appropriate column. Weinstein-Evron (1987b), in a detailed statistical analysis of sampling and subsampling techniques, showed this method, of smoothing the curves by grouping samples together, to be quite reliable. This way of presentation also makes easier correlations with other sections, since it eliminates minor local influences and differences, showing more readily the general palynostratigraphic and paleoenvironmental trends.

9.2.2. *Palynostratigraphy and paleoenvironments*

The late Cenozoic paleogeography of Israel was studied in detail by Garfunkel and Horowitz (1966), Gvirtzman (1970) and many others, summarized in Horowitz (1979). Rates of accumulation for the coastal plain sediments are given in Gvirtzman (1970) and in Derin and Reiss (1973), following detailed biostratigraphic studies, especially of the Neogene sequences. Rates of accumulation and subsidence for the Jordan–Dead Sea Rift and its several subbasins are discussed in detail by Horowitz (1987c, 1989c), based on palynostratigraphic studies.

The Neogene sequence was subdivided in Israel into five palynozones characterized by different pollen assemblages, (Horowitz and Derin, 1987; Horowitz, 1990). These palynozones have been encountered in both continental sediments drilled in the Dead Sea Rift and marine sequences penetrated by boreholes offshore Israel; foraminifera assemblages of the latter are used as basis of the stratigraphy. All Neogene palynozones have been encountered in the offshore Bravo 1 borehole, where sedimentation is considered to have been continuous throughout this period, except for a break within the middle Miocene. The Neogene sequences of the Dead Sea Rift are not always continuous, resulting from differential faulting and troughs formation in various localities during this period (Horowitz, 1987a), but since parts of the sections overlap it was possible to construct a cumulative sequence for the continental Neogene.

The Neogene palynozones are Ma, of latest Oligocene–early Miocene age; Mb of late early through middle Miocene; and Mc of late Miocene; Pa of early through middle Pliocene; and Pb of late Pliocene age. The Quaternary sequence was subdivided by Horowitz and Horowitz (1985) into ten palynozones: QI and QII of Preglacial Pleistocene age; QIII through QIX of Glacial Pleistocene age; and QX representing the Holocene.

Palynological and paleoenvironmental characteristics of the Quaternary palynozones are summarized in Table 9.2.1. For rates of accumulation of Neogene palynozones in the various subbasins of the Dead Sea Rift please refer to Fig. 8.5.1.; for the Quaternary, to Fig. 8.5.4.

Pollen diagrams presented here for the Neogene and Quaternary palynozones, are taken for each from the borehole in which maximum thickness of the specific palynozone was encountered. Occasionally diagrams are brought from such boreholes that best display the palynozone characteristics. Paleogeographic and palynological descriptions of the palynozones follow in stratigraphic order.

A. The Miocene (Ma, Mb and Mc)

I. Palynozone Ma (latest Oligocene through early Miocene). Sea level was quite low at that time (Fig. 9.2.2); it seems that the landscape consisted of a relatively elevated flatland, which could have been up to 800–900 meters above sea

TABLE 9.2.1

Main characteristics of the Quaternary palynozones in the present day Mediterranean domain of Israel

Palynozone	Vegetation	Climate	Age
			QUATERNARY
			Holocene
QX	Evergreen oak maquis	Dry Mediterranean	
			Glacial Pleistocene
QIX	Deciduous oak forest	Wet Mediterranean	Würm
QVIII	Garigue	Dry Mediterranean	R/W
QVII	Deciduous oak forest	Wet Mediterranean	Riss
QVI	Evergreen oak and pine stands	Dry Mediterranean	M/R
QV	Deciduous oak forest	Wet Mediterranean	Mindel
QIV	Oak and pine stands	Dry Mediterranean	G/M
QIII	Oak, pine and spruce forest	Wet Mediterranean	Günz
			Preglacial Pleistocene
QII	Oak and spruce stands	Dry Mediterranean?	
QI	Oriental spruce forests	Wet, warm temperate	

Reprinted with permission from Horowitz (1989b).

level, as evidenced by canyons cut to this depth during that period (Neev, 1960). The area presently occupied by the Dead Sea Rift was slowly subsiding during Ma times, but no sign of rifting or trough formation is seen until the late Miocene (Horowitz, 1987c).

The northern part of the Dead Sea Rift — the central Jordan Valley and the Hula — was drained in Miocene times toward the Persian Gulf (Michelson, 1972); the southernmost tip — the Elat region — was drained southward toward the forming Red Sea, as it does today, while the rest of Israel's area was drained to the Mediterranean (Garfunkel and Horowitz, 1966). Marine sediments are rare for Ma times and the only locality in which such had been accumulated is the offshore, studied palynologically only from Bravo 1 borehole. Gravel had accumulated in southern Israel in the slowly subsiding, most probably intermediate basins on rivers' way to the Mediterranean.

It is very difficult to draw any figures for rates of subsidence during Ma palynozone, since it is only known from one locality in the southern Dead Sea basin, Ami'az 1 borehole. However, in this locality about 400 meters of sediments had been accumulated during approximately six million years (Horowitz and Derin, 1987; Horowitz, 1987c), which is a figure typical in Israel for a syncline, rather than a rift. Boreholes drilled in other localities in the Dead Sea Rift did not go deep enough to encounter sediments of this palynozone, which may have accumulated in other subbasins in the subsiding Dead Sea region.

Palynozone Ma comprises pollen spectra (Fig. 9.2.3) rich in non-arboreal components, of which Compositae are the most common. The arboreal part

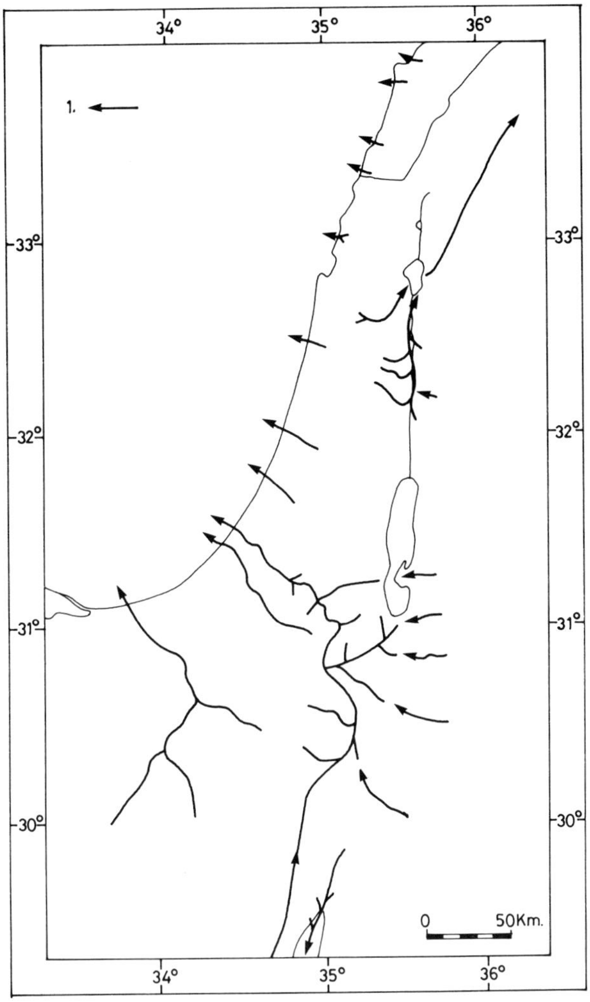

Fig. 9.2.2. Paleogeography, palynozone Ma. *1* = Main rivers. Reprinted with permission from Horowitz (1979).

consists of pollen derived from trees of the families Juglandaceae, Betulaceae and Pinaceae. The very high percentages of Compositae within regional components of the pollen diagrams indicate that, in general, the environment was very dry and desertic. The arboreal pollen indicate that some conifers, probably quite rare, may have grown on hills, or brought by winds from distant localities. Some lowland environments, probably marshes, gave place to growth of trees of the Betulaceae and Juglandaceae. Pollen of steppe plants, such as *Artemisia* and *Ephedra*, are quite rare through Ma times. It seems that there was either no suitable biotope for development of this vegetational formation, maybe not only because of the

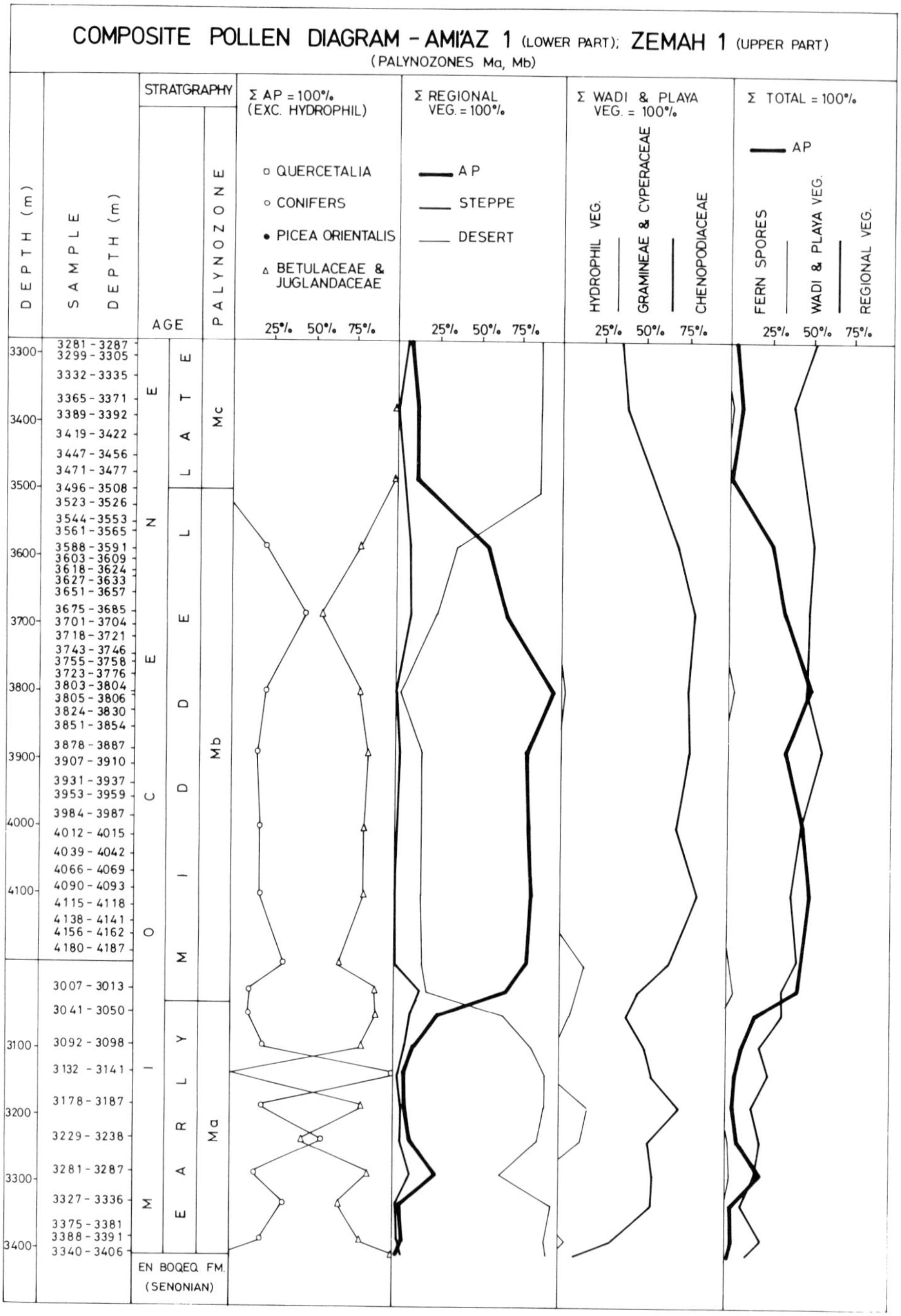

Fig. 9.2.3. Pollen diagram, palynozones Ma and Mb.

climate, but also as a result of the topography at that time; or, that the region was extremely-arid, quite far away from any steppe.

The pollen assemblages indicate a warm, dry to extremely-arid, subtropical environment for late Oligocene through early Miocene times in Israel. Palynozone Ma is correlative to the biostratigraphic foraminifera zones N4, N5 and N6.

II. Palynozone Mb (latest early through middle Miocene). Sediments of this palynozone had been deposited concurrently with the global middle Miocene sea level rise, which caused deposition of several hundred, up to a thousand meters of marine sediments in the offshore and part of the onshore in Israel. This invoked filling up of the previous late Oligocene–early Miocene channels, thus forming wide, meandering rivers with rather large, intermediate lakes occupying the slowly subsiding synclines (Fig. 9.2.4). Some volcanic activity is also known for that period, from both the coastal plain and northern parts of Israel. Sediments of Mb are known from all along the Dead Sea Rift, both from boreholes and outcrops, attaining several hundred meters in thickness. As for the previous, Ma palynozone, no endoreic basins are known within the Dead Sea Rift and most of Israel was drained toward the Mediterranean, except for the northeast, which was still draining toward the Persian Gulf, while the southernmost part toward the Red Sea.

Mb palynozone is characterized (Fig. 9.2.3) by arboreal pollen of types quite similar to those of the underlying Ma, namely conifers, as yet unidentified, Betulaceae and Juglandaceae. Typical arboreal pollen shares for Mb are around 50% of the pollen derived from regional vegetation, while the rest of this group is chiefly originated from desert plants such as Compositae, accompanied by some pollen of steppe plants, especially in the lower part. The lower half and the upper half of Mb have almost similar shares of arboreal pollen, but within the lower half Betulaceae and Juglandaceae prevail, probably indicating wetter lowlands, formed by the regional subsidence and marine transgression, while the upper part is typified by much higher percentages of conifers, most probably indicating development of a drier highlands forest.

The much higher percentages of arboreal pollen in Mb as compared to Ma seem to indicate a higher humidity, or the turning of the dry, subtropical desert of Ma into a wet, subtropical savannah, or sometimes even forest during Mb times. However, this change is gradual; it seems that the higher percentages of Betulaceae and Juglandaceae — the lowland marsh floras (Horowitz and Derin, 1987) at the beginning of Mb — are mainly a result of the rising sea level, that created suitable biotopes for these trees. Only later on, the increasing humidity all over the region permitted development of a highlands forest, which mainly consisted of conifers. This is also supported by the relatively high percentages of steppe elements in the lower part of Mb, indicating that the regional vegetation was probably a steppe, while only the lowlands were occupied by trees. Toward the upper part of Mb

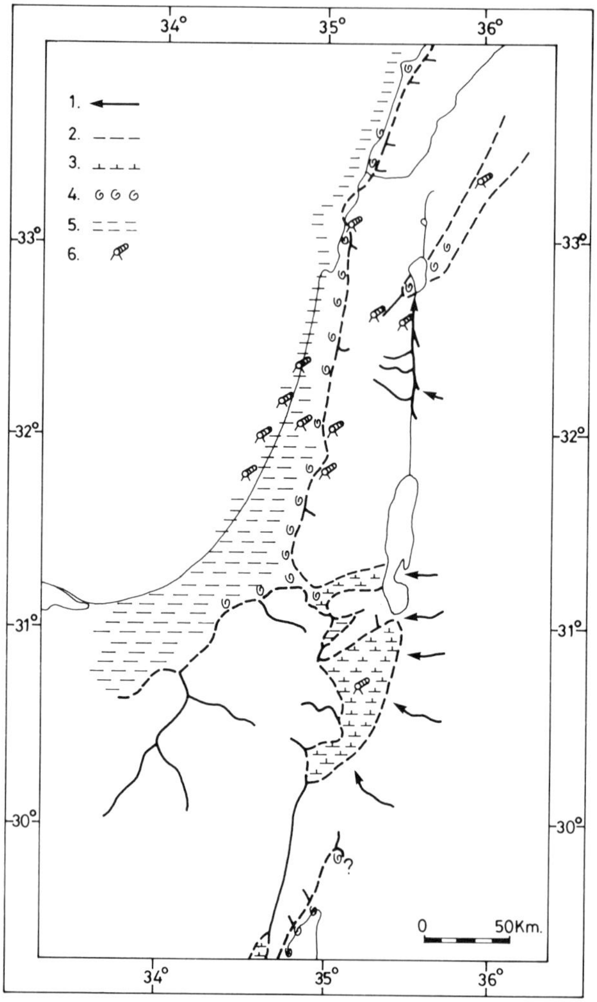

Fig. 9.2.4. Paleogeography, palynozone Mb. *1* = Main rivers. *2* = Shoreline. *3* = Freshwater lakes. *4* = Oyster banks. *5* = Fine marine clastics. *6* = Volcanos. Reprinted with permission from Horowitz (1979).

the steppe elements declined sharply and it may well be that a major part of the Compositae, which are generally regarded as indicative of desert environments, comprised in fact, in the upper part of Mb, undergrowth of the highlands forest. To sum up, it seems that the relatively high percentages of arboreal pollen in Mb represent a combination of rising sea level and climate, which only gradually became wetter.

It is quite interesting to note that arboreal pollen shares are higher in Mb sequences obtained from southern Israel, as compared with those from the north.

This may indicate that rains at that period came from the south, most probably connected with the monsoon system which had commenced to develop at that time, affecting, at least for a while, the Near East.

The climate was quite humid during Mb times, maintaining subtropical savannahs in which elephants, giraffes and other creatures roamed. Water bodies seem to have been heavily infested with crocodiles (Savage and Tchernov, 1968). Palynozone Mb is correlative to foraminifera zones N7 through N14.

III. Palynozone Mc (late Miocene). In late Miocene (Mc) times (Fig. 9.2.5) sea level had dropped to its lowest point during the late Cenozoic, causing erosion and channelling of previously laid sediments, mainly toward the Mediterranean and the Red Sea, in which evaporites had been accumulated. Connection of northeastern Israel with the Persian Gulf seems to have ceased at that time, probably as a result of the formation of troughs in the Dead Sea Rift region. Such troughs (Horowitz, 1987c, 1989c) are known from the southern Dead Sea region, south of the present Dead Sea, from the central Jordan Valley and probably also from the northern Jordan Valley, in the Hula region. These troughs accumulated extensive thicknesses of Mc palynozone sediments — almost two kilometers encountered in Arava 1 borehole south of the Dead Sea, close to 1,750 meters at Zemah 1 borehole in the central Jordan Valley. Only 500 meters of Mc sediments are known from the Hula basin. The southern Dead Sea basin was, during Mc times, a terminal erosion base level which had captured the drainage from its close vicinity, receiving clastic sediments especially from the northern Arava, to the south. The trough in the central Jordan Valley may well had also been a terminal base level, but information from this region is not sufficient for a firm statement.

This palynozone is typified (Fig. 9.2.6) by a characteristic regional vegetation pollen, which is represented in Mc almost exclusively by Compositae, with very low values of arboreal pollen, lower even than those encountered for Ma times. This indicates rather extreme desert conditions for Mc times. The climate was hot and extremely-arid, maintaining only a sparse desert vegetation, in a manner similar to Ma times, but apparently even slightly drier. Palynozone Mc corresponds to foraminifera zones N16 and N17.

B. The Pliocene (Pa, Pb)

The transition from Miocene to the Pliocene was marked in the entire Mediterranean area by considerable structural changes, which had also affected Israel. Several troughs had been opened in the Dead Sea Rift region, connected to the sea through the Yizre'el Rift Valley, which cut the country diagonally in a northwesterly direction, linking the Dead Sea Rift region with the Mediterranean (Fig. 9.2.7). As a result, much of the drainage pattern had changed by capturing many rivers toward the forming rift valleys. The Piacenzian and Tabianian transgressions, which inundated the forming rift valleys (Fig. 9.2.8), deposited

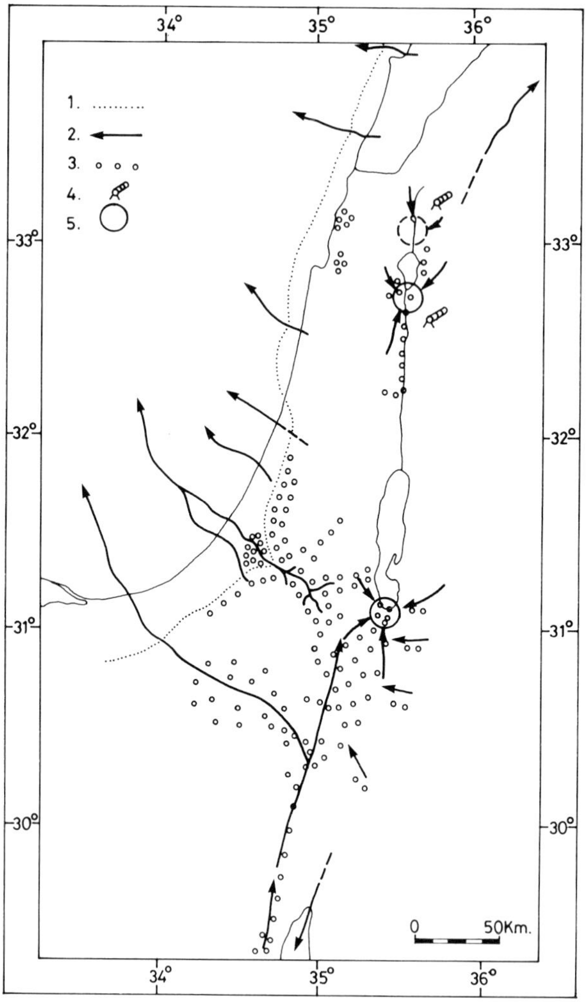

Fig. 9.2.5. Paleogeography, palynozone Mc. *1* = Eastern limit of evaporites. *2* = Main rivers at the maximum regression. *3* = Gravels, mainly fluviatile. *4* = Volcanos. *5* = Endoreic taphrogenic basins. Reprinted with permission from Horowitz (1979).

brackish sediments to the north, where seawater had been diluted by freshwater flowing in rivers, while saline to hypersaline lagoons had been formed to the south, where freshwater supply was limited during the Pliocene.

Sediments had been accumulated in great thicknesses in Pliocene troughs of the Dead Sea Rift — up to 2,000 meters of rocksalt, gypsum and clastic rocks are known from the southern Dead Sea basin. Some 1,500 meters of evaporites and clastics are known from the central Jordan Valley, but only 400 meters from the northern Jordan Valley. Some 600 meters of Pliocene deposits are known from the offshore, an area which was also subject to a considerable subsidence during this period. Toward the end of the Pliocene the sea regressed, followed by rivers depositing gravel (Fig. 9.2.9).

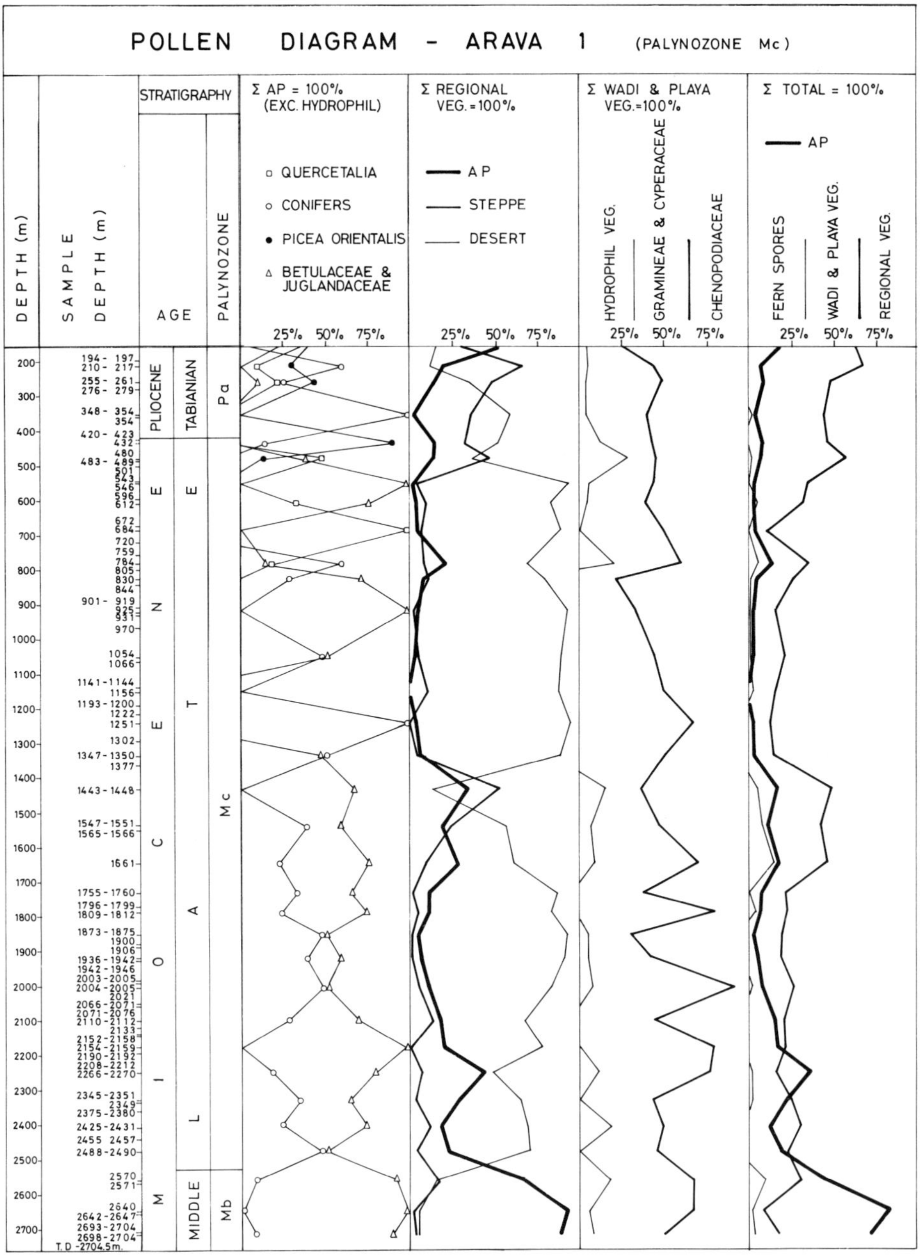

Fig. 9.2.6. Pollen diagram, palynozone Mc.

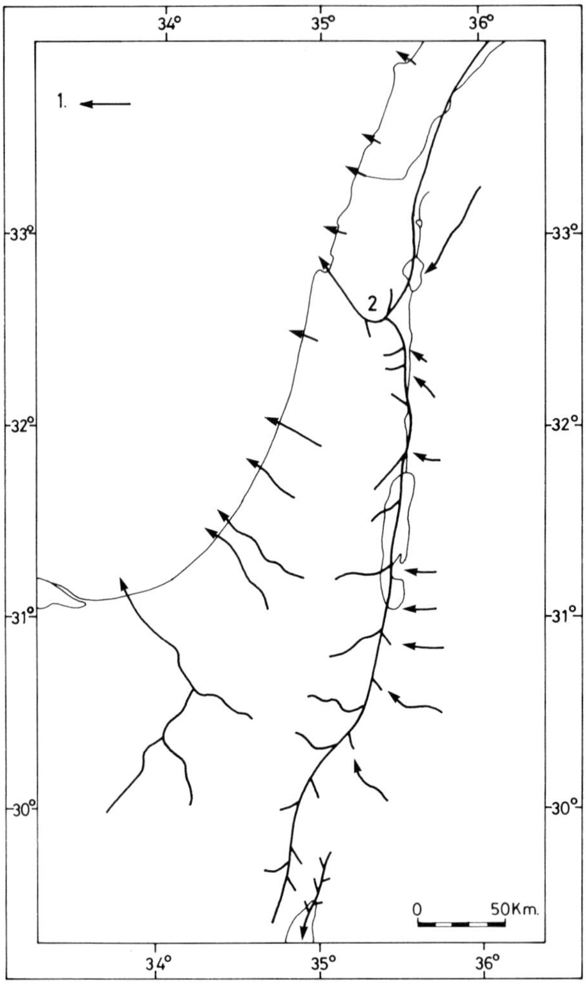

Fig. 9.2.7. Paleogeography, Mc–Pa transition. *1* = Main rivers. *2* = Yizre'el Rift Valley. Reprinted with permission from Horowitz, 1979.

The Pliocene sequence was subdivided (Fig. 9.2.10) into two palynozones: Pa, of lower through middle and Pb, of late Pliocene age, which bear pollen spectra considerably different from those of underlying Miocene sediments, the "M" palynozones. Dominating arboreal pollen are derived from two groups of trees, the *Quercetalia* and a conifer, which most probably was *Picea* (*orientalis*?). It is quite difficult to be exact about this specific definition: of all the conifer pollen examined in reference collections, the Pliocene grains looked very much like those presently produced by *P. orientalis*. It must be stressed that Ms. N. Drivaliari (Laboratoire de Palynologie, U.S.T.L., Montpellier, personal communication, 1989) had assigned

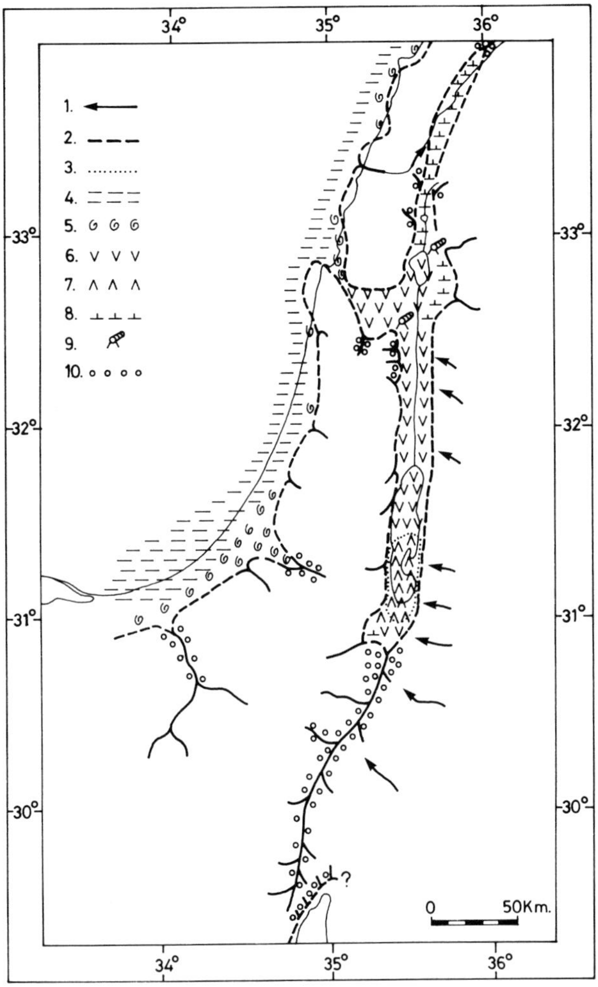

Fig. 9.2.8. Paleogeography, maximum of the Tabianian (Pa) and Piacenzian (Pb) transgressions. *1* = Main rivers. *2* = Shoreline. *3* = Evaporitic basin of the southern Dead Sea. *4* = Fine marine clastics. *5* = Biogenic limestones. *6* = Brackish water bodies. *7* = Hypersaline lagoons. *8* = Freshwater lakes. *9* = Volcanos. *10* = Fluviatile gravel. Reprinted with permission from Horowitz (1979).

these grains to *Abies*, or to the type genus *Cathaya*, which may as well be correct.

At any rate, composition of the arboreal pollen group is essentially different than that of the Miocene, probably representing initial stages in the developing Mediterranean domain. Palynozone Pa is richer in arboreal pollen as compared to the overlying Pb. Shares of up to 70% AP from the pollen produced by regional vegetation are typical for Pa, while Pb is characterized by figures of 20–30% for that group. There is no essential difference in components of non-arboreal

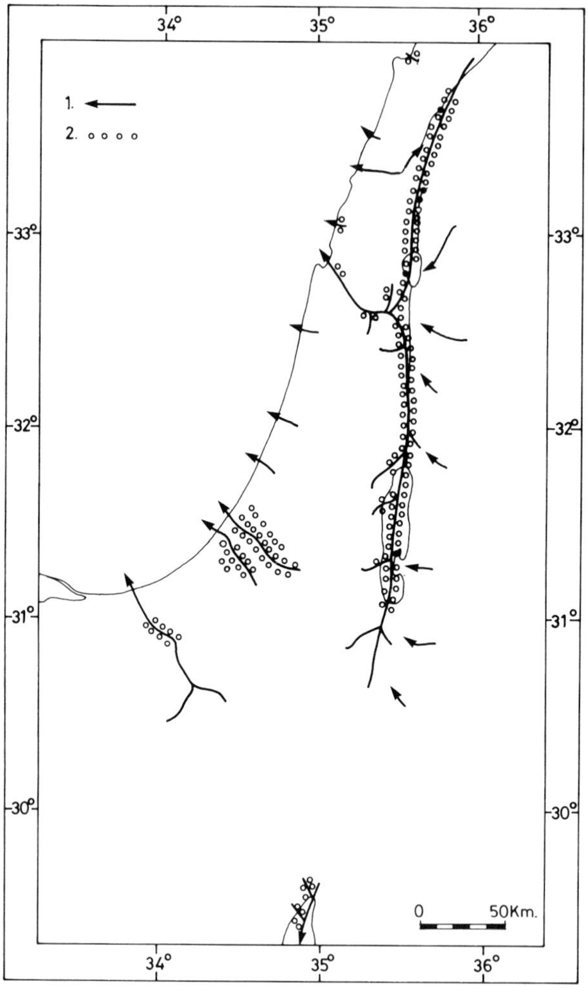

Fig. 9.2.9. Paleogeography, maximum regression at end of Pb. *1* = Main rivers. *2* = Fluviatile gravel. Reprinted with permission from Horowitz (1979).

pollen groups when the Pliocene and Miocene are compared and indeed the same NAP continue to the present day, with only minor changes, except for temporary prevalence of certain groups, under specific environmental conditions. Part of this apparent continuity may be a result of problems involved with identification, since most of the non-arboreal pollen have been identified only to the family level, except for *Artemisia*, *Ephedra* and several other minor constituents. There might have been changes within the non-arboreal pollen which are not reflected in the identifications, changes that may have been on less than genus, or occasionally family level.

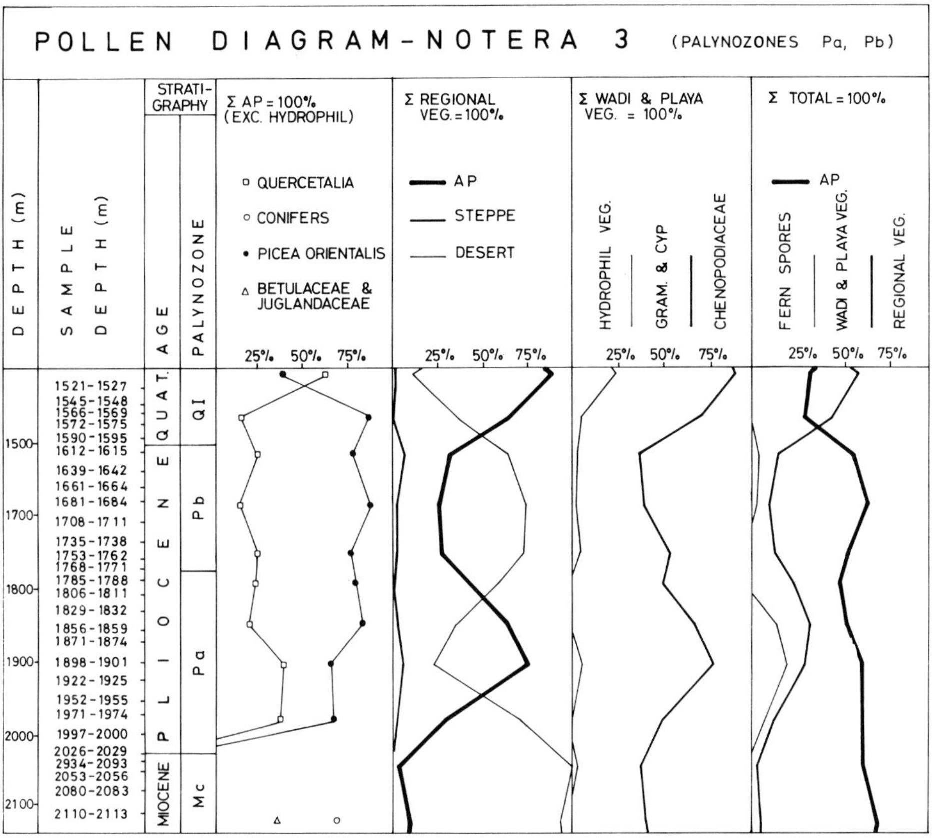

Fig. 9.2.10. Pollen diagram, palynozones Pa and Pb.

The combination of pollen produced by oaks and *Picea* (or *Abies*) indicates a temperate environment for the Pliocene, which was probably wetter in Pa and dryer in Pb times, but in both cases representing environments for which the present day counterparts can be found in places such as Turkey, the Balkan and central to northern Italy.

Arboreal pollen are more abundant during the Pliocene in sediments from northern Israel as compared to the south, opposed to the situation during Miocene times, indicating a major change in the climate regime. Rains must have come during Pliocene times from the north, rather than from the south as was concluded for the Miocene, pointing to a major shift of climatic belts and influence exerted upon Israel by the European temperate type climate, which had replaced the previously monsoon-controlled climate of the Miocene.

During the Pliocene, offshore sections show lower arboreal pollen percentages than those obtained from the Dead Sea Rift region and are also richer in oak grains. The reason for over-representation of non-arboreal pollen in the offshore sequences is probably tied up with activity of the River Nile, which had commenced at the early Pliocene, carrying with it and depositing in the offshore considerable amounts of non-arboreal pollen derived from its bank vegetation, especially Gramineae and Cyperaceae, accompanied by some tropical spores. The area of Israel constituted during the Pliocene a lowland, in which most probably the oaks grew, while to the northeast, where the landscape became more elevated, conifer forests occupied the hilly terrains.

The climate in Pa times was quite humid, as evidenced by wealth of arboreal pollen in the spectra. It is in some contrast to the extensive accumulations of evaporites and even seems more so, since the landscape appears to have been quite flat, with no considerable relief which could have formed a suitable biotope for arboreal vegetation. It seems however that solar insolation was enough to evaporate the seawater and, aided by the very restricted connection to the sea, deposit evaporites. The climate during Pb is somewhat drier than in the preceding Pa, but it still maintained the temperate characteristics.

C. The Preglacial Pleistocene (QI, QII)

I. Palynozone QI. The transition from the Pliocene to the Quaternary, some 2.65 my ago, was accompanied by considerable volcanic activity in northern and northeastern Israel, which sealed the previous, Pliocene connection of the Dead Sea Rift region with the Mediterranean (Fig. 9.2.11). However, no trough formation and rifting processes are known for this period from the Dead Sea Rift and a system of rivers leading from east to west, from the present day Transjordan to the Mediterranean, had developed in Preglacial Pleistocene times (Horowitz, 1979; Levin and Horowitz, 1987).

The country was rather flat, low-lying and crossed by several rather wide, meandering rivers. The Dead Sea Rift at that time showed only slight subsidence, which made some of its parts intermediate lakes on the way of these rivers from Transjordan to the Mediterranean. The most pronounced of these intermediate basins was in the southern Dead Sea basin, where up to 870 meters of sediments had been accumulated in Melekh Sdom 1 borehole during QI times, while in other boreholes drilled within the Dead Sea Rift typical figures are 400–600 meters. However, even though subsidence was considerable in the region, the sediments do not show any signs of having been deposited in a terminal basin.

Palynozone QI (Fig. 9.2.12) was subdivided into three parts: the lower, QIa, is typified by very high shares of arboreal pollen, sometimes exceeding 80% of the total, almost exclusively derived from *cf. Picea orientalis*; the middle, QIb, is characterized by lower arboreal pollen shares as compared with the underlying and overlying sequences, but higher percentages of steppe elements and Chenopodiaceae. Pollen derived from desert vegetation are also somewhat more abundant as compared to QIa. Usually, *cf. Picea orientalis* is the common arboreal pollen, though in some boreholes, *Quercetalia* are quite common, especially in the lower part; the upper, QIc, resembles QIa by the very high arboreal pollen values, again dominated by *cf. Picea orientalis.*

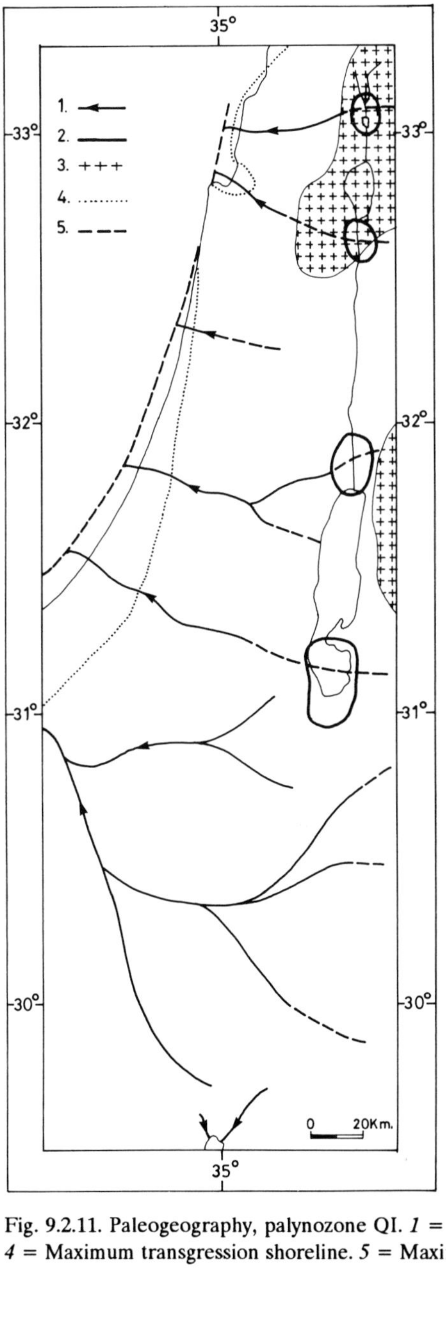

Fig. 9.2.11. Paleogeography, palynozone QI. *1* = Main rivers. *2* = Intermediate lakes. *3* = Lava flows. *4* = Maximum transgression shoreline. *5* = Maximum regression shoreline.

The high percentages of pollen from *Picea orientalis* (if identification is correct, a precaution that should be remembered all along this chapter), which presently only grows in some secluded parts of northeastern Turkey, in typical euxine environment,

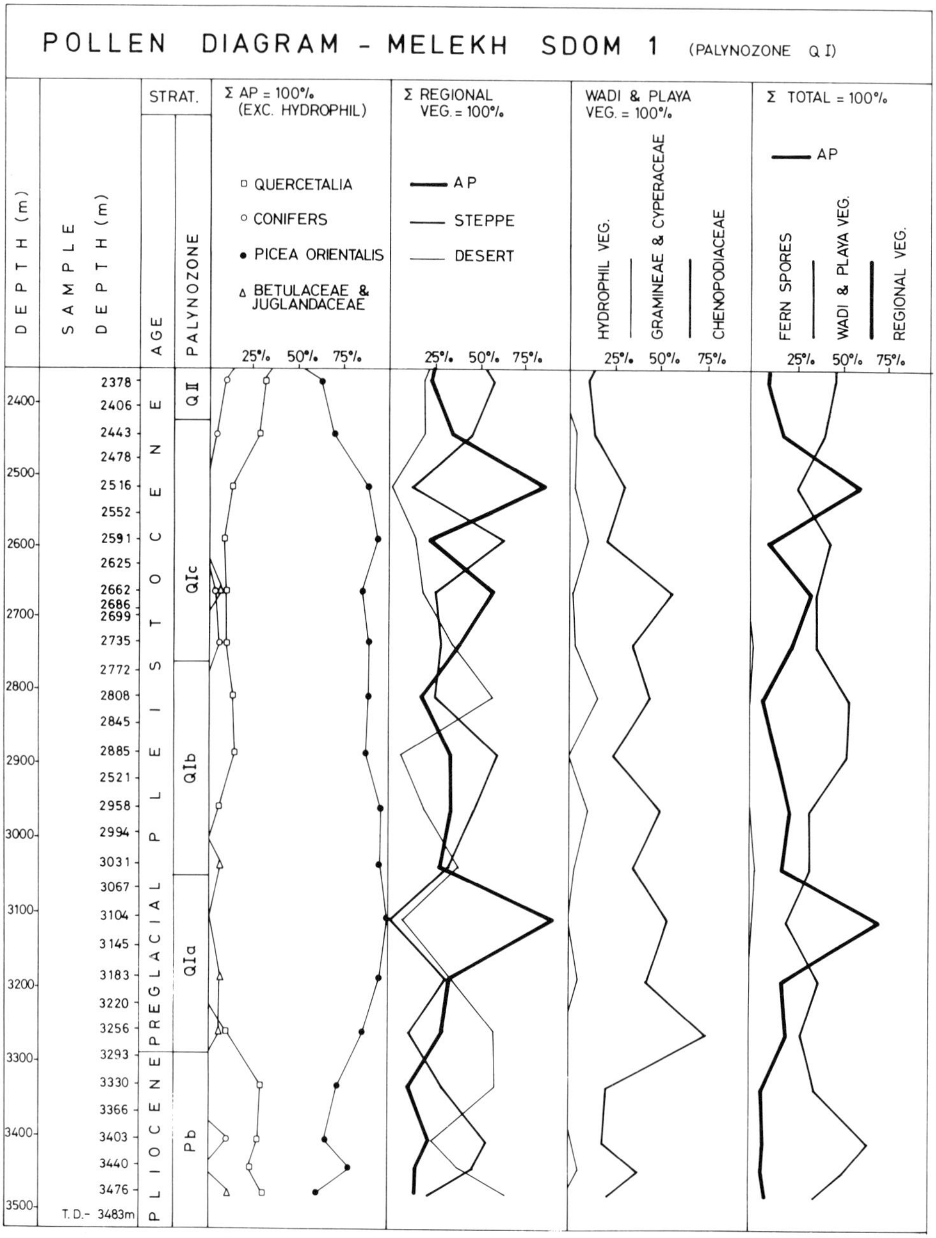

Fig. 9.2.12. Pollen diagram, palynozone QI.

seem to indicate a cool, temperate environment for QI palynozone times, which oscillated from rather wet during QIa to somewhat dryer in QIb, back to wetter at QIc. This wetter climate stands in contrast to the preceding Pb palynozone and also the overlying QII, both characterized by pollen spectra indicating rather dry climates.

An interesting feature of QI is that, in contrast to the Pliocene palynozones, it displays somewhat higher arboreal pollen shares to the south, where *Picea* pollen are also accompanied by some Betulaceae and Juglandaceae, which were typical for the Miocene wet lowland vegetation. This situation changes at QIc and arboreal pollen are more abundant to the north, a trend which increasingly develops toward the present day. These findings may indicate that in QIa times the monsoon rains again hit Israel from the south, while toward QIc the present day regime of rains, arriving from the northwest, commenced developing again, reminding of the Pliocene climates.

Same as was observed for the preceding, Pliocene palynozones, oak pollen are more abundant in the offshore sediments, as are also non-arboreal pollen, apparently again due to similar reasons, namely that oaks grew in the lowlands to the west, while *Picea orientalis* grew on the highlands to the east and northeast. At the same time a considerable amount of non-arboreal pollen were contributed by the Nile.

QI climate was rather wet, temperate, with a somewhat drier period in its middle part, which served to subdivide the palynozone into three subpalynozones: QIa, QIb and QIc. It corresponds to foraminifera zone N22, which also includes QII in its upper part.

II. Palynozone QII. Palynzone QII times mark a considerable change in the morphotectonic regime of the region, when the Dead Sea Rift commenced its development toward the present day shape, of an endoreic drainage system, for which the Dead Sea had served ever since QII and up to the present day as terminal erosion base level (Fig. 9.2.13). At the same time the central parts of Israel had been longitudinally uplifting, a process which continues to the present day, that resulted in clear subdivision of the country into the western low lying coastal plain, elevated hilly central backbone and the Jordan–Dead Sea Rift to the east.

The incipient rifting during QII times resulted in development of a trough in the southern Dead Sea basin, where almost 900 meters of sediments had been accumulated in Melekh Sdom 1 borehole during QII, which was relatively a rather short stage, spanning less than 200,000 years (Fig. 8.5.4). This considerable rate of subsidence caused capturing of previous QI rivers into the newly developing terminal basin, a capturing process which continues until the present day.

Palynozone QII is typified (Fig. 9.2.14) by very low arboreal pollen shares, comprising mainly *Picea orientalis*, with some oak and rare occurrences of small triporate grains, possibly derived from one of the Juglandaceae. Pollen of desert

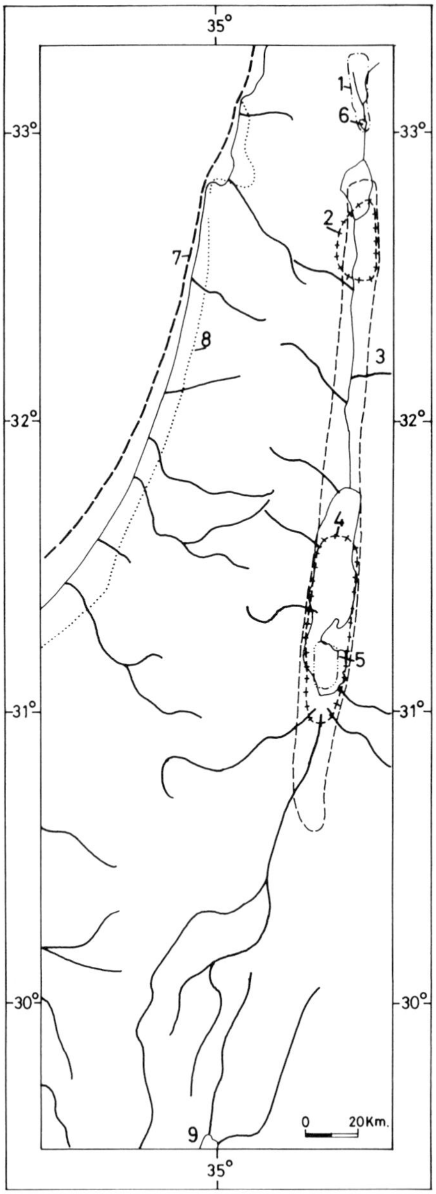

Fig. 9.2.13. Paleogeography, palynozones QII through QX. *1* = Maximum extension of lakes in the Hula sub-basin. *2* = The central Jordan Valley sub-basin. *3* = Maximum extension of lakes in the Jordan–Dead Sea region. *4* = The southern Dead Sea sub-basin. *5* = Minimum extension of water bodies in the southern Dead Sea sub-basin. *6* = Minimum extension of water bodies in the Hula sub-basin. *7* = Pluvial regressive shoreline (schematic). *8* = Interpluvial transgressive shoreline (schematic). *9* = Bay of Elat (Aqaba).

vegetation, mainly Compositae, slightly increase as compared with the underlying QI, but among the regional vegetation the steppe elements predominate.

The rarity of arboreal pollen and composition of the pollen flora seem to indicate gradual onset of Mediterranean interpluvial arid climate for the first time (Horowitz, 1987a). The environment in QII times was somewhat similar to that of QIb, but apparently much dryer. The predominance of pollen derived from steppe vegetation indicates that this flora grew not far away, possibly in a manner quite similar to the present day (Fig. 9.3.3). The Mediterranean climate of QII times remained of the same nature until the present day. Oscillations, of dry interpluvials and wet pluvials, both interrupted by interstadials, characterize the period from QII, still within the Preglacial Pleistocene, through to the present day, including the Glacial Pleistocene and Holocene.

D. The Glacial Pleistocene–Holocene (QIII through QX)

Ever since its incipience in QII times, the Dead Sea Rift became a prominent feature in Israels' geography throughout the Glacial Pleistocene and Holocene (Fig. 9.2.13). This is even more accentuated by the contemporaneous longitudinal uplifting of both its shoulders, in Israel and Transjordan (Horowitz, 1979). Five subbasins, in the form of deep troughs, had developed along the Dead Sea Rift — the Hula to the north; the central Jordan valley; the northern Dead Sea basin; the southern Dead Sea basin; and the Bay of Elat to the south, all of which had accumulated great thicknesses of sediments, up to almost 4,000 meters in the terminal southern Dead Sea basin. Rates of subsidence in the five subbasins (Fig. 8.5.4) differ in each and are also variable along different periods, so that thicknesses of various Glacial Pleistocene and Holocene palynozones considerably change from one borehole to another.

The Quaternary palynozones are well known only from boreholes drilled in various subbasins of the Dead Sea Rift and not from the coastal plain, where only fragmentary evidence is available due to incomplete sampling of the offshore sequences (Rossignol, 1969b). The coastal plain sections are not complete because during low sea levels (corresponding to the glacials), the area was subject to erosion and sediments had only accumulated during transgressive, interglacial phases (Horowitz, 1979). Thus the coastal plain sections only serve to complement knowledge gathered from the Dead Sea Rift boreholes and sequences.

In Glacial Pleistocene–Holocene times, Israel seems to have constantly been under a dominant influence of the Mediterranean type climate. Three end members have been defined for these paleoclimates (Horowitz and Horowitz, 1985; Horowitz, 1987a): a pluvial, corresponding to higher amounts of rainfall, some of it during the summer time, that gave rise to development of a rather rich vegetation in which trees take a considerable share (Fig. 9.3.4); interpluvial, dryer than the present day climate, during which the region was subject to only occasional thunderstorms arriving from the northwest, consequently supporting only a very poor vegetation

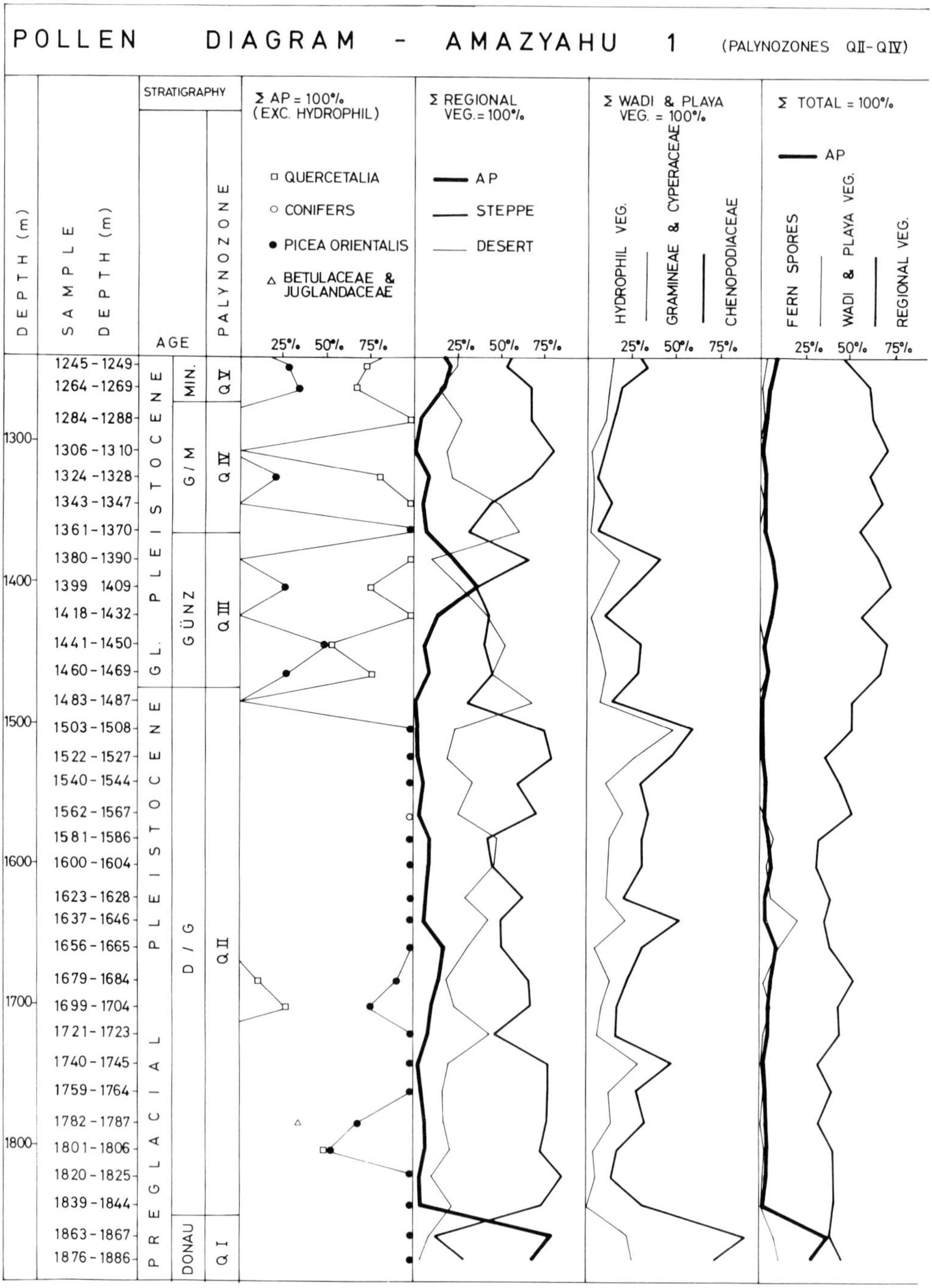

Fig. 9.2.14. Pollen diagram, palynozones QII, QIII and QIV.

(Fig. 9.3.5); and interstadial, of the present day character, when the area was subject to winter rains with a long, hot, dry summer, permitting development of seasonal non-arboreal vegetation during the wet winters and growth of such trees which could withstand summer drought, notably the evergreen oak *Quercus calliprinos* (Fig. 9.3.3).

I. Palynozone QIII. This palynozone (Fig. 9.2.14) displays higher arboreal pollen shares than the preceding QII, with oak as the dominant constituent. AP percentages are in the order of the present day, or somewhat higher. The increase in arboreal pollen is accompanied at the base of the palynozone by a similar trend in Compositae, probably derived from undergrowth of the submesic maquis, which is replaced upward by a peak in the steppe curve, indicating some drying. The prevalence of oak pollen grains probably indicates a developed Mediterranean type environment. It seems that during QIII times the region enjoyed somewhat higher amounts of rainfall as compared with the present day, the climate probably oscillating between pluvial and interstadial.

II. Palynozone QIV. Palynozone QIV is characterized (Fig. 9.2.14) by a drop in arboreal pollen and a rise in the desert vegetation pollen. Toward the top the latter decreases to give place to an increase of the steppe elements. Again, oak is the main constituent of the arboreal pollen spectra. The climate is most probably of an interpluvial nature.

III. Palynozone QV. This palynozone is typified (Fig. 9.2.15) by considerable peaks of the arboreal pollen curve, among the highest along the entire Glacial Pleistocene. The AP are dominated by oak, of the winter deciduous type, but pine also makes its appearance. Some occurrences of *cf. Picea orientalis* are still recorded at the base of this palynozone. Typically, higher pine shares are encountered when the arboreal pollen curve drops at the middle part of the palynozone, where desert vegetation pollen curve also increases. The rest of QV sequence shows rather low values of desert plants pollen. Steppe vegetation pollen maintain high values of the top of the preceding QIV palynozone through the base of QV and diminish upward, to reach moderate shares. The climate was essentially pluvial, separated by interstadial phases.

IV. Palynozone QVI. Characterized (Fig. 9.2.15) by relatively low AP shares along the sequence, interrupted by several somewhat higher peaks. The main arboreal pollen is the evergreen oak, *Quercus calliprinos*, with subordinate *Pinus halepensis*. Desert plants pollen are quite abundant, while steppe vegetation shows median values. The climate was interpluvial, possibly interrupted by some interstadials.

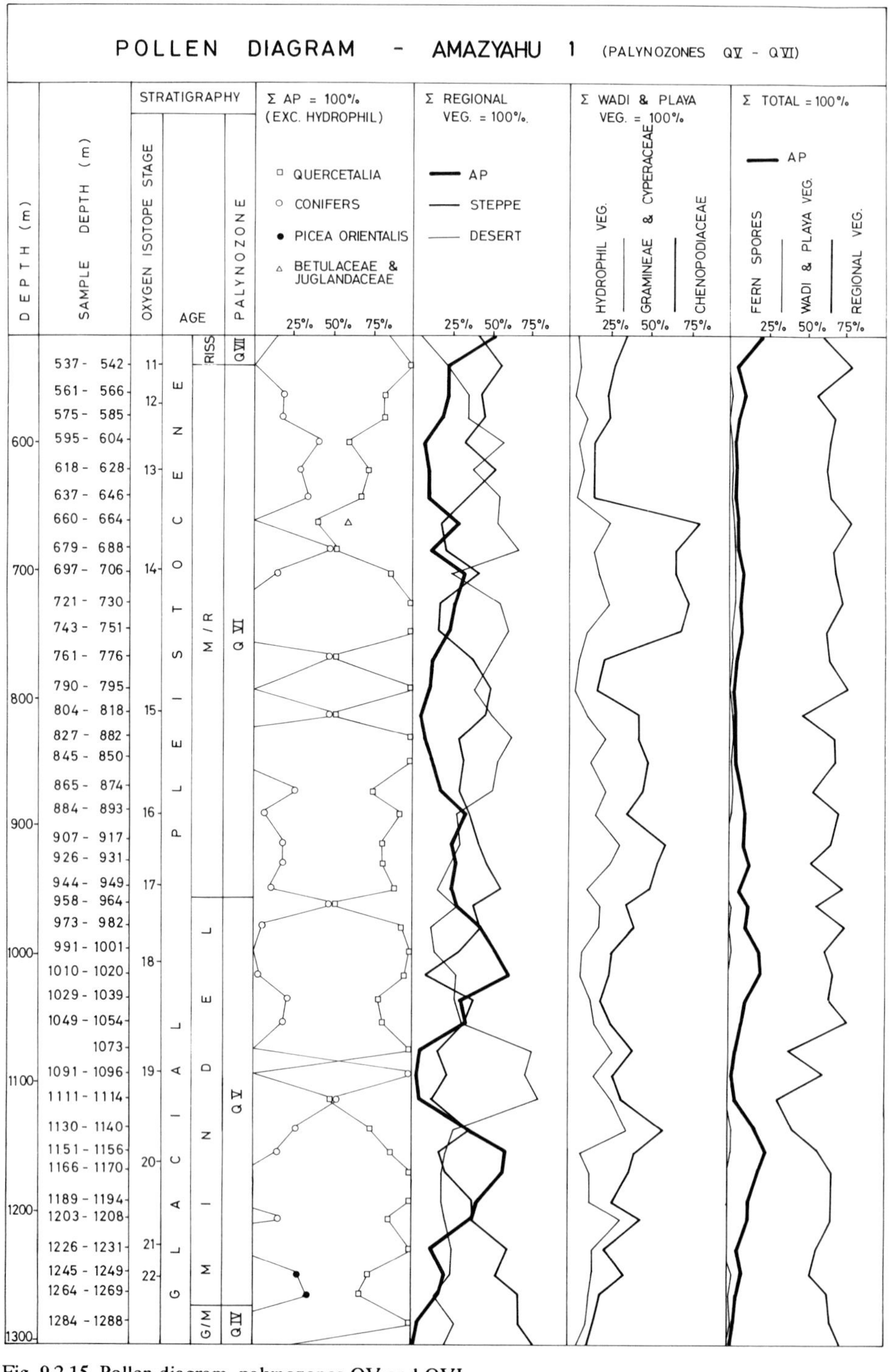

Fig. 9.2.15. Pollen diagram, palynozones QV and QVI.

V. Palynozone QVII. Typified (Fig. 9.2.16) by two distinct arboreal pollen peaks, at the lower and upper parts. Oak pollen dominate when AP values are high, replaced by a mixture of oak and pine following declines of this curve. The middle part of the palynozone shows low AP shares, with a considerable increase of pollen produced by steppe vegetation. The climate is pluvial, with an interstadial separating the wetter peaks.

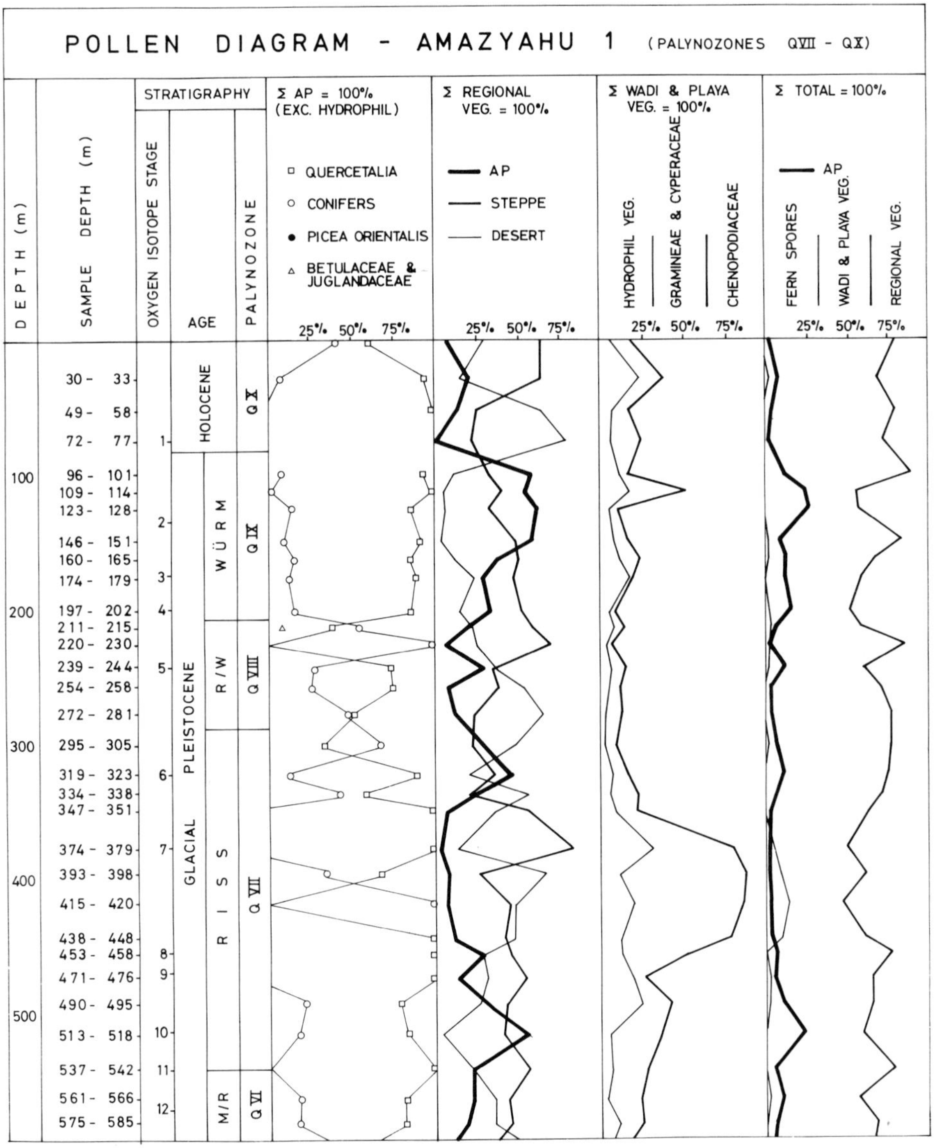

Fig. 9.2.16. Pollen diagram, palynozones QVII through QX.

VI. Palynozone QVIII. this palynozone is characterized (Fig. 9.2.16) by rather low arboreal pollen shares, consisting of both oak and pine. The desert plants pollen curve shows a peak which decreases toward the upper part, to be replaced by higher shares of steppe elements. The climate was interpluvial.

VII. Palynozone QIX. Characterized (Fig. 9.2.16) by very high arboreal pollen percentages, mainly consisting of the winter deciduous oak *Quercus ithaburensis*, accompanied by the evergreen *Q. calliprinos* and by some pine grains. Pollen of desert plants take a relatively small share of the spectra, while those originating in steppe vegetation show medium values. The climate was mostly pluvial, separated by interstadials. The pluvial phases seem to be progressively drier upward. The upper part of QIX is further detailed in Chapter 10.

VIII. Palynozone QX. This palynozone is characterized (Fig. 9.2.16) by very low AP and very high desert plants pollen values at the bottom. Both curves change direction toward the middle part of QX, which is followed by a drop of the AP and an increase of the steppe elements, to reach their present day values. A slight increase toward the top is also seen in the desert plants pollen curve, but present day values are considerably lower than those at the base of the palynozone. The arboreal pollen peak at the middle part is accompanied by decrease of local vegetation pollen shares, probably indicating expansion of the water bodies. This peak is dominated by the winter deciduous *Quercus ithaburensis*. The climate was interpluvial at the beginning of the palynozone, changing to interstadial later on, which persisted with slight fluctuations until the present day. A wetter phase is recorded for the Atlantic period, some 7,000 to 4,500 years ago, when summer rains have most probably occurred (Horowitz and Gat, 1984). QX is further detailed in Chapter 10.

To sum up (Figs. 9.3.3–5), QIII through QX show alternations of humid and dry Mediterranean climates, with peaks of interpluvials when almost the entire area of Israel was a desert; interstadials during which only the southern part of Israel was a desert, as is the present day situation; and pluvial climates when even the southern part of the country became covered by steppe vegetation, while at the same time oak forests grew to the north, most probably dominated by winter deciduous oaks.

9.3. Development of the desert system in Israel

Desert environments have been recorded in late Cenozoic pollen sequences from Israel by the very low percentages, or sometimes total absence of arboreal pollen from the regional components diagrams, accompanied by dominance of pollen grains produced by desert plants, notably Compositae. These desert phases, or indeed the entire sequence of climatic fluctuations, can be subdivided into two styles: long term fluctuations, in the order of several million years of duration, such as deserts of the late Oligocene–early Miocene (Ma) and the late Miocene (Mc); and arid phases which only persisted for tens, or maximum 200–300 ky, typical for the Quaternary.

9.3.1. The Neogene

Paleoenvironmental and paleoclimatic development of the region throughout the Neogene (Horowitz, 1990) can thus be summarized as follows (see Fig. 9.4.1, also for correlations with the foraminifera zones): a desert, occasionally oscillating between dry savannah and extremely-arid environment, existed during palynozone Ma times, latest Oligocene through early Miocene. During this time southern Israel had enjoyed more rainfall than the north. Amount of rainfall gradually increased in Mb times, latest early Miocene through middle Miocene, when the lowlands were covered with wet savannah of Juglandaceae and Betulaceae, probably also other trees which are not represented by pollen, while conifers grew on the drier highlands. The north continued to be drier than the south. This environment had entirely changed toward Mc times, late Miocene, when hardly any arboreal pollen are at all recorded. Pollen diagrams are dominated by Compositae, indicating arid to extremely-arid environments, probably a Sahara type desert for that period. The climatic cycles are in the order of six million years duration, with only subordinate fluctuations.

In Pliocene times climatic cycles became shorter, in the order of 1–1.5 million years and the entire region was subject to temperate conditions, climatically affected from the northern, European direction. Arboreal pollen spectra are dominated by such trees as *Picea orientalis* (or *Abies*?) and oaks, while the environmental gradient became drier southward. Thus Pa times, early through middle Pliocene, enjoyed a rather cool, humid, temperate climate, while during Pb, late Pliocene, the temperate environment became drier, possibly somewhat warmer.

9.3.2. The Quaternary

In contrast to the frequency of Neogene climatic and environmental cycles, which are in the order of millions of years, Quaternary oscillations are much shorter in duration and seem to become progressively shorter when the present day is approached. The reason for the latter phenomenon may well be artifactual, due to much more data acquired for the later part of the Quaternary, as compared with its earlier times. It is however notable that oxygen isotope curves for the last 3.5 million years (Shackleton and Hall, 1984) also seem to show longer durations of the earlier oscillations.

Several periods are recorded for the Quaternary, during which almost the entire area presently occupied by Israel had turned to desert. These arid phases characterize palynozones QII, QIV, QVI, QVIII and QX, separated by humid pluvials and occasionally interrupted by somewhat more humid interstadials. The Quaternary desert phases of Israel seem to correlate quite well with the interglacials as recorded in the deep sea cores (Horowitz and Weinstein-Evron, 1986; Cheddadi, 1988; Horowitz, 1989b) and the Alps (Horowitz, 1979; Horowitz and Horowitz, 1985).

The information available on Quaternary paleoclimates in Israel and its neighboring countries is much more extensive than for the Neogene, coming from several sources which seem to corroborate each other such as geology, pedology, geomorphology, paleontology, palynology and isotope analyses.

Three types of climates have been discerned (Horowitz, 1979) for the Quaternary of Israel: interpluvial, very dry; pluvial, rather humid; and interstadial, of the present day characteristics. All comparisons are made with the present day environments, taken as reference (Table 9.3.1).

Pollen diagrams obtained from Israel for the last 150 ky, radiocarbon dated for their upper parts and uranium series dated for the lower (Horowitz and Weinstein-Evron, 1986), have shown quite clearly that pluvial climates correspond to the European and oceanic glacial phases; interpluvials to the interglacials, while interstadials to both stadials within interglacials and interstadials within glacials. North of Israel, however, especially in the higher elevations of northern Syria, Iran and Turkey, the glacials are recorded in pollen diagrams as seemingly

TABLE 9.3.1

Summary of the pluvial, interpluvial and interstadial climates characteristics for the Quaternary in Israel

	Pluvial	Interpluvial	Interstadial
Prevailing pollen of	Deciduous oak forest	Desert and steppe vegetation	Evergreen oaks and Mediterranean maquis
Rift Valley lakes	Extensive	Very limited	Like that at present
Wadi and river sediments	Aggradation	Strong erosion	Erosion
Soils	Well leached, red	Hardly any	Gray, calcareous
Deuterium excess of rain	d = 7‰ (average)	?	d = 22‰ (average)
$\delta^{18}O$ of rain	−7.6‰	?	−5.0‰
Rain regime	Moderate rains in winter and summer	Rare, probably winter thunderstorms	Winter thunderstorms only
Mediterranean water balance	Positive, freshwater from the North	Positive, freshwater from the Nile	Negative
Origin of rain	Atlantic Ocean	?	Mediterranean

Reprinted with permission from Horowitz (1989b).

dry periods (Van Zeist and Bottema, 1988), concluded from poverty in arboreal pollen. However, El Moslimany (1986) indicates that trees growing in the Zagros mountains, Iran, are more sensitive to heavy snowfall than to drought, thus concluding that the last glacial was not arid, but on the contrary, characterized by a considerable winter precipitation, which in higher elevations caused heavy snow accumulation that may have hindered the growth of oaks in that region.

Admittedly, several researchers had claimed quite the opposite for Israel, namely that glacials correspond to dry climates, (Gat and Magaritz, 1980), or that interglacials were humid (Livnat and Kronfeld, 1985). The former authors observed a dry climate phase in Israel at 18 ky ago, which indeed was also seen in pollen diagrams from the Hula (Horowitz, 1979; Weinstein-Evron, 1990). This was a very short phase however, at the peak of the glacial, that may reflect influence of the very cold climate to the north. Weinstein (1976) had pointed out that this last phase of the Pleistocene was markedly colder than most preceding pluvial periods. It is possible that short, dry phases would occur in Israel at peaks of glacials, but in general periods recorded in Israel by low sea levels enjoyed humid, pluvial climates. Since these short dry phases (or short wet phases within an interpluvial) are very limited in duration they are here considered interstadials.

Weinstein (1976) and Weinstein-Evron (1983, 1990) also showed that climatic stages, both pluvials and interpluvials, are frequently interrupted by stadials and interstadials in Israel, which would inflict an almost opposite paleoclimate for a given phase. It is therefore essential that conclusions be drawn from continuous sequences, rather than point evidence, as seems to be the basis for all those studies that appear to contradict the general paleoclimatic trends, which are almost always interrupted by opposing minor oscillations. A detailed discussion of this problem is given in Horowitz (1979, 1987a, b, 1989b).

Correlation of the Israeli interpluvials with odd numbered oxygen isotope stages and pluvials with even numbered stages, was suggested by Weinstein-Evron (1983, 1990), Horowitz and Weinstein-Evron (1986), Cheddadi (1988) and Horowitz (1989b) for this region. These studies mainly dealt with the late Quaternary, the last 150 ky, while others (Fuji and Horowitz, 1989; Horowitz, 1989b) extended these correlations through the Brunhes chron and further down to 3.5 million years ago. It seems that a good correlation exists between the climatic curves for Israel, based on relative distributions of pollen components of the regional vegetation and oxygen isotope curves obtained for that period from the northern Atlantic ocean (Shackleton and Hall, 1984). These climatically based correlations are brought in Figs. 9.3.1 and 9.3.2.

A. Interstadial

Interstadial climates, of which the present day is representative, are known from the Quaternary as phases within both pluvial and interpluvial periods. The present day conditions are characterized by pollen spectra which display some 10–20%

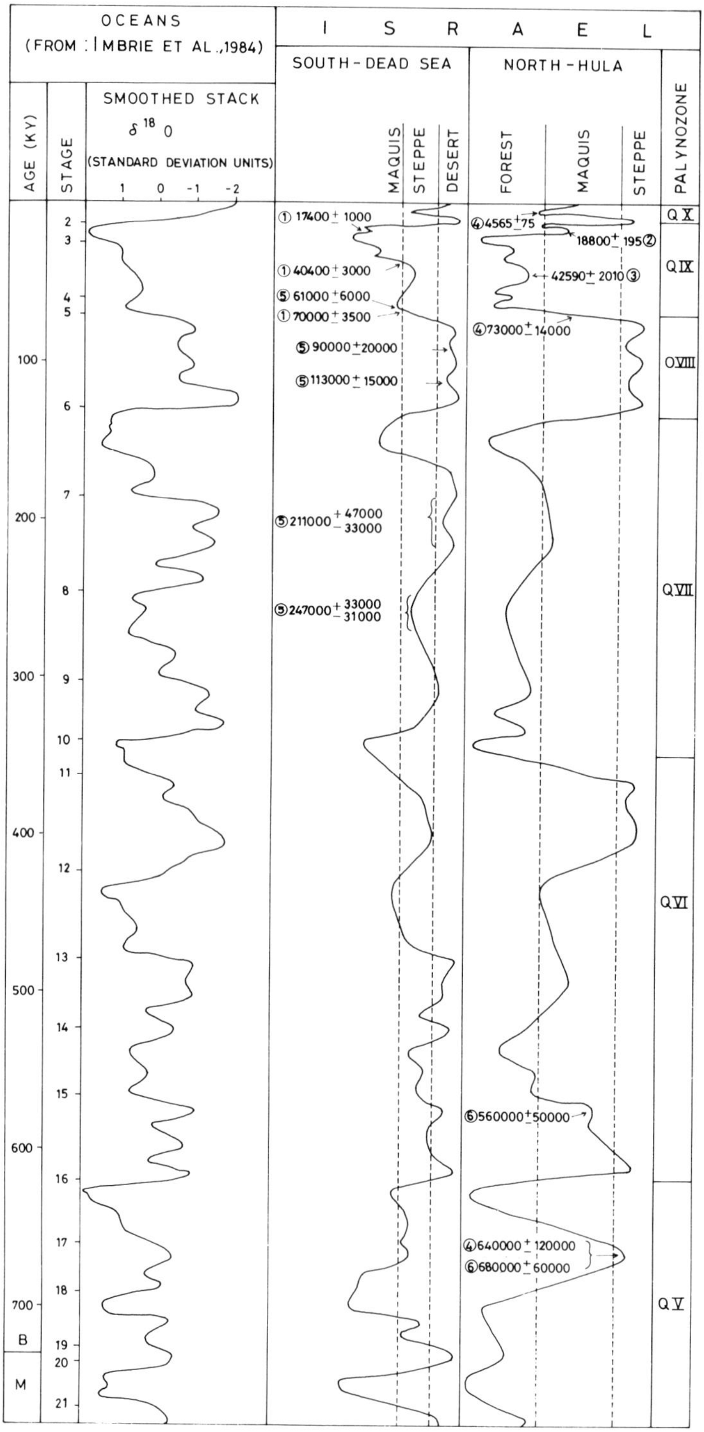

Fig. 9.3.1. Correlation of Quaternary paleoenvironments in Israel and the oxygen isotope ocean sequence: a model for the Brunhes epoch, with radiometric datings. Each point of the paleoenvironmental curves comprises at least 2–4 samples of similar palynological characteristics, for clarity and noise reduction. Ages (years BP) are from: 1. Vogel, in Horowitz (1979); 2. Horowitz (1979); 3. Weinstein-Evron (1983); 4. Siedner and Horowitz (1974); 5. Livnat and Kronfeld (1985); 6. Siedner, in

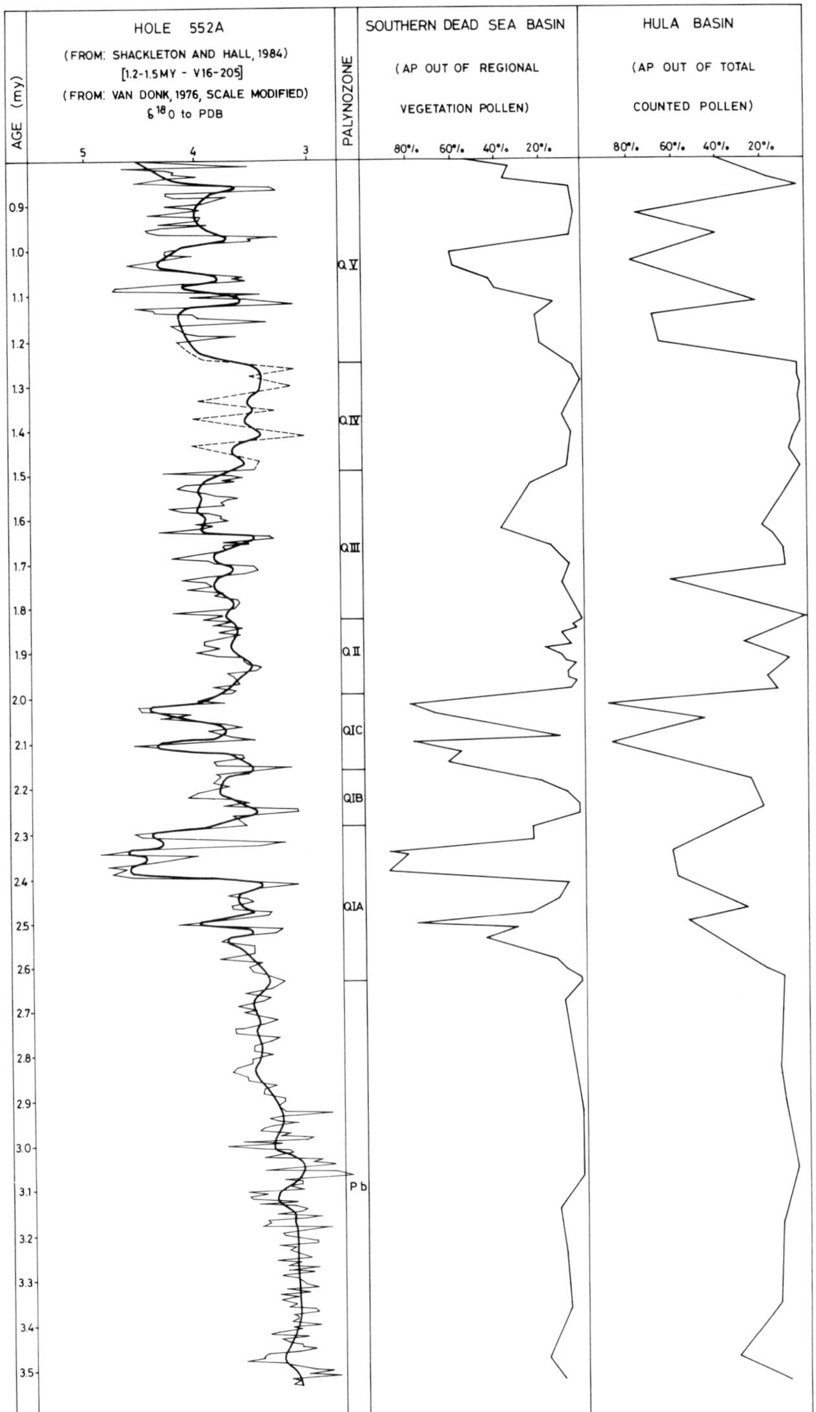

Fig. 9.3.2. Correlation of the Quaternary paleoenvironments in Israel and the ocean sequence from the base of the Brunhes down to 3.5 my. Each point of the paleoenvironmental curves comprises at least 2–4 samples of similar palynological characteristics, for clarity and noise reduction. Reprinted with permission from Horowitz (1989b).

arboreal components to the north of the country, where Mediterranean conditions prevail, diminishing southward, to the Negev, where no AP occur at all, reflecting the desert conditions in that region (Fig. 9.3.3). The situation, it should be noted,

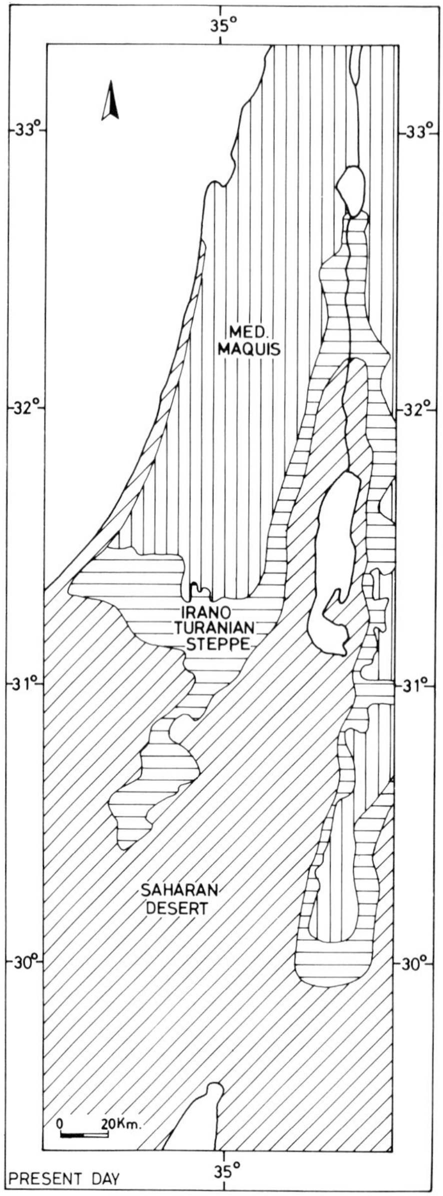

Fig. 9.3.3. Present day (interstadial) vegetation zones of Israel.

is somewhat different in the Dead Sea, which although surrounded by a desert, yields pollen spectra typical for the Mediterranean domains, due to long range transportation by the winds and Jordan River (Rossignol, 1969a).

The major contributor to the recent arboreal pollen is evergreen oak, *Quercus calliprinos*, partly replaced by pine in the coastal plain, depending on the prevailing wind directions during flowering seasons (Weinstein, 1979). Lakes are present in the Dead Sea Rift in three localities: the Hula to north, Lake Kinneret at the central Jordan Valley and the Dead Sea to the south, all three rather limited in area (Fig. 9.2.13). Typical soils of the present day are gray, lime-rich, with a rather shallow profile and very little organic matter in the Mediterranean domain, while to the south bare rocks make most of the landscape. Wadis are presently subject to erosion, caused by the frequent winter floods, as indeed is almost the entire area of Israel. Loess is presently wind deflated from its former accumulations in southern Israel and in no place any significant new deposits of this type are being formed. Eolian sediments that do accumulate at present are sand dunes, at the coastal plain.

B. Pluvial

Pluvial climates are characterized by pollen spectra considerably richer in arboreal components, as compared to the present day ones. Arboreal pollen also occur in sediments from the south of the country, in shares similar to the present day northern regions, while to the north the percentage in pluvial times is more than double than at present. The main increase in arboreal pollen is derived from winter deciduous oaks, growing in the much expanded Mediterranean environment (Fig. 9.3.4).

Mediterranean sea levels were much lower than at present, thus continental sediments are found far west of the recent coastline. Lakes in the Dead Sea Rift occupied considerably larger areas than today, sometimes covering the entire valley floor. Their sediments are known from many localities all over the Jordan Valley, both from outcrops and boreholes (Fig. 9.2.13). Paleosols are quite abundant, comprising mainly red loams, also known from the southern region of Israel (Magaritz, 1986), where hardly any soil is formed today. Gravel and silt were being accumulated in the wadis in considerable quantities (Goldberg, P., 1976, 1986), filling and silting up previous channels which had been cut by the floods during preceding interpluvials. This process is known from all over Israel, extending southward to southern Sinai and northward up to Turkey. Loess was accumulated in sequences several tens of meters thick (Ginzbourg, 1963), mainly in the northwestern Negev, but extending to the central Negev as well, in thinner sheets as one goes south. The loess was being subject to pedogenic processes, quite extensive in the northern Negev, less so to the south.

The pluvial climate must have been considerably more humid than the present day, with rains which also reached the south of the country. Rains mostly originated from warm fronts, while thunderstorms seem to have been quite rare. Isotopic compositions of paleowaters dated to the last pluvial phase (Horowitz and Gat, 1984) indicate that rains at that time originated in the Atlantic Ocean, rather than the Mediterranean at present, some arriving at Israel also in summer time.

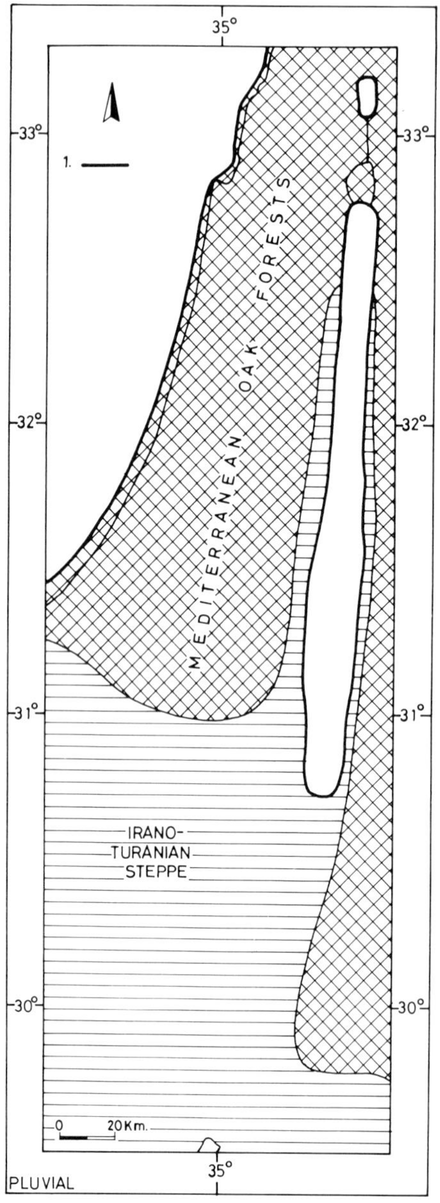

Fig. 9.3.4. Reconstruction of the pluvial vegetation zones of Israel, from the pollen data. *1* = Coastline and Jordan Valley lakes.

C. Interpluvial

Interpluvial climate is defined in Israel by pollen spectra which, while compared to the present day ones, are extremely poor in arboreal constituents, none to the south and only some 3–5% to the north of the country. The prevailing components

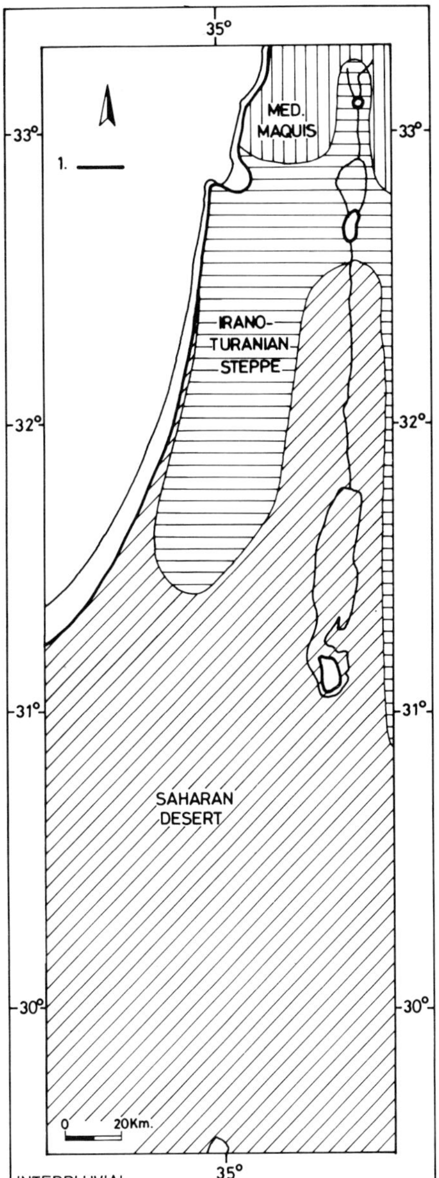

Fig. 9.3.5. Reconstruction of the interpluvial vegetation zones of Israel, from the pollen data. *1* = Coastline and Jordan Valley lakes.

of interpluvial assemblages are steppe elements to the north and desert pollen to the south, reflecting distribution of their parent vegetations (Fig. 9.3.5).

The coastal plain was covered by sand dunes, which were pushed eastward deep into the hinterland by high sea levels of the interglacial Mediterranean. The Dead Sea Rift lakes were restricted to shallow marshes to the north and a terminal hypersaline lake in the Dead Sea region, both very small in area, their sediments only known from boreholes (Fig. 9.2.13). Hardly any paleosols, wadi deposits or loess are known for the interpluvials, due to erosion that had prevailed throughout the entire region. The interpluvial paleoclimate responsible for this suite of phenomena must have had very little rains, the rare ones which did reach the region brought by occasional thunderstorms only to the northern part of the country.

9.4. Paleoclimatic models

Any attempt at modelling the late Cenozoic climatic development of Israel must account for both the long range variations during the Neogene and, on the other hand, the short duration ones typical for the Quaternary.

9.4.1. The Neogene

While many and various models have been suggested for the global Quaternary paleoclimates, especially for the alternating glacials and interglacials, very little is in fact known about Neogene paleoclimates. The main reason for that may be the poverty of data acquired from continuous continental sequences, relative to the Quaternary. Most studies of Neogene paleoclimates concentrated on marine sediments and faunas, while pollen diagrams from uninterrupted continental (or even marine, for that matter) Neogene sections are quite rare. It is however very difficult and sometimes quite hazardous, to assess continental climates exclusively from marine faunal records. The basis for many problems in such attempts is that the sequences only reflect marine environments, which may considerably differ from conditions on land.

Just an example is the occurrence of coral reefs, generally thought of as representing tropical to subtropical conditions. Tropical, in many instances, is tied up almost automatically with rain forests and the like, but one must remember that corals are happily flourishing in places where the surrounding continent is a bare desert, for instance the extensive reefs of the Red Sea.

Attempts have been done to acquire data and conclude on Neogene paleoclimates from fossil plants, which seem to pose yet another problem. Most localities, if not all, of fossil plants analyzed, are intimately connected with freshwater sources, namely either travertines or marsh deposits. Plants typical for such environments, both arboreal and non-arboreal, are much less sensitive to the general climate as long as water is available, thus when successions of spring deposits or marsh sediments were analyzed, conclusions were that throughout the Neogene there was hardly any climatic change (Gregor and Velitzelos, 1987), or that such changes had been subordinate (Velitzelos and Gregor, 1990).

It seems that the main difference in global circulation pattern between the Neogene and Quaternary is the onset, during the latter period, of such critical conditions as would amplify the variations in solar insolence (the Milankovitch curve), toward concurrent formation of glaciers on the one hand and planetary deserts on the other. A possible trigger for such a mechanism could be (Horowitz, 1977b, 1979) the global uplift of the Alpine orogeny (Guillaume and Guillaume, 1982), which barred the way of hot, humid, tropical air beyond certain latitudes and created Hadley cells, typical for today's atmospheric circulation, which resulted in the formation of global climatic belts.

It is therefore suggested that causes for the Neogene environmental shifts, of rather long duration, are not relevant for the Quaternary, with climatic oscillations best explained by the Milankovitch curve. It seems that the causes for the Neogene shifts should be sought for with the late Tertiary–Quaternary gradual development of global climatic belts, when the globe changed from a rather uniform, subtropical environment (Velitzelos and Gregor, 1987) toward the present day configuration of prevalent climatic belts.

Thus during the Miocene, Israel was still within the subtropical monsoon system, alternating from dry to wet, finally to a Sahara type desert during Mc. The middle Miocene amelioration of desert conditions in Israel and the Near East was probably a result of global cooling at that time, corresponding to the initial glaciation of Antarctica (Van Zinderen Bakker and Mercer, 1986), rather than caused by migrating climatic belts, as thought before. This assumption is mainly based on the subtropical environments typical for the northern and northwestern Mediterranean regions in southern Europe at that time (Bessais and Suc, 1987; Benda and Meulenkamp, 1990) and on the similar characteristics of Levantine floras (Horowitz, 1990).

In the course of the Pliocene, when the present day type climatic belts had been further established, the region was under influence of the southern European temperate domain, wetter at the beginning, Pa, becoming dryer toward the late Pliocene, Pb (Fig. 9.4.1). Pollen analyses of northern Mediterranean regions indicate a pronounced cooling for the Pliocene (Suc, 1989).

9.4.2. The Quaternary

Climatic models suggested for the Quaternary of Israel (Horowitz, 1987a, 1989b) are based on previous studies and suggestions (Horowitz, 1977b, 1979; Horowitz and Assaf, 1981; Horowitz and Gat, 1984; Horowitz and Horowitz, 1985; Horowitz

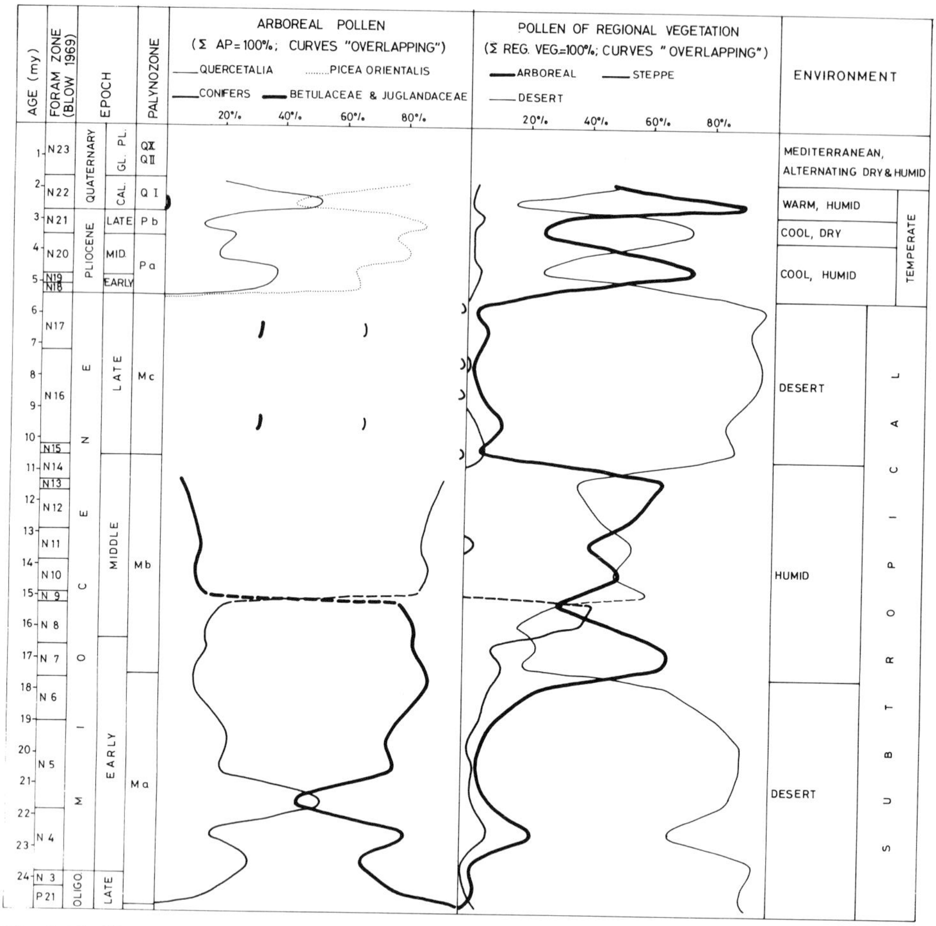

Fig. 9.4.1. Neogene paleoclimates of Israel, from the pollen data. Reprinted with permission from Horowitz (1990).

and Weinstein-Evron, 1986; and others). The point of reference is the present day conditions, for which the climate can be observed and measured; inferences to the past are based on the proxy-data described above, for which the basis is the pollen spectra and diagrams of the various palynozones.

Rain falls in Israel presently only in winter time, precipitating annually up to 800 millimeters to the north of the country, diminishing gradually to the south and southeast, until in southernmost Israel the total yearly amount does not usually exceed 30–40 mm. Consequently, the environment is Mediterranean to the north, while southward, the desert gradually takes the upper hand (Fig. 9.3.3). Precipitation originates as cyclonic thunderstorms in the eastern Mediterranean during the winter, as a result of combined several factors: high barometric pressure

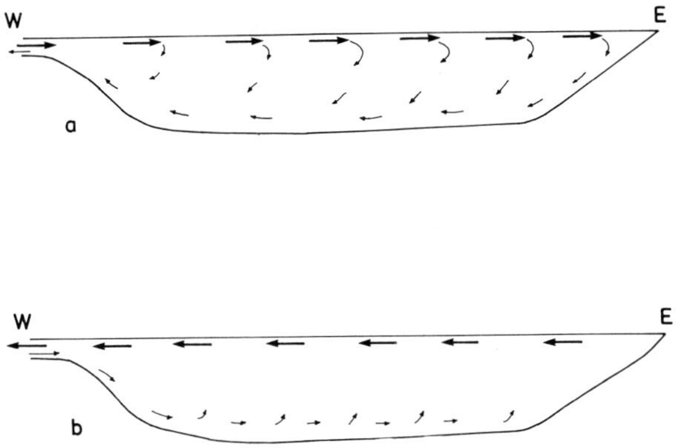

Fig. 9.4.2. Mediterranean water balance. a. Negative; b. positive. Reprinted with permission from Horowitz (1987a).

over the Sahara keeps any rains from reaching the region in summer time; at the same time solar radiation causes warming up of the eastern Mediterranean seawater; the only connection of the Mediterranean with the ocean is the restricted path of the Straits of Gibraltar; freshwater supply to the Mediterranean by rivers is not sufficient to compensate for evaporation, thus the basin has a negative water balance; as a result the Atlantic Ocean seawater flow into the Mediterranean, becoming warmer and saltier as they propagate eastward due to evaporation, until becoming heavy enough to sink to the bottom; consequently, the Mediterranean is quite well mixed, storing solar energy in its entire water column (Fig. 9.4.2a).

Solar energy thus stored in the Mediterranean during the long, hot summer is only released to the atmosphere upon contact with cold air masses arriving from the north, which can reach the Mediterranean region only in winter, when the Saharan high barometric pressure lessens and the polar high becomes more active. The combination of cold fronts and stored energy creates cyclonic thunderstorms. The confluence of northern, polar winds and western winds over southern Europe and the Mediterranean, the latter also being pushed southward during the European winter, defines the paths of thunderstorms and accounts for the present day distribution of rains in the Levant (Fig. 9.4.3). This climatic system is also considered responsible for rains in an interstadial climate.

In pluvial times the situation must have been different. Northern climatic belts had been located south of their present day position (Horowitz, 1977b); tropical regions had received less insolation (Rossignol-Strick, 1983), resulting in decreased rain quantities (Gasse et al., 1980; Hamilton, 1982; and many others), also causing the Saharan high pressure belt to be much less developed; concurrently, the western winds belt was quite active over the Mediterranean, pushed there by high

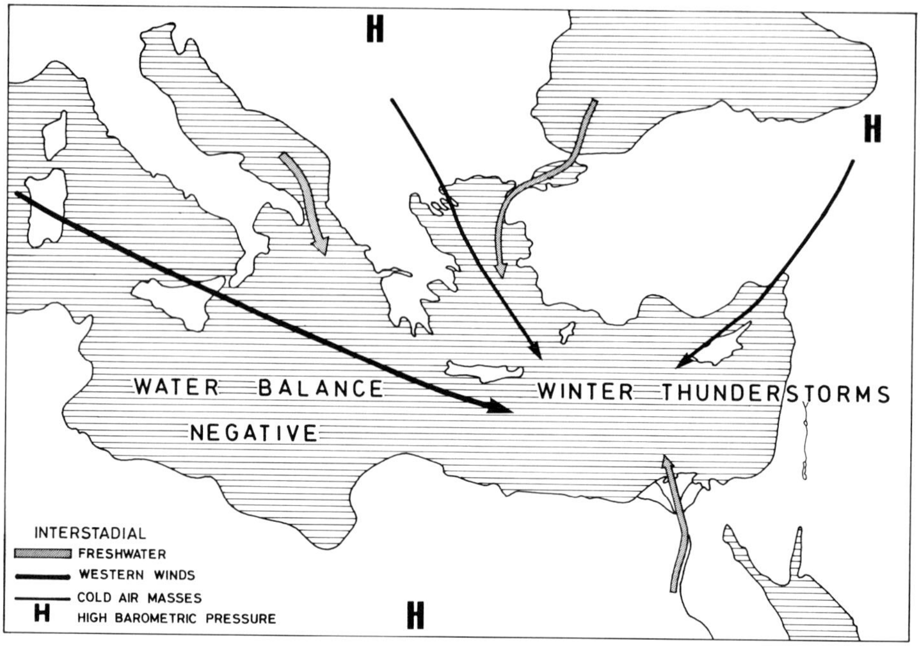

Fig. 9.4.3. Quaternary paleoclimatic model: interstadial, similar to the present day. Reprinted with permission from Horowitz, 1987a.

pressure over Europe and the Atlantic, created by the extensive distribution of glacial ice. Consequently, the Mediterranean water balance became more positive, due to rains coming in summer time from the Atlantic, much less evaporation because of increased summer cloudiness, with some increase in inflow of rivers from southern Europe; decrease in the Nile water quantity typical for the glacial periods apparently did not alter the water balance. In times of positive water balance the Mediterranean became stratified, which prevented the bulk of seawater from acting as solar energy reservoir (Fig. 9.4.2b), thus diminishing considerably the extent of cyclonic thunderstorms genesis.

On the other hand, the western winds had brought rains from the Atlantic ocean to the Levant (Fig. 9.4.4) and indeed also to much of the present day Sahara (Van Campo, M., 1975; Williams and Faure, 1980; and many others). The rain regime of the Levant in pluvial times was therefore characterized by rather gentle rains, some also falling during summer time, spreading southward considerably more than at present, probably in somewhat larger annual averages.

It should however be noted that the most significant factor in pluvial climates seems to be the style and more even annual distribution of rains, rather than the absolute quantity. Such rain regime seems to explain the pluvial characteristics of the Levant, the rich flora, its composition and distribution, the depositional

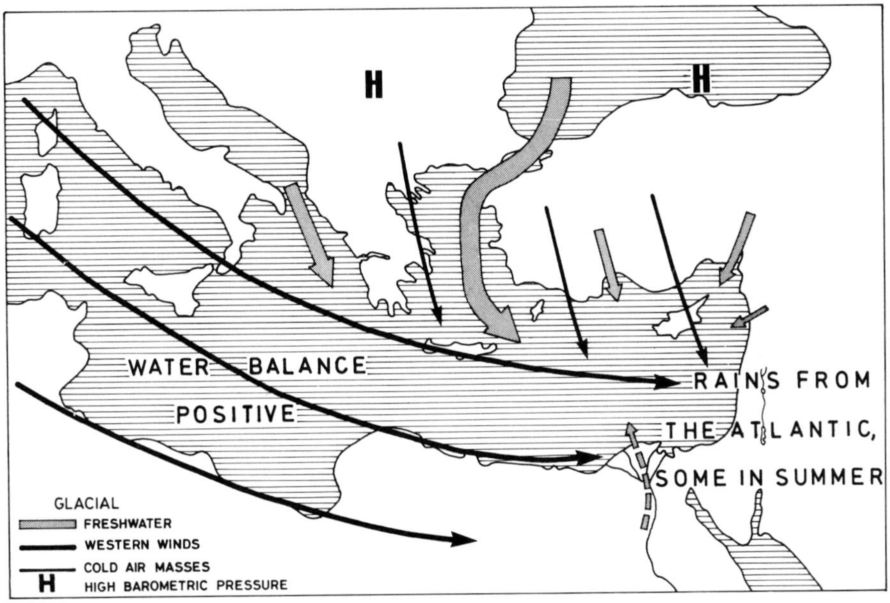

Fig. 9.4.4. Quaternary paleoclimatic model: glacial (pluvial). Reprinted with permission from Horowitz (1987a).

processes, the isotopic composition of paleowaters, the extensive pedogenesis resulting in widespread paleosols and the distribution of lakes and loess deposits. The occurrences of all these pluvial characteristics in Israel in times of low Mediterranean sea levels indicate quite clearly that pluvial climates correspond to the glacial phases.

The situation is again completely different in interpluvial times. The global climatic belts moved to the north, the equatorial monsoons were much more active (Rossignol-Strick, 1983), while the western winds belt and cold polar fronts hardly reached the Mediterranean (Fig. 9.4.5). The southern Levant became an integral part of the Saharan high pressure system throughout most of the year. The Mediterranean again had a positive water balance, this time due to considerable increased activity of the Nile, probably also due to rivers coming from Europe, benefitting from glaciers meltwater. The positive water balance had again prevented solar energy from being stored in the Mediterranean, thus the source for winter rains had considerably diminished during interpluvials. It seems that only rare thunderstorms could have reached the Levant under these conditions, most probably from the north.

The northern origin is suggested because, dry as the region was, the north was wetter than the south (Horowitz, 1979). The idea which was occasionally expressed, that monsoons had reached as far as Israel in interglacial times, has

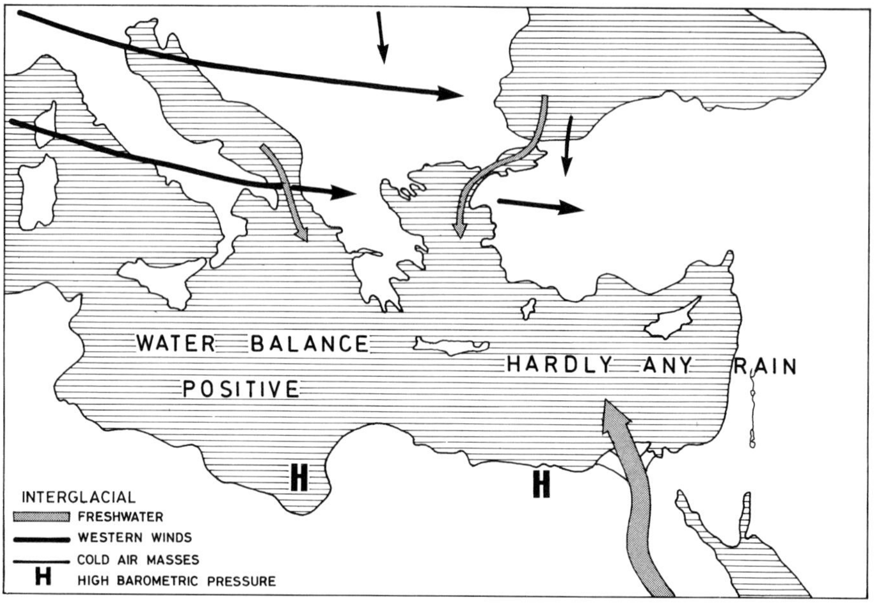

Fig. 9.4.5. Quaternary paleoclimatic model: interglacial (interpluvial). Reprinted with permission from Horowitz (1987a).

therefore no support in the evidence, including Sinai. Our model seems to explain the phenomena observed for interpluvial climates in Israel, discussed above. These phenomena, which occur at times of high Mediterranean sea levels, seem to justify correlation of interpluvials with the interglacials.

9.4.3. Conclusion — paleoclimates

It seems that uplift of the Alpine orogeny was already sufficient to initiate formation of global climatic belts in late Oligocene times. This created a tropical, monsoon controlled belt in the nowadays tropics, a planetary deserts belt, which is so typical for Ma times in Israel and a temperate climate to the north, getting somewhat cooler as one goes northward.

A global cooling in middle Miocene times was probably sufficient to create an ice cap over Antarctica, due to its higher elevation as compared with the north pole area, where no ice is known for this time. This global cooling resulted in amelioration of the planetary desert conditions which had affected the Near East, causing spreading of trees in the entire region. This situation again changed in Mc times, when extreme desert conditions prevailed in the Near East. Indeed, mangroves are known from southern and southwestern Europe in middle Miocene times (Bessais and Suc, 1987).

Another global cooling took place in Pliocene times, most probably resulting from accentuation of the climatic belts characteristics, by further uplift of the Alpine orogenic belt. Consequently, a temperate climate of northern origin spread southward to the Near East, causing coverage of the region by pontic type forests. Long term oscillations of this kind of climate resulted in Pa being more humid, while Pb less so.

During the Quaternary Israel was mostly under influence of the Saharan climatic regime, which was interrupted by pluvial climates, initiated by the glacials to the north, or interstadial climates characterized by winter rains, resulting from a negative water balance of the Mediterranean.

Chapter 10

Pollen analysis of archeological sites in arid lands

Archeological research of arid lands is going hand in hand with pollen analysis for a long time now, almost anywhere around the world. Evidently, the principal cause for this fruitful cooperation is, because the study of pollen embedded within deposits of archeological context can almost readily yield information about environmental conditions and human activities in a given site. In some contrast to other methods which could provide similar information about past environments connected with human habitations, such as geology, geomorphology and other branches of the earth sciences, pollen analysis requires much less field work, which facilitates logistics when studying remote sites. While application of many other sciences comprises tedious, long stays in the field, palynology mainly involves collecting samples, as the rest of the work can be conveniently carried out in the airconditioned laboratory. Incidentally, this also means less pressure on the usually tight budgets.

It seems that through the ages people had almost always inhabited arid and especially semi-arid environments on Earth. The main reason could be that such environments combine a favorable climate, offering a relatively comfortable weather and possibilities of obtaining enough edible plants for gathering, or game for hunting. This facilitated life and made basic subsistence available, especially for the so termed "primitive" societies. This is quite evident from the fact that temperate regions had been occupied by man considerably later than the dry ones. One only has to realize that most Paleolithic, especially early Paleolithic sites are found in semi-arid and Mediterranean regions, while those parts of the Earth which are nowadays densely populated, like western Europe, had been settled quite late in the history of mankind. This situation makes palynological research of archeological sites in dry lands of prime importance and significance for understanding evolution of man and his cultures, in connection with the natural environment. The only seemingly exception for this rule may be the Americas, but even in this part of the world, when it had finally been reached by man, it was the drier parts which had first been inhabited, much before the tropical and temperate.

Reconstructions of paleoenvironments where these early people could dwell are therefore most important for understanding modes applied for achieving food supplies. Incidentally, this can also shed light on the influence those people had exerted on the environment, by their activities such as agriculture, grazing and gathering of plant material for various needs.

It seems that the pioneers in routinely including pollen analysis in almost any archeological excavation are those groups of anthropologists studying the American Southwest. Already in 1963, Paul S. Martin had presented a synthesis of the Holocene environments of SW USA, based on pollen profiles studied by himself and others (see the historical overview of palynological research in this region in Bryant and Holloway, 1983), a work that had and still has many followers. A glance at the bibliography of Southwestern Quaternary palynology (Davis and Hall, 1988), which covers hundreds of references to publications and reports dealing with palynology of almost any archeologic site which was explored or excavated, only proves it beyond any doubt.

This trend of archeopalynology was followed by many others, especially the French who carried out a great deal of pollen analysis in the Sahara and east Africa, connected with human habitation and hominid sites (Van Campo, M., 1975; Bonnefille, 1976 and many following publications; and many others). It seems that presently there is hardly any archeological excavation in arid or semi-arid regions which is not escorted and aided by pollen analysis, usually carried out concurrently with the digging activities.

Pollen analysis was used in arid lands archeology for various purposes, such as establishing the stratigraphy and chronology of prehistoric sites (Horowitz, 1988a), reconstructing paleoenvironments and structural changes (Bonnefille et al., 1987), establishing environments of early man (Mehringer and Haynes, 1965), discussing diet and medicinal plant usage (Holloway, 1985) and so on.

Needless to say, archeopalynology is also practiced in fertile lands (Leroi-Gourhan and Renault-Miskovsky, 1977; Dimbleby, 1985; Renault-Miskovsky et al., 1985; Behre, 1986; and a wealth of publications by other researchers). But it is the arid, where hardly any other paleoenvironmental indicators are available and subsistence critically depends on environment, that archeology and anthropology are greatly benefitting from palynology. Since both the problems involved and methods applied are different for arid lands, pollen analysis in dry regions had to be treated in its special way, which would have to furnish answers to both the technical and scientific questions.

A summary of the possibilities of pollen analysis pertaining to human habitations is given in Bryant and Holloway (1983). These authors summarize (p. 217): "Archeologists should carefully consider the value of fossil pollen data when they conduct their excavations. In the past, palynologists have been able to help archeologists determine past environmental changes, prehistoric diets, the use of native and cultivated plants, graveside rituals, intersite and intrasite dating, aboriginal use of certain categories of artifacts, utilization patterns for rooms at pueblo sites, and most recently, a method for pretesting sites to determine their archeological potential. In addition, there are often instances where the fossil pollen record will provide precise answers to questions that often cannot be answered by the artifact record alone. For example, the determination of the exact onset of the Neolithic in Denmark, the precise beginning of maize agriculture in Ontario, and how a series of buffalo jumps at Bonfire Shelter perfectly correlated with the expansion of grasslands into southwestern Texas, represent only a few of many such examples."

10.1. The significance of samples collected from archeological sites in arid lands

Archeologic and prehistoric sites located in arid lands are subject to a variety of processes, usually alien to temperate and tropical regions. These may significantly affect the composition of pollen spectra and their bearing on reconstructing paleoenvironments concerned with a certain settlement. Fundamentally, these processes could be divided into three groups: the first concerns processes which directly affect the stratigraphic connection of an analyzed sample with the horizon of interest; the second is the possible loss of certain taxa from the pollen spectra; and the third involves the addition of taxa to the assemblages. Evidently, this discussion is only pertinent if utmost care and cleanliness had been observed while collecting the samples and no inclusion of artifactual contaminations is suspected.

10.1.1. Stratigraphic considerations

The stratigraphic context of samples collected from archeological sites in arid lands and the chronologic relations between the obtained spectra and occupation levels, are quite frequently hard to define. This is particularly a result of the strong winds and floods activity in the dry regions, causing both rapid sediment accumulation and considerable erosion, which may swing from one to the other quite frequently at the same locality.

Strong winds may cause deflation of the finer fraction of sediments (Fig. 3.1.2.1), in which artifacts are embedded. This phenomenon was especially observed for many prehistoric occupations in the Negev Desert, Israel, where younger implements were found mixed with older ones in originally multi-layered sites. Archeologic tells, which are quite common in the dry lands, are higher than their close surroundings, therefore very strongly prone to wind deflation. When a multi-layered site is studied such a mixture is usually easy to discern, thus it would be straightforward to define inadequacy of samples collected from the artifact bearing bed, for any stratigraphic and environmental implications concerned with the time the implements had been buried. The problem becomes much more complicated if a single-culture site, quite commonly found in arid regions (Marks, 1976, 1977, 1983), is explored or excavated. Two possibilities may occur with such sites: either the implements are directly lying on bedrock, in which case samples are anyway not available for palynology; or the artifacts are found embedded in a seemingly contemporaneous deposit, but may have been brought there by vertical downward migration through deflation of the original fine sediment (Fig. 3.1.2.1). It may take a lot of study and attention in the latter case, to correctly assess the contemporaneity of the sediment and its enclosed implements.

These considerations may even be further complicated with sites which had been subject to wind deflation at a certain stage in times of, or following, its occupation, but was subsequently covered again by eolian or other sediments. Such processes of repeated denudation and deposition phases are well encountered for the late Quaternary in Israel (Horowitz, 1979) and had indeed affected many prehistoric sites in arid regions of the country (Goldberg, P., 1986). In such cases it may seem to unaware excavators that the site is sealed and the artifacts are embedded in contemporaneous deposits, which had been sheltered from deflation.

Experience shows that these are the most tricky sites in arid lands and unfortunately, it seems that they are more common than usually believed. This happens especially among older prehistoric sites which had underwent a rather long geomorphological history, including several alternating stages of erosion and deposition, following the period of settlement.

Similarly to consequences of eolian processes, sites found near or in wadi beds are affected by the erosional–depositional regime at a certain time, which may depend on either climate changes or structural disturbances, or both. Again, in the arid region of southern Israel alternations between deposition and erosion of wadi sediments at a certain point in the wadi course are well known for the Quaternary (Goldberg, P., 1976, 1986; Horowitz, 1979) and had affected prehistoric and archeologic sites. Just as an example, wadi sediments have at places been accumulated in the Roman period, some 2,000 years ago, preceded and followed by phases of erosion.

Dust accretion can be very rapid in arid and semi-arid regions, but is usually not continuously active. Thus an abandoned site may be quite readily covered by a blanket of dust or sand, which could sometimes be considered synchronous with the occupation period, because in many cases when it covers structures dust enters into cavities or empty ceramic vessels, found in seemingly intimate connection with the archeological remains. Similar phenomena have also been observed for wadi deposits. In terms of their significance, pollen spectra recovered from such dust or silt would not be exactly contemporaneous with the occupation, but somewhat younger, which evidently would preclude the possibilities of reconstructing conditions in time of habitation.

Such problems of wind or flood erosion and deposition intervention with pollen spectra could be satisfactorily solved in sites where mud (adobe) bricks had been used for construction (O'Rourke, 1982, 1983; Drori and Horowitz, 1988–89; and see Chapter 4.6). It is only unfortunate that the use of mud bricks does not date back more than a few millennia, thus precluding their application for older human settlements.

Complications concerned with eolian deposition have also been observed in caves and rockshelters, where two main points should be attended to. The first concerns the fact that pollen deposition is not uniform in all parts of a cave and some preferences are evident, depending on size, direction and shape of the opening. Such differential deposition of sediments and pollen within a cave may cause apparent differences between various layers when their respective spectra are compared, if samples have not been collected in similar relations to the opening. The problem is that both size and shape of this opening usually change with time, thus it may be found quite difficult to obtain optimal spectra. Although the direction in which the opening faces does not usually change with time, a change in wind direction may have a parallel effect. Another problem may arise when spectra from cave sediments are compared with a reference sequence collected from another type of site or sedimentary basin, as is quite frequently the case for stratigraphic studies. In such cases a specific environment could be differently manifested in the two localities.

Dissimilarities between regional pollen spectra for a given period and those obtained from caves and rockshelters may also result from the direction into which the opening faces (see Chapter 2.2). Such discrepancies may lead to wrong conclusions, both in stratigraphic and paleoenvironmental interpretations, whenever comparisons are attempted between a cave site and a reference section, even if obtained from the same region. Stratigraphic problems are evidently more complicated when sites become older, which may occasionally lead to more than a single plausible solution. Thus Weinstein-Evron (1983), while studying palynostratigraphy and paleoenvironments of Middle Paleolithic sites in Israel, could not arrive at a single definite conclusion and suggested two possible interpretations for the chronology of these cultures. Initially the chronologic discrepancies resulted from several sets of radiometric datings, when each group of researchers arrived at different "absolute" ages. In this case however palynology could not solve the problem one way or another.

Stratigraphic discrepancies could also evolve from burrowing activity (Fig. 2.2.14) and underground storage of food reservoirs by animals, which are quite commonly found in archeologic sites due to the usual friability of the sediments. Thus pollen accumulations (Bottema, 1975) by burrowing Apidae (bees and wasps) are abundant in excavations (Fig. 2.2.15). Recent burrowings can usually be readily distinguished in the course of excavation, but can easily be overlooked whenever boring samples are analyzed, or samples collected by unaware archeologists. It is however much more difficult to discern burrowings that took place a long time ago, in former periods.

Stratigraphic disturbances are known from infiltration of younger pollen and their incorporation within older layers, probably aided by rainwater circulation through friable sediments or fissures, down the section. This phenomenon was observed in both cave sediments (Scott and Bonnefille, 1986) and open sites (Weinstein-Evron, 1986). Evidently, it is quite easy to follow such infiltrations of recent pollen if they are not oxidized or otherwise deteriorated (Weinstein-Evron, 1986). It should be remembered that such processes acted also in the past, in which case the state of preservation may occasionally be helpful for discriminating the older from younger palynomorphs (Horowitz, 1975), but is not necessarily applicable in all cases. Downward movement of pollen is aided in caves by bats and other animals, whose acid excrements help dissolve sediments, thus enabling pollen infiltration into older strata. Such movements, incidentally, also affect radiocarbon dating.

Inclusion of younger pollen can also occur whenever a site, which is not a rare phenomenon with archeological sites in Israel, is burnt down, due to accidental fire, or intentional, following siege and occupation. In such cases, which could usually be identified by higher occurrences of charcoal particles in the slides prepared for palynological study (Clark, 1982), the original pollen contents of the layers is largely destroyed by the fire. Pollen which arrive at the following years find a rather favorable environment for embedding and preservation and are, in most cases, also washed down into the older sediments by percolating rainwater. The younger pollen seemingly become an integral part of the previously burnt horizon, thus may easily be mistaken for being contemporaneous. Although the time span between occupation of the site and subsequent inclusion of younger pollen may be minimal, the habitation would not be fully reflected in the obtained spectra, but frequently only in the higher shares of weeds, which persist for a much longer period after a site is abandoned.

Careful considerations must therefore be made for two aspects, while attempting pollen analysis of anthropogenic sediments in arid lands, to ascertain the significance of pollen spectra recovered from sediments in which artifacts are found. The first is to determine contemporaneity of the sediments and implements, while the second would be to establish synchroneity of the pollen assemblages and sediments from which they are extracted. Admittedly, this task may turn quite difficult too often and could involve extensive geologic and stratigraphic studies.

10.1.2. Preferential inclusion and deficiency of taxa in anthropogenic deposits

Preferential inclusion or deficiency of a certain taxon or taxa in sediments collected from prehistoric and archeologic sites in arid lands results from its over- or under-representation, either by its own shares or in comparison to percentages of other constituents of the pollen spectra. This may be caused either by differential preservation, or by addition or omission of a certain taxon by anthropogenic induced processes.

The problem of differential preservation (or destruction, which amounts to the same thing) is discussed by many authors (see Chapter 6.1.1) but has not yet received a systematic treatment, which could present solutions such as by offering appropriate correction factors that could be used in all particular cases. The problem is much more pertinent to archeological sites in arid lands than elsewhere, mainly due to the relative high Ph of the sediments, even when some organic material is present, which results in pollen extracted from such samples to appear in all different states of preservation (Bryant and Holloway, 1983; Holloway and Bryant, 1986). It seems that the problem has to be tackled separately for each analyzed site and its various environments of deposition. One of the possible solutions would be to compare pollen spectra obtained from an archeological site with subrecent ones from the same region, in approximately similar environments of deposition. It does not seem that at the present state of knowledge a better solution can be offered. The problem must however be considered, since such processes do exist and affect the pollen spectra.

Initial additions to pollen spectra from sites are by those plants cultivated or introduced by man and by the group of opportunistic, ruderal plants constituting weeds, which find a suitable biotope close to human habitations. Occurrences of these pollen taxa do help to reconstruct economy, subsistence and living conditions of the site's inhabitants. However, in some cases, shares of these pollen could amount to such values that could distort or even completely mask regional contributions to the spectra. Such is usually the case near water reservoirs, which are very frequently used in arid lands throughout history, trash pits and so on, which make excellent biotopes for weeds due to the higher amounts of available water and organic materials. Jelinek (1966) found out that percentages of pollen produced by plant families that make a large share of weeds, such as Chenopodiaceae and Amaranthaceae, differ in connection with various archeological contexts such as bone, shard or stone, thus suggested the use of correction factors to account for such differences in arid SW USA.

Other anthropogenic induced contributions for preferential pollen representations could be made by animals. This may especially be the case in caves and rockshelters, into which the inhabitants frequently herd their livestock for protection from the elements. In such cases the excrements are mingled with other deposits. Pollen produced by plants consumed by the animals are thus highly over-represented.

Similar preferences hold true also for animal and human coprolites, occasionally analyzed for pollen from various sites (see Chapter 2.2.5), in which pollen produced by those plants usually consumed by the inhabitants are over-represented, rather than natural components of the flora, whose pollen are only accessories in such samples (Williams-Dean and Bryant, 1975; Mead et al., 1986; Sobolik, 1988).

All above mentioned contributors and possibly also others, to archeological sites spectra, may occasionally mask the pollen produced by regional vegetation, but may also mask those pollen taxa important for understanding subsistence. Many such taxa are produced by economic plants which are insect pollinated, or otherwise do not produce large quantities of grains, or may have been cultivated some distance from the locality where samples had been collected, thus can be severely under-represented. To solve this problem Clary (1989) suggested extended counts of grains from archeological samples, so that economic pollen could be represented in significant numbers, for interpreting diet and subsistence of the inhabitants.

Exotic or strange substances are occasionally found in archeological excavations and proved in some instances to carry pollen grains, which could be helpful for identification and interpretations of usage. These pollen spectra, which differ from the rest extracted from other sediments of the same site, are usually dominated by a certain group of taxa thus giving the desirable clues.

Quite often such unique pollen spectra could be used to identify substances which appear as black lumps. One such little black lump was analyzed from a Roman site near the Dead Sea and yielded a rich variety of pollen produced by insect pollinated, melliferous plants, which indicated that the lump could originally be honey (Horowitz, 1988b). This stands in contrast to other black pieces collected from the same site, which contained a mixture of pollen typical for hydrophil and halophil vegetations of the Dead Sea area, in addition to palynomorphs of Triassic age. This mixture is typical for asphalts in the region (Horowitz and Langozky, 1965), with additions of recent pollen.

Pollen analysis of jar contents could yield, by the prevalence of a certain taxon, information concerning the original contents of the jars, such as higher representation of olive pollen in jars used for oil, or various flowers pollen in some perfume bottles. The internal (again black) coating of Roman wine jars excavated in Israel was found to contain a majority of pine pollen, suggesting that its resin was used for that purpose, serving for both sealing and preservation. Incidentally it apparently gave the wine a taste of retsina (Horowitz, 1988b).

Amber also contains pollen grains which may indicate its provenance, but the main problem seems to be its dissolution, in order to obtain the grains for analysis. In its classical sense, amber is a fossil resin widely known from western Europe, at a certain time also imported to the Near East. In its wider sense, all kinds of resins which are hard enough are used for jewelry, including those that may be collected from a variety of local trees.

Pollen extracted from an Egyptian mummy (Girard and Maley, 1987) gave clues to the substances used for mummification processes. The use of flowers for ceremonial burial was suggested by Leroi-Gourhan (1975), who discovered considerable amounts of pollen produced by various flowers just below a Neanderthal skeleton at Shanidar Cave, Iraq, concluding that flowers collected on the hills had been used for the ceremony.

Maybe the most extreme case of anthropogenic intervention in pollen spectra is by the intentional import of pollen itself to the site, usually for ceremonial purposes, a habit quite common with several Indian tribes in North America. These rituals have been revealed for several pueblo sites in Southwestern USA by the excess in shares of otherwise lesser represented pollen types. Hill and Hevly (1968), who analyzed the Broken K Pueblo in Arizona had noted the probable ceremonial use of *Sphaeralcea* and pine pollen in an infant's burial; Holloway and Bryant (1986) report on results of pollen analysis of Antelope House Pueblo by Bryant and Morris (Unpubl.), who suspected a similar use of corn (maize) pollen at a graveside ceremony. These two studies suggest that pollen or flowers were used deliberately in such rituals.

10.2. Recognition of anthropogenic and livestock activities

Anthropogenic activity hardly exerted any noticeable influences on natural vegetation and ensuing pollen spectra before the inception of agriculture and pastoralism, that is not until the beginning of the Neolithic, in its wide sense, in various parts of the world. The appearance of human and livestock influence is a result of the beginning of agriculture, which had changed natural biotopes closely surrounding human habitations and pastoralism, affecting larger areas used for grazing grounds. The anthropogenic effect is further felt due to the increase in number of occupants and size of sites of many habitations, from that time onward. It seems that even if some changes of the natural vegetation composition have indeed taken place before agriculture was practiced, they would be very difficult to discern in the pollen spectra, since any such effect was most probably restricted to a very limited area. A detailed account of man's role in the shaping of the eastern Mediterranean landscape is brought in Bottema et al. (1990).

Anthropogenic influence on pollen spectra dates back in some arid regions for quite a long time, such as in northern Mali, where human activity was interpreted for pollen spectra dating to more than 5,650 years ago (Cour and Duzer, 1976); in Rajasthan Desert, India, where human influence was disclosed for the last 9,000 years (Singh, 1980); or in Qatar (Bonnefille and Riollet, 1988), where anthropogenic effects are recorded in sites 6,000–8,000 years old.

Various aspects of anthropogenic and livestock activities are recorded in pollen spectra obtained from archeological sites, such as agriculture and subsistence, occurrences of imported plants used for food, building and so on, development of ruderal weeds and the influence on natural vegetation, both intentional and unintentional.

10.2.1. Cultivated plants

Agriculture and husbandry of plants used for food has two aspects in the pollen spectra. The first is seen in cultivated plants that had underwent genetic changes throughout the process of domestication, which had also affected the shape of their pollen, such as corn and cereals. These can thus be easily recognized in the spectra and differentiated from natural Gramineae. In this way cultivation of cereals could easily be followed in the Middle East (Leroi-Gourhan, 1974; Horowitz, 1977a, 1979) and intentional growing of corn for food was traced in the Americas (Martin and Schoenwetter, 1960; Fish, 1984; Byrne and Horn, 1989), or for the more recent past also in Egypt (Mehringer et al., 1979).

In contrast to pollen grains produced by cereals and corn, which can be readily recognized, there stands a relatively large group of plants that, although being cultivated, do not seem to have undergone such genetic changes that significantly affected the morphology of their pollen. In this case the grains are similar to those produced by the wild counterparts of these plants. Some plants are foreign to the region and could be identified as imported, while most such plants do grow in the same region, thus their husbandry or use may be identified in pollen spectra only by implying certain assumptions.

Suspicion for possible cultivation may be raised when a certain pollen type displays unnaturally high percentages in the archeological samples in comparison with its shares elsewhere, such as *Ephedra* being one of the major contributors to fossil pollen spectra between 10,000 and 7,000 years BP at Bonfire Shelter site, west Texas (Bryant, 1978); or by the relatively more frequent occurrence of lumps of grains produced by a certain taxon and found in archeologic sediments, which indicates that the plants grew in an intimately close vicinity of the site (Fish, 1984), suggesting their intentional husbandry or use.

Baruch (1986; 1990) indicates large scale olive cultivation in areas surrounding Lake Kinneret, Israel, during the second and first millennia BC, by the dominance of their pollen grains in the spectra, as opposed to sequences analyzed from other localities in Israel (Horowitz, 1979), which do not show any increase of *Olea europaea* pollen for these periods.

Bryant (1983), in a study of the Pratt Cave, New Mexico, found rather high percentages of pollen produced by *Agave*, *Opuntia* (prickly pear), Liliaceae and Malvaceae, interpreted as representing probable introduction by man. Bryant further noted that many prehistoric groups had frequently utilized flowers as a source of food, concluding that if indeed the people using Pratt Cave were flower eaters, then some of the pollen from these flowers could easily have become dislodged during food preparation and thus incorporated into the cave sediments.

In some cases even low percentages of trees and other plants used for subsistence, in a region in which they do not otherwise grow, are reliable indicators for such a way of human activity. Such could be low percentages of *Amygdalus* (almond tree) and *Olea europaea* (olive tree) pollen found in Iron Age sites in the Negev desert, Israel (Kempinski, in press); or pollen of plants such as beans, cucurbits and cotton, which are hardly ever represented in the natural pollen assemblages, unless cultivated (Fish, 1984).

Trees which do not naturally grow in the region, or which may usually occur only in limited numbers, may occasionally be imported by inhabitants and grown in an archeological site or its close vicinity, for use of their timber for construction. Pollen spectra from an Early Bronze site near Tel Aviv, Israel (Horowitz, 1979), yielded unusually high pollen percentages, up to 25, of cypress (*Cupressus sempervirens*), which only rarely occur in synchronous deposits of this region, never exceeding 1–2% anywhere in Israel at that time. It was therefore suggested that cypresses have been grown on or near the site, for use as raw material for building. Similar high percentages of cypress were also found in an Iron Age site in the northern Negev (Kempinski, in press), where the same interpretation was proposed.

Unusual percentages of pollen produced by insect pollinated flowers are sometimes a result of either collecting their mother plants for food, or maybe even growing them on the site for aesthetic purposes. Such may be the case with finds of up to 14% of *Althaea hirsuta* (hollyhock) pollen extracted from Iron Age sediments of Masos, a site at the Negev, Israel (Kempinski, in press), where it was difficult to decide for which purpose the flowers had been brought, or cultivated on the locality, since the present day local Bedouins use the seed and leaves of these plants for food, but also occasionally collect and grow them for beautification.

It is not always clear whether a certain plant is natural to the region, or imported by man long ago. An example is papyrus (*Cyperus papyrus*) that, for a long time was considered the most typical plant of the Hula swamps, in northern Israel (Zohary, 1959). Combined palynological and isotopic studies indicated (Bein and Horowitz, 1986) that these plants did not grow in this area before some 4,500 years ago, which raised a suggestion that papyrus was indeed introduced by the Egyptians, since the date coincides too well with the beginning of their presence in northern Israel.

10.2.2. Ruderal plants

Ruderal plants are part of the natural vegetation, which seem to adapt much better and considerably increase their populations in cultivated fields, along irrigation canals and close to human habitations. They utilize the better conditions offered by anthropogenic activity, such as increase in amounts of water and organic materials. These processes are known also from forested regions, but seem to be much more important in arid, where the local vegetation is rather poor. It is quite common to find nowadays, around settlements in arid regions, that almost the entire flora is dominated by ruderals. Incidentally, most of the ruderal plants, such as *Typha*, *Plantago*, Cheno-Ams and others are also important pollen producers, their grains frequently occurring in high proportions in sediments affected by anthropogenic activity. The problem is that there is no way to differentiate by pollen between ruderal and natural plants of the same genus or family, thus exact calculations of the relative amounts of ruderals are impossible.

Pollen of ruderal plants occur in almost any archeological site which was analyzed, undoubtedly indicating anthropogenic interference with the natural conditions. This was found in the Middle East (Leroi-Gourhan, 1980), in the Americas (Davis et al., 1985b; Byrne and Horn, 1989) and indeed almost anywhere in arid regions around the world.

Higher occurrences of pollen produced by ruderal plants in certain anciently cultivated fields in central Arizona, in comparison with other contemporaneous deposits at the same region, led Fish (1984) to suggest that the former had been irrigated. Increased percentages of *Typha* pollen seem to denote the onset of perennial irrigation in Birket Qarun, Egypt (Mehringer et al., 1979) some 100 years ago.

Davis et al. (1985b) had analyzed sediment cores from Pecks Lake and Tevasci Marsh in Arizona and found out that during the Historic interval *Juniperus* pollen percentages doubled, *Prosopis* and *Larrea* percentages reached maxima, while pollen of introduced weeds first appear. During the Prehistoric interval, concurrent with the occupation of nearby Tuzigoot Pueblo, Cheno-Ams pollen reached maximum percentages. During both intervals pollen of other weeds are present, however in small amounts. The authors concluded that differences between the two intervals probably result from the more widespread disturbance, the introduction of grazing livestock and the presence of introduced weeds during the Historic interval.

Byrne and Horn (1989) found out that a sequence from Vera Cruz, Mexico, contained corn pollen, especially common in the middle part of a core dating to the first millennium AD. This section is also characterized by rather low arboreal pollen percentages, but rich in herbaceous taxa, particularly in pollen of *Ambrosia*. It is interesting to note that this weed is ruderal in more than one environment and abundantly accompanies human settlements also in temperate North America.

Horowitz et al. (1981) found out that percentages of pollen produced by Chenopodiaceae and Amaranthaceae, considered to be ruderals, in pueblo sites near El Paso, southwestern Texas, seem to increase the longer the site is inhabited. Based on this, reconstruction of the order of building was suggested, namely which was the first structure to be constructed in the site and which one the last. Construction times could accurately be thus reconstructed, since pollen was analyzed directly from their adobe bricks.

Thus the weeds, although useless and annoying for early men, seem quite useful for the present day palynologist who tries to follow and reconstruct man's activities by the pollen spectra obtained from archeological sites.

10.2.3. Man induced changes in the natural vegetation

The intentional introduction, domestication and cultivation of plants and livestock, together with the ensuing expansion of opportunistic ruderal vegetation, make immediate contributions to pollen spectra recovered from an archeological site. Besides this intimately connected influence, man's and livestock activities can in many cases affect also natural vegetation and its composition in the surrounding regions. The expansion of ruderal weeds can also be on the expense of natural vegetation, thus such disturbances in the pollen spectra have been used (Fish, 1984) to assess ancient land use.

The use of fires by man, for clearing forests and preparing suitable areas for agriculture, is well known from western and northern Europe, but had also been practiced in the arid and semi-arid regions for quite a long time, in fact beginning long before this is known from the temperate lands. The considerable use of fire for clearing natural forests of the temperate regions resulted in reduction of arboreal pollen shares in the sediments and increased appearance of pollen produced by weeds, which rapidly occupied the cleared lands. This was extensively used in Europe to assess processes connected with agriculture (Faegri and Iversen, 1964).

In the dry regions, it is mainly savannah and steppe vegetation which had been put to fire, to clear land for agriculture and especially for pasture. Thus the decline in pollen produced by savannah shrubs and trees was interpreted in Rajasthan, India (Singh, 1980) as being caused by fires, intentionally set by inhabitants throughout most of the Holocene. Similarly, use of fire was concluded in some Mediterranean countries by Pons and Thinon (1987). Moore (1979) had detected clearings resulting from fires set near sites in Iraq more than 10,000 years ago, in order to gain areas for agriculture and stock breeding, in a pre-Neolithic farmers village on the Euphrates River.

It is usually quite rare to observe fires set intentionally by humans in arid and extremely-arid lands, but it seems that in the semi-arid this practice was quite common through the last millennia, at least in the dry regions of the Near East. It should be remembered that natural fires are quite common in these regions, especially during the long, hot, dry summer. There seems to be no conclusive way to establish, palynologically, which fire was artificially set, or which one came from natural causes. Some researchers (Naveh, 1990) consider fire, at least in the eastern Mediterranean region, to be an important driving force for renovation of the natural vegetation. Clearly however fires cause change in the succeeding plant associations, that are sure to affect pollen spectra.

Considerable influence can be exerted on the expansion and composition of the natural vegetation in dry regions by man's and livestock activities. Clason and Clutton-Brock (1982), summarizing such effects for archeologic periods in the Near East, maintain that (p. 146): "by protecting the herds of domestic caprines from predators and other competing ungulates, man initiated the denudation and even the desertification of much of the landscape of western Asia. Furthermore, in addition to overforaging by domestic livestock, man had other destructive activities, as is known from modern ethnographic parallels, that cause a rapid depletion of vegetation. These are the cutting of wood for fuel and for the manufacture of charcoal, by the extensive cutting of thorns for fencing livestock corrals and human habitations. There was also the wide-scale destruction of pine trees for the tapping of turpentine, that has been carried out at least since the classical periods".

The combined effect of these activities can become so pronounced as to convert a region from semi-arid steppe into an almost bare desert (Fig. 1.2.12).

In some arid regions, such as parts of northern Africa or the American Southwest, higher hills or mountains occur, enjoying more rains which can maintain trees. Pollen spectra obtained from the lower localities could display the influence of anthropogenic activities on the mountain vegetation, by people living in the drier lowlands. Opinions vary as to the possible causes for changes which may inflict an arboreal decline. Reille (1979), or Fall et al. (1981) have interpreted such decline as having been caused by the changing climate, while others (Davis and Turner, 1986; Veblen and Markgraf, 1988) concluded that changes in the distribution of forests adjacent to arid lands have been caused by man's and livestock activities, especially grazing. Such changes, incidentally, are not always in the way of reducing the forest areas as would initially seem, but could, in certain cases (Davis and Turner, 1986) have caused expansion of the forest due to grazing activities.

Such activities of man and livestock, when coming out of control and replenishment possibilities of the natural environments, may cause desertification quite rapidly. This is shown in Horowitz (1979) for the desert fringe regions in Israel, by a-index Potter and Krenetsky (1967) for the American Southwest and by Lézine and Edorh (1991) for Senegal, based on palynological studies. It should however be noted that not all researchers agree with putting the blame for desertification on man's activities, as Petit-Maire (1990) states (p. 281): "The evidence for climatic changes in the North African arid belt thus confirms the predictive model and does not imply any responsibility of man, although deforestation and erosion locally emphasize the trend".

Processes of human induced desertification are quite rapid and may occur in a matter of years, depending on how sensitive is the environment and what are the type and extent of human activity. In a similar manner, recovering of the natural vegetation may also be quick. This can be seen in Fig. 1.2.12, photographed in 1971, which shows a dark rectangle in northeastern Sinai, where the natural vegetation is renovated, only four years after the area was fenced for military purposes and protected from grazing and plant collection. Potter and Krenetsky (1967) have shown that after 25 years of protection from grazing, grassland plots tripled in percent of ground cover of grasses.

It should however be stressed that apparently not the entire damage could be blamed on man or his livestock alone, as stated by Potter and Krenetsky (p. 145): "The retrogressions of vegetational cover and composition cannot be attributed entirely to grazing, for no single factor can be all-destructive or singly effective".

The contributions made by man to such processes can however tilt the natural trends to the worse.

10.3. Archeopalynology in Israel

Pollen analyses accompany archeological excavations in Israel for the last 25–30 years, routinely carried out in several archeologic institutions in the country. The palynological research is geared into two main directions, to provide stratigraphic and chronologic framework for the numerous prehistoric sites in the country and to reconstruct paleoenvironments for both prehistoric and archeologic sites, throughout the long history of human settlement in this part of the world.

Stratigraphy and chronology are particularly important for the prehistoric cultures, for which hardly any other dating methods are available; the environmental studies are of significance since the country, situated between the desert to the south and east and the Mediterranean environments to the north, had been subject throughout the entire Quaternary to climatic fluctuations which have caused considerable latitudinal shifts of the vegetational belts. The desert had spread to the north during the dry interglacials, while the Mediterranean domain moved to the south during the wet glacials, changing previous desert regions into habitable land. For details please refer to Chapter 9. Please note that the terms "glacial" and "interglacial" are used here in their stratigraphic sense, which does not imply to paleoclimates (see also discussion in Horowitz, 1979).

Human habitations are known in Israel from as long as some two million years ago, commencing at the transition from the Preglacial to the Glacial Pleistocene (Horowitz, 1979, 1989d; Braun et al., 1991). Ever since, human settlements existed in the country, but their remains are frequently found in secluded outcrops, which are not situated within a long continuous section, nor are they intimately connected with volcanic rocks or bear any other witness that could be used to assess their ages. Severe erosion and tectonic movements only help confusion. Just as an example, the important Early Acheulian Ubeidiya site was attributed ages ranging from some 2.5 million years by some authors, up to some 300,000 by others. Incidentally, the present views are around 1,0 to 1,4 my.

Fortunately, continuous sequences are now available for the entire Quaternary of Israel from the Jordan–Dead Sea Rift, palynologically analyzed in several boreholes drilled at that region (see Chapter 9). Correlations of pollen assemblages recovered from the sites' sediments with the continuous sequences, the latter dated by both radiometric methods (Heimann and Steinitz, 1989; Heimann, 1990) and by correlations with the oxygen isotope curves (Horowitz, 1989b) resulted in a chronostratigraphic framework for the prehistory of Israel (Horowitz, 1989d). It should however be noted that exact datings of many prehistoric cultures are still in great dispute, for that matter not only in Israel. The chronologic scheme adopted here is therefore only of a general nature, while other opinions may as well be relevant.

The chronology of sites dating from the last stages of Paleolithic cultures through the present day, could better be resolved by radiocarbon datings, thus palynology is usually not used for this purpose. However, especially in this time span, when agriculture have been developed, the study of paleoenvironments is of crucial importance. This stage is detailed below, in Chapter 10.3.2.

When paleoenvironments and their connection with human habitation patterns are concerned, the history of settlements in Israel (and in most other localities) can be divided into two epochs. The first, which in terms of time takes the major part of human history, is characterized by a passive man, who had to follow appropriate environments (or occasionally microenvironments) suitable for his needs of hunting and food gathering. The second, commencing with the initiation of agriculture, is a period of active behavior toward the environment. The early stages only cause minor changes in the close vicinity, with the introduction and husbandry of plants and animals, or the unintentional development of weeds. With developing habitation, progressively pronounced changes are felt in the natural environments, until at present they are in places changed beyond recognition.

These anthropogenic induced environmental changes, when affecting sensitive boundary environments on the desert margins, could result, depending on the type of activity, in either progressive desertification or fertility (see Fig. 1.2.12).

10.3.1. Chronology and paleoenvironments of prehistoric industries in Israel

The Glacial Pleistocene sequence was subdivided in Israel into seven palynozones, QIII through QIX (topped by the Holocene QX), which house the prehistoric industries found within sites and find spots in the country. The stratigraphic and paleoenvironmental characteristics of these palynozones are detailed in Chapter 9; locations of the more important early prehistoric sites in Israel and the vicinity are shown in Fig. 10.3.1; the chronological framework for the Glacial Pleistocene palynozones, their correlation with the oxygen isotope curve, radiogenic ages and the contained prehistoric industries is displayed in Fig. 10.3.2. The later cultures are dealt with in more detail in Chapter 10.3.2.

Early and Middle Acheulian industries are found embedded in sediments of both the coastal plain and Jordan Valley lowlands. There seems to be no occurrence of implements made by these early cultures in the hilly highlands of Israel, apparently due to the fact that no sediments have been sheltered from subsequent erosion in the elevated regions, or maybe because the inhabitants had not chosen to live in these places.

Some of the prehistoric sites are found in highly oxidized sediments, either primarily such as the red paleosols (locally termed "hamra") of the coastal plain, or due to deep oxidation of outcrops, as is the case with many formations in the Jordan Rift Valley. In these cases detailed fieldwork could occasionally locate single samples from representative sites which did yield pollen spectra, or, in the absence of such, stratigraphic correlations with less oxidized, such as borehole

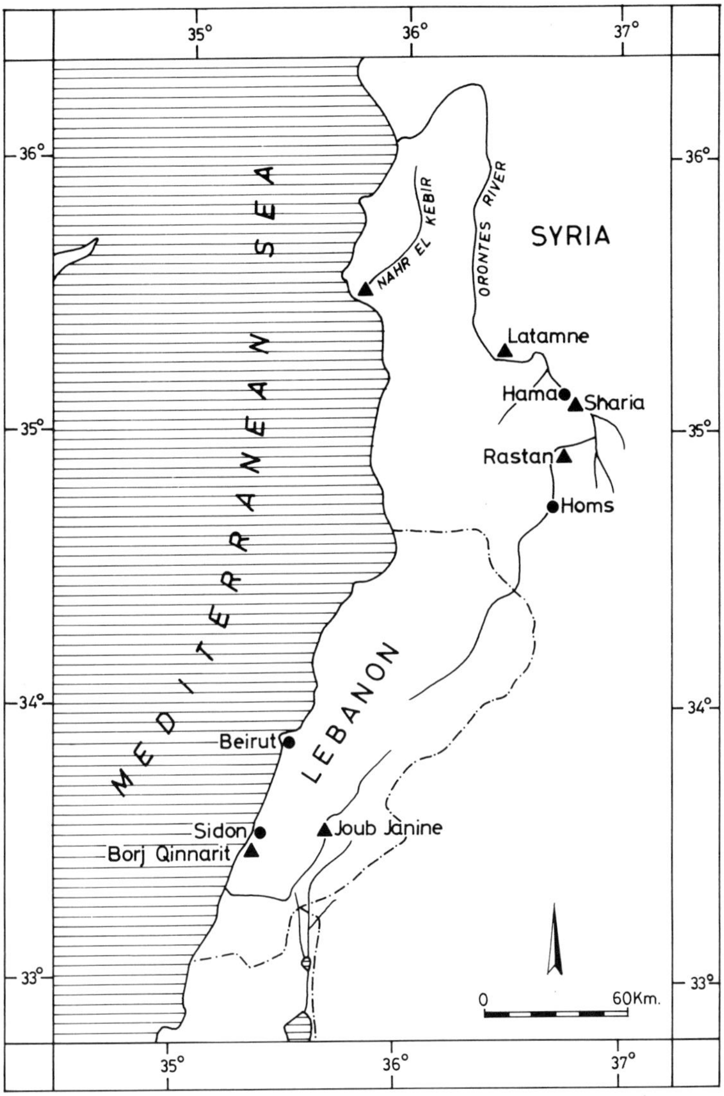

Fig. 10.3.1. Location map for the mentioned paleolithic sites. a. Syria and Lebanon.

sequences, supplied the necessary material for analysis. The sites had been assigned to the corresponding palynozones, based on both pollen analyses and stratigraphic considerations (Horowitz, 1979, 1987a, 1988a, 1989d; Weinstein-Evron, 1983, 1988, 1990).

Absolute dating of the palynozones (see also Chapter 9) was done both by correlating the Israeli sequences with the oxygen isotope curve (Horowitz, 1989b) and by potassium/argon or $^{39}Ar/^{40}Ar$ radiometric datings of basalt flows.

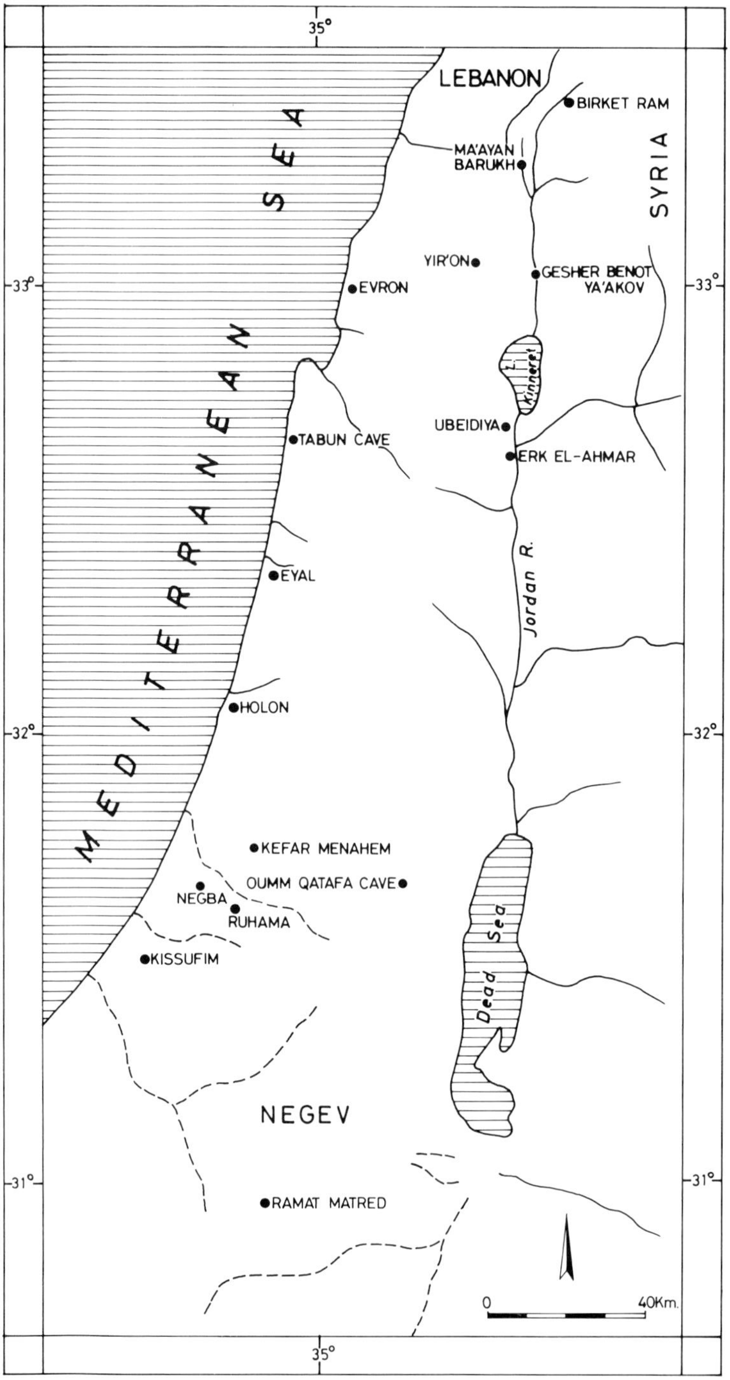

Fig. 10.3.1 (continued). b. Israel.

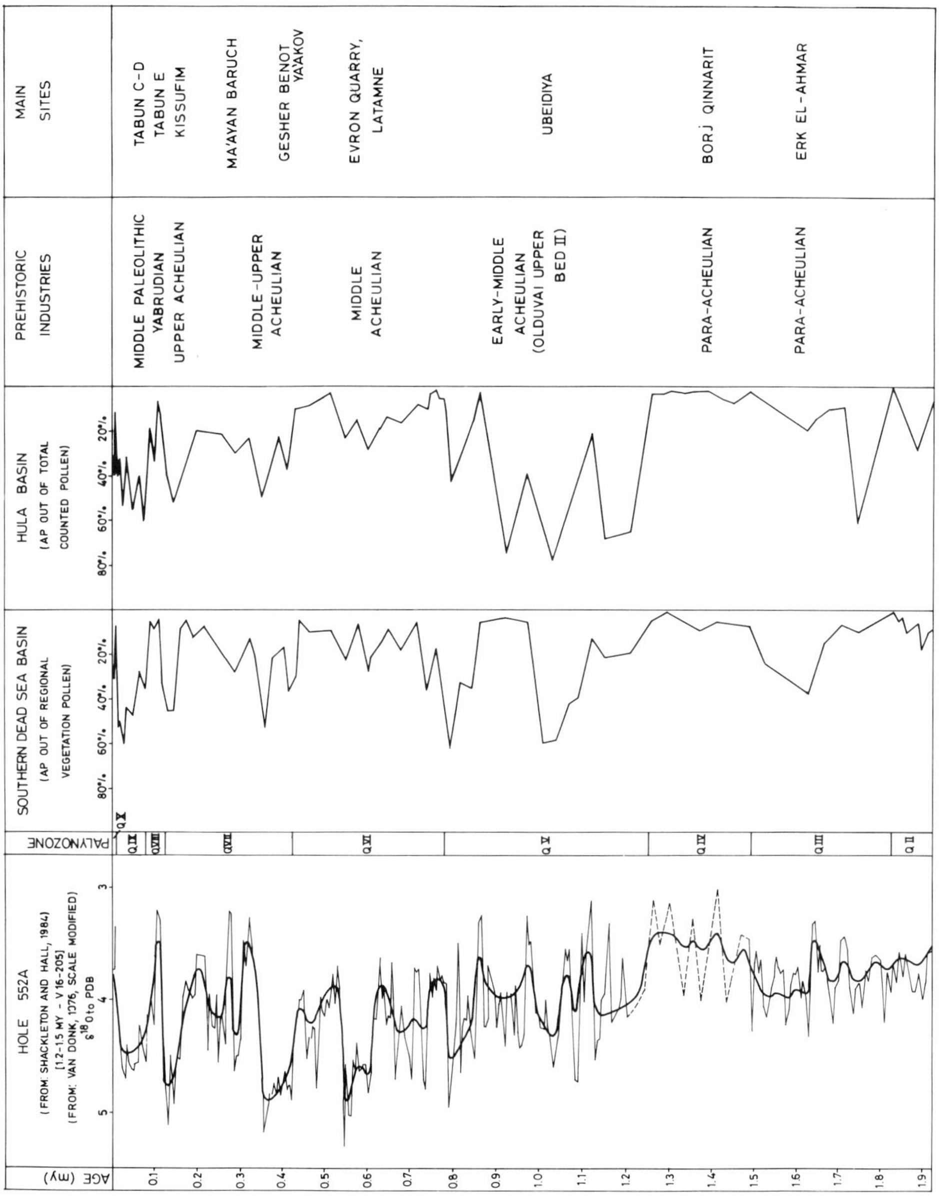

Fig. 10.3.2. Correlation of the Glacial Pleistocene palynozones in Israel, the oxygen isotope curve and prehistoric industries of the Levant.

These volcanics interfinger with the palynologically analyzed borehole sequences (Heimann and Steinitz, 1989; Heimann, 1990) and outcrops (Horowitz et al., 1973; Siedner and Horowitz, 1974).

The oldest artifacts had been recovered from Erk el-Ahmar formation in the central Jordan Valley and from the Hesi member of Gaza formation in the southern coastal plain, chiefly comprising crude flakes and primitive chopping tools, made on pebbles with minimum blows. No pollen was recovered from the oxidized Hesi member fluviatile sediments, but several horizons of the Erk el-Ahmar formation yielded decent pollen spectra. An average of these reads as follows:

Quercus spp. — 23%
Pinus sp. — 6%
Picea orientalis — 4%
Betulaceae — 0.5%
Total AP (excluding hydrophil trees) — 33.5%
Hydrophil trees — 3.5%
Chenopodiaceae — 9%
Gramineae — 24%
Cyperaceae — 3%
Typha — 2%
Compositae — 7%
Artemisia — 6%
Fern spores — 4%

and occurrences of: *Ephedra*, *Scabiosa*, Papilionaceae, Liliaceae, Malvaceae and Umbelliferae

This composite pollen spectrum for Erl el-Ahmar outcrops resembles very much those obtained from sediments of QIII palynozone in the borehole sequences. This correlation is mainly based on the considerable shares of arboreal pollen, indicating wet, pluvial conditions, with their composition of oaks, pine and *cf. Picea orientalis*. Combined occurrences of these three taxa together is typical for QIII. The reconstruction of QIII paleoenvironments is based on detailed information obtained from continuous pollen diagrams of drillings, where the material is not oxidized.

The only two occurrences of artifacts in QIII sediments are intimately connected with water, being found embedded within fluvio-lacustrine sediments at Erk el-Ahmar and in river sands and gravel of the Hesi member. It seems that inhabitants chose to settle in close vicinity to a stream or lake at that time.

The age assigned to QIII palynozone by its correlation with the oxygen isotope curve is in the range of 1.8 up to 1.5 million years ago. Paleomagnetic measurements recently carried out for Erk el-Ahmar outcrops in the central Jordan Valley (Braun et al., 1991) may indicate that the base is somewhat older. It is quite difficult at the present state of study to be more specific as to the exact age of the implements within this range.

Two occurrences of industries described from Syria (Hours, 1975) are assigned to strata correlative to palynozone QIII in Israel (Horowitz, 1979). The industries had been unearthed from Sharia, south of Hama, where very few artifacts, two spheroids, some irregular globular cores and small flakes, but no bifaces had been recovered. The other occurrence is at Rastan, on a high terrace of the Orontes River, between Homs and Hama. This assemblage also includes pebble cortex flakes, with no cores and bifaces. Hours (1975) had assigned these sites to the Para-Acheulian industries and correlated the layers with the end of the Günz glacial.

Sites correlative to palynozone QIV are not known from Israel, but do occur to the north, in the Lebanon (Hours et al., 1973). The industries had been found near Sidon, at Borj Qinnarit, in a fossil beach 95 meters above the present day sea level, assigned by the excavators to the Günz/Mindel interglacial times. The industry consists of two pebble tools, two cores, a single large anvil, two heavy flakes and ten small preparation flakes. This industry is also considered by Hours (1975) as of a Para-Acheulian affinity, which is thought to be younger than the African Olduvan and may be broadly correlative with the European Clactonian. The correlation with QIV is only based on stratigraphic considerations (Horowitz, 1979), with no pollen spectra available for the Lebanese site. The age of the site is in the range of 1.5 to 1.25 million years ago. The climate in Israel during QIV times was dry, interpluvial, as can be seen from its pollen spectra, rather poor in arboreal components.

Early Acheulian industries occur in several localities in Israel (Ronen, 1979; Bar Yosef, 1987), both in the coastal plain and the Jordan Valley. The best known by far is the site of Ubeidiya, where in various deposits that had been laid down on a lakeshore, at least 14 lithic assemblages comparable to the Olduvai Upper Bed II industry, have been encountered (Goren, 1981). Chopping tools form the most common and characteristic implement, accompanied in some of the assemblages by spheroids, large trihedral picks and/or primitive handaxes. Quite similar industries are known from a few sites in the coastal plain, such as Kefar Menahem and Negba and also at other localities, in the Jordan Valley. Sites such as Joub Janine in the Lebanon and Nahr el-Kebir in Syria, bearing similar characteristics had been assigned by Hours (1975) to the Early Middle Acheulian, correlative to the Mindel glacial.

Pollen spectra recovered from sediments of the Ubeidiya formation at the excavation site (Horowitz, 1979) yielded arboreal pollen in the range of 80–85%, of which oaks, particularly the winter deciduous ones, comprise the majority, accompanied by occurrences of pine, juniper and olive grains. These spectra undoubtedly represent pluvial conditions. The striking prevalence of winter deciduous oak pollen over all other constituents serves for correlating the Ubeidiya formation and its coastal plain counterparts to palynozone QV.

The age range for this palynozone is quite wide, commencing at some 1.25 million years ago and continuing up to some 800,000 years. This age is based on both correlations with the oxygen isotope curve and radiometric datings. Ages attributed recently to Ubeidiya formation on the basis of its contained faunal remains are somewhat older, in the range of 1.4 to 1.0 my (Tchernov, 1987). It seems that the main assemblage of industries could be correlated with the considerable peak of arboreal pollen at the middle part of palynozone QV, which recurs in many analyzed sequences, dating to approximately one million years ago.

The following phase of lithic industries is rarely found in Israel and seems to be present only in a single locality, the Evron Quarry (Ronen, 1979), at the northwestern tip of the country, on the coastal plain. Several artifact horizons are found there, embedded in red sandy loams and in paludine deposits. The principal tools are handaxes of great size, with as few as 12–15 flakes per tool. A characteristic subgroup comprises bifacial cleavers, with narrow edges. The chopping and flake tools are accompanied by a few racloirs and awls. The flake tools are small and thin, with a quite fine workmanship.

Evron Quarry seems to have been a hunting and butchering site located in a marshy area, probably not far from a river. The site broadly ties up, stratigraphically, with the site at Latamne, Syria. The pollen spectra, from correlative layers in boreholes, are poor in arboreal components, among which pine prevails, thus seem to generally indicate an interpluvial, rather dry climate for this site and the QVI palynozone. It is probable that the two main occupation level in Evron had been settled during somewhat more humid stages, possibly interstadials.

QVI palynozone took place from approximately 780,000 years ago, up to about 420 ky. It is however quite difficult to exactly define the stratigraphic position of Evron within QVI. It seems plausible to assume (Horowitz, 1979) that in Syria, possibly also in northernmost Israel and the Lebanon, occupation took place almost continuously throughout QVI palynozone times.

The succeeding Middle Acheulian industries are typified by presence of bifaces, use of a soft hammer and absence of Levallois technique. Otherwise, each of the specific sites located in several places in Israel has its own distinctive features. Except for a single site, Gesher Benot Ya'akov, most others are located in paleosols of no direct stratigraphic context, thus it is very difficult to assess internal time relations within the various Middle Acheulian variants.

The site at Gesher Benot Ya'akov yielded pollen spectra with relatively high shares of arboreal components, up to 45%, of which oak comprises the majority, typical for the lower part of QVII. Small amounts of *Fagus* pollen (3–4%) seem to further justify correlation of the site with palynozone QVII, for which they are typical (Horowitz, 1973a), hardly ever found in any other palynozone. The time range for the lower part of QVII is between 425–340 ky ago (Horowitz, 1989d). An attempt to study pollen from soils in which the Middle Acheulian Yir'on site was embedded (Weinstein-Evron, 1986), in northern Israel, failed due to impossibility to find credible assemblages.

The Upper Acheulian is known from many sites and find spots all over Israel, many of which are just surface finds, even in regions that are presently a hardly habitable desert. The industry is characterized by the finest workmanship of handaxes, as indicated by their symmetry, thinness and small size.

The Upper Acheulian industries comprise in Israel three main phases, accompanied by several minor variants. The two early ones are found within sediments of QVII palynozone, while the later stage is encountered within deposits of QVIII. Ma'ayan Barukh, in northern Israel and Oumm Qatafa D2, in the Judean Desert, are assigned to the middle part of QVII, with median arboreal pollen values, between 300 and 220 ky ago. Frequent occurrences of the Kissufim type Upper Acheulian all over the country are correlated with the uppermost pluvial phase of QVII, dated at 160–130 ky ago. During this later phase of QVII, characterized by high arboreal shares as compared with the previous, people had inhabited the entire area of Israel, including the Negev Desert and even Sinai.

No sites are known from Israel for the drier interstadials within QVII palynozone.

The youngest Upper Acheulian phase is only known from several localities in northern Israel, such as Tabun Cave. The sites are embedded in sediments which yielded pollen spectra of interpluvial affinity, very poor (2–3%) in arboreal taxa (Jelinek et al., 1973), corresponding to palynozone QVIII (oxygen isotope stage 5), dated from 125 up to approximately 75–80 ky ago.

The palynology of palynozone QIX, in connection with prehistoric settlement of the country, was studied in detail by Weinstein-Evron (1983, 1987c, 1988, 1990). It seems that, broadly speaking, Middle Paleolithic settlements occupied the entire area of Israel, down to the Negev, during the early pluvial phase of QIX.

The early humid phase of QIX is further subdivided by Weinstein-Evron into three subphases of higher humidity. Attempts at exact correlations of Middle Paleolithic cultures in Israel with substages of the oxygen isotope curve, through the use of her pollen diagrams, are presented in 1983 and 1988. It seems that the very humid initial subphase is probably contemporaneous with the Negev early Middle Paleolithic depositional stage and associated industries (Marks, 1976, 1977, 1983; Goldberg, P., 1986). It is also possible that the Middle Paleolithic commenced somewhat earlier, within the end of QVIII. The erosional phase which follows the early Middle Paleolithic in the Negev probably reflects a somewhat drier stage, as compared with the former, although quite humid by itself. An erosion phase is apparent also in the caves of northern Israel, but is thought to have been caused by a more humid stage, accompanied by increased spring activity (M. Weinstein-Evron, personal communication). A distribution map of Middle and Upper paleolithic settlements in Israel, superimposed on paleogeographic and environmental reconstruction, is brought in Fig. 10.3.3.

The lack of later Middle Paleolithic sites in the central Negev seems better explained by the local erosional regime (Goldberg, P., 1986), rather than by a long and extremely-arid phase, for which there is no evidence in the pollen diagrams

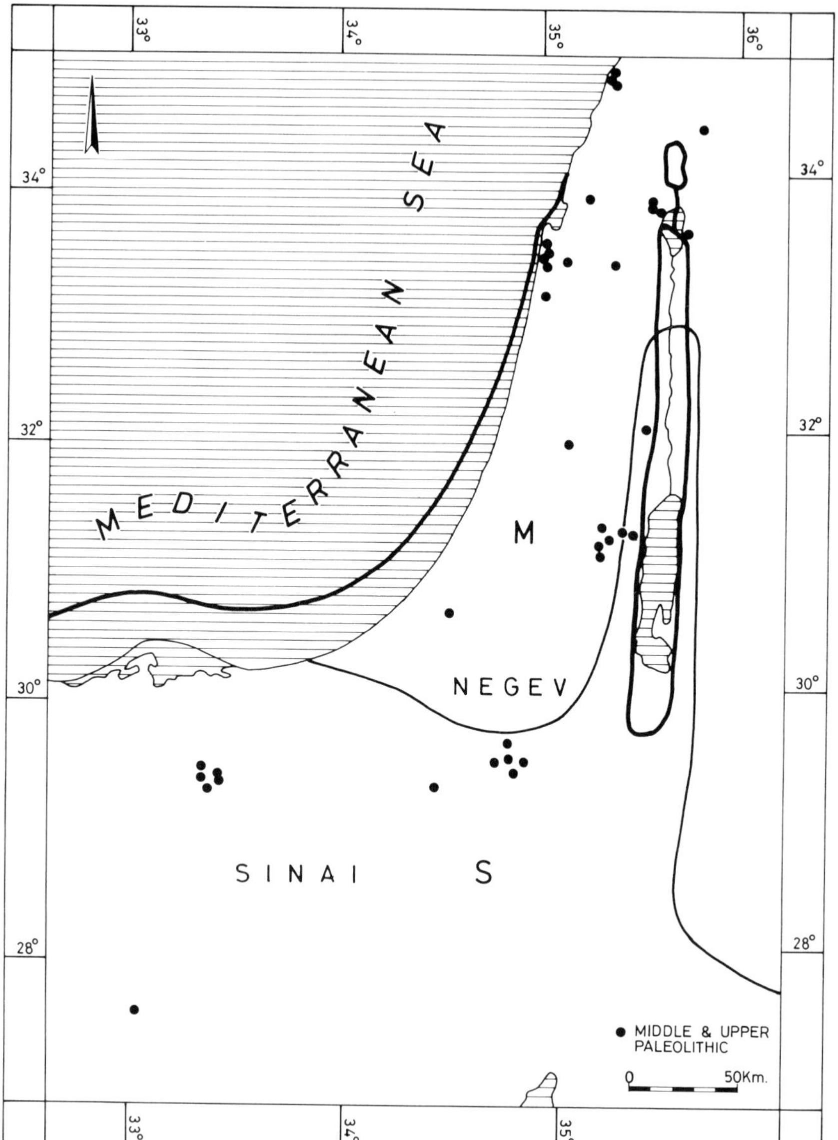

Fig. 10.3.3. Middle and Upper Paleolithic settlement pattern in Israel, within the reconstructed environment. *M* = Mediterranean. *S* = Steppe. Thick lines denote approximate extension of Mediterranean coast line and Dead Sea Rift Valley lakes. Location of sites from Marks (1976), Bar Yosef (1987, 1989).

for that period (Weinstein-Evron, 1988). Evidently, since erosion prevailed to the south, climatic inferences had to be based on pollen diagrams from northern Israel. However, pollen spectra recently obtained from the southern Dead Sea (Horowitz and Horowitz, 1990) seem to strongly corroborate Weinstein-Evron's views.

The last phase of the Pleniglacial, from ca. 30,000 up to ca. 22,000 years ago, is characterized by relatively high arboreal pollen shares (Weinstein-Evron, 1990). During this period, in much the same way as in Middle Paleolithic times, Upper Paleolithic settlements spread south (Fig. 10.3.3) to the Negev and Sinai (Marks, 1976, 1977, 1983). Following comes a dry period, that extended up to some 14 ky ago, when settlements, mainly of the Kebaran cultures, are restricted to the north (Fig. 10.3.5). The Late Glacial is again typified by increase in AP, followed by a spread of sites to the south (Figs. 10.3.6, 10.3.7).

The transition from QIX to QX, again broadly speaking, is characterized by appearance of the Neolithic cultures and the beginning of agriculture (Bar Yosef, 1987) in this part of the world. A more detailed discussion of settlement pattern in relation to environments, for the later prehistoric cultures, is brought below, in Chapter 10.3.2.

A. Conclusion

In terms of paleoenvironments, it is quite clear that most prehistoric industries and sites found in Israel are embedded in sediments bearing pollen assemblages which almost unanimously indicate wetter conditions, during palynozones QIII, QV, QVII and QIX, as compared with the present day.

At first glance this seems to be a natural response of man living at the desert boundary to environmental constraints. Whenever the desert, during the pluvial stages, retreated to the south, man followed and settled in the newly available fertile territories, down to the present day deserts of the Negev and Sinai. Fertility enabled habitation and accordingly, sites are found in the now desolate lands. Interpluvial times sites, of palynozones QIV, QVI and QVIII, seem only to be found to the north, abundant in the Lebanon and Syria, but only rarely occurring in northern Israel. Again, a straightforward explanation would tend to connect the northward retreat of Mediterranean environments with similar movements of ancient populations.

On its face, this seems a very reasonable explanation to the behavior of man and his settlement patterns in the Levant throughout prehistory, when no means had been available to change natural conditions. Some caution must however be applied before this simplistic conclusion is adopted and accepted. Evidently, it is true that the more fertile the environment is, the more suitable it is for habitation. On the other hand, it should be remembered that during most of the interpluvials erosion prevailed in Israel, rather than deposition in times of the pluvials (Horowitz, 1979). The Jordan–Dead Sea Rift lakes shrank during the dry periods, in comparison with their extent in humid times.

This may well explain the abundance of sites in deposits accumulated during the wetter palynozones, as a simple outcome of the widespread distribution of such sediments, as compared with the rarity of outcrops of layers laid down throughout the dry palynozones. The latter are restricted to the coastal plain of Israel, where only sand dunes are known for the drier periods, which could hardly form any settling place for man. Interpluvial sediments are also quite abundant in the Jordan Rift Valley, but due to the lesser extent of its lakes are only known from boreholes (Horowitz, 1979, 1987a).

While settlements could indeed exist at the lake shores also during the dry interpluvials, there is no way to find them, since they are now buried under overlying extensive cover of pluvial lakes sediments, deposited during the succeeding wetter palynozones. The same holds true for the wadis and hilly regions of Israel, where extended erosion wiped out any possible exposures of sediments rarely laid down during the earlier, dry palynozones. In fact, even sediments of the rather recent QVIII palynozone are quite rare in Israel.

Sites corresponding in age to the last interglacial, palynozone QVIII, are known from several caves in Israel, where sediments had been accumulated and sheltered by the cave structure from subsequent erosion. Again, it may well be the case that sites corresponding to preceding dry palynozones existed in such caves, but had been long washed away. Erosional phases of cave sediments, however short for the more recent deposits, are quite well known from Israel (Weinstein-Evron, 1988).

To sum up, it seems plausible to assume that human settlement patterns during the Glacial Pleistocene closely followed paleoclimatic and paleoenvironmental fluctuations. The possibility cannot however be ruled out that, although the number of settlements may have been considerably reduced, such had indeed existed in this region while dry, for people could have found refuge in places where locally water, game and edible plants were still available in sufficient quantities. Our chances of finding such sites are quite slim, since they either are long eroded, or buried deep inside the rift valley or the coastal plain, where suitable conditions were to be found even during the drier periods.

10.3.2. Archeopalynolgy of late QIX and QX palynozones in Israel

The peak of oxygen isotope stage 1 is expressed in Israel by a sharp decline of arboreal pollen, most probably resulting from the extensive northward propagation of the desert regime, up to the northern part of Israel, some 10,500 years ago. This AP decline is indicated both in continuous pollen diagrams obtained from various basins of the Dead Sea Rift (Horowitz, 1987a) and from pollen analyses of Late Natufian sites in a region just north of the Dead Sea (Alon, 1976; Darmon, 1984, 1987, 1988; Leroi-Gourhan and Darmon, 1987). Following this extremely-arid episode, Neolithic industries are found in many sites in Israel and vicinity.

Although it is usually maintained that the Neolithic cultures correspond to the beginning of agriculture, it is noteworthy that cereal type Gramineae pollen are found already in the preceding, Natufian culture sites (Darmon, 1987) north of the Dead Sea. Ever since, cereal pollen have been found in almost every archeological bed from the arid regions of the Levant which was subject to palynological analysis, testifying that cereal agriculture, however in initial stages, was practiced even in the dry regions for at least the last 12–13,000 years.

It is however quite evident that the Neolithic cultures, at the beginning of QX times, had settled in a variety of environments in which a slow deterioration of climate was almost continuous since the peak of the last Pleniglacial, at the end of QIX palynozone, when the region was settled by people of the Kebaran cultures.

Pollen analyses of late QIX–early QX sites enabled Darmon (1988) to reconstruct paleoclimates of the southern Jordan Valley, north of the Dead Sea, during the Epipaleolithic–Early Neolithic periods, between 19,000 and 9,500 years ago.

The present day values for arboreal pollen north of the Dead Sea, within continental sediments, are around 1–2% (Alon, 1976). Compared with these, AP obtained from Kebaran settlements sediments, 19,000–14,500 years BP, at the same locality, are in the range of 5–12% (however, drier phases may have occurred during this time span); the Geometric Kebaran, 14,500–12,500 years BP, yielded values in excess of 10%; the Early Natufian (12,500–11,000 years BP) 8–10%; the Late Natufian (11,000–10,750 years BP) 2–4%; while during the Terminal Natufian (around 10,500 years ago) AP values drop down to zero; following this extreme stage the Pre-Pottery Neolithic A (10,200–9,200) shows arboreal pollen values of 10–15%, increasing up to 20% for the Pre-Pottery Neolithic B, from approximately 9,200 up to 8,000 years BP (Darmon, 1988).

Kebaran cultures seem to have enjoyed environments which had been only slightly more humid as compared with the present day. Humidity increased during the following, Geometric Kebaran, until approximately 12,500 years ago. The Natufian shows gradual desiccation, attaining a peak during the Terminal Natufian, corresponding to oxygen isotope stage 1, in which arboreal pollen are entirely absent from the sites' deposits. During the beginning of the Neolithic settlement humidity increased, which is correlated by Darmon with similar trends at the beginning of the Holocene in Europe.

A study of goat coprolites from an Early Neolithic site in the Judean Desert (Darmon, 1989) indicated that during this time the environment was significantly more humid as compared with the present day. Truly, the coprolites reflect mainly the goats' diet, but the environment is also reflected in this component of the flora. Similar paleoenvironmental trends for this region and period were also previously reported, based on preliminary pollen analyses of some of the sites analyzed by Darmon, in Alon (1976).

Fig. 10.3.4 is an attempt at presenting arboreal pollen factors for the last 18–19 ky in Israel. Since no continuous habitation is known for the arid parts of the country in a single locality, a combined diagram was constructed, in which each archeologic horizon is represented by the proportion of average arboreal pollen shares to the present day values, at the same locality. Thus, although the curve is based on pollen analyses from many sites, the present day percentages are always

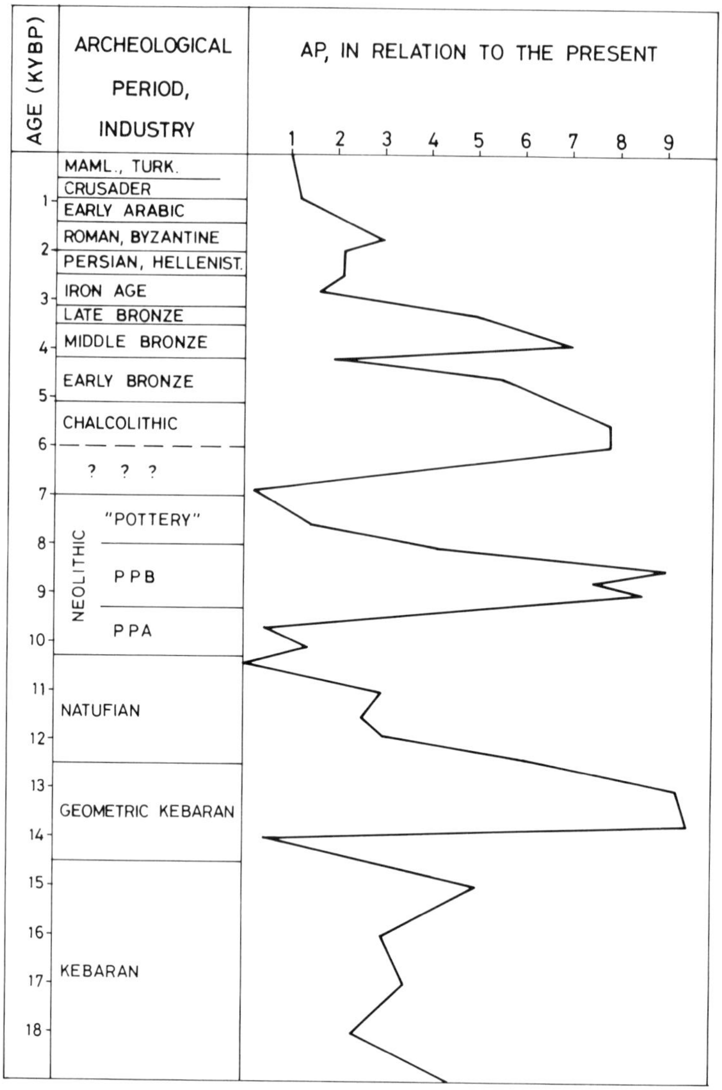

Fig. 10.3.4. Synthetic diagram of arboreal pollen factors for various sites, as compared with the present day, from late QIX through QX times. Sites have been palynologically analyzed from all over Israel and the recent values for each region are taken as a unit, with which the AP shares of the settlement period are compared. Where no spectra are available for a certain period, the factor is calculated from equivalents in synchronous samples from pollen diagrams.

taken as a single unit, with which the arboreal factor in times of settlement is compared.

The settlement patterns of these cultures (Figs. 10.3.5 through 10.3.11) seem to follow quite parallel the climatic curve based on the arboreal pollen factors,

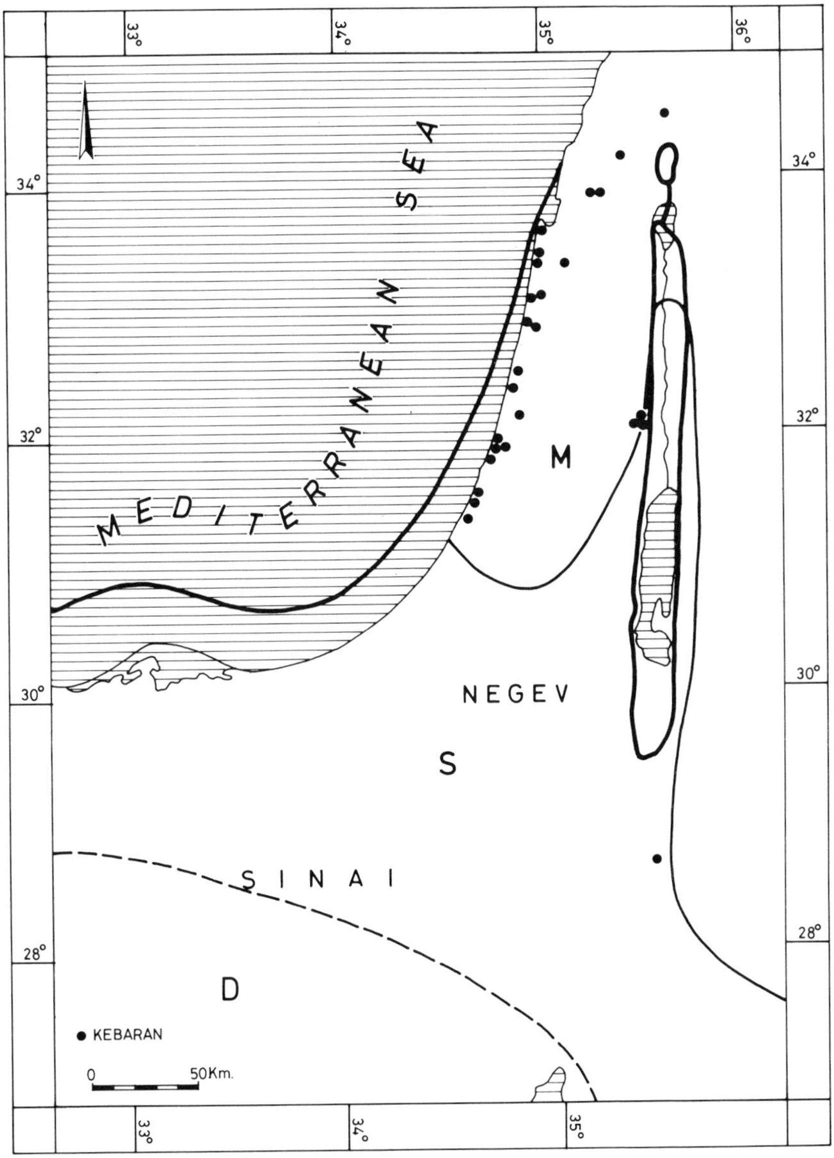

Fig. 10.3.5. Kebaran settlement pattern in Israel, within the reconstructed environment. *M* = Mediterranean. *S* = Steppe. *D* = Desert. Thick lines denote approximate extension of Mediterranean coast line and Dead Sea Rift Valley lakes. Location of sites from Goring-Morris (1987), Bar Yosef and Belfer-Cohen (1989).

which also coincides with pollen diagrams obtained for this time span from the Hula basin and the Dead Sea (see Chapter 9). Kebaran settlements (AP factor 2.5–3) are mainly found in the present day Mediterranean region of Israel, quite

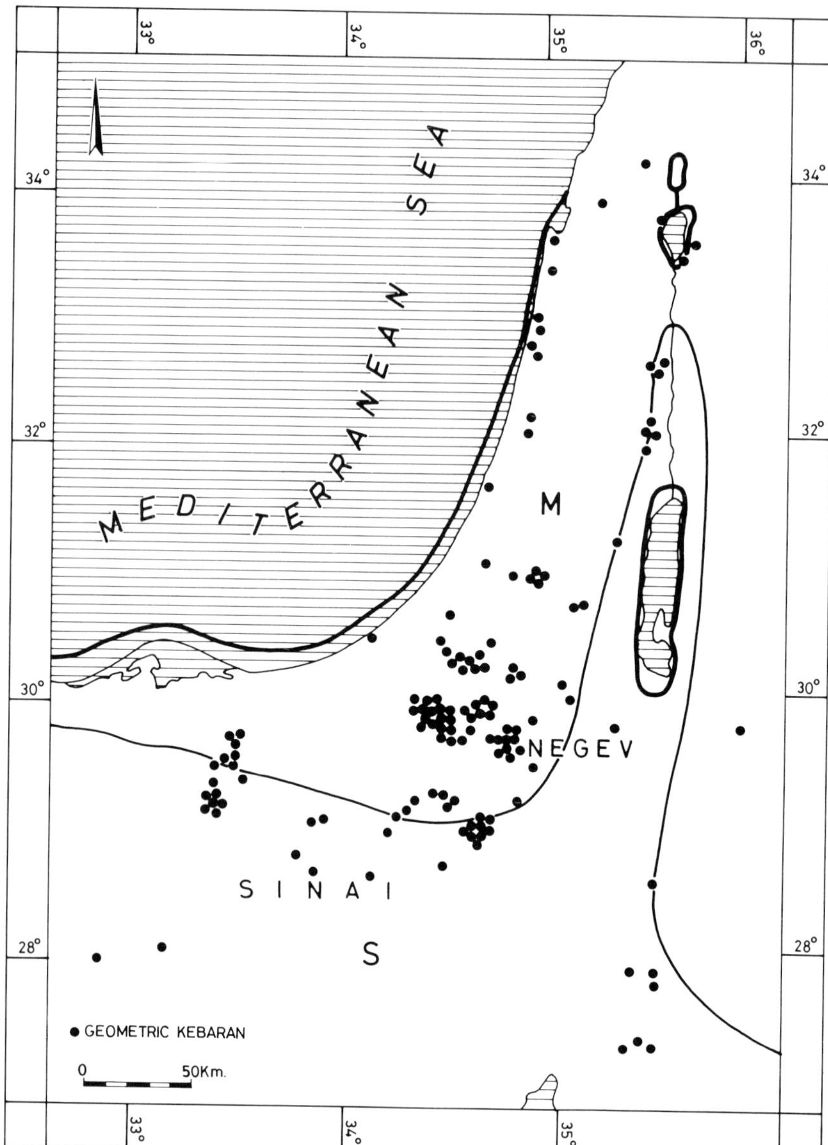

Fig. 10.3.6. Geometric Kebaran settlement pattern in Israel, within the reconstructed environment. *M* = Mediterranean. *S* = Steppe. Thick lines denote approximate extension of Mediterranean coast line and Dead Sea Rift Valley lakes. Location of sites from Goring-Morris (1987); Bar Yosef and Belfer-Cohen (1989).

rare to the south; Geometric Kebaran (AP factor 9–10) sites are found in many localities in the present day deserts of the Negev and northern Sinai (Horowitz, 1979); Natufian sites (AP factor 2.8–3.2) generally maintain the Geometric Kebaran

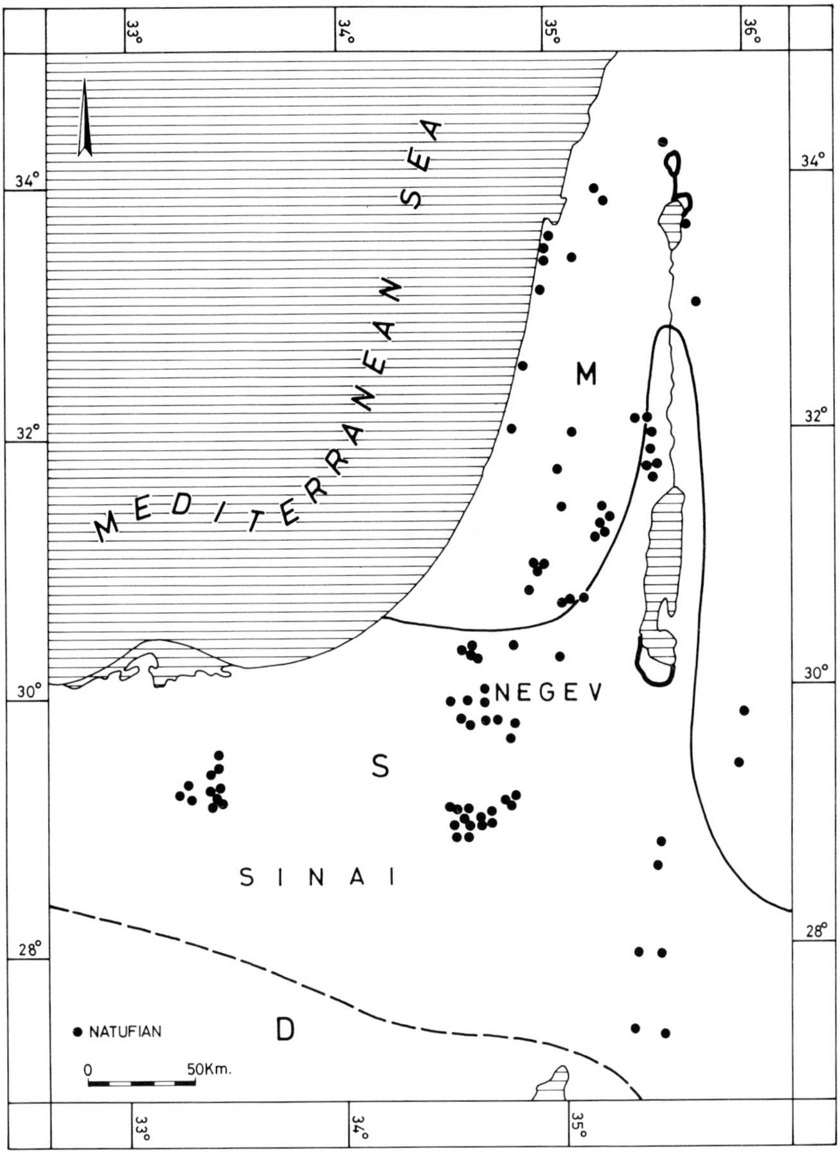

Fig. 10.3.7. Natufian settlement pattern in Israel, within the reconstructed environment. *M* = Mediterranean. *S* = Steppe. *D* = Desert. Thick lines denote approximate extension of Dead Sea Rift Valley lakes. Location of sites from Goring-Morris (1987), Bar Yosef and Belfer-Cohen (1989).

pattern, somewhat more abundant in northern Israel, while those sites found to the south are either intimately connected with perennial springs in the Jordan Valley, such as Jericho and Fatza'el or somewhat smaller than the preceding, Geometric

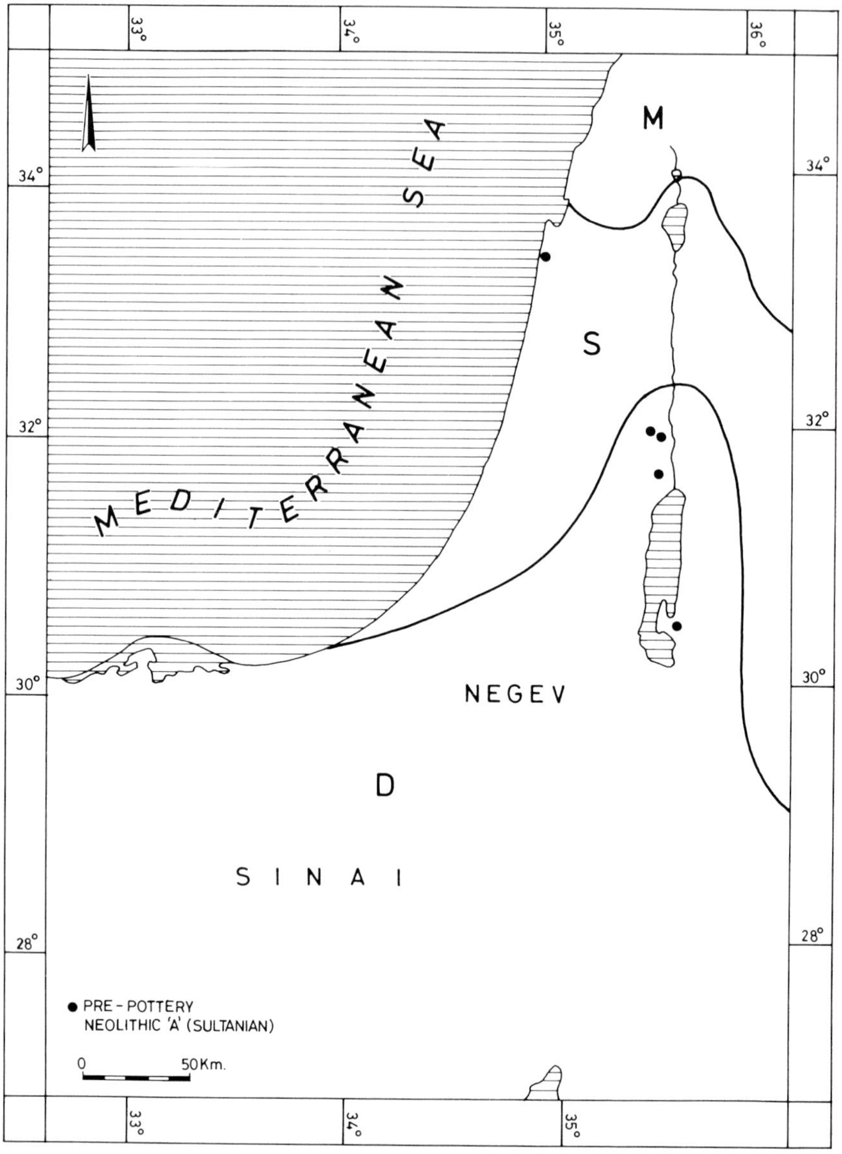

Fig. 10.3.8. PPNA settlement pattern in Israel, within the reconstructed environment. *M* = Mediterranean. *S* = Steppe. *D* = Desert. Location of sites from Goring-Morris (1987), Bar Yosef (1989).

Kebaran sites; Pre-Pottery Neolithic A sites, with AP factor less than 1, are hardly found even to the north of the country; Pre-Pottery Neolithic B sites again spread to the south, enjoying AP factors of 8–9. The following, "Pottery" Neolithic (pottery was already in use before, see Bar Yosef, 1987) is restricted to the northern part

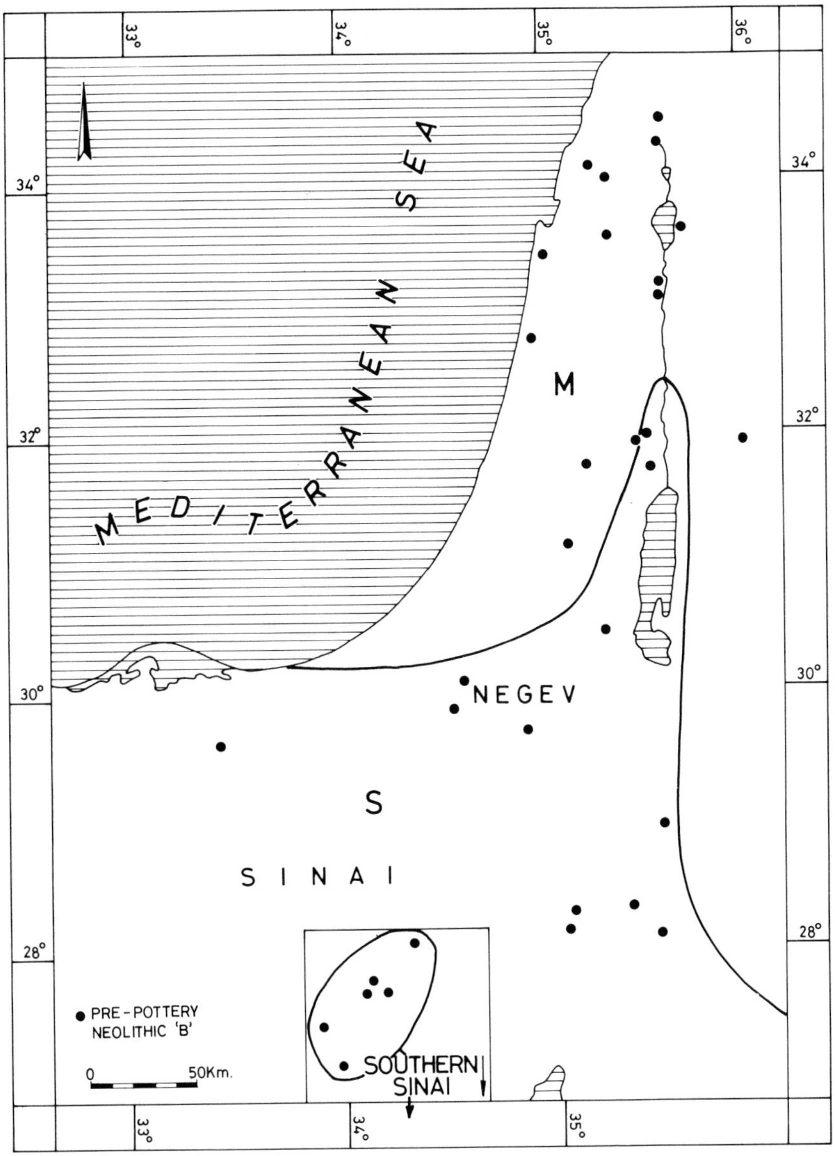

Fig. 10.3.9. PPNB settlement pattern in Israel, within the reconstructed environment. *M* = Mediterranean. *S* = Steppe. Location of sites from Goring-Morris (1987), Bar Yosef (1989).

of Israel and is typified by pit houses, dug into the soil, maybe to cope with the extremely-arid climate, with AP factors less than 1. Such pit houses were also found in Jericho, where the large spring was continuously active even during the driest phases.

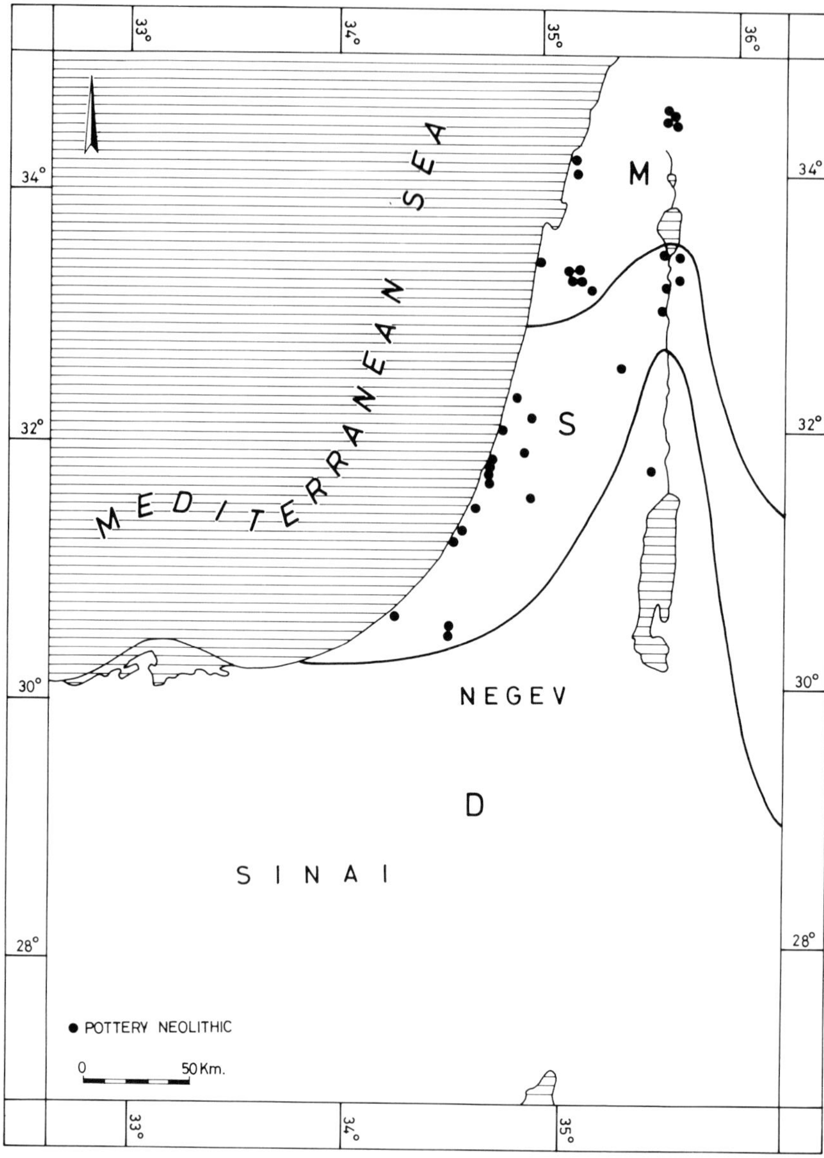

Fig. 10.3.10. Pottery Neolithic settlement pattern in Israel, within the reconstructed environment. *M* = Mediterranean. *S* = Steppe. *D* = Desert. Location of sites from Bar Yosef (1989).

The transition to Chalcolithic cultures in the Near East seems to follow a "break" in settlement, which extended over some 1,000–1,500 years (Bar Yosef and Mintz, 1979). This break is usually explained as a result of climatic deterioration, which is also seen by a corresponding drop in the arboreal pollen shares in continuous

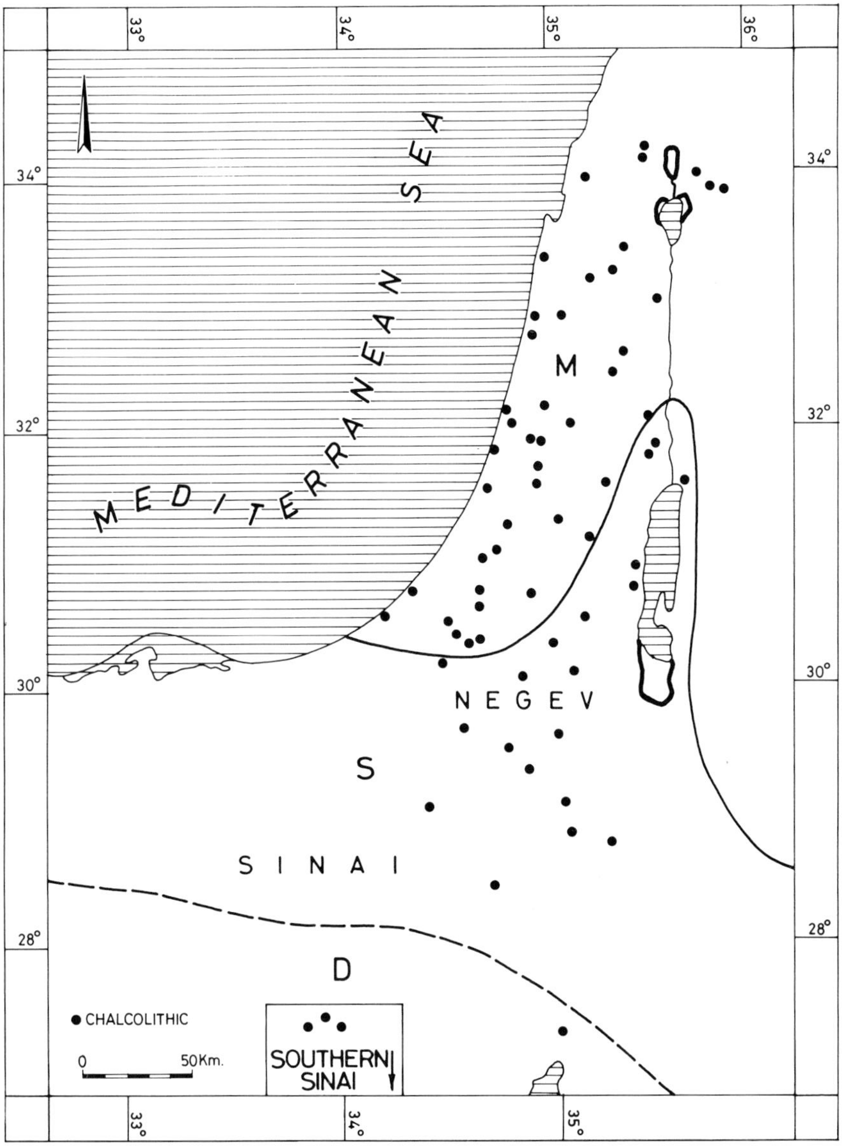

Fig. 10.3.11. Chalcolithic settlement pattern in Israel, within the reconstructed environment. *M* = Mediterranean. *S* = Steppe. *D* = Desert. Thick lines denote approximate extension of Dead Sea Rift Valley lakes. Location of sites from Cohen (1986), Gilead (1989), Gonen (1989).

sequences. Not many sites corresponding to this "break" are known and excavated from Israel, thus it is quite difficult to assess the exact impact of the dry climate on the few remaining settlers.

Pollen analysis of submerged sites from the Carmel coast in northwestern Israel, a region presently under Mediterranean conditions (Galili and Weinstein-Evron, 1985), which corresponds to this period, with radiocarbon ages of $6,830 \pm 60$ and $6,830 \pm 80$ yr BP, yielded several pollen assemblages very poor in arboreal components as compared with the present day. The range of AP for samples collected from the site is 1.1–4.5%, while the recent values are 36–66%. Although it is quite clear that the present day values are high because of the inclusion of imported and cultivated trees such as *Eucalyptus* and pines, the natural arboreal components are still more abundant than in the site's samples.

The pollen spectra therefore indicate environmental conditions similar to the present day northern Negev Desert, as far north as the Carmel, a distance of some 200 kilometers, for the transition from the Neolithic to the Chalcolithic, a period which also involved a fundamental change in culture, lithic assemblages and settlement pattern. the Neolithic was separated from the Chalcolithic by a relatively long dry period, for which hardly any settlements are known from Israel, none from the southern parts of the country, where extremely-arid climate dominated.

Pollen spectra for the Pre-Pottery Neolithic B and the Late Natufian settlements in the central Negev highlands (Horowitz, 1976, 1977a) show considerably higher percentages of arboreal components as compared with the present day, including oaks, cypress, olive, pistachio and almond trees. The Chalcolithic settlements in the central Negev show similar tendencies (Horowitz, 1979), but with somewhat lower arboreal pollen shares (compared to none at present). It seems that Chalcolithic people have found favorable conditions in their times, in places which are now extremely-arid. The wealth of sites corresponding to the wetter periods in the present day deserts seem to testify the intimate connection of the settlers with pleasant semi-arid environments during these periods, in the Negev and Sinai (Horowitz, 1979).

Aridity probably also prevailed in periods separating the occurrences of sites in southern Israel, particularly in places where no large springs were available to maintain settlements. This aridity was also accompanied by severe erosion, thus the permanent question arises, whether people had not possibly stayed in the region during dry phases, but moved closer to the few remaining springs, to such localities that have probably been washed away by subsequent floods.

Following the Chalcolithic culture historic times began, during which the settlement pattern had not depended solely on natural environments, but also on other considerations. Sites are found in desert regions either for reasons in which the environment had not played any role, such as state security, guarding of highway and commerce routes, exploitation of mines (which already began in Chalcolithic times) and so on; or by applying advanced technologies such as damming of wadis for agriculture; storage and transport of water, digging wells and water reservoirs, all of which served to extend the limits of habitable regions, sometimes far into the desert. An example is the extensive Nabatean farming in the central Negev highlands some 2,000 years ago, in an environment only slightly more humid than the present day (Fig. 7.2.5).

The transition from Chalcolithic, mainly pastoral cultures, to the Early Bronze period is characterized by the beginning of a large scale process of urbanization in the Near East and a completely different organization of societies and agriculture. This transition is not expressed in the pollen diagrams obtained from basins far from sites of this period, except for a slight decrease in arboreal pollen, which seems insignificant in terms of the overall environment. It was therefore suggested (Horowitz, 1980) that this change was purely cultural, not affected by environment.

The climate was slowly deteriorating (or the increased number of inhabitants caused progressive desertification?) until the present day, with some oscillations toward better conditions. These are seen connected with the Middle Bronze II, the Late Bronze and the Byzantine periods, which were also times of extensive settlement in southern Israel. The drier phases separating favorable times are characterized by massive abandonments of settlements and by the northward invasions of nomad tribes, especially in the southern part of Israel.

It is quite clear that environmental conditions had always determined the boundary between the settled area, in which agriculture could be practiced and arid environments, where the only possible subsistence was by adopting the nomad way of life, such as the Bedouins. Here also comes up one of the more nagging, yet unanswerable questions, of which comes first, the desert or the nomads. It is quite well established that the nomads cause desertification, at least to a certain degree, by the grazing, wood collection and other activities (see Fig. 1.2.12 and Chapter 10.2). Could it be that it was not climatic deterioration which had caused the regression in agricultural settlements and spreading of nomads, but the nomads came first, pushed the previous settlers out and turned the land desolate?

Incidentally, the higher percentages of arboreal pollen at the Late Bronze period correspond to the time when the country was first occupied by the Israelites, for which the Land of Israel is described in the Bible as a land "of milk and honey". Unfortunately, this paradise did not last too long, either due to natural or anthropogenic causes, or a combination of both.

The other problem for which no satisfactory solution is yet available is concerned with the so termed "cultural breaks" in arid lands archeology and especially prehistory. Any attempt at reconstructing relations of human settlements with paleoenvironments through the ages has to consider the possibility that sedimentary sequences in which sites are found are usually far from being complete, continuous ones. The only possible indications for missing parts of sequences can be obtained from palynological analyses of continuous sections of the studied periods and correlating the sites with these.

Unfortunately, such continuous sequences which could be used for reference hardly ever crop out, thus usually the only possible way of acquiring them is from boreholes. Palynology, to date, seems the only useful method for correlating sites with such boreholes, giving indication of missing parts of the section (reflected in the industries or settlements as "cultural breaks") and specifying the length of such hiata.

The question whether during such dry periods settlements had existed in the region and their remains either washed away or covered by younger deposits, is still left as a matter of pure speculation. A search for such sites should however be conducted along the paleoenvironmental gradients typical for each period and region, toward the more favorable territories for any given time span.

Chapter 11

Pre-late Cenozoic palynology of arid lands

Two methods have been principally adopted for palynologically founded reconstructions of pre-late Cenozoic (or, as frequently termed, “pre-Quaternary”) arid lands paleoenvironments. One is based on species (or, in paleopalynology, rather taxa) diversity and dominance factors and gradients (Stehli, 1968a, b; Horowitz, 1972), while the other (Visscher and Van Der Zwan, 1981) on occurrences of index species considered typical for aridity. The two methods are detailed below, each with its specific limitations. Constraints common to both approaches are delineated in the following discussion.

11.1. Constraints and precautions

First and foremost, it must be established whether the reconstructions of pre-late Cenozoic environments are based on continental or marine plant microfossils, since conditions are entirely different for each, within the same region, especially in arid lands. As an example, one should only refer to the present day seas surrounding the warm regions where, regardless whether the land climate is verily tropical or extremely-arid, marine life is equally flourishing under best of conditions.

Unicellular algae such as hystrichospheres, acritarchs and others, which become an important part of pre-late Cenozoic marine palynomorph assemblages, may occur as well developed communities, rich in number of species and individuals. This would usually indicate a warm sea, but would not say anything about conditions on the adjacent land, which could maintain any environment from very humid tropical through to extremely-arid desert, as long as the seawater temperature is kept warm. Thus marine plant microfossils can be used as temperature indicators, with bearing on the neighboring land, but can hardly say much in terms of terrestrial humidity.

Terrestrial palynomorph assemblages extracted from marine sediments, particularly so in seas adjacent to arid lands, are especially prone to long distance transportation (see Chapter 2.2), which could readily produce a false picture of the land vegetation. This is indeed evident while analyzing present day pollen assemblages obtained from marine sediments in extremely-arid and arid regions (Cross et al., 1966; Horowitz, 1966; Cheddadi, 1988). An opposed view is held by Kremp (1977), who maintains that long distance transportation does not essentially intervene with the palynomorph assemblages, thus would not affect the reconstructions of regional terrestrial paleoenvironments, which he presented for the Permian and Triassic periods.

Caution should however preferably be exercised, especially when attempts at reconstructing terrestrial arid paleoenvironments are made, since composition of the relevant spectra is usually affected by transportation. This may be the reason why in the reconstructions presented by Kremp, arid environments are only vaguely determined by any typical palynomorph assemblages, while terms such as "extremely-arid" are never used. Kremp explains the lack of typical arid lands spectra by the absence of ice caps in late Permian times, which resulted in a less pronounced development of climatic belts at that time. Another explanation is here suggested, that since many analyzed samples are of marine origin, the long distance transported palynomorphs somewhat averted the spectra by inclusion of elements brought from the more humid environments. These may comprise a considerable part of the assemblages, mainly due to the low masking factor of palynomorphs produced by the relatively poor xerophytic vegetations.

There are two main differences between pre-late Cenozoic and late Cenozoic (or Quaternary) palynology. The first depends on natural constraints, the other unfortunately originates from techniques traditionally applied in the field of pre-late Cenozoic palynology.

The natural constraints and restrictions are concerned with the fact that hardly ever the regional paleoenvironments where plants had been growing are securely known. This is especially pronounced in palynology, when the palynomorphs are studied, in the majority of cases, after they had been detached from their parent plants. Thus quite frequently the exact botanical affinity of the grains is largely questionable. Identifications are usually based solely on their morphology, which gave rise to the use of terms such as "morphospecies" or "morphogenus".

The other problem pertains to the plants themselves. Even when sporomorphs are extracted from these plants, which is unfortunately quite a rare case, their known living environments are frequently useless for any regional reconstructions. This is because the only exception to the rule, that the palynomorph–plant connection is not known, usually occurs with those plants which became fossilized in their natural environments such as marshes, peat bogs, playas, springs and the like. All of these constitute biotopes which are fundamentally controlled by local hydrologic conditions, less by the regional pattern.

These plants and consequently their pollen and spores, can therefore provide useful information concerning their intimate environments, but usually no more than that. Thus while trying to reconstruct regional vegetations and paleoclimates, one is faced with the problem that palynomorphs produced by plants growing in such local environments as mentioned above become of hardly any use. This is because similar types of local environments tend to exist under a variety of regional conditions, not necessarily confined to a certain type or another. Evidently, to that is added the fact that in many cases it is unknown which of the pre-late Cenozoic palynomorphs originated from a local, or from a regional plant association. Thus the reconstruction of regional paleoenvironments from the plant remains, including palynomorphs, may quite frequently turn to be of doubtful significance, if relying exclusively on those restricted to hydrologically controlled biotopes.

An example can be brought from Gregor and Velitzelos (1987), who had studied fossil plants from spring deposits in Europe throughout the Neogene, concluding that there was no climatic change for the entire period. This conclusion is understandable when analysis is carried out on plants intimately attached to microenvironments, that would normally not follow regional climatic fluctuations, as long as water supply is available. In addition, if such a spring would cease its output, the typical vegetation would simply disappear, but without indications of changes in its composition. If it so happens that spring activity is renewed at the same locality, the vegetation would essentially continue to grow in a similar way, superimposed on the former. The break in sedimentation is usually hard to tell in such type of sediments as spring deposits, let alone define its duration.

A straightforward example can be seen in the present day, when trees such as poplars or willows are growing along flowing water courses, springs and streams, in almost any regional environment around the world, warm or cool, arid and humid, thus giving no evidence whatsoever regarding general conditions, except for the occurrence of running freshwater. Consequently, even when the connection of pre-late Cenozoic palynomorphs with their parent plants is well established, it would frequently turn of hardly any use for the reconstruction of regional conditions. This results from the fact that mostly, such connections could only be safely secured for plants living in intimate connections with water bodies.

The "human" factor in pre-late Cenozoic palynology refers to the tradition by which spectra from earlier strata are in most cases not counted. Stratigraphy is usually based on occurrence/non-occurrence factors, or on first appearances and extinctions of index taxa. In some cases authors would publish indications such as "present", "rare", "frequent" or "abundant" and so on, but hardly with specific numeric data, let alone a numerically treated pollen diagram. The reason for omission of numeric data, although regrettable, is quite understandable, since pre-late Cenozoic stratigraphy is based on evolution and extinctions, thus the appearance of new taxa usually makes a good marker, because the time spans involved are considerably longer. With late Cenozoic stratigraphy, when periods become shorter and shorter, ecostratigraphy must be exercised, which deserves numeric data.

This drawback, which dates far back into the history of palynology, is unfortunately largely practiced up to the present day by pre-late Cenozoic palynologists. This makes almost futile all attempts at precisely reconstructing regional climatic conditions, let alone global, based on the published data, which are fundamentally insufficient for such treatments. Evidently, it is strongly urged that palynologists, regardless of the age of studied samples, would routinely present numeric data of the palynomorphs assemblages, for any possible future use, as a sound basis for paleoenvironmental reconstructions, or other needs which may arise from the development of new concepts and techniques. The extra amount of work in this case is negligible, while future gains may be considerable.

Fundamentally, two ways of paleoenvironment reconstructions are practiced. The more common treatment is temporal, when sequences from a certain locality are studied and the paleoenvironmental development of a restricted area is presented, in ascending order of time or geologic periods. The second approach is spatial, when a relatively large region (sometimes even the entire Earth) is dealt with and paleoenvironments are presented for a certain slice of time. When data are sufficient, which is still a rare case at the present state of art, the two approaches may be combined to present a paleoenvironmental development of a considerable part of the globe, throughout the studied periods.

Temporal paleoenvironment reconstructions seem to be the simpler of the two ways, mainly since they are done in a geographically restricted area, which facilitates contacts between the involved palynologists. One should not necessarily agree with the principle laid down for the reconstruction, but it is at least clear what the approach is, while its consistent use for a given sequence ensures no hazardous combinations.

The situation is frequently different with spatial reconstructions, based on studies from different locations. The first possible discrepancy may involve as basic a datum as the age of studied spectra, which directly bears on the regional reconstruction. As an example, Kremp (1977) had indicated that palynostratigraphic correlations are being done across considerable distances by tying up similar palynomorph assemblages. However, if the movement of Pangea is taken into consideration (Fig. 11.1.1), spectra of similar nature may be of different ages, following migration of the vegetation belts, which tend to maintain their positions in relation to the geographic pole. Needless to say, continents do not have to be moved for such discrepancies to evolve — the changing positions of climatic belts such as so frequently happened throughout the Quaternary, with no continental drift, would be sufficient for diachronous correlations. There is indeed no reason to believe that climatic oscillations of belts had not occurred in pre-late Cenozoic times.

It is thus clear that, in order to safely present reconstructions of paleoenvironments on a larger geographical scale, it is essential to initially secure accurate ages for the spectra. These ages should preferably be also based on data which have no connection with vegetation, or with other fossils that may have been affected by possible movements of paleoclimatic belts. How to obtain such stratigraphic data is yet another problem, which seems far from an ideal solution. Incidentally, this is also the source of great debates on issues such as where to put exactly the Permian/Triassic, or other boundaries. Such debates are typical for periods when climatic belts were better developed, possibly also oscillating.

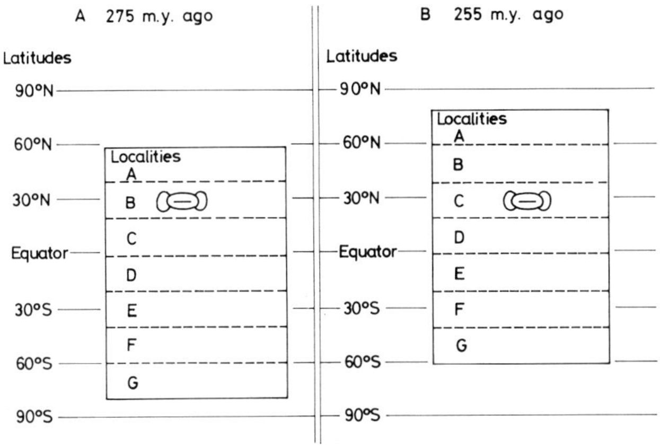

Fig. 11.1.1. Possible miscorrelations by migrating climate belts due to possible plant migration during Pangea's northward drift. It may be assumed that: A. a new plant group evolved at locality *B* (left) of the supercontinent about 275 my ago; B. during the following 20 my, Pangea drifted continuously northward while this new plant group, in order to stay in the same climate belt, had migrated to locality *C* (right). It would thus be erroneous to assume that strata from the two localities, which contain similar earliest finds of the same plant group, are of the same age. Reprinted with permission from Kremp (1977).

The other problem concerned with regional widescale reconstructions involves the subjective personal approach of researchers. Thus a certain publication can define specific palynomorph assemblages as of an arid affinity, while another, based on somewhat different concepts regarding the significance of this very spectrum, would suggest another paleoenvironment. Euramerian Permian floras and palynomorph assemblages were regarded by Hart (1965) as of a warm, dry character; by Horowitz (1972) as temperate (probably erroneously); while by Kremp (1974) as indicative of tropical paleoenvironments, grading to semi-arid to the north and south. These different views are not brought here with any preference, just to indicate the possible difficulties involved before any attempt at suggesting global paleoenvironmental reconstructions is done.

Another problem which is faced with, while trying to present global paleoclimates, involves the differentiation between warm and cold arid paleoenvironments. When the plants themselves are not known, as is the usual case with ancient geological periods, the palynomorph spectra would bear almost similar characteristics for both cold and warm dry lands. Be it the low taxa diversity, the prevalence of a certain taxon over the others, or the use of an index species, it would at any rate be very difficult to tell temperatures from the terrestrial palynomorph assemblages. Needless to say, such confusion may even occur in Quaternary palynology, such as when spectra rich in *Artemisia* typify both cold and warm steppes. In such cases, when studying continental deposits, sedimentological criteria should be consulted.

Such may be interfingering with tillites in cold deserts, or occurrences of evaporites and red beds, indicating higher temperatures.

The problem of paleotemperature is however greatly reduced when spectra are obtained from marine sediments, since the population compositions of various algae are usually significant indicators to seawater temperatures. In such cases, the rule of thumb is that well developed communities, rich in number of species and individuals, grew in tropical and subtropical warm seas; while poor speciation and prevalence of a single taxon would be typical for cold seas. Care should however be exercised, because if samples had been collected from restrictive, unfavorable warm marine environments such as hypersaline lagoons, with a heavy environmental stress, the ensuing algal spectra could resemble in character the cold seas. A sedimentological analysis would be mandatory in this case, which frequently accompanies palynological studies concerned with the Permian (Visscher, 1971), the Triassic (Visscher and Van Der Zwan, 1981), as well as other geologic periods.

11.2. Identifying paleoaridity by species diversity

11.2.1. Species diversity and dominance — the recent model

Application of the species diversity concept for reconstructing past environments had been outlined by Stehli et al. (1967) and Stehli (1968a, b; and other publications), based on a straightforward principle: the more favorable the environmental conditions are, the higher is the overall number of taxa in almost any group of plants or animals, both marine and continental. Some researchers add to the number of species another factor, which is the relative importance of each species (Prance, 1982), measured in a variety of ways by different authors. It seems that the importance factor can hardly be measured for the past, but anyway is non-applicable at all to palynology, which deals with dispersed palynomorphs. The dominance factor of certain species over the others should be added to diversity and importance, a factor which in some way may parallel the importance factor. The dominance factor increases as the environmental conditions become less and less favorable (Walton, 1964).

Attempts had also been done to use species diversity of animals for reconstructing paleoenvironments of archeologic sites (Avery, 1982; and others) in southern Africa. Diversity is applied in the same manner as by Stehli, while importance factor, as it seems, concerned the size of dominant animals, which became larger under less favorable conditions.

Initial explanation for the two combined phenomena, diversity and dominance, observed in most groups of living organisms is that, under favorable environmental conditions competition is low, thus many taxa can find sufficient subsistence to enable their existence. When environmental stresses progressively develop to become more and more restrictive, competition is becoming crucial, so that the first species within a given biotope to develop better adaptation would be the one to multiply and flourish. Quite rapidly this taxon would overrule others and become the dominant species in its environment, pushing the others into minor niches, thus causing disappearance of those many taxa that could not adapt fast enough. Taxa that disappear during this process are not necessarily such that could not live in the new environments, but seem to be less adapted, or slower to adapt, than the one that spread first at the crucial time, thus left without sufficient supply of basic needs. Figure 11.2.1 presents both species diversity and dominance factor displayed

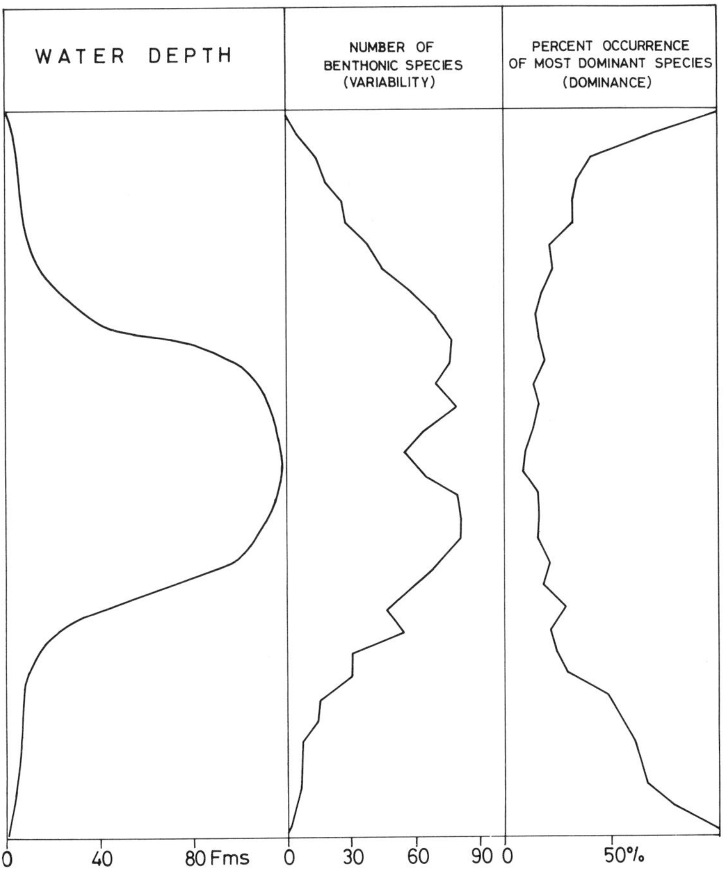

Fig. 11.2.1. Species diversity and dominance in benthonic foraminifers from the Gulf of Mexico, as a function of water depth. Optimal environments are in the range of 40–100 Fms., where the number of species is maximal and dominance minimal. Shallower environments are more restrictive, expressed by a lower number of species and dominance of a certain taxon. Drawn from data in Walton (1964).

one against the other, for recent foraminifers distribution in the Gulf of Mexico, as a function of water depth.

Another reason for the greater variety of species in favorable environments, especially the tropics, is that these seem to provide the best conditions for development of new species. This was observed in both the spatial and temporal planes, in pollen diagrams from tropical South America (Van Der Hammen, 1957).

A theory discussed by several authors in a volume edited by Prance (1982), maintains that the tropical regions became refuge areas for plants and animals whenever environmental conditions deteriorated during the Quaternary, thus "collecting" more and more species which had invaded the area at some stage or another. Not all seem to agree (Endler, 1982) with the refuge theory which, if true, could shed doubt on the validity of paleoenvironment reconstructions in periods when such climatic changes as known for the Quaternary had not been taking place.

Be the reason what it may, it seems the data indicate quite clearly that favorable environments are typified by a greater variety and number ("diversity") of species, but a lesser dominance of certain taxa over the others. This, in contrast to less favorable environments in which, as conditions become progressively adverse, the tendency toward development of oligotypic, or sometimes even monotypic communities is much stronger.

There are certain groups of plants and animals for which the favorable environmental conditions are far from the tropics in both marine and continental

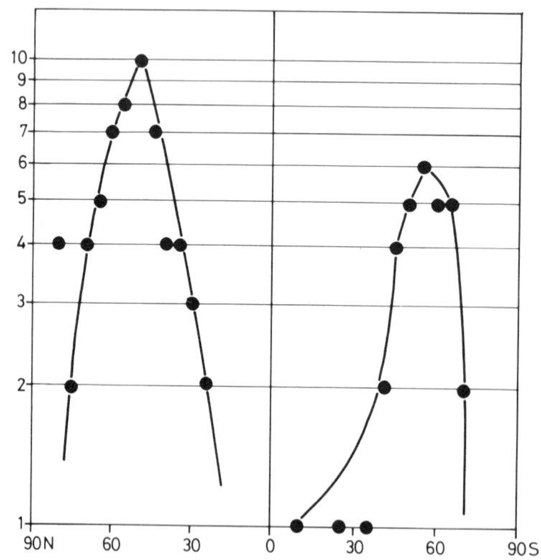

Fig. 11.2.2. Species diversity in penguins (southern hemisphere, right) and alcids (northern hemisphere, left), as a function of latitude. The two exhibit atypical distribution patterns, sometimes encountered among small groups occupying only a portion of the total ecological niche available to the larger group to which they belong. Reprinted with permission from Stehli (1968a).

environments, thus such groups flourish in biotopes considered stressed for others, both in number of species and individuals. In this way their species diversity patterns would contradict the general trend, centering around their most favorable environments. If such groups are treated by themselves, conclusions pertaining to global climatic belts may turn misleading. This is no problem with the present day or the Quaternary, when these groups are known, but may cause confusion in earlier periods. Examples for the present day are penguins in the southern hemisphere and alcids of the northern, both birds which developed highly specialized adaptations to cold environments (Fig 11.2.2); lizards, snakes, cacti, thorn trees and other plant and animal groups of the arid lands. These, or other such minor deviations from the ideal do not have a significant effect on the overall picture, when large bodies of data are considered. It is therefore suggested to refrain from species diversity analysis of a single group, but rather regard the overall data.

The picture is much clearer with marine, especially shelf and littoral environments, where ecological conditions are primarily determined by the amount of absorbed solar energy (Stehli, 1968a), which determines the surface temperature of seawater and its worldwide gradients, being high near the equator, diminishing toward the poles. Terrestrial environments, affected by numerous diverse factors, are more complicated in terms of species diversity, but in general the principle seems to remain essentially reliable.

A. Diversity and dominance in marine environments

The recent model for species diversity gradients of marine invertebrates was discussed in Stehli et al. (1967) and Stehli (1968a, b; and other publications). The model is constructed and based on benthonic invertebrate groups living in the photic zone of shallow seas, which is the marine biotope most significant for reconstructing environments of earlier periods, due to the fact that most marine rocks exposed for these ages had been deposited on shelves of epicontinental seas.

The studies of Stehli and his collaborators centered around attempts at reconstructing the paleomagnetic pole of Permian times, based on the assumption that there is a connection between location of the pole and pattern of climatic belts on Earth, a connection clearly shown for the present day. This may also serve as a sound foundation for similar reconstructions of land environments, relying on the same principle of species diversity.

There is however an essential difference between climatic belts as expressed in marine or shelf environments and their counterparts on land. Marine environments are dependent on a smaller number of physical climatic factors than terrestrial, thus are much more homogeneous. Marine habitats are therefore quite uniform in terms of environmental conditions, thus more regular and gentle with regard to ecological gradients, than the terrestrial. Some coastal features, such as lagoons or enclosed bays of restricted circulation with the open sea, may occasionally present locally stressed environments as compared with the open marine biotopes. These are however quite subordinate in areal distribution and can easily be identified by their peculiar deposits. It is however noteworthy that such coastal features are widespread in arid regions, which calls for further attention when marine plant microfossils are concerned.

The main factor largely controlling the type of marine environments anywhere on Earth is the water temperature which, besides certain parts of the sea affected by either cold or warm currents, depends on solar energy absorption by the sea. Since the albedo of seawater does not change much all over the globe, the average seawater temperature is a function of the length of time that any given part of the sea is exposed to solar insolation and also, the angle by which the sun rays hit the area. In other words, there is a direct dependence of seawater temperature on geographic latitude.

By analyzing the species diversity in certain recent benthonic groups Stehli (1968b) demonstrated (Fig. 11.2.3) that almost invariably, the highest number of species is centered around the equator, diminishing both northward and southward. Stehli had applied this principle to Permian brachiopod species, to reconstruct climatic belts of this period, as indicated by species diversity gradients of these fossils.

This basic principle could readily be applied to palynomorphs of marine origin, such as hystrichospheres, acritarchs and others, for reconstructing paleolatitudinal belts at any given period. Naturally, since seawater temperatures do not have direct connections with rainfall regime on adjacent continents, species diversity of marine plant microfossils could only be used to determine seawater temperatures, certainly not the degree of aridity on land.

This however may turn helpful while distinction should be made between a cold and warm arid land, adjacent to a locality where marine samples are collected. While such samples are palynologically analyzed it is habitual that both marine and terrestrial palynomorphs are studied. In this way, a combination of high species diversity and non-dominance of marine components, together with a low number of species of terrestrial with very few taxa prevailing, would indicate a warm arid land. The combination of poor species diversity and dominance of certain components in both marine and continental fractions, would point to a cold aridity of the adjacent land (unless the marine environment was restrictive, which should be ascertained from sedimentology of the analyzed samples).

B. Diversity and dominance in terrestrial environments

Terrestrial environments are controlled and affected by many more factors as compared with marine, thus much more varied in the number and extent of different ecologic biotopes and niches, available for both plants and animals. Factors such as winds and their different velocities, extreme temperatures and their much steeper gradients, or different substrates, have pronounced effects on terrestrial environments, determining the possibilities of life in each biotope.

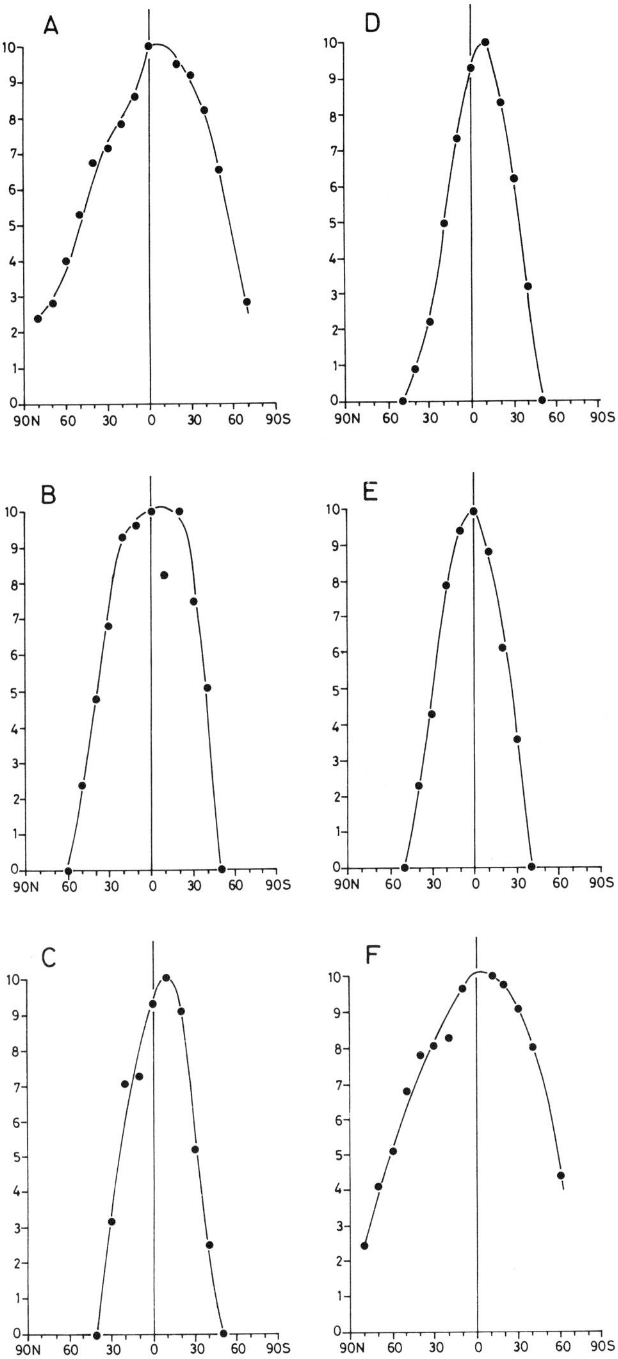

Fig. 11.2.3. Taxonomic diversity plots against present day latitude for recent benthonic marine invertebrates. A. Pelecypods genera. B. Species of Cypraeid gastropods. C. Species of *Strombus* (Gastropoda). D. Species of the family Pinnidae (Pelecypoda). E. Families of Pelecypoda. F. Genera of hermatypic corals. Reprinted with permission from Stehli (1968b).

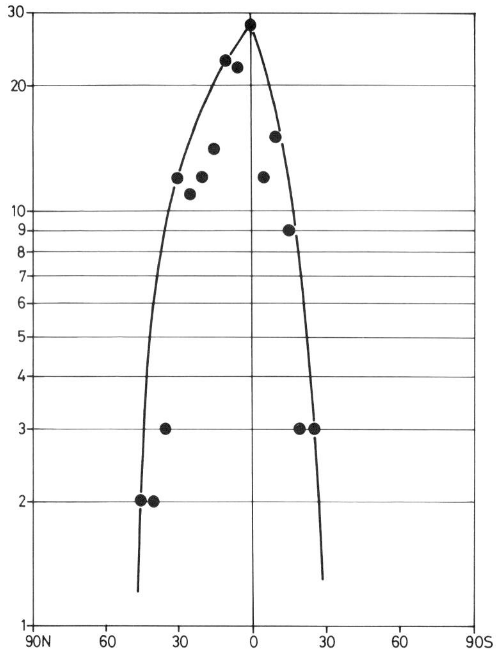

Fig. 11.2.4. Genetic diversity of palms, a leptokurtic distribution pattern found in many tropical–subtropical forms. Reprinted with permission from Stehli (1968a).

Consequently, types of biotopes are varied and numerous on land, even within the limits of a specific regional environment, each supporting different assemblages of species. Thus the diversity becomes greater and much more complicated in comparison with the rather simple pattern shown for the marine fauna and flora. It is however possible to observe trends, similar to the marine record, also in various groups of terrestrial plants (Fig. 11.2.4).

It would therefore be considerably more difficult to define climatic belts from continental data of species diversity, thus it seems more appropriate to refer to terrestrial environments in terms of a biotope, or complex of biotopes, being either favorable or less so. A look at the global species diversity distribution for plants in general, would immediately display the wealth in number of species of the tropical regions (Fig. 11.2.5) in comparison with other environments such as deserts and tundras, where the number of plant species is minimal. Thus a general outline, which would ignore local, restricted occurrences that may change the total number of species, would display highest species diversity in the tropical domain; sharp decrease in planetary deserts; rising again in Mediterranean subtropical environments; diminishing in the temperate; and falling back to minimum values closer to the cold polar regions (Fig. 11.2.5).

This ideal curve may be used as a fundamental guideline for reconstructing past

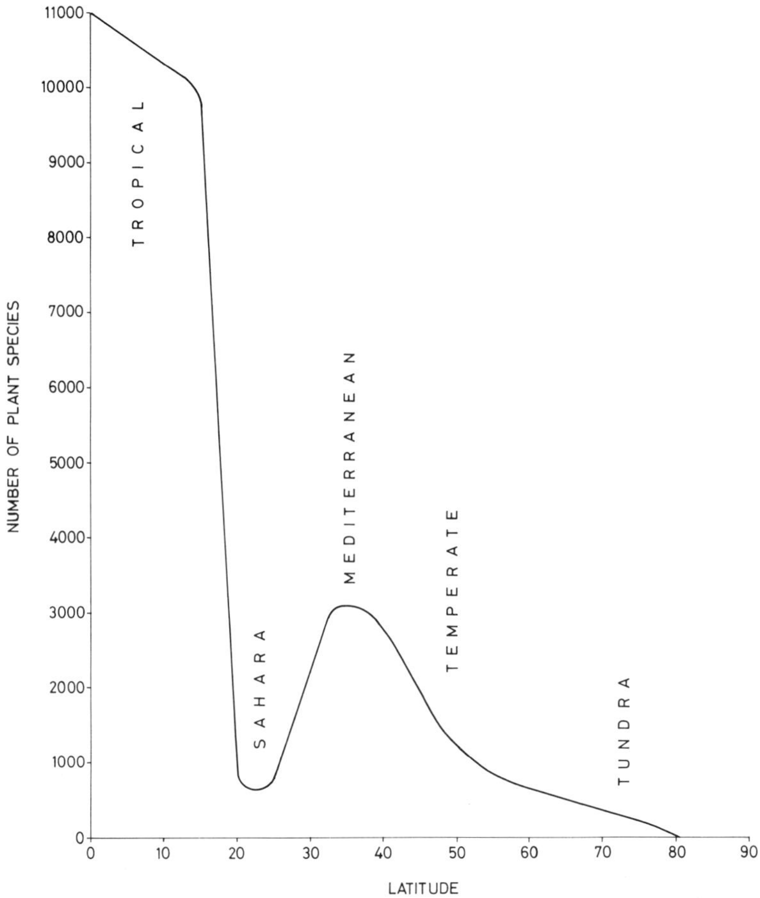

Fig. 11.2.5. Species diversity in terrestrial plants, from the equator to the north pole. From various sources.

environments by their species diversity. It should be taken into consideration that factors such as the variability of substrates in a certain arid environment would immediately cause an increase in the number of species represented, as can be seen in deserts where both sand dunes and rocky soils occur, which almost doubles the number of species, due to the fact that each substrate maintains its own typical vegetation.

Another factor which should be considered is whether a desert is geologically young, or had a long history of aridity. The latter category also includes these arid lands which for a long time oscillated in position, but had always have access to a permanent stock of xerophytic vegetation, which had time enough for the development of adaptations to the dry conditions. Such are the arid regions of Southwestern USA and northern Mexico, which show a much higher species diversification as compared with younger deserts, or those latitudinal dry lands which had been cut from any stock of xerophytic flora during humid phases. Thus, the Sonoran desert in northern Mexico exhibits species diversity which is much higher (Pianka, 1979) as compared with the Sahara.

The relief has also a pronounced influence on the number of species which could survive in a certain arid land, since altitudinal vegetation belts develop and each of these maintains different plants (Spaulding, 1990), thus readily adding to the species diversity. This, incidentally is yet another cause for the greater variety of plant species in Sonora (Pianka, 1979).

Arid regions maintain quite frequently microenvironments used as refuge niches by both plants and animals, usually living in different environments, which had at some former stage dominated the now desolate land. Such are many Mediterranean plants presently growing in the Negev and Sinai deserts (Danin, 1972), where this type of environment is known for the late glacial times; or tropical plants attached to springs at the vicinity of the now extremely-arid zone surrounding the Dead Sea (Zohary, 1959). These may also eventually add to the species diversity.

It is however quite clear, that even when all these and probably some other additions are taken into consideration, the total number of species which can grow in a desert, either cold or warm, is very limited in comparison with more favorable environments. The Sahara, with its 800–900 species altogether, spread over an area of some 12–13 million square kilometers, seems quite a convincing example, as compared with adjacent Mediterranean Israel with more than 2,000 plant species, growing within its mere 20,000 square kilometers.

Since all these components are averaged in the pollen spectra, in addition to contributions made by eolian and fluviatile long distance transportation, the ensuing species diversity of an arid land could be considerably higher than primarily predicted by its environment. This poses no problems with the present day model, or in most Quaternary paleoenvironments, because the allochthonous contributions are usually easily recognized as such; it may however cause considerable confusions with older periods, where such clear distinctions can hardly be made.

An additional variable which should be attended to concerns the quantitative dominance of a certain taxon or a small number of taxa over all others, which is apparent in most environmentally stressed biotopes. When both the species diversity and dominance factors are considered together, it seems possible to use these criteria for the reconstruction and definition of pre-late Cenozoic paleoenvironments. It would be useful, in such cases, to have at hand as many palynomorph spectra as possible, preferably collected from different contemporaneous depositional environments.

It is clear from the discussion brought above, concerning both species diversity and dominance factors, that the only way paleoenvironments could be reconstructed with some degree of credibility, when these principles are applied, is by comparisons with other, different environments. These could be either of the same period, carried out over long horizontal traverses, or along time in a given, long enough sequence, where comparisons could be made with either underlying or overlying spectra (or preferably, both). In this way spatial reconstructions can be presented as a paleoenvironmental map, while temporal analysis would display paleoenvironmental evolution at a given locality.

11.2.2. Species–taxon relationships in paleopalynology

One of the principal problems with which palynologists studying pre-late Cenozoic spectra are faced is the definition of "taxon" ("morphospecies", "morphogenus", etc.) in relation to the actual biological definition of "species", as applied to living plants. This problem becomes crucial with paleopalynology, since in times preceding the late Cenozoic almost all plant microfossils are determined solely by their morphology, usually even in such cases when their connection with parent plants is secured.

The concepts on which this technique is based maintain, along with other claims, that even if a palynomorph of a certain shape is physically derived from a fossil plant, it is not guaranteed that this had been the only plant producing such forms. Thus the use of fossil plant name for its own palynomorph may be misleading and could especially hamper paleovegetation reconstructions. Indeed, such similarities between pollen produced by altogether different plants is well known for the present day. A knowledge of the possibilities for resemblance, which can be acquired for palynomorphs of living plants, should be sufficient for preventing possible mistakes. Since palynomorph-plant ties are known for only a small number of pre-late Cenozoic taxa, this precaution cannot be overlooked.

The fact that most pre-late Cenozoic palynological studies are based on dispersed sporomorphs makes this branch of paleontology different than most others, which actually deal with large parts the organisms themselves. This fundamental difference has to be considered whenever any conclusions are to be based on palynomorphs as representatives of plants, such as vegetation, environment or climate reconstructions. This limitation does not apply to "pure" palynostratigraphy, for which palynomorph shapes are a sufficient basis, which may explain the widespread successful use of palynology in the oil industry.

Identification by morphology alone presents two main problems, one evolves from natural constraints, while the second results from subjective attitude taken by various researchers toward the definition and use of morphogenera and species. Natural constraints involve the fact that although palynomorphs differ from plant to plant, such differences are not always visible or identifiable, even more so with the fossil material. The most common present day example of this kind of drawback is probably portrayed by Gramineae, a family rich in number of genera and species, of a cosmopolitan distribution. Taking into account even a very small area such as the 20,000 square kilometers of Israel, the number of Gramineae genera (Zohary, 1980) exceeds 100, while the number of species reaches 240. All of these are regarded by most palynologists as a single taxon, mainly for objective lack of possibility for further identification. Incidentally, only a very restricted part of the Gramineae species grow in the arid parts of Israel, while the majority occupies the Mediterranean environments. There is of course no possibility whatsoever to discuss species diversity among the Gramineae, when only their pollen are available.

A similar situation can also be seen with the Chenopodiaceae in Israel, with 23 genera and 71 species, again usually treated as a single taxon, to which Amaranthaceae are also added, with additional species. The situation with Compositae is not much different. These groups are brought as an example, for they make a considerable part of pollen spectra typical for the present day arid regions.

Much as we hate to think of this possibility, which is frequently simply wiped under the carpet, ignored and hardly ever discussed, there seems no reason to assume that such situation could not have existed also during the remote past. In this way, while pre-late Cenozoic palynology is considered, it is evident that a single morphogenus or morphospecies could in fact represent many plant genera or species. A single morphospecies may even represent an entire plant family. Incidentally, it seems that the better candidates for such suspicions are those common palynomorphs of worldwide distribution, such as the Mesozoic *Cyathidites australis*, the late Paleozoic–early Mesozoic *Platysaccus queenslandi* or *Vittatina costabilis*, the late Permian *Aculeisporites variabilis* and many others (Fig. 11.2.6).

When the present day model of species diversity is concerned, the wider margins of pollen taxa do not pose any problems in interpretation of stressed environments. On the contrary, since most arid regions plants pollen cannot be determined further than the family or genus level, the ensuing number of taxa is reduced for the dry lands, which helps in accentuating the "species" diversity analysis. On the other hand, both temperate and tropical vegetations are represented in the recent pollen spectra by a considerably greater variety of taxa, partly thanks to better resolution in identification, due to the greater morphological variability of the concerned grains. Consequently, pollen spectra from the present day tropical regions show great "species" diversity, the temperate display medium values, while the arid, both hot and cold, are typified by low to very low diversity.

The applicability of species diversity analysis based on dispersed palynomorphs, in those periods when their ties with the plants are far from being satisfactorily secured, calls therefore for an additional assumption, namely that the present day situation, in terms of morphological variability of the palynomorphs, had also existed in the past. This assumption places some limitations on reconstructions, that however still seem valid. It should be remembered that this last assumption does not have at present any broad factual basis, thus it cannot be guaranteed that indeed the situation was not, at some time, inverted. Imagine tropical vegetations

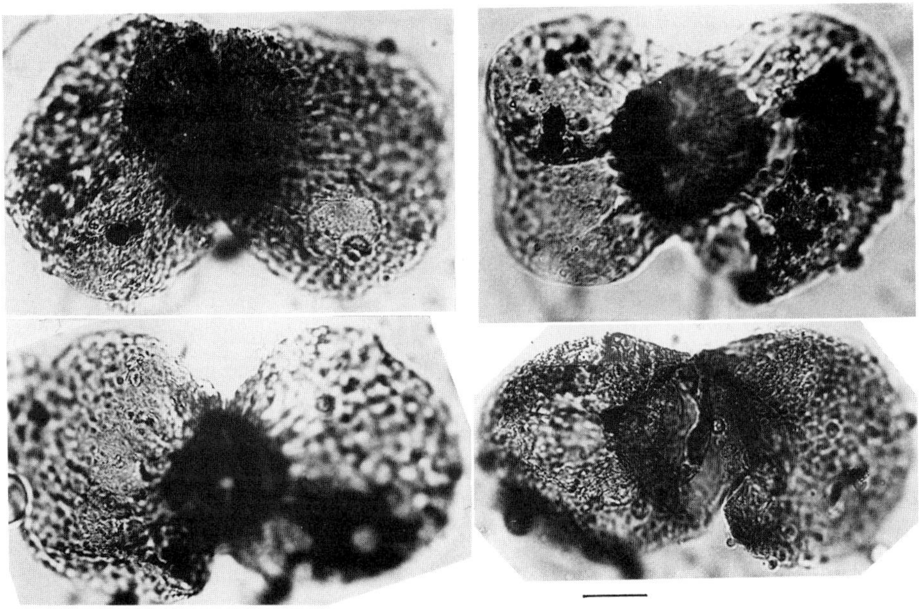

Fig. 11.2.6. *Platysaccus queenslandi*, a Triassic morphospecies which may represent several plant taxa. Zuq Tamrur 2 borehole, near the Dead Sea, Israel, core 1, 1,850 m. Courtesy of Y. Eshet. Scale bar: 10 μm. For location see Fig. 11.3.2.

rich in number of plant species, but belonging to only a few families, where palynologic distinction is quite difficult within the family level; or an arid land vegetation comprising plants from many different families which can easily be identified by their pollen.

The only way presently available for clearing up this situation is to add other, independent criteria for reconstruction of paleoenvironments, in conjunction with the palynological study. Such can be paleogeomorphology, lithology, isotope geochemistry, occurrences of particular rock types such as evaporites, tillites and so on. Cross examinations, wherever carried out, seem to have supported the validity of the above mentioned assumption (Visscher and Van Der Zwan, 1981; Eshet, 1987), so it seems that the use of species diversity methods may be justified for reconstructing paleoenvironments.

A self evident outcome of the above discussion is that the term "species" should not actually be used, but rather replaced by "taxon" in palynological analyses of diversity. Incidentally, this applies not only to paleopalynology, but even through to the present day.

Another factor, which has a considerable effect on the possibility of taxa diversity analysis, is the subjective approach taken by almost any paleopalynologist while the definition of a taxon is concerned. Or, in other words, how much different is different enough for differentiation between two taxa? Certain morphological features of a given palynomorph, in comparison to another, may be considered by one researcher sufficient for attaching different names for the two, while another would regard the two as a single systematic entity, with only some minor variations.

Examples from the recent pollen are numerous, like some Chenopodiaceae which show a wider variation in the number of pores within a certain plant species, than within close genera. From paleopalynology, the variety of morphospecies attributed to *Vittatina* by certain researchers, as opposed to others considering almost all variations as a single systematic treat, is only one of many examples. In such cases it would probably be more appropriate, for the reconstruction of past environments based on taxa diversity, to rely on morphogenera rather than morphospecies. Although some disagreement is also known on that systematic level between various authors, it probably is much less than for morphospecies, where many contradicting views are frequently published.

When comparing the number of taxa from one region to another, or along a vertical sequence, in studies made by different authors, an initial examination of the individual approaches to systematics seems to be fundamental for further conclusions. A preference for detailed systematic subdivision by one student, as opposed to a general approach by another, could well lead to false conclusions regarding paleoenvironments. Thus exclusive use of morphogenera for taxa diversity analysis may provide a better balanced body of basic data.

This problem is evidently more pertinent to regional reconstructions, where almost invariably one has to rely on studies by various authors, each of whom may had adopted a different attitude. It is of less significance when reconstructions are made along sequences in restricted areas, where usually sections are analyzed by a single palynologist or team, thus approaches are consistent.

11.2.3. Paleoaridity reconstructions by taxa diversity and dominance gradients

Arid environments predating the late Cenozoic seem to be quite rare in the geological history of the Earth, as compared with the Neogene and Quaternary. A variety of explanations was suggested for this phenomenon, such as the connection of planetary deserts with formation and uplift of orogenic mountain ranges, which induced limitations on the global atmospheric circulation, thus creating climatic belts (Horowitz, 1977b). A refined version (Ruddiman and Kutzbach, 1991) blames the uplift of large continental plateaus such as the Tibetan and the North American, for the creation of large scale aridity, by blocking possible paths of humid air to the dry regions. Still others maintain that aridity is a function of continental drift which, for these periods when dry regions existed, had placed the continents in an appropriate position on Earth for the development of these environments (Kremp, 1974, 1977).

There seems to be an intimate chronologic connection between the occurrence of glacial deposits and large scale aridity, for both the late Paleozoic and late Cenozoic periods. Strong indications are also for such connections in the late Precambrian and the Siluro-Devonian. All these periods are also characterized by stronger than usual orogenic activities.

The closest large scale aridity to the late Cenozoic, in terms of time, is evident for the Permian and to some extent the Triassic periods, by the considerable geographical extent of evaporites and red beds (Robinson, 1973; Kremp, 1977; Pearson, 1978). Explanations for the late Paleozoic aridity mostly concentrate on the possible occurrence of climatic belts in the modern style (Kremp, 1977; Horowitz, 1977b); while for the Triassic, when the global climate seems to have been more uniform, without pronounced climatic belts, aridity is explained as a result of the large size of the Triassic continent, which had caused dry air circulation in the equatorial regions (Robinson, 1973).

Bearing this in mind, I had initially planned to present a global reconstruction of the late Paleozoic and Triassic arid lands, based on palynological evidence. Instead, I can only come up with an apology: although Permo-Triassic palynology is well known from many localities around the globe, with a considerable body of literature published on the topic, hardly any of these publications presents quantitative data, which would permit paleoenvironmental reconstructions based on taxa diversity and dominance gradients. When "quantitative" data are at all present, they in most cases constitute semi-quantitative representations such the use of circles in varying diameters to represent abundance ranges, or the use of terms such as "abundant" and so on, with relatively wide numeric margins (10–40%, for example). Evidently, no sound reconstructions could be based on such data.

Consequently, only some spot reconstructions are shown here for several chosen localities, just to indicate the possible validity of the method. A more detailed analysis of the Triassic aridity in Israel is presented in Chapter 11.3, where the use of both index fossils, combined with taxa diversity and dominance, is applied for reconstructing the paleoenvironments.

The global distribution of Permian plant remains, both megafossils and microfossils, had suggested that they grew in four principal floristic provinces: the Euramerian (or Euramerican, a term preferred in North America), Angaran, Gondwanian and Cathaysian. The worldwide pattern of plant megafossils was outlined by Gothan and Weyland (1954), by Horowitz (1972) and by Kremp (1977), among others. The latter two also added data from the palynological literature (Fig 11.2.7).

The late Permian Euramerian microflora is characterized by high shares of striatiti palynomorphs, of which the most abundant are *Lueckisporites* and *Protohaploxypinus*. The first is the most typical microfossil of this floristic province. Trilete and other pteridophyte sporomorphs are rather poorly represented, by a relatively small number of taxa with low to very low shares. The diversity and total

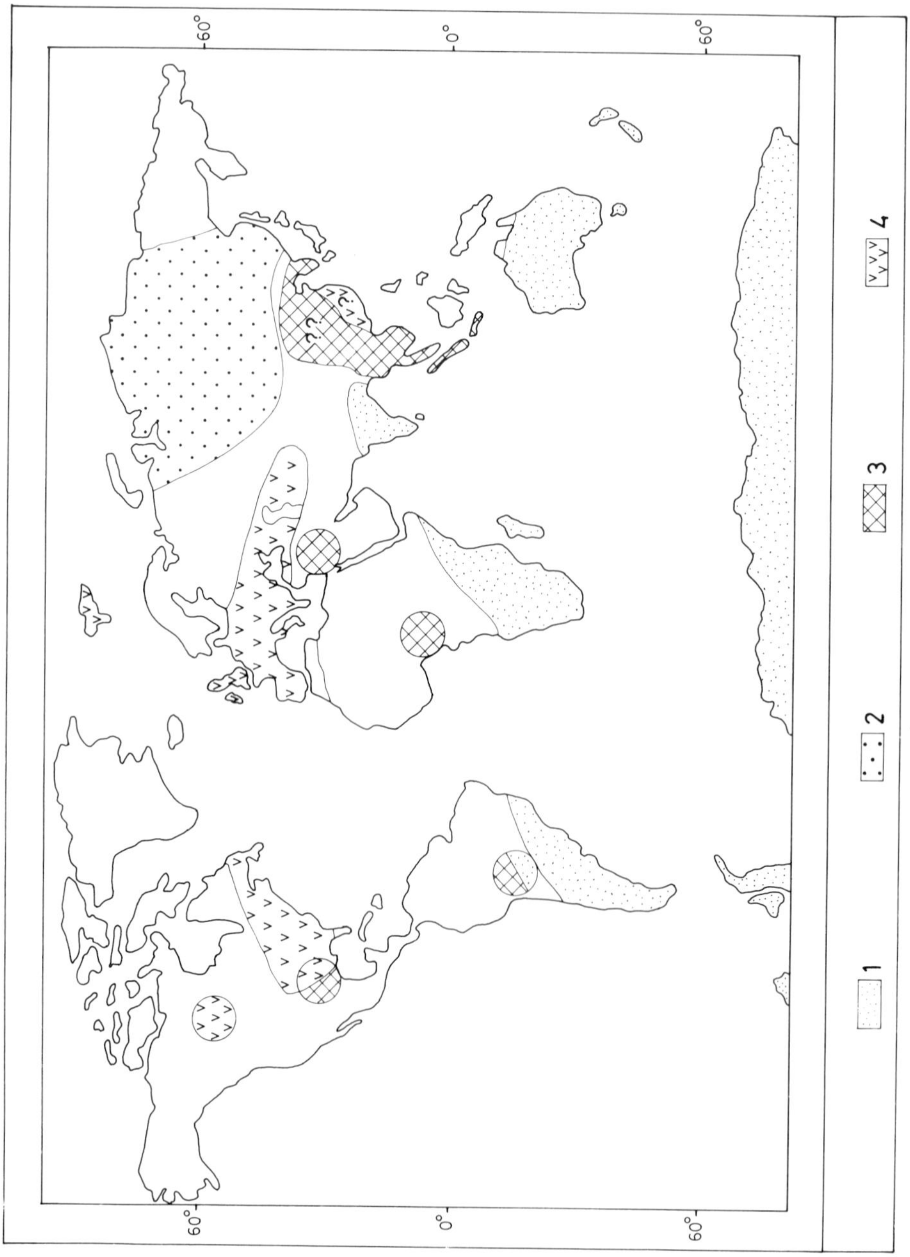

Fig. 11.2.7. Global distribution of late Permian floras and microfloras. *1* = Gondwanian. *2* = Angaran. *3* = Cathaysian. *4* = Euramerian. Data from Hart (1965, 1970), Horowitz (1972), Kremp (1977).

number of taxa is limited. This assemblage occurs presently in Europe and the northern parts of North America. Hart (1965, 1970) assumed that this assemblage represents a warm, dry climate, which is also supported by the concepts brought above for taxa diversity and dominance.

The Angaran microflora is typified by monocolpate sporomorphs and monosaccate palynomorphs, somewhat richer than the Euramerian in trilete spores. It is presently known from regions east of the Ural mountains — Siberia and northeast Asia. The suggested environment for development of this association is thought to had been temperate to subtropical.

The Gondwanian microflora is quite similar to the Euramerian in its general composition, although dominated by different morphogenera. Some possible similarities with the Angaran character had also been pointed out. The most common are various forms of striatiti, accompanied by relatively high shares of monosaccate palynomorphs. Trilete spores are not too abundant in the assemblages, usually constituting up to some 20% of the spectra. This microflora is known from all the continents thought to comprise Gondwana–Africa south of the equator, India, Australia, central and southern South America and Antarctica. It most probably represents a temperate, grading to subarctic environments. Attempts had been done to subdivide this extensive floral province into subprovinces.

The Cathaysian microflora, for which an example is discussed from the late Permian of Israel (Horowitz, 1972, 1973b), encompasses great abundance and diversity of various trilete sporomorphs, relatively limited shares but a variety of taxa of saccate palynomorphs, with occurrences of relics from earlier, Carboniferous and early Permian times. Of the 65 sporomorph taxa known from Israel, *Aculeisporites variabilis* palynodem is the more common, attaining shares of up to 60% in some of the samples (Horowitz, 1974b). This dominance seems to be a local phenomenon connected with vegetation of the depositional basin, as synchronous sediments from other localities in Israel show considerably lower shares (Eshet, 1983). It may also be quite safely assumed that the taxon was produced by at least several plant species or genera, as expressed in its great variability of shapes (hence the morphospecies name *variabilis*).

Aculeisporites variabilis is known in small quantities from Carboniferous sediments (or early Permian, according to Eshet, 1983) in Israel (Horowitz, 1974b) and other localities, also continuing into the Triassic. Most of the saccate forms of the Israeli late Permian, which occur in small percentages, are of Euramerian affinity, such as *Klausipollenites schaubergeri*, *Sulcatisporites kraeuseli*, *Falcisporites zapfei* and others. Several palynomorphs which occasionally occur in small numbers, such as *Nuskoisporites gondwanensis*, indicate remote connections with the Gondwanian province. These have most probably travelled to the region by winds or rivers, over considerable distances. The relatively warm seawater temperatures during the late Permian in Israel are evidenced by rich and varied acritarchs populations of this time (Horowitz, 1974c).

Cathaysian plant micro- and megafossils are known from China, while palynological assemblages were reported from the Middle East and central Africa. "Mixed" Euramerian-Cathaysian palynomorph assemblages are known from Oklahoma, USA, (Clapham, 1970), while Gondwanian-Cathaysian are recorded from Bolivia, South America (Cousminer, 1965). The Cathaysian assemblages most probably represent tropical environments, concluded from composition of the flora and abundance of pteridophytes, with its great variety of taxa and usually low dominance factors. This assumption is further supported by the occurrences of relics from former times, which probably found refuge in the favorable tropical conditions, that helped to postpone their extinction in these parts of the world.

Assuming that climatic belts of late Permian times had observed rules approximately similar to the present day ones, one would look for planetary widescale arid environments on both sides of the occurrences of tropical, Cathaysian paleoenvironments. The immediate inclination would be to look both north and south, if such could be safely located for that time. The problem in locating such paleogeographic positions is concerned with different views on how should the late Permian world look like, views which considerably differ among authors. It is not for us to judge in the present context, although a thorough reconstruction of paleoenvironments could have been of great help in this issue. However, as mentioned above, regrettably the palynologic data are not yet sufficient.

The Euramerian province lies to the present day north of the Cathaysian, previously considered arid or semi-arid by many authors. It seems therefore the first candidate for applying the taxa diversity and dominance gradients to analyze the validity of previous paleoenvironmental assignments in the light of these methods, validity which is also largely supported by the common types of rocks deposited during the late Permian in these regions (Bambach et al., 1980).

There is however a problem in applying taxa diversity criteria for the Euramerian (and, incidentally also the Gondwanian) province, which results from the spreading pattern of the present day western civilization. Both these floristic provinces are now located in places where numerous palynological laboratories are centered: Europe and North America on the one hand, South Africa, Australia and India on the other. Consequently, the amount of palynological information available for the Euramerian and Gondwanian provinces greatly exceeds, probably by several orders of magnitude, the data bank for both Cathaysian and Angaran microfloras.

A typical outcome of this situation is found in Traverse (1988), who observed (p. 203): "Late Permian, from the Kungurian stage on, is palynologically a time of saccates: monosaccates or bisaccates, and especially of striate bisaccates, surely bespeaking conifer or conifer-like gymnosperm dominance (plus glossopterids in Gondwanaland)".

The implications for palynostratigraphy are, since most sequences are dated and correlated by their saccate components, that correlations over large distances could easily cross time lines and be diachronous, in fact only following the location of climatic belts. This problem was pointed out by Kremp (1977; and see also Chapter 11.1 and Fig. 11.1.1), but unfortunately such correlation methods are still in wide use (see Eshet, 1983; Ecke, 1986; and others).

Thus, the number of taxa recorded throughout the years of research by various students of the Euramerian province became quite considerable (which applies as well to the Gondwanian) and, added to that, are various opinions which result in a seemingly even larger number of taxa. This stands in some contrast to the poorly studied Angaran and Cathaysian microfloras. Since the taxa diversity gradients used for reconstructions of past environments are based on comparisons of assemblages, this artifactual point should be remembered.

As can be seen from Fig. 11.2.7, regions south of the late Permian equator where arid environments could be looked for, are situated in countries in which very little palynological research of this period was carried out, for several reasons. Many such areas, as large parts of the Sahara and the northeastern part of South America are eroded down to the Precambrian; in others, Permian sediments are deeply buried, or when exposed, highly oxidized. This leaves only borehole material for analysis, which I am sure took place in oil companies, only they usually do not publish their information. Yet other localities may be presently covered by sea. What remains in the available literature is very little; a locality in southeastern China, where arid conditions may have prevailed in late Permian times, is discussed below.

Visscher (1971), while analyzing the late Permian palynomorph assemblages of the Kingscourt Outlier in Ireland, found out that the number of taxa was rather restricted, with an abundance of *Lueckisporites virkkiae*. This taxon is abundant in most late Permian deposits of Europe and North America, almost always displaying a clear dominance over the rest. His results (although numeric data are not given) can be summarized as follows, for 14 late Permian assemblages:

"*Cycadopites*" sp.	present in 5
"*Striatisaccus*" spp.	present in 3, rare in 10, frequent in 1
"*Vittatina*" spp.	present in 6, rare in 1
Klausipollenites schaubergeri	present in 3, common in 7, abundant in 4
Stellapollenites sp.	present in 3
Guttulapollenites sp.	present in 8, rare in 1, frequent in 1
Lueckisporites virkkiae	present in 3, rare in 1, frequent in 2, common in 1, abundant in 7
Jugasporites delasaucei	present in 3, common in 11
"*Limitisporites*" *moersensis*	present in 3, common in 9, abundant in 2
"*Perisaccus granulatus*"	present in 2, rare in 10
Nuskoisporites dulhuntyi	present in 2, rare in 3
"*Rhizomaspora*" sp.	present in 4

Present = percentage not known; rare = 1–5%; frequent = 6–15%; common = 16–25%; abundant = >25%; not mentioned when absent.

These assemblages clearly represent an arid environment, by the dominance of one or two taxa and the poor diversity, for the late Permian of this region.

Ecke (1986) studied palynomorphs from the late Permian Zechstein evaporites in Germany and recorded 35 taxa in 14 horizons. Of these taxa, seven only rarely occur in a single horizon, others rarely occur in several horizons, while only eight taxa are recorded continuously along the sequence, with a marked prevalence (although no numeric data are presented) of *Lueckisporites virkkiae* and *Klausipollenites schaubergeri*. This assemblage again marks aridity in late Permian times for Germany, by the poor taxa diversity and clear dominance of certain types.

Doubinger et al. (1990) described assemblages of 27 taxa from the late Permian and Triassic of the southeastern Iberian ranges, Spain. The late Permian assemblages are poorer in number of taxa as compared with the overlying Triassic, but not markedly so, again dominated by the saccate *Lueckisporites virkkiae* and occasionally by *Nuskoisporites* and *Paravesicaspora* (Fig. 11.2.8). This dominance, accompanied by poor taxa diversity, seem to be indicative of the late Permian Zechstein arid or semi-arid environments.

The relatively poor taxa diversity and dominance of *Camerosporites secatus* and occasionally *Ovalipollis ovalis* and *Triadispora aurea*, seem to indicate that aridity, although maybe to a lesser extent than during the late Permian, had also continued into the Triassic in this part of the world. This conclusion is also suggested in a study by Visscher and Van Der Zwan (1981).

The indication of aridity in Meishan, near Shanghai, southeastern China, in a study by Ouyang and Utting (1990) seems quite striking, when comparing the late Permian with the overlying early Triassic assemblages. Although the authors present the relative shares of terrestrial and marine plant microfossils as parts of the total spectra and not separately for each group, while numeric data which would permit recalculations are omitted, it is evident that the number of late Permian terrestrial palynomorphs is very much lower than the early Triassic (Fig. 11.2.9) ones. This seems to indicate that the rather restrictive, most probably arid environment of the late Permian had considerably ameliorated through the transition to Triassic times, to become almost tropical in nature, displayed by the considerably higher taxa diversity in the latter times.

Among the late Permian terrestrial assemblages the gymnosperm, saccate palynomorphs show marked dominance over others. Some trilete sporomorphs which do occur, such as *Cyclogranisporites* and *Azonotriletes*, may have derived from coastal or marsh vegetation. The warm arid environment postulation is further supported for the late Permian by the great wealth, diversity and abundance of marine plant microfossils, especially Leiosphaeridia and acritarchs in the spectra, testifying to a warm sea in this region, wherever it was located at that time. The decrease in marine microfossils following transition to the Triassic was most probably caused by a change in the local environment, as also evidenced by a marked change in lithology from late Permian limestone to Triassic shales and

LATE PERMIAN, CUENCA, SPAIN					
				•	*Converrucosisporites eggeri*
		•		•	*Coridaccites alutas*
	•				*Corisaccites* sp.
		•			*Cycadopytes* sp.
				•	*Densoisporites nejburgii*
		•			*Endosporites* sp.
	•				*Falcisporites nuthualensis*
•		•		•	*Falcisporites zapfei*
	•				*Falcisporites* sp.
		•			*Fimbriaesporites fimbriatus*
				•	*Gardenasporites oberrauchii*
	•			•	*Jugasporites perspicuus*
•	•	•	•	•	*Klausipollenites schaubergeri*
		•			*Limitisporites* sp.
				•	*Lueckisporites microgranulatus*
•	•	•	★	•	*Lueckisporites virkkiae*
				•	*Lundbladispora* sp.
★		★	•	•	*Nuskoisporites dulhuntyi*
	•		•		*Nuskoisporites* sp.
			•	★	*Paravesicaspora splendens*
	•			•	*Platysaccus papilionis*
•	•				*Playfordiaspora crenulata*
•					*Protohaploxypinus microcarpus*
				•	*Protohaploxypinus sevardi*
		•			*Protohaploxypinus* sp.
			•		*Triadispora staplinii*
•					*Trizonaesporites grandis*
		•	•	•	Indetermined Bisaccates

• < 15% ★ 15 - 30%

Fig. 11.2.8. The relative poverty of late Permian taxa, with dominance of only three saccate palynomorphs, in sediments from the southeastern Iberian ranges, near Cuenca, Spain. Reprinted with permission from Doubinger et al. (1990).

argillaceous limestones (Ouyang and Utting, 1990, Fig. 3). This change does not seem to have any connection with climate, as opposed to the above discussed changes of the terrestrial vegetation.

It is also quite interesting to compare the late Permian assemblages from Meishan, considered here of arid character, with other late Permian palynomorph spectra from China. This comparison is further facilitated because these other assemblages are also analyzed by Ouyang (1982; and other publications), so that comparisons are quite free from the effect of different subjective approaches. In eastern Yunan, southern China, the late Permian assemblages are of a typical Cathaysian nature, with some 160 morphospecies, belonging to 73 genera; pteridophyte spores predominate, while saccate gymnosperm palynomorphs comprise only

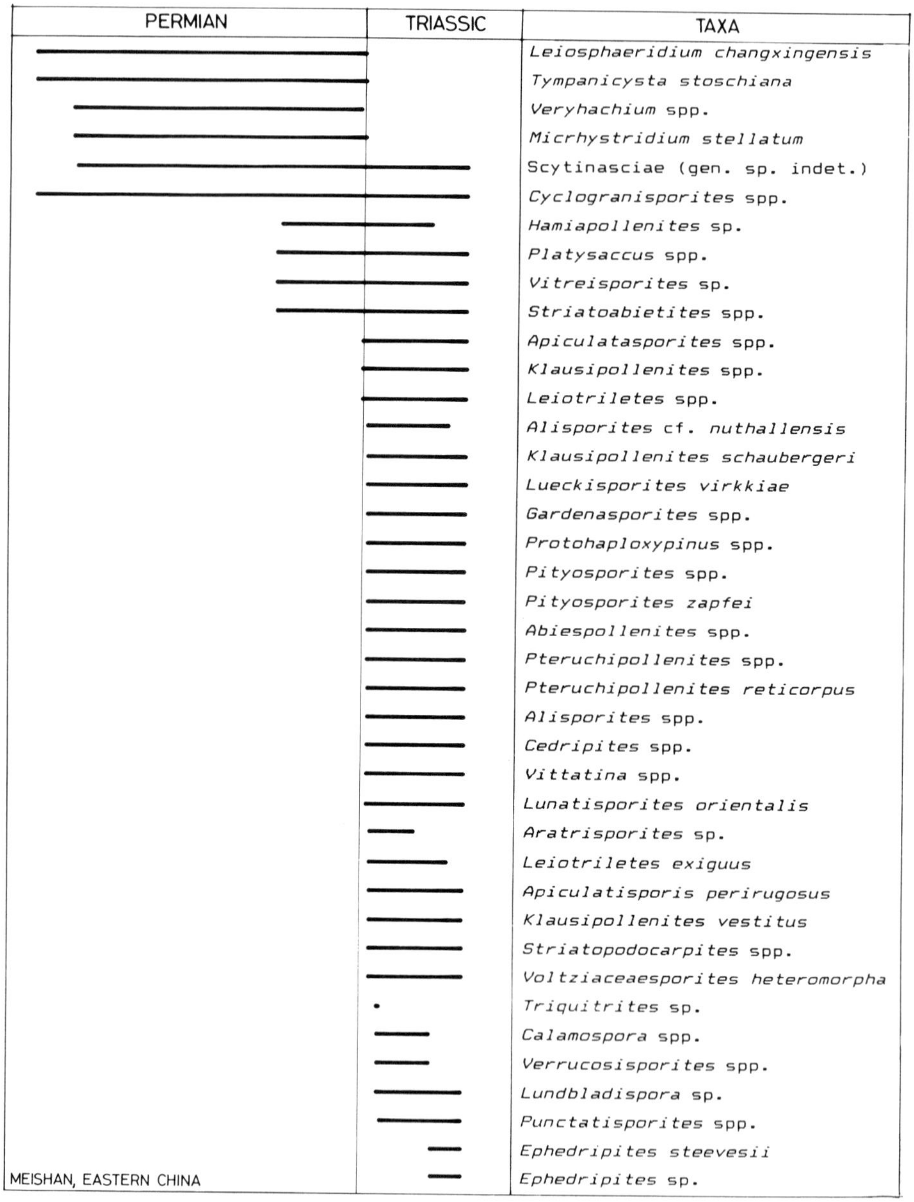

Fig. 11.2.9. Vertical ranges of selected taxa from the late Permian and early Triassic of Meishan, southeastern China. The late Permian terrestrial palynomorphs diversity is considerably lower than the early Triassic, but marine plant microfossils are abundant and varied, represented by the uppermost five taxa, which probably represent a greater number of species. Not shown here, but mentioned by the authors, is the dominance of gymnosperm palynomorphs together with *Azonotriletes* and *Cyclogranisporites*. Reprinted with permission from Ouyang and Utting (1990).

a minor part of the spectrum; Cathaysian elements play an important role in the spectra, with many forms also known from former periods, or extend to Triassic beds. This assemblage seems to represent a tropical environment, in which coal was deposited on land.

In contrast, Ouyang (1982) describes palynological assemblages from northern China (although no exact location is specified) in which palynomorphs of gymnosperms dominate, with abundant bisaccate striatiti totalling 80%, accompanied by low shares of pteridophyte sporomorphs. This assemblage seems to indicate a temperate, or dry subtropical environment.

It would be premature to reconstruct climatic belts from these spot findings, but the differences in late Permian environments of China seem quite clear. Arid to the (present day) southeast, tropical to the south and temperate or dry subtropical to the north. Further studies may delineate the late Permian climatic belts.

It seems plausible to assume that the arid to semi-arid belt north of the late Permian tropical, Cathaysian province, is to be found in the Euramerian province, as seen from the few examples brought above. To the south(?), since only a single indication of possible aridity was found, no paleogeographic reconstruction can be proposed. The seemingly direct contact of Gondwanian temperate(?) elements with Cathaysian tropical ones in Bolivia (Cousminer, 1965) may either indicate the absence of arid belt separating the two provinces at this locality, or that some transportation agent should be blamed. On the other hand, the possibility that some part of the Gondwanian province was arid or semi-arid to some extent, should not be altogether overlooked. Again, for lack of data this hypothesis cannot be further advanced.

Interfingerings of horizons bearing both Euramerian and Cathaysian microfloras, reported from Oklahoma by Clapham (1970), may be attributed to either some minor oscillations of the climatic belts, or to causes connected with transportation, such as periods of stronger winds or rivers activity, carrying palynomorphs from the arid lands to the tropical, or vice versa. Examples from the present day situation indicate that such processes are possible (see Chapter 2.2).

To the north (if indeed it was north at that time), Euramerian semi-arid environments gave rise to different plant taxa, but still with a clear dominance of small number of elements, among which the most common are *Vittatina* and *Protohaploxypinus*, in localities such as Greenland (Balme, 1979) and Alaska (Jansonius, 1962). These may represent cold dry, subpolar environments, or cool temperate habitats. Again, unfortunately the data are not sufficient for further elaboration or sounder conclusions.

11.3. Reconstructing paleoaridity from index taxa *

The use of index taxa (a term preferable to "index species" for reasons discussed above) for reconstruction of paleoaridity is quite common for a long time. Usually, attribution of taxa to paleoenvironments follows trends observed in the present day vegetation. Thus, it is widely agreed that pteridophyte spores are connected with humidity, while saccate palynomorphs, probably representing conifers, are characteristic of drier environments (*cf.* Hart, 1970; Kremp, 1977; Traverse, 1988; and others). Analysis of the late Permian palynofloras, based on taxa diversity and dominance, seems to support these general lines for that period. It should however be remembered that even at the present day there are some exceptions to the rule, although admittedly not too many. Such are conifers typical of tropical environments, as *Podocarpus*, or those growing in marshes, as *Taxodium*. Very few pteridophytes are known from arid biotopes, such as *Ophioglossum* growing on sand dunes. A combination of ferns and conifers growing side by side is however quite common, as in Europe, where *Pteridium aquilinum* accompanies pine forests.

The index taxa chosen as aridity indicators may present an interesting problem, also known for the present day typical dry environments pollen, such as Chenopodiaceae. These plants are very abundant in arid regions, regarded by many authors as characteristic for these environments (*cf.* El Moslimany, 1990). However, they only occupy biotopes which are secondarily enriched in water such as wadi beds or playas, which are less typical for the regional conditions. Thus (see Chapter 9 for an example) some authors prefer to refrain from using Chenopodiaceae for concluding about regional conditions, leaving them for defining the local environments. The point is, when pre-late Cenozoic palynomorphs are defined as xerophytic indicators, there is hardly any way to know whether they represent regional aridity, or local environments typical to arid regions, but which may also occur elsewhere.

An example for a possible misuse of Chenopodiaceae pollen can be seen with the early-middle Pliocene evaporites in Israel, known from the Dead Sea region. Pollen spectra obtained from rocksalt and anhydrite of Pa palynozone are quite rich in Chenopodiaceae pollen (see Chapter 9 for details), which evidently is not surprising. However, when the regional fraction of the spectra is analyzed, its composition and shares of arboreal pollen suggest a rather wet, temperate regional environment. This reconstruction is even clearer where Pa sediments are not evaporitic, when Chenopodiaceae are indeed subordinate. The overlying, late Pliocene Pb palynozone is typified by regional spectra indicating a drier climate, but Chenopodiaceae shares are also lower. This is because paleogeography had changed, which almost terminated evaporites accumulation in the Dead Sea region.

Thus, if the Chenopodiaceae were taken as aridity indicators, the paleoenvironment reconstruction would have been reversed for Pa and Pb. The paleoenviron-

---

* By Yoram Eshet, Geological Survey of Israel, 30, Malkhe Israel Street, Jerusalem, Israel.

ments of the Israeli Pliocene were rather easy to determine since both evaporites and open sea sediments are available for analysis. It would be much more difficult to tell for earlier periods whether the accumulation of evaporites had occurred under truly arid conditions, or resulted from paleogeography, when no samples of other synchronous rock types are available for comparison.

The case presented below is based on the principles of index taxa, which had been defined from evaporites, considered typical for a dry climate, against others from coals, characteristic of humid environments. These definitions should however be carefully viewed, considering the above mentioned limitations.

Robinson (1973) presented a theoretical model for the global Triassic climate, based on the paleogeographic configuration. According to this model, humid climate occurred at high latitudes, whereas at or near the paleoequator, in regions far from the ocean, arid climate developed due to the winds, which were blowing over great land masses. This paleoclimatic pattern is supported by both the geographic distribution of fossil plant associations and types of sediments, known for various parts of the world (Visscher and Van Der Zwan, 1981) for that period. Considering this theoretical model and assuming that the plant record would represent climatic belts, one may expect humid (hygrophytic) vegetation at high latitudes north and south of the paleoequator, whereas an arid (xerophytic) vegetation belt is expected near or at the paleoequator, within the nowadays Mediterranean region. The model further suggests aridity increase toward the western Mediterranean, as compared with its eastern side. This assumption also receives support from the geographic distribution of Carnian rock types in the Mediterranean region, where limestones are more common to the east and coal to the north, whereas evaporites are widespread to the west, especially in southwestern Europe and northwestern Africa.

The first step in attempting paleoclimatic reconstructions based on index taxa of an extinct vegetation, is to create reference assemblages which would represent the various environments for any given period of time. This approach was applied by Visscher and Van Der Zwan (1981) for Ladinian and Carnian strata of the Mediterranean region, in comparison with the west European. They considered coal a typical rock of humid environments, while evaporites (anhydrite and gypsum) as indicators of arid ones. With the use of the term "environment" rather than "climate", they referred to the possibility that humid environments such as oases, can locally occur in arid climatic belts and vice versa.

Visscher and Van Der Zwan subdivided the late Triassic assemblages into 15 palynological groups, seven of which considered typical of hygrophytic environments, five characteristic of xerophytic, while three either unknown or indifferent. It should be pointed out that in terms of diversity, the hygrophytic group contains more taxa than the xerophytic.

The groups are as follows:

Hygrophytic elements:
- Group A: Monolete acavate spores (e.g. *Punctatosporites*)
- Group B: Trilete acavate laevigate or apiculate spores (e.g. *Carnisporites*, *Deltoidospora*)
- Group C: Trilete acavate muronate spores
- Group D: Trilete cingulate and zonotrilete spores (e.g. *Camerozonotriletes*)
- Group E: *Aratrisporites* group
- Group F: *Porcellispora* complex
- Group G: Monosulcate pollen grains (e.g. *Cycadopites*)

Xerophytic elements:
- Group K: Taeniate bisaccate pollen grains (e.g. *Lunatisporites*, *Protodiploxypinus*)
- Group L: *Triadispora* complex
- Group M: Vesicate pollen grains (e.g. *Vallasporites*, *Enzonalasporites*)
- Group N: Monosaccate pollen grains (e.g. *Patinasporites*)
- Group O: Circumpolles sporomorphs (e.g. *Paracirculina*, *Camerosporites*, *Praecirculina*, *Duplicisporites*).

Groups of no apparent paleoenvironmental significance:
- Group H: *Ovalipollis*
- Group I: Alete bisaccate pollen grains (e.g. *Alisporites*, *Falcisporites*)
- Group J: *Samaropollenites*

The results, summarized in Fig. 11.3.1, seem to provide support to the Triassic paleoclimatic model of Robinson (1973), since they demonstrate quite clearly the predominance of xerophytic elements to the present day south and west, whereas hygrophytic elements are more common to the north. Data are also added to this figure from Israel, discussed below.

11.3.1. Triassic aridity in Israel

During the late Triassic Israel was located somewhat south, but quite close to the paleoequator (Fig. 11.3.1), thus according to Robinson's model, is expected to have been well within the arid climate belt. This assumption is lithologically supported by predominance of evaporites in the late Triassic Mohilla formation.

In order to test this hypothesis by reconstructing Triassic paleoclimates in Israel, the phytogeographic classification of Visscher and Van Der Zwan was adopted for analysis of selected middle and late Triassic formations (Eshet, 1990). Locations of these sequences are shown in Fig. 11.3.2. This study presents new information on the phytogeography of the Triassic in Israel, with emphasis on gradual temporal climatic changes that could be concluded from the palynological record.

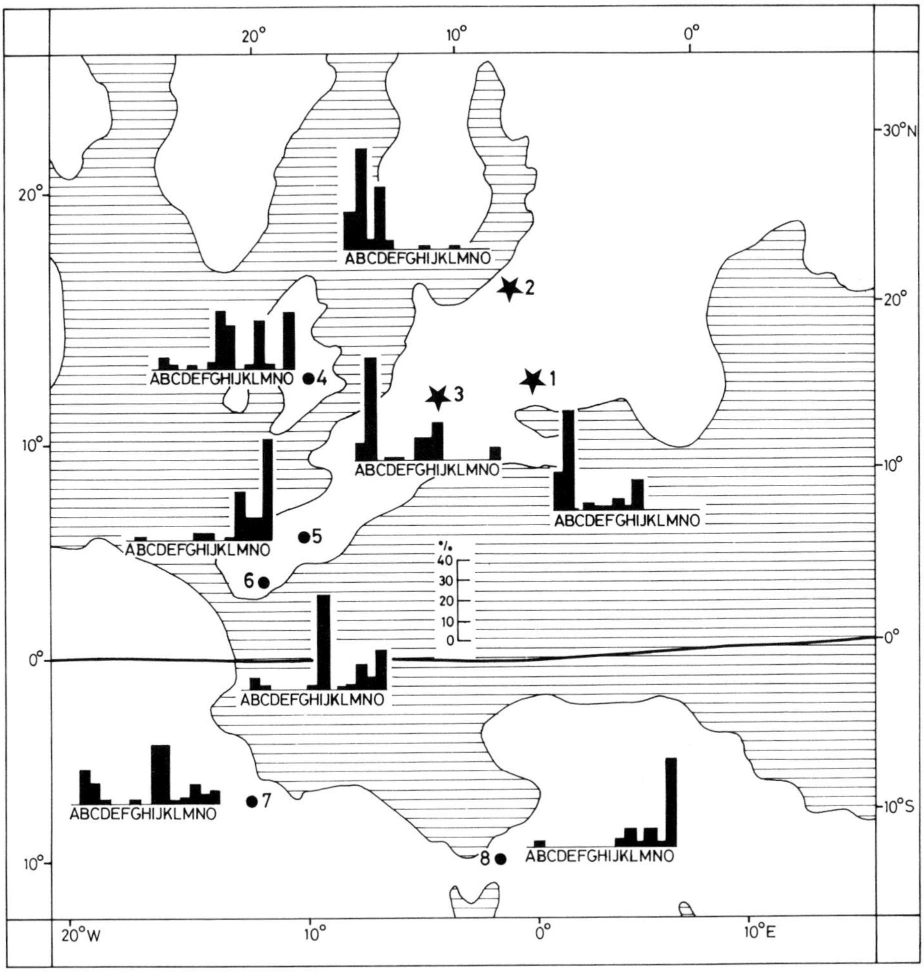

Fig. 11.3.1. The Mediterranean–west European region in Carnian times: palynomorph assemblages, grouped for environments. Hygrophytic groups *A* through *G*; non-conclusive groups *H*, *I*, *J*; xerophytic groups *K* through *O*. For details of assemblages please see text. *1* = Austria. *2* = Poland. *3* = Germany. *4* = England. *5* = Northern Spain. *6* = Southern Spain. *7* = Lybia. *8* = Israel. Black circles denote dry localities, stars indicate humid. Western part reprinted with permission from Visscher and Van Der Zwan (1981), eastern part from Eshet (1990).

A. General geology

The Triassic sequence in Israel was deposited in a variety of environments, including nearshore marine, tidal flats, evaporitic basins and terrestrial (Druckman, 1974; Weissbrod, 1981). The paleogeographic configuration during most of the Triassic was governed by the Arabian Craton to the south and the Tethys extension to the north. Continental sediments, mainly sandstones and shales, had

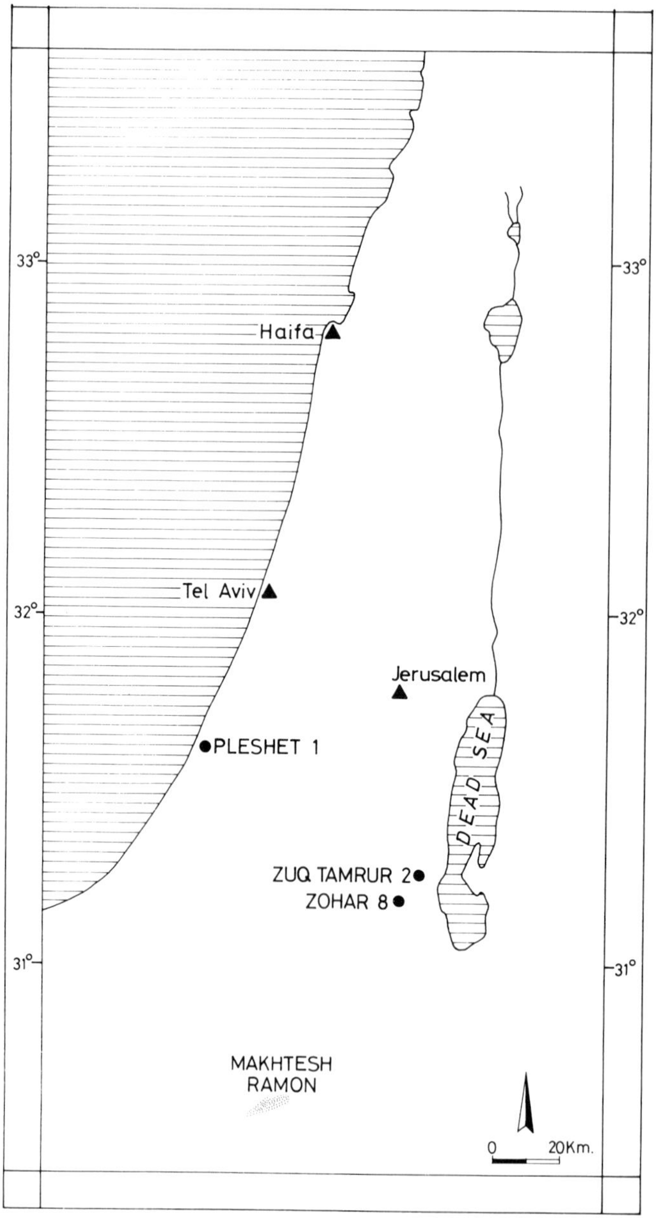

Fig. 11.3.2. Location map for the described Triassic sequences in Israel.

accumulated on land, while carbonates, clastics and evaporites had been deposited in the neighboring sea. Transgressions appear to have invaded generally from the northwest. Continental detritus, including sporomorphs and other types of organic matter, was transported to the sea from the continent, to the south and southeast. Consequently, the sedimentary facies to the southeast are more clastic, while those to the northwest, away from the source of detritus, are poor in organic and mineral allochthonous material. Epeiric movements of the Arabian Craton are thought to have triggered the transgression–regression regime (Druckman, 1974; Goldberg M. and Friedman, 1974) at that time.

A general outline of the Triassic sequence in southern Israel is shown in Fig. 11.3.3. The sequence comprises three major sedimentary cycles, each commencing with a transgression and terminating by regression (Druckman, 1974). The first cycle began with a transgression that resulted in deposition of both carbonates of the upper part of the early Triassic Yamin formation and lower part of the early Triassic Zafir formation. The cycle ended with a regression, represented by clastic sedimentation of the upper part of Zafir formation. The second cycle began with a transgression, forming carbonates of the Anisian Ra'af formation and terminated with a regression, that had deposited clastics of the Anisian Gevanim formation. The transgression beginning the third cycle, which appears to have been the most extensive Triassic transgression in the Middle East, is represented by carbonates of the Anisian-Ladinian Saharonim formation. The cycle ended with a large scale regression, invoking deposition of evaporites of the Carnian Mohilla formation.

Middle and upper Triassic strata crop out in southern Israel in Makhtesh Ramon (Fig. 11.3.2) and have also been penetrated by a dozen boreholes, three of which provided palynological material for the present discussion. The upper part of the Triassic succession, the Norian-Rhaetian interval, is missing due to unconformity that truncated the top of the section.

B. Materials and methods

Core and well-cutting samples from three boreholes that penetrated the Triassic succession in southern Israel (Fig. 11.3.2) were selected. Samples were processed according to standard palynological procedures which included digestion in concentrated HCl and HF, followed by fractionation and sieving of the organic residue (Doher, 1980). No oxidation was used during the sample treatment. All quantitative results presented here are based on the counting of all sporomorphs along ten randomly selected traverses of the microscope slide.

C. Palynological zonation

The Permo-Triassic succession of Israel was subdivided by Eshet and Cousminer (1986) and by Eshet (1990) into seven palynozones, three of which are included within the time interval dealt with in the present discussion:

Anisian: Species of *Aratrisporites* predominate, together with a variety of bisaccate pollen grains (*Triadispora* spp., *Angustisulcites klausii*, *Platysaccus queenslandi*, *P. leschikii*, *Voltziaceaesporites heteromorphus*, *Lunatisporites pellucidus* and *Alisporites grauvogeli*) and trilete spores (*Densoisporites nejburgii*, *Keuperisporites*

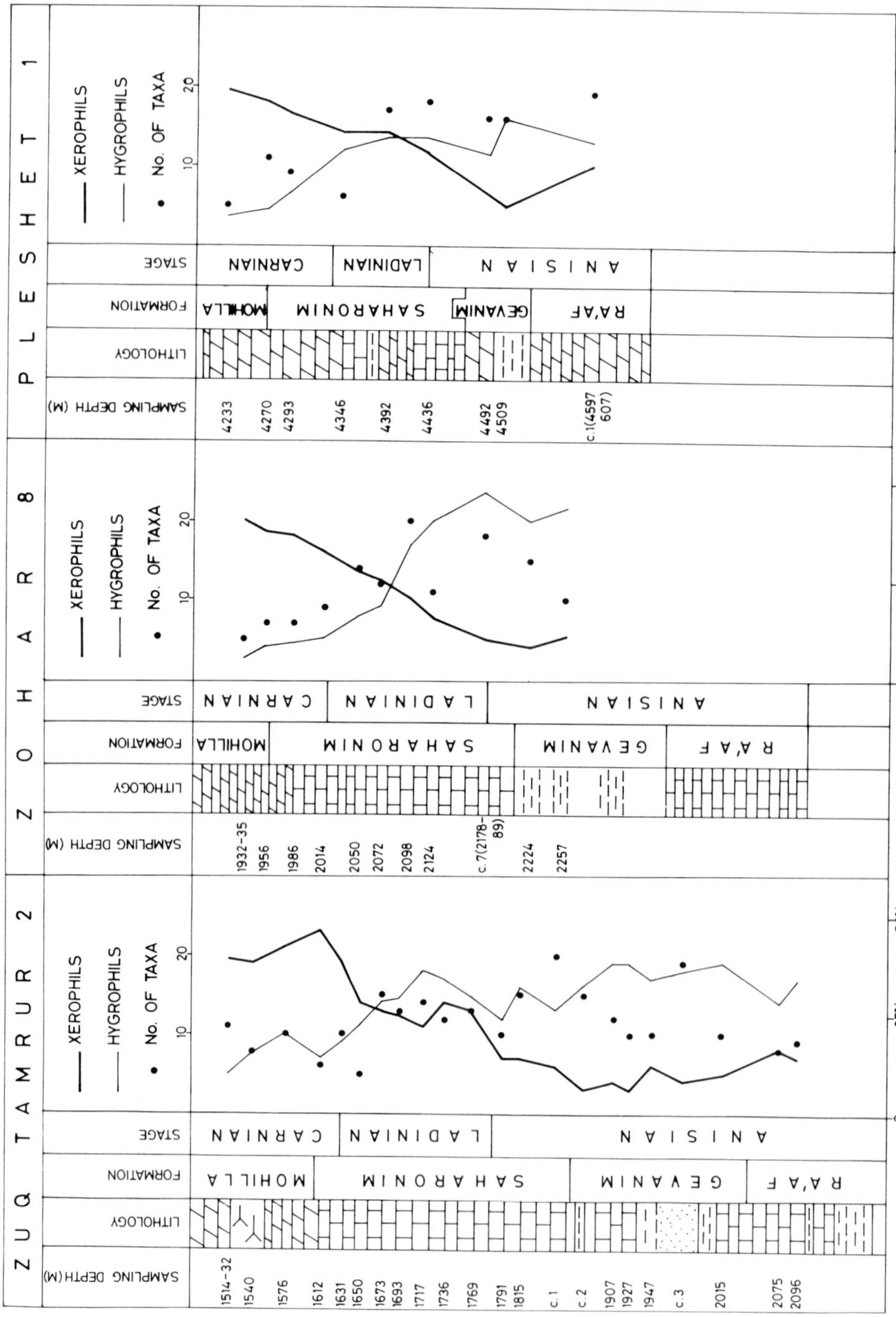

Fig. 11.3.3. Triassic lithostratigraphy in three analyzed boreholes from Israel, xerophil *vs.* hygrophil elements curves and taxa diversity in the studied sections. For details see text.

baculatus, Spinotriletes echinoides and *Cristianisporites triangulatus*). Also present are *Kuglerina meieri, Stelapollenites thiegartii* and *Enzonalasporites vigens*. Acritarchs are common in most samples.

Ladinian: Characterized by a decrease in *Aratrisporites*. Many of the Anisian forms continue into this stage. New taxa are *Echinitosporites iliacoides, Podosporites amicus, Eucommiidites microgranulatus* and species of the Circumpolles group such as *Duplicisporites granulatus*. Acritarchs decrease or disappear.

Carnian: A drastic palynological change marks the Ladinian/Carnian transition. Most Ladinian taxa disappear, but the Circumpolles group becomes dominant. This includes *Paracirculina* spp., *Praecirculina* spp., *Duplicisporites granulatus* and *Camerosporites secatus*.

D. Results

To enable paleoenvironmental interpretation, sporomorphs in the studied samples were classified into the same 15 environmentally significant palynological groups of Visscher and Van Der Zwan (1981), as detailed above.

Each studied sample was analyzed for the relative abundance of those groups which are present (some of the groups do not occur at all in Israel). Representative samples are shown in Fig. 11.3.4. These histograms show that samples from the Anisian stage (Gevanim and Ra'af formations) are characterized by predominance of both hygrophytic (groups E, A and B) and xerophytic elements (groups K, L, M). During the Ladinian (Saharonim formation), the hygrophytic elements decrease and xerophytic groups increase (especially group O). In the Carnian stage, almost only xerophytes are present in the samples.

The histograms therefore suggest coexistence of the hygrophytes and xerophytes, with a slight predominance of the hygrophytes during the Anisian. This may have resulted from either the coeval occurrence of humid and arid environments in southern Israel, much in the same way local humid environments exist in oases, or along a river crossing a desert; or, alternately, that the region was arid, but received a considerable supply of humidity indicating sporomorphs by long distance transportation, most probably fluviatile. Sporomorphs from both Anisian environments were carried by the well developed fluviatile systems of the Gevanim formation (Druckman, 1974) to the sea. The decrease in hygrophytes and increase in xerophytes in Ladinian times indicates that arid environments became more dominant, while the almost exclusive occurrences of arid elements suggest a complete "aridization" of the area during the Carnian.

The aridization trend from the Anisian to the Carnian can be exemplified by plotting the quantitative distribution of selected groups along the strata of this time span (Fig. 11.3.3). The xerophytes curve is the percentage of xerophytes (groups K–O) calculated from the total sum of counted sporomorphs, while the hygrophytes curve represents the shares of groups A–G. Throughout the Anisian, in all the studied sections, the hygrophytic elements are more common than the xerophytes.

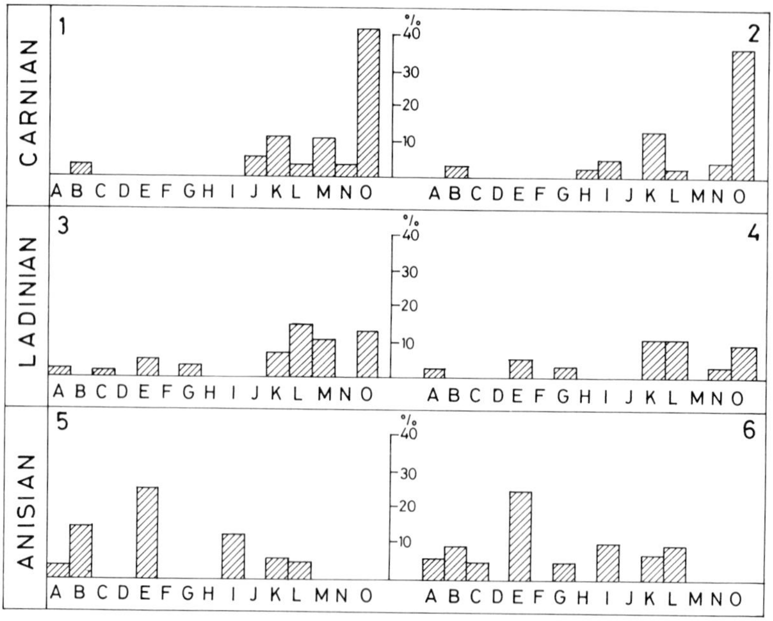

Fig. 11.3.4. Quantitative distribution of palynomorph assemblages in selected samples from the Triassic in Israel, indicating increased aridity from Anisian through Carnian times. For details of assemblages please see text. 1. Zuq Tamrur 2, 1492 m. 2. Zuq Tamrur 2, 1532 m. 3. Zohar 8, 2050 m. 4. Pleshet 1, 2392 m. 5. Zuq Tamrur 2, 1815 m. 6. Zohar 8, 2257 m. Reprinted with permission from Eshet (1990).

These relationships between arid and humid floras continue unchanged at the beginning of the Ladinian, with a slight decrease in the hygrophytes and increase of xerophytes.

From the middle or upper part of the Ladinian, the xerophytes become dominant and hygrophytes decrease drastically. The overall floral changes from the Anisian to the Carnian, as represented by the palynomorph groups relations in Fig. 11.3.4, indicate a regional aridization which is manifested by replacement of the hygrophytic flora, as the environments change. This trend of aridization reaches its maximum at the Carnian, with the predominance of Circumpolles taxa. The Carnian aridity is also well documented throughout the northern and western Mediterranean regions, where Circumpolles taxa are dominant (Fig. 11.3.1). This belt is referred to as the "*Camerosporites* Phase" (Visscher and Krystyn, 1978).

11.3.2. Taxa diversity and dominance in the triassic of Israel

It would, in this respect, be interesting to compare the above conclusions arrived at by the index taxa approach, to data concerning taxa diversity and dominance along the same sequences (Fig. 11.3.3).

The only borehole in which the number of taxa gradually and linearly decreases from the Anisian through the Carnian, indicating the onset of arid conditions, is Pleshet 1, drilled in the present day coastal plain of Israel, some 70 km northwest of the other two. Zohar 8, drilled near the Dead Sea, displays a gradual decrease in the number of taxa from the beginning of the Ladinian through the end of the Carnian, but the top of the Anisian spoils the picture with a lower than expected number. Zuq Tamrur 2, drilled somewhat further east, presents a general lower number of taxa for the Carnian as compared with the underlying strata, but the picture is somewhat fuzzy and the trend not as clear.

When the sedimentary facies of the three boreholes are compared, it is evident that Zuq Tamrur 2 is the most affected by the continent, Zohar 8 somewhat less and Pleshet 1 is almost entirely dominated by marine environments of deposition (Druckman, 1974). These facies changes may explain the inconsistencies observed for the taxa diversity gradients. Zuq Tamrur 2, with its marked terrestrial stamp, is closer to various microenvironments than Pleshet 1, that probably had received an average pollen supply from a wider region, being located further seaward. The microenvironments did not contribute large quantities of palynomorphs, but however added to the number of taxa some minor constituents, which apparently only grew there, affecting Zuq Tamrur 2 in a more pronounced way due to its closeness to these local occurrences. Thus the marine environment is most probably a faithful representative of the average pollen flora, while the continental is also subject to local donations of palynomorphs.

This explanation becomes even more convincing when the dominance factor is concerned (Fig. 11.3.4). The arid Carnian assemblages are quantitatively dominated by Group O (Circumpolles) even in Zuq Tamrur 2, proving that indeed the somewhat higher number of taxa, observed in this borehole, is of restricted paleoenvironmental significance.

It seems therefore that all three approaches — index taxa, taxa diversity and dominance — are applicable for identifying paleoaridity. It would naturally be beneficial if all three tests are applied together, preferably with the aid of sedimentology. There is not too much additional work involved with this joint approach, it only needs the numeric data, which are easy to get while the slides are analyzed.

References

Alimen, H., Beucher, F., Lhote, H., Delibrias, G., 1968. Les gisements Néolithiques de Tan-Tartait et d'I-n-Itinen Tassili-n-Ajjer (Sahara Central). Bull. Soc. Préhistorique Française 65(1): 421–458.

Almagor, G., 1976. Physical properties, consolidation processes and slumping in recent marine sediments in the Mediterranean continental slope off southern Israel. Unpubl. PhD Dissertation, Dept. of Geology, The Hebrew Univ. of Jerusalem. 1: 139 pp.; 2: 120 pp. (in Hebrew, English summary).

Alon, G., 1976. Pollen analyses of the Fatza'el sites, southern Jordan Valley. Unpubl. MA Thesis, Dept. of Archaeology, The Hebrew Univ. of Jerusalem, 45 pp. (in Hebrew).

Amit, R., 1990. Salts as a factor in gravel weathering in reg soils. Ran Gerson Memorial Volume, Dept. of Geography, The Hebrew Univ. of Jerusalem, 20–21.

Avery, D.M., 1982. Micromammals as palaeoenvironmental indicators and an interpretation of the late Quaternary in the southern Cape Province, South Africa. Ann. S. Afr. Mus., 85: 183–374.

Bagnold, R.A., 1941. Physics of blown sand and desert dunes. Methuen, London, 265 pp.

Ballais, J.L., Cohen, J., 1985. Problèmes de fossilization et d'interprétation des pollens d'un travertin actuel de Sidi-Masmoudi (Aures — Algerie). C. R. Soc. Biogéogr., 61(4): 118–128.

Balme, B.E., 1979. Palynology of Permian-Triassic boundary beds at Kap Stosch, east Greenland. Medd. Grönland, 200(6): 1–37.

Bambach, R.K., Scotese, C.R., Ziegler, A.M., 1980. Before Pangea: the geographies of the Paleozoic world. Scientific American, 68: 26–38.

Baruch, U., 1986. The late Holocene vegetational history of Lake Kinneret (Sea of Galilee), Israel. Paléorient, 12/2: 37–48.

Baruch, U., 1990. Palynological evidence of human impact on the vegetation as recorded in late Holocene lake sediments in Israel. In: Bottema, S., Entjes-Nieborg, G., Van Zeist, W. (Editors), Man's role in the shaping of the eastern Mediterranean landscape. A.A. Balkema, Rotterdam, 283–293.

Bar Yosef, O., 1987. Prehistory of the Jordan Rift. Israel J. Earth Sci., 36(1–2): 107–120.

Bar Yosef, O., 1989. The Neolithic period. The Open Univ. of Israel, Tel Aviv, 52 pp. (in Hebrew).

Bar Yosef, O., Belfer-Cohen, A., 1989. The origins of sedentism and farming communities in the Levant. J. World Prehistory, 3(4): 447–459.

Bar Yosef, O., Mintz, E., 1979. Epipaleolithic and Neolithic industries. In: Horowitz, A., The Quaternary of Israel, Academic Press, New York, 307–319.

Beaumont, P., 1968. Salt weathering on the margin of the Great Kavir, Iran. Geol. Soc. Am. Bull., 79: 1683–1684.

Behre, K.E. (Editor), 1986. Anthropogenic indicators in pollen diagrams. A.A. Balkema, Rotterdam, 232 pp.

Bein, A., Feinstein, S., 1988. Late Cenozoic thermal gradients in Dead Sea Transform basins. J. Petr. Geol., 11(2): 185–192.

Bein, A., Horowitz, A., 1986. Papyrus — a historic newcomer to the Hula Valley, Israel? Rev. Palaeobotan. Palynol., 47: 89–95.

Benda, L., Meulenkamp, J.E., 1990. Biostratigraphic correlations in the eastern Mediterranean Neogene. Newsl. Stratigr., 23(1): 1–10.

Bessais, E., Suc, J.P., 1987. Climate control over the Neogene geographic and chronologic variations in the west Mediterranean vegetations. Interim Colloquium, RCMNS, Montpellier, p. 14.

Betancourt, J.L., Van Devender, T.R., Martin, P.S. (Editors), 1990. Packrat middens: the last 40,000 years of biotic change. Univ. of Arizona Press, Tucson, 467 pp.

Beucher, F., 1975. Étude palynologique de formations Néogènes et Quaternaires au Sahara nord-occidental. Éditions CNRS, Paris, Sér. Géologie, 20, 285 pp.

Birks, H.J.B., Birks, H.H., 1980. Quaternary palaeoecology. E. Arnold, London, 289 pp.

Birks, H.J.B., Gordon, A.D., 1985. Numerical methods in Quaternary pollen analysis. Academic Press, London, 317 pp.

Bonnefille, R., 1976. Palynological evidence for an important change in the vegetation of the Omo basin between 2.5 and 2 million years ago. In: Coppens, Y., Howell, F.C., Isaac, G.L., Leakey, R.E.F. (Editors), Earliest man and environments in the Lake Rudolf Basin: stratigraphy, paleoecology and evolution. Univ. of Chicago Press, London, 421–431.

Bonnefille, R., Riollet, G., 1988. Palynologie des sédiments holocènes de sites archéologiques du Qatar. In: Inizan, M.L. (Editor), Préhistoire à Qatar. Éditions Recherches sur les Civilizations, Paris, 2: 137–145.

Bonnefille, R., Gasse, F., Azema, C., Denefle, M., 1980. Palynologie et interprétation paléoclimatique de trois niveaux Pléistocène supérieur d'un sondage du Lac Abhe (Afar, Territorie de Djibuti). Mém. Mus. Nat. d'Histoire Naturelle, Paris, B(XXVII): 149–164.

Bonnefille, R., Vincens, A., Buchet, G., 1987. Palynology, stratigraphy and paleoenvironment of a Pliocene Hominid site (2.9–3.3 m.y.) at Hadar, Ethiopia. Palaeogeogr., Palaeoclimatol., Palaeoecol., 60: 249–281.

Bonnefille, R., Roeland, J.C., Guiot, J., 1990. Temperature and rainfall estimates for the past 40,000 years in equatorial Africa. Nature, 346: 347–349.

Bottema, S., 1975. The interpretation of pollen spectra from prehistoric settlements (with special attention to Liguliflorae). Palaeohistoria, XVII: 17–35.

Bottema, S., 1984. Pollen stratigraphical investigations in the Mediterranean area (with special emphasis on the problems of sampling and interpretation of the results. Webbia, 38: 465–472.

Bottema, S., Entjes-Nieborg, G., Van Zeist, W. (Editors), 1990. Man's role in the shaping of the eastern Mediterranean landscape. A.A. Balkema, Rotterdam, 352 pp.

Boyd, W.E., 1990. Quaternary pollen analysis in the arid zone of Australia: Dalhousie Springs, central Australia. Rev. Palaeobotan. Palynol., 64: 331–341.

Bradley, R.S., 1985. Quaternary paleoclimatology. Allen and Unwin, Winchester, 472 pp.

Braun, D., Ron, H., Marco, S., 1991. Magnetostratigraphy of Erk el-Ahmar formation in the northern Dead Sea Rift. Abstract, Israel Geological Society Ann. Mtg. Abs., Akko, 16–17.

Brown, C.A., 1960. Palynological techniques. Private Edition, Baton Rouge, 188 pp.

Brun, A., 1985. La couverture steppique en Tunisie au Quaternaire supérieur. C.R. Acad. Sci. Paris, 301(II, 14): 1085–1090.

Bryant, V.M. Jr., 1974a. Prehistoric diet in southwest Texas: the coprolite evidence. American Antiquity, 39(3): 407–420.

Bryant, V.M. Jr., 1974b. Pollen analysis of prehistoric human feces from Mammoth Cave. In: Watson, P.J. (Editor), Archeology of the Mammoth Cave area. Academic Press, New York, 203–209.

Bryant, V.M. Jr., 1975. Pollen as indicators of prehistoric diets in Coahuila, Mexico. Bull. of the Texas Archeological Society, 46: 87–106.

Bryant, V.M. Jr., 1978. Late Quaternary pollen records from the east-central periphery of the Chihuahuan Desert. In: Wauer, R.H., Riskind, D.H. (Editors), Transactions of the Symposium on the Biological Resources of the Chihuahuan Desert Region, United States and Mexico. US Dept. of the Interior Transactions and Proceedings Ser., 3: 3–21.

Bryant, V.M. Jr., 1983. Pollen analysis of Pratt Cave. In: Schroeder, A.H. (Editor), The Pratt Cave studies. The Artifact, 21(1–4): 161–166.

Bryant, V.M. Jr., Holloway, R.G., 1983. The role of palynology in archaeology. Advances in archeological method and theory, 6: 192–224.

Bryant, V.M. Jr., Holloway, R.G., 1988. Fossil pollen preservation under different lithological conditions. Abstract, 7th Intl. Palynological Congr., Brisbane, p. 20.

Bryson, R.A., Swain, A.M., 1981. Holocene variations of monsoon rainfall in Rajasthan. Quat. Res., 16: 135–145.

Buchet, G., 1982. Transport des pollens dans les fleuves Omo et Awash (Ethiopie): étude de vases actuelles. Unpubl. PhD Dissertation, Univ. des Sciences et Techniques du Languedoc, Montpellier, 171 pp. + App.

Byrne, R., Horn, S.P., 1989. Prehistoric agriculture and forest clearance in the Sierra de los Tuxtlas, Veracruz, Mexico. Palynology, 13: 181–193.

Caratini, C., Bellet, J., Tissot, C., 1976. Étude microscopique de la matière organique: palynologie et palynofaciès. In: Arnould, M., Pelet, R. (Editors), Géochimie organique des sédiments marins profonds, Orgon III, Mauritanie–Sénégal–Iles du Cap-Verts. Éditions CNRS, Paris, 215–264.

Caratini, C., Bellet, J., Tissot, C., 1978. Étude microscopique de la matière organique: palynologie et palynofaciès. In: Arnould, M., Pelet, R. (Editors), Géochimie organique des sédiments marins profonds, Orgon IV, Golfe d'Aden, Mer d'Oman. Éditions CNRS, Paris, 265–307.

Caratini, C., Tissot, C., Fredoux, A., 1988. Caractérisation des aérosols désertiques à Niamey (Niger) par leur contenu pollinique. Inst. Fr. Pondichéry, Trav. Sec. Sci. Tech., XXV: 251–268.

Caulton, E., Gibson, W., 1988. Étude du contenu pollinique des matières fécale d'herbivores sélectionnées. Inst. Fr. Pondichéry, Trav. Sec. Sci. Tech., XXV: 301–308.

Cerceau-Larrival, M.T., Carbonnier-Jarreau, M.C., 1980. Caractéristiques tectales de certaines Ombellifères indiennes et hymalayennes. Relation avec le climat et la biogéographie. Mém. Mus. Nat. d'Histoire Naturelle, Paris, B(XXVII): 39–44.

Cerceau-Larrival, M.T., Roland-Heydacker, F., 1980. Adaptation de l'ultrastructure de l'exine de quelques Ombellifères australes en fonction des conditions biogéographiques. Mém. Mus. Nat. d'Histoire Naturelle, Paris, B(XXVII): 31–38.

Cheddadi, R., 1988. Paléoclimats au nord de la Mediterranée oriental depuis 250,000 ans. Analyse pollinique et stratigraphie isotopique de quatre carottes marines. Unpubl. PhD Dissertation, Univ. des Sciences et Techniques du Languedoc, Montpellier, 116 pp.

Cheddadi, R., Van Campo, E., 1985. Image pollinique des steppes à *Artemisia* au nord de la Mer d'Arabie: application à l'étude des carottes marines. Actes du 110e Congr. Nat. Soc. Savantes, II: 203–209.

Clapham, W.B., 1970. Permian miospores from the Flowerpot formation of western Oklahoma. Micropaleontology, 16(1): 15–36.

Clark, R.L., 1982. Point count estimation of charcoal in pollen preparations and thin sections of sediments. Pollen et Spores, XXIV(3–4): 523–535.

Clary, K.H., 1989. Reducing ambiguity in the archeological pollen record by improving recovery of economic pollen. Palynology, 13: 281.

Clason, A.T., Clutton-Brock, J., 1982. The impact of domestic animals on the vegetation during the first phases of animal husbandry in the Mediterranean and Near East. BAR Intl. Ser., 133(II): 146–148.

Coetzee, J.A., 1967. Pollen analytical studies in east and southern Africa. Palaeoecology of Africa, 3: 1–146

Cohen, R., 1986. The settlement of the central Negev in light of archaeology and literary sources during the 4th–1st millenia B.C.E. Unpubl. PhD Dissertation, Dept. of Archaeology, The Hebrew Univ. of Jerusalem, 356 pp. (in Hebrew, English summary).

Cour, P., 1974. Nouvelles techniques de détection des flux et des retombées polliniques: étude de la sédimentation des pollens et des spores a la surface du sol. Pollen et Spores, XVI(1): 103–141.

Cour, P., Duzer, D., 1976. Persistance d'un climat hyperaride au Sahara central et méridional au cours de l'Holocène. Rev. Géogr. Physique Géol. Dynamique , XVIII(2–3): 175–198.

Cour, P., Duzer, D., 1978. La signification climatique, édaphique et sédimentologique des rapports entre taxons en analyse pollinique. Annales des Mines de Belgique, 7–8: 155–164.

Cour, P., Duzer, D., 1980. Action actuelle des courants anmosphériques sur la dissémination des pollens au Sahara et dans la régions climatiques avoisinantes le long d'un transect Oran–Abidjan. Mém. Mus. Nat. d'Histoire Naturelle, Paris, B(XXVII): 66–80.

Cour, P., Guinet, Ph., Cohen, J., Duzer, D., 1973. Reconnaissance des flux polliniques et de la sédimentation actuelle au Sahara nord-occidental. In: Palynology and medicine, Proc. 3rd Intl. Palynological Congr., Moscow, 41–58.

Cousminer, H.L., 1965. Permian spores from Apillapampa, Bolivia. J. Paleontology, 39: 1097–1111.

Cross, A.T., Thompson, G.C., Zaitzeff, J.B., 1966. Source and distribution of palynomorphs in bottom sediments, southern part of the Gulf of California. Mar. Geol., 4: 467–524.

Cushing, E.J., 1967. Evidence for differential pollen preservation in late Quaternary sediments in Minnesota. Rev. Palaeobotan. Palynol., 4: 87–101.

Dan, J., 1981. Soil formation in the arid regions of Israel. In: Dan, J., Gerson, R., Koyumdjisky, H., Yaalon, D.H. (Editors), Aridic soils of Israel. The Volcani Center, Bet Dagan, Israel, Sp. Publ. 50: 17–50.

Dan, J., Gerson, R., Koyumdjisky, H., Yaalon, D.H. (Editors), 1981. Aridic soils of Israel. The Volcani Center, Bet Dagan, Israel, Sp. Publ. 50, 353 pp.

Danin, A., 1972. Mediterranean elements in rocks of the Negev and Sinai deserts. Notes of the Royal Botanical Garden, 31(3): 437–440.

Danin, A., 1989. The use of biogenic weathering of limestone in paleoecologic research. In: Spanier, E. (Editor), Environmental quality and ecosystem stability. ISEEQS Publ., Jerusalem, IV/B: 53–64.

Danin, A., 1990. Dust and blue-green algae as stabilizers of dunes in the western Negev. Ran Gerson Memorial Volume, Dept. of Geography, The Hebrew Univ. of Jerusalem, 25–26.

Danin, A., Ganor, E., 1991. Trapping of airborne dust by mosses in the Negev Desert, Israel. Earth Surface Processes and Landforms, 16(2): 153–162.

Danin, A., Gerson, R., Garti, Y., Marton, K., 1982. Patterns of limestone and dolomite weathering by lichen and blue-green algae and their paleoclimatic significance. Palaeogeogr., Palaeoecol., Palaeoclimatol., 37: 221–233.

Danin, A., Gerson, R., Garti, Y., 1983. Weathering patterns of hard limestones and dolomite by endolithic lichen as an indicator of the eolian origin of terra rossa. Soil Sci., 136: 213–217.

D'Antoni, H.L., 1978. Palinologìa del perfil del Alero del Cañadòn de las Manos Pintadas. Relaciones Soc. Argentina de Antropologìa, XII: 249–262.

D'Antoni, H.L., Markgraf, V., 1977. Dispersìon del polen actual en el oeste arido Argentino. IANIGLA (CONICET) Ann., 4: 125–150.

D'Antoni, H.L., Nieto, A., Mancini, M.V., 1984. Pollenanalytic stratigraphy of Arroyo Las Brusquitas profile (Buenos Aires Province, Argentina). Zbl. Geol. Paläont. Teil I, H. 11/12: 1721–1729.

Daoud, A., 1984. Spectres polliniques de l'atmosphère et d'échantillons de surface en Tunisie: rélations avec la végétation. Unpubl. PhD Dissertation, Univ. des Sciences et Techniques du Languedoc, Montpellier, 94 pp.

Darmon, F., 1984. Analyses polliniques de deux sites de la basse Vallée du Jourdain: Fazaël VIII et Salibiya IX. Paléorient, 10/2: 106–110.

Darmon, F., 1987. Analyses polliniques de trois sites natoufiens (ancien, récent, final) dans le région de Salibiya- Fazaël. Paléorient, 13/1: 121–129.

Darmon, F., 1988. Essai de reconstitution climatique de l'Épipaléolithique au début du Néolithique ancien dans la région de Fazaël-Salibiya (basse Vallée du Jourdain) d'après la palynologie. C. R. Acad. Sci. Paris, 307(II): 677–682.

Darmon, F., 1989. Étude de l'environnement de la grotte de Nahal Hemar dans le Désert de Judée au Néolithique ancien d'apres l'analyse des coprolithes de chevres. C. R. Acad. Sci. Paris, 308(II,19): 1759–1764.

Davis, O.K., 1983/1984. Pollen frequencies reflect vegetation patterns in a Great Basin (USA) mountain range. Rev. Palaeobotan. Palynol., 40: 295–315.

Davis, O.K., 1990. Caves as sources of biotic remains in arid western North America. Palaeogeogr., Palaeoclimatol., Palaeoecol., 76: 331–348.

Davis, O.K., Anderson, R.S., 1987. Pollen in packrat (*Neotoma*) middens: pollen transport and the relationship of pollen to vegetation. Palynology, 11: 185–198.

Davis, O.K., Hall, S.A., 1988. A bibliography of Southwestern Quaternary palynology. AASP Data Committee Contr., Tucson, 62 pp.

Davis, O.K., Turner, R.M., 1986. Palynological evidence for the Historic expansion of juniper and desert shrubs in Arizona, USA. Rev. Palaeobotan. Palynol., 49: 177–193.

Davis, O.K., Anderson, R.S., Fall, P.L., O'Rourke, M.K., Thompson, R.S., 1985a. Palynological evidence for early Holocene aridity in the southern Sierra Nevada, California. Quat. Res., 24: 322–332.

Davis, O.K., Hevly, R.H., Foust, R.D., 1985b. A comparison of historic and prehistoric vegetation change caused by man in central Arizona. In: Jacobs, B.F., Fall, P.L., Davis, O.K. (Editors), Late Quaternary vegetation and climates of the American Southwest. AASP Contr. Ser., 16: 63–76.

Derin, B., Gerry, E., 1971. Micropaleontological report: Joshua No. 2. Israel Inst. of Petroleum Rep. 5/71, Tel Aviv, 12 pp.

Derin, B., Reiss, Z., 1973. Revision of marine Neogene stratigraphy in Israel. Israel J. Earth Sci., 22: 199–210.

De Vernal, A., Richard, P., Jetté, H., 1983. Sur les effets de la technique de Lüber utilisée en analyse pollinique pour éliminer la matière organique non décomposée. pollen et Spores, XXV(3–4): 541–548.

Dimbleby, G.W., 1985. The palynology of archaelogical sites. Academic Press, London, 176 pp.

Dodson, J.R., 1983. Pollen recovery from organic lake clays: a comparison of two techniques. Pollen et Spores, XXV(1): 131–138.

Doher, I., 1980. Palynomorph preparation procedures currently used in the paleontology and stratigraphy laboratories. US Geol. Surv. Circ. 830, 29 pp.

Doubinger, J., López-Gómez, J., Arche, A., 1990. Pollen and spores from the Permian and Triassic sediments of the southeastern Iberian ranges, Cueva de Hierro (Cuenca) to Chelva- Manzanera (Valencia–Teruel) region, Spain. Rev. Palaeobotan. Palynol., 66(1/2): 25–45.

Drori, I., Horowitz, A., 1988–89. Tell Lachish: environment and subsistence during the Middle Bronze, Late Bronze and Iron ages. Tel Aviv, 15–16(2): 206–211.

Druckman, Y., 1974. The Stratigraphy of the Triassic sequence in southern Israel. Geol. Surv. Israel Bull., 64: 1–92.

Dumait, P., 1962. Le vibroséparateur. Pollen et Spores, IV(2): 311–316.

Dumait, P., 1964. Nouvelles techniques physiques d'enrichissement pour l'analyse pollinique des sédiments. Unpubl. PhD Dissertation, Fac. des Sciences, Univ. de Paris (Centre d'Orsay), 91 pp.

Dupont, L.M., Hooghiemstra, H., 1989. The Saharan–Sahelian boundary during the Brunhes chron. Acta Bot. Neerl., 38(4): 405–415.

Ecke, H.H., 1986. Palynologie des Zechsteins und Untere Buntsandsteins im Germanischen Becken. Unpubl. PhD Dissertation, Georg-August Univ., Göttingen, 117 pp., 12 plates.

El-Moslimany, A.P., 1982. The late Quaternary vegetational history of the Zagros and the Taurus mountains in the regions of Lake Mirabad, Lake Zeribar and Lake Van — a reappraisal. BAR Intl. Ser., 133(II): 343–351.

El-Moslimany, A.P., 1986. Ecology and late Quaternary history of the Kurdo-Zagrosian oak forest near Lake Zeribar, western Iran. Vegetatio, 68: 55–63.

El-Moslimany, A.P., 1988. The climatic significance of common herbaceous pollen, examples from the Middle East. Abstract, 7th Intl. Palynological Congr., Brisbane, p. 44.

El-Moslimany, A.P., 1990. Ecological significance of common non- arboreal pollen: examples from dry lands of the Middle East. Rev. Palaeobotan. Palynol., 64: 343–350.

Elsic, W.C., 1966a. Degradation of arci in a fossil *Alnus* pollen grain. Nature, 209: 825.

Elsik, W.C., 1966b. Biologic degradation of fossil pollen grains and spores. Micropaleontology, 12(4): 515–518.

Endler, J.A., 1982. Pleistocene forest refuges: fact or fancy? In: Prance, G.T. (Editor), Biological diversification in the tropics. Columbia Univ. Press, New York, 641–657.

Erdtman, G., 1952. Pollen morphology and plant taxonomy. Angiosperms: an introduction to palynology. Almqvist and Wiksell, Stockholm, XII + 539 pp.

Erdtman, G., 1971. Notes on the resistance and stratification of the exine. In: Brooks, J., Grant, P.R., Muir, M., Gijzel, P. van, Shaw, G. (Editors), Sporopollenin. Academic Press, London, 249–255.

Eshet, Y., 1983. Palynostratigraphy, thermal alteration index and kerogen analysis of the Permo-Triassic succession in Makhtesh Katan 2 well, Negev, Israel. Geol. Surv. Israel, Jerusalem, Rep. S/12/83, 172 pp.

Eshet, Y., 1987. Palynologic aspects of the Permo-Triassic sequence in the subsurface of Israel. Unpubl. PhD Dissertation, City Univ. of New York, 193 pp.

Eshet, Y., 1990. Palynological aspects of the Permo-Triassic sequence in the subsurface of Israel. Bull. Geol. Surv. Israel, 81: 1–57.

Eshet, Y., Cousminer, H.L., 1986. Palynozonation and correlation of the Permo-Triassic succession in the Negev, Israel. Micropaleontology, 32: 193–214.

Faegri, K., Iversen, J., 1964. Textbook of pollen analysis. Munksgaard, Copenhagen, 237 pp.

Faegri, K., Kaland, P.E., Krzywinski, K., 1989 (4th edition, previously by Faegri, K. and Iversen, J.). Textbook of pollen analysis. John Wiley and Sons, Chichester, 328 pp.

Fall, P.L., 1985. Holocene dynamics of the subalpine forest in central Colorado. In: Jacobs, B.F., Fall, P.L., Davis, O.K. (Editors), Late Quaternary vegetation and climates of the American Southwest. AASP Contr. Ser., 16: 31–46.

Fall, P.L., 1987. Pollen taphonomy in a canyon stream. Quat. Res., 28: 393–406.

Fall, P.L., Kelso, G., Markgraf, V., 1981. Paleoenvironmental reconstruction at Canyon del Muerto, Arizona, based on principal- component analysis. J. Archaeol. Sci., 8: 297–307.

Fall, P.L., Lindquist, C.A., Falconer, S.E., 1990. Fossil hyrax middens from the Middle East: a record of paleovegetation and human disturbance. In: Betancourt, J.L., Van Devender, T.R., Martin, P.S. (Editors), Packrat middens: the last 40,000 years of biotic change. Univ. of Arizona Press, Tucson, 408–427.

Faure, H., 1969. Le Quaternarie du Sahara. Mitt. Internat. Verein. Limnol., 17: 131–146.

Fish, S.K., 1984. Salt–Gila aqueduct project pollen analysis. In: Teague, L.S., Crown, P.L. (Editors), Hohokam archeology along the Salt–Gila aqueduct, central Arizona project, VII: Environment and subsistence. Arizona State Museum Archeological Ser., Tucson, 150: 9–16; 39–51; 111–138.

Frumkin, A., 1990. Branching caves as information sources for the morphotectonic development of Israel. Ran Gerson Memorial Vol., Dept. of Geography, The Hebrew Univ. of Jerusalem, pp. 7–8.

Fuji, N., Horowitz, A., 1989. Brunhes Epoch paleoclimates of Japan and Israel. Palaeogeogr., Palaeoclimatol., Palaeoecol., 72: 79–88.

Galili, E., Weinstein-Evron, M., 1985. Prehistory and paleoenvironments of submerged sites along the Carmel coast of Israel. Paléorient, 11/1: 37–52.

Garfunkel, Z., Horowitz, A., 1966. The upper Tertiary and Quaternary morphology of the Negev, Israel. Israel J. Earth Sci., 15: 101–117.

Garner, G.E., Bryant, V.M. Jr., 1973. Preparation of modern palynomorphs for scanning Electron microscopy by the critical point drying method. Geoscience and Man, VII: 83–88.

Gasse, F., Rognon, P., Street, F.A., 1980. Quaternary history of the Afar and Ethiopian rift lakes. In: Williams, M.A.J., Faure, H. (Editors), The Sahara and the Nile. A.A. Balkema, Rotterdam, pp. 361–380.

Gat, J.R., Galai, A., 1982. Groundwaters of the Arava Valley: an isotopic study of their origin and interrelationships. Israel J. Earth Sci., 31: 25–38.

Gat, J.R., Magaritz, M., 1980. Climatic variations in the eastern Mediterranean sea area. Naturwissenschaft, 67: 80–87.

Gerson, R., 1981. Geomorphic aspects of the Elat mountains. In: Dan, J., Gerson, R., Koyumdjisky, H., Yaalon, D.H., (Editors), Aridic soils of Israel. The Volcani Center, Bet Dagan, Israel, Sp. Publ. 50: 279–296.

Gerson, R., Amit, R., 1987. Rates and modes of dust accretion and deposition in an arid region — the Negev, Israel. In: Frostick, L., Reid, I. (Editors), Desert sediments: ancient and modern. Geol. Soc. Am. Sp. Publ. 35: 157–169.

Gilead, I., 1989. Grar: a Chalcolithic site in the northern Negev, Israel. J. Field Archaeology, 16(4): 377–394.

Ginzbourg, D., 1963. Petrography of the loess in the Beer Sheva basin. Unpubl. MSc Thesis, Dept. of Geology, The Hebrew Univ. of Jerusalem, 126 pp. (in Hebrew, English summary).

Girard, M., Maley, J., 1987. Étude palynologique. In: David, L., Mourer, R. (Editors), Autopsie d'un mommie Egyptienne du Musée de Lyon. Nouv. Arch. Mus. Hist. Nat. Lyon, Fasc. 25: 103–110.

Glennie, K.W., 1970. Desert sedimentary environments. Elsevier Science Publ., Amsterdam, 222 pp.

Goldberg, M. and Friedman, J.M., 1974. Paleoenvironments and paleogeographic evolution of the Jurassic system in southern Israel. Geol. Surv. Israel Bull., 61: 1–44.

Goldberg, P., 1976. Upper Pleistocene geology of the Avdat/Aqev area. In: Marks, A.E. (Editor), Prehistory and paleoenvironments in the central Negev, Israel. I. The Avdat/Aqev area. SMU Press, Dallas., 25–56.

Goldberg, P., 1986. Late Quaternary environmental history of the southern Levant. Geoarchaeology, 1(3): 225–244.

Goldsmith, N.F., Hirsch, F., Friedman, G.M., Tchernov, E., Derin, B., Gerry, E., Horowitz, A., Weinberger, G., 1988. Rotem mammals and Yeroham crassostreids: stratigraphy of the Hazeva formation (Israel) and the paleogeography of Miocene Africa. Newsl. Stratigr. 20(2): 73–90.

Gonen, R., 1989. The Chalcolithic period. The Open Univ. of Israel, Tel Aviv, 64 pp. (in Hebrew).

Goren, N., 1981. The lithic assemblages of the site of Ubeidiya, Jordan Valley. Unpubl. PhD Dissertation, Dept. of Archaeology, The Hebrew Univ. of Jerusalem, 202 pp.

Goring-Morris, A.N., 1987. At the edge: Terminal Pleistocene hunters–gatherers in the Negev and Sinai. BAR Intl. Ser., 361, 526 pp.

Gothan, W., Weyland, H., 1954. Lehrbuch der Paläobotanik. Akademie Verlag, Berlin, 535 pp.

Gregor, H.J., Velitzelos, E., 1987. Evolution of Neogene Mediterranean vegetation and the question of a dry upper Miocene period (Salinity Crisis). Ann. Inst. Geol. Publ. Hung., LXX: 489–496.

Guillaume, A., Guillaume, S., 1982. L'érosion dans les Alpes au Plio-Quaternaire et au Miocène. Eclogae Geol. Helv., 75/2: 247–268.

Gvirtzman, G., 1970. The Saqiye group in the coastal plain and Hashefela regions, Israel. Unpubl. PhD Dissertation, Dept. of Geology, The Hebrew Univ. of Jerusalem, 170 + X pp. (in Hebrew, English summary).

Hafsten, U., 1959. Bleaching + HF + acetolysis, a hazardous preparation process. Pollen et Spores, I(1): 77–79.

Hall, S.A., 1981. Deteriorated pollen grains and the interpretation of Quaternary pollen diagrams. Rev. Palaeobotan. Palynol., 32: 193–206.

Hamilton, A.C., 1982. Environmental history of East Africa. Academic Press, London, 328 pp.

Hart, G.F., 1965. The systematics and distribution of Permian miospores. Witwatersrand Univ. Press, Johannesburg, 252 pp.

Hart, G.F., 1970. The biostratigraphy of Permian palynofloras. Geoscience and Man, I: 89–131.

Havinga, A.J., 1984. A 20-year experimental investigation into the differential corrosion susceptibility of pollen and spores in various soil types. Pollen et Spores, XXVI(3–4): 541–558.

Heimann, A., 1990. The development of the Dead Sea Rift and its margins in northern Israel during the Pliocene and the Pleistocene. Geol. Surv. Israel, Jerusalem, Rep. GSI/28/90, 83 pp.

Heimann, A., Steinitz, G., 1989. $^{40}Ar/^{39}Ar$ total gas ages of basalts from Notera 3 well, Hula Valley, Dead Sea Rift: stratigraphic and tectonic implications. Israel J. Earth Sci., 38: 173–184.

Heusser, L.E., Stock, C.E., 1984. Preparation techniques for concentrating pollen from marine sediments and other sediments with low pollen density. Palynology, 8: 225–227.

Hill, J.N., Hevly, R.H., 1968. Pollen at Broken K Pueblo: some new interpretations. American Antiquity, 33: 200–210.

Holloway, R.G., 1981a. Pollen exine deterioration and preservation in Brady Reservoir. National Parks Service Tech. Rep., Santa Fe, 93 pp.

Holloway, R.G., 1981b. Preservation and experimental diagenesis of the pollen exine. Unpubl. PhD Dissertation, Texas A&M Univ., College Station, 317 pp.

Holloway, R.G., 1985. Diet and medicinal plant usage of a Late Archaic population from Culberson County, Texas. Bull. of the Texas Archeological Society, 54: 319–329.

Holloway, R.G., 1989. Experimental mechanical pollen degradation and its application to Quaternary age deposits. Texas J. Sci., 41(2): 131–145.

Holloway, R.G., Bryant, V.M. Jr., 1986. New directions of palynology in ethnobiology. J. Ethnobiol., 6(1): 47–65.

Hooghiemstra, H., 1988a. Palynological records from northwest African marine sediments: a general outline of the interpretation of the pollen signal. Phil. Trans. R. Soc. Lond., B318: 431–449.

Hooghiemstra, H., 1988b. Changes of major wind belts and vegetation zones in northwest Africa 20,000–5,000 YR B.P., as deduced from a marine pollen record near Cap Blanc. Rev. Palaeobotan. Palynol., 55: 101–140.

Hooghiemstra, H., 1989. Variations of the NW African trade wind regime during the last 140,000 years: changes in pollen flux evidenced by marine sediment records. In: Leinen, M., Sarntheim, M., (Editors), Paleoclimatology and paleometeorology: modern and past patterns of global atmospheric transport. Kluwer Academic Publ., Dordrecht, 733–770.

Hooghiemstra, H., Agwu, C.O.C., 1986. Distribution of palynomorphs in marine sediments: A record for seasonal wind patterns over NW Africa and adjacent Atlantic. Geol. Rund., 75/1: 81–95.

Hooghiemstra, H., Agwu, C.O.C., 1988. Changes in the vegetation and trade winds in equatorial northwest Africa 140,000–70,000 YR B.P. as deduced from two marine pollen records. Palaeogeogr., Palaeoclimatol., Palaeoecol., 66: 173–213.

Hooghiemstra, H., Agwu, C.O.C., Beug, H.J., 1986. Pollen and spore distribution in recent marine sediments: a record of NW African seasonal wind patterns and vegetation belts. Meteor Forsch. Ergebn., C40: 87–135.

Hooghiemstra, H., Bechler, A., Beug, H.J., 1987. Isopollen maps for 18,000 years B.P. of the Atlantic offshore of northwest Africa: evidence for paleowind circulation. Paleoceanography, 2(6): 561–582.

Horowitz, A., 1966. Tropical and northern pollen and spores in Recent sediments from the Bay of Elat

(Aqaba). Israel J. Earth Sci., 15: 125–130.

Horowitz, A., 1968. Application of palynological methods in solving hydrodynamic problems (Abstract). Israel J. Earth Sci., 17: 46.

Horowitz, A., 1969. Recent pollen sedimentation in Lake Kinneret, Israel. Pollen et Spores, XI(2): 353–384.

Horowitz, A., 1970. Palynological tracing of saline water sources in Lake Kinneret region (Israel). J. Hydrol., 10: 177–184.

Horowitz, A., 1971. Climatic and vegetational developments in northeastern Israel during upper Pleistocene–Holocene times. Pollen et Spores, XIII(2): 255–278.

Horowitz, A., 1972. Probable palaeogeographic implications of the global distribution of the late Permian Cathaysian microflora. An. Acad. Brasil. Cienc., 44: 173–177.

Horowitz, A., 1973a. Development of the Hula Basin, Israel. Israel J. Earth Sci., 22: 107–139.

Horowitz, A., 1973b. Late Permian palynomorphs from southern Israel. Pollen et Spores, XV(2): 315–341.

Horowitz, A., 1974a. Some pollen spectra from the Neogene of Israel. Pollen et Spores, XVI(1): 59–65.

Horowitz, A., 1974b. Palynostratigraphy of the subsurface Carboniferous, Permian and Triassic in southern Israel. Geoscience and Man, IX: 63–70.

Horowitz, A., 1974c. Espèces du genre *Veryhachium* du Permo-Trias du sud d'Israël. Rev. Micropaléontologie, 17(2): 75–80.

Horowitz, A., 1975. Preliminary palaeoenvironmetal implications of pollen analysis of Middle Breccia from Sterkfontein. Nature, 258: 417–418.

Horowitz, A., 1976. Late Quaternary paleoenvironments of prehistoric settlements in the Avdat/Aqev area. In: Marks, A.E. (Editor), Prehistory and paleoenvironments of the Central Negev, Israel. I: The Avdat/Aqev area. SMU Press, Dallas, 57–68.

Horowitz, A., 1977a. Pollen spectra from two early Holocene prehistoric sites in the Har Harif (west central Negev). In: Marks, A.E. (Editor), Prehistory and paleoenvironments in the Central Negev, Israel. II: The Avdat/Aqev area, (part 2) and the Har Harif. SMU Press, Dallas, 323–326.

Horowitz, A., 1977b. Climatic changes: atmospheric reaction to astronomic and geotectonic processes. In: Horie, S. (Editor), Paleolimnology of Lake Biwa and the Japanese Pleistocene. Kyoto Univ. Press, 5: 302–312.

Horowitz, A., 1979. The Quaternary of Israel. Academic Press, New York, 394 pp.

Horowitz, A., 1980. Palynology, climate and settlement pattern in Israel. Qadmoniot, XIII(3–4): 80–86 (in Hebrew).

Horowitz, A., 1987a. Subsurface palynostratigraphy and paleoclimates of the Quaternary Jordan Rift Valley fill, Israel. Israel J. Earth Sci., 36: 31–44.

Horowitz, A., 1987b. Travertines of the arid regions, oxygen isotope stages, and late Quaternary climates of Israel. Quat. Res., 27: 103–105.

Horowitz, A., 1987c. Palynological evidence for the age and rate of sedimentation along the Dead Sea Rift, and structural implications. Tectonophysics, 141(1–3): 107–116.

Horowitz, A., 1988a. Synthèse de la stratigraphie et des paléoclimats quaternaires du rift de la Mer Morte, Israël. L'Anthropologie, 92(3): 753–767.

Horowitz, A., 1988b. Palynology. In: Drinkard, J.F., Mattingly, G.L., Maxwell-Miller, J. (Editors), Benchmarks in time and culture. Scholars Press, Atlanta, 261–278.

Horowitz, A., 1989a. Adaptation of pollen production in arid environments. In: Spanier, E., Steinberger, Y., Luria, M., Environmental quality and ecosystem stability. ISEEQS Publ., Jerusalem, IV/B: 81–83.

Horowitz, A., 1989b. Continuous pollen diagrams for the last 3.5 m.y. from Israel: vegetation, climate and correlation with the oxygen isotope record. Palaeogeogr., Palaeoclimatol., Palaeoecol., 72: 63–78.

Horowitz, A., 1989c. Palynological evidence for the Quaternary rates of accumulation along the Dead Sea Rift, and structural implications. Tectonophysics, 164: 63–71.

Horowitz, A., 1989d. Prehistoric cultures of Israel: correlation with the oxygen isotope scale. In: Bar Yosef, O., Vandermeersch, B. (Editors), Investigations in South Levantine prehistory. BAR Intl. Ser., 497: 5–17.

Horowitz, A., 1990. Neogene palynostratigraphy, paleoflora and paleoclimatic reconstruction in Israel. Paléobiologie Continentale, XVII: 245–257.

Horowitz, A., Assaf, G., 1981. The pluvial and interpluvial climates of the Levant. Beih. Tübinger Atlas Vord. Orients, A8: 110–120.

Horowitz, A., Derin, B., 1987. Palynological correlation of continental and marine Neogene sequences in Israel. Ann. Inst. Geol. Publ. Hung., LXX: 503–507.

Horowitz, A., Gat, J.R., 1984. Floral and isotopic indications for possible summer rains in Israel during wetter climates. Pollen et Spores, XXVI(1): 61–68.

Horowitz, A., Horowitz, M., 1985. Subsurface late Cenozoic palynostratigraphy of the Hula Basin, Israel. Pollen et Spores, XXVII(3–4): 365–390.

Horowitz, A., Horowitz, M., 1990. Neogene — recent paleogeography of the southern Dead Sea. Unpubl. Rep., Israel National Oil Co., Tel Aviv, 153 pp.

Horowitz, A., Langozky, Y., 1965. Preliminary palynological study of hydrocarbons in Israel. Geologie en Mijnbouw, 44: 59–63.

Horowitz, A., Weinstein-Evron, M., 1986. The late Pleistocene climate of Israel. Bull. Assoc. Fr. Étude Quaternaires, 25–26: 84–90.

Horowitz, A., Zak, I., 1968. Preliminary palynological analysis of an evaporitic sequence from Mount Sedom, Israel. Rev. Palaeobotan. Palynol., 7: 25–30.

Horowitz, A., Siedner, G., Bar Yosef, O., 1973. Radiometric dating of the Ubeidiya formation, Jordan Valley, Israel. Nature, 242: 186–187.

Horowitz, A., Weinstein, M., Ganor, E., 1975. Palynological determination of dust storms' provenances in Israel. Pollen et Spores, XVII(2): 223–231.

Horowitz, A., Sampson, C.G., Scott, L., Vogel, J.C., 1978. Analysis of the Voightspost site, O.F.S., South Africa. S. Afr. Archaeol. Bull., 33: 152–159.

Horowitz, A., Gerald, R.E., Chaiffetz, M.S., 1981. Preliminary paleoenvironmental implications of pollen analyzed from Archaic, Formative, and Historic sites near El Paso, Texas. Texas J. Sci., XXXIII(1): 61–71.

Hours, F., 1975. The Lower Paleolithic of Lebanon and Syria. In: Wendorf, F., Marks, A.E. (Editors), Problems in prehistory: North Africa and the Levant. SMU Press, Dallas, 249–271.

Hours, F., Copeland, L., Aurenche, O., 1973. Les industries paléolithiques du Proche-Orient, essai de correlation. L'Anthropologie, 77: 229–280; 437–496.

Jacobs, B.F., 1985. Identification of pine pollen from the Southwestern United States. In: Jacobs, B.F., Fall, P.L., Davis, O.K. (Editors), Late Quaternary vegetation and climates of the American Southwest. AASP Contr. Ser., 16: 155–168.

Jacobs, B.F., Fall, P.L., Davis, O.K. (Editors), 1985. Late Quaternary vegetation and climates of the American Southwest. AASP Contr. Ser. 16, 185 pp.

Jansonius, J., 1962. Palynology of Permian and Triassic sediments, Peace River area, western Canada. Palaeontographica, B110: 35–98.

Jelinek, A.J., 1966. Correlation of archeological and palynological data. Science, 152: 1507–1509.

Jelinek, A.J., Farrand, W.R., Haas, G., Horowitz, A., Goldberg, P., 1973. New excavations at the Tabun Cave, Mount Carmel, Israel, 1967–1972: a preliminary report. Paléorient, 1(2): 151–183.

Kempinski, A., in press. Masos. Institute of Archaeology, Tel Aviv University.

Kershaw, A.P., Nix, H.A., 1988. Quantitative paleoclimatic estimates from pollen data using profiles of extant taxa. J. Biogeogr., 15: 589–602.

Köppen, W., 1923. Die klimate der Erde — Grundriss der Klimakunde. Walter de Gruyter, Berlin, X + 369 pp.

Koreneva, E.V., 1971. Spores and pollen in Mediterranean bottom sediments. In: Funnel, B.M., Reidel, W.R. (Editors), The micropaleontology of oceans. Cambridge Univ. Press, Cambridge, 361–371.

Kremp, G.O.W., 1974. A re-evaluation of global plantgeographic provinces of the late Paleozoic. Rev. Palaeobotan. Palynol., 17: 113–132.

Kremp, G.O.W., 1977. The positions and climatic changes of Pangaea and five southeast Asian plates during Permian and Triassic times. Paleo Data Banks, Palynodata, Tucson, 7: 1–21.

Kummel, B., Raup, D., 1965. Handbook of paleontological techniques. Freeman, San Francisco, 852 pp.

Lavee, H., Wieder, M., 1990. Salt and gypsum crusts and interflow in the Judean Desert. Ran Gerson Memorial Volume, Dept. of Geography, The Hebrew Univ. of Jerusalem, 5–6.

Leffingwell, H.A., Hodgkin, N., 1970. Techniques for preparing fossil palynomorphs for study with the scanning and transmission electron microscopes. Rev. Palaeobotan. Palynol., 11: 177–199.

Leroi-Gourhan, A., 1974. Les possibilités de l'analyse pollinique en Syrie at au Liban. Paléorient, I: 39–47.

Leroi-Gourhan, A., 1975. The flowers found with Shanidar IV, a Neanderthal burial in Iraq. Science, 190: 562–564.

Leroi-Gourhan, A., 1980. Analyses polliniques à Tell Aswad. Paléorient, V: 170–176.

Leroi-Gourhan, A., 1982. Palynological research of Near Eastern archaeological sites. BAR Intl. Ser., 133(II): 353–356.

Leroi-Gourhan, A., Darmon, F., 1987. Analyses palynologiques de sites archéologiques du Pléistocène final dans la Vallée du Jourdain. Israel J. Earth Sci., 36: 65–72.

Leroi-Gourhan, A., Renault-Miskovsky, J., 1977. La palynologie appliquée a l'archéologie. In: Approche écologique de l'Homme fossile, Suppl. Bull. Assoc. Fr. Étude Quaternaire, 35–49.

Levin, N., Horowitz, A., 1987. Palynostratigraphy of the early Pleistocene QI Palynozone. Israel J. Earth Sci., 36(1–2): 45–58.

Lézine, A.M., 1988. New pollen data from the Sahel, Senegal. Rev. Palaeobotan. Palynol., 55: 141–154.

Lézine, A.M., 1989. Le Sahel: 20,000 ans d'historie de la végétation. Bull. Soc. Géol. France, 8(V,1): 35–42.

Lézine, A.M., Edorh, T.M., 1991. Modern pollen deposition in West African Sudanian environments. Rev. Palaeobotan. Palynol., 67: 41–58.

Lézine, A.M., Hooghiemstra, H., 1990. Land–sea comparisons during the last glacial–interglacial transition: pollen records from West Tropical Africa. Palaeogeogr., Palaeoclimatol., Palaeoecol., 79: 313–331.

Liu, T.S. (Editor), 1987. Aspects of loess research. China Ocean Press, Beijing, 447 pp.

Liu, T.S. (Editor), 1991. Loess, environment and global change. Science Press, Beijing, 288 pp.

Livingstone, D.A., 1975. Late Quaternary climatic change in Africa. Ann. Rev. of Ecology and Systematics, 6: 249–280.

Livnat, A., Kronfeld, J., 1985. Paleoclimatic implications of U- series dates for lake sediments and travertines in the Arava rift valley, Israel. Quat. Res., 24: 164–172.

Lobreau-Callen, D., 1980. Variations morphologiques du pollen du genre *Thessium* (Santalaceae). Mém. Mus. Nat. d'Histoire Naturelle, Paris, B(XXVII): 23–30.

Magaritz, M., 1986. Environmental changes recorded in the upper Pleistocene along the desert boundary, southern Israel. Palaeogeogr., Palaeoclimatol., Palaeoecol., 53: 213–229.

Maher, L.J., 1972. Nomograms for computing 0.95 confidence limits of pollen data. Rev. Palaeobotan. Palynol., 13: 85–93.

Maley, J., 1972. La sédimentation pollinique actuelle dans la zone du Lac Tchad (Afrique Central). Pollen et Spores, XIV(3): 263–307.

Maley, J., 1973. Les variations climatiques dans le bassin du Tchad durant le dernier millenaire: essai d'interpretation climatique de l'Holocène Africain. C. R. Acad. Sci. Paris, D276: 1673–1675.

Maley, J., 1981. Études palynologiques dans le bassin du Tchad et paléoclimatologie de l'Afrique nord-tropicale de 30,000 ans à l'époque actuelle. O.R.S.T.O.M., Paris, 586 pp.

Maley, J., 1989. 20,000 ans d'évolution des climats du Sahara central aux savanes tropicales humides. In: Centre Culturel Français d'Abidjan (Editor), Sud Sahara–Sahel Nord, 35–52.

Maley, J., Cohen, J., Faure, H., Rognon, P., Vincent, P.M., 1970. Quelques formations lacustres et fluviatiles associées a differentes phases du volcanisme au Tibesti (Nord du Tchad). Cah. O.R.S.T.O.M., Ser. Géol., 2(1): 127–152.

Marcus, E., Slager, J., 1985. The sedimentary-magmatic sequence of the Zemah 1 well (Jordan–Dead Sea Rift, Israel) and its emplacement in time and space. Israel J. Earth Sci., 34: 1–10.

Markgraf, V., 1985. Late Pleistocene faunal extinctions in southern Patagonia. Science, 228(4703): 1110–1112.

Marks, A.E. (Editor), 1976. Prehistory and paleoenvironments in the central Negev, Israel. I: The Avdat/Aqev area, part 1. SMU Press, Dallas, 383 pp.

Marks, A.E. (Editor), 1977. Prehistory and paleoenvironments in the central Negev, Israel. II: The Avdat/Aqev area, part 2, and the Har Harif. SMU Press, Dallas, 356 pp.

Marks, A.E. (Editor), 1983. Prehistory and paleoenvironments in the central Negev, Israel. III: The Avdat/Aqev area, part 3. SMU Press, Dallas, 349 pp.

Martin, P.S., 1963 (1970 — second printing). The last 10,000 years. Univ. of Arizona Press, Tucson, 87 pp.

Martin, P.S., Schoenwetter, J., 1960. Arizona's oldest cornfield. Science, 132: 33–34.

Martin, P.S., Sables, B.E., Shutler, D., 1961. Rampart Cave coprolite and ecology of the shasta ground sloth. Am. J. Sci., 259: 102–127.

McKee, E.D., Crosby, E.J., Berryhill, H.L., 1967. Flood deposits, Bijou Creek, Colorado, June 1965. J. Sediment. Petrol., 37(3): 829–852.

Mead, J.I., O'Rourke, M.K., Foppe, T.M., 1986. Dung and diet of the extinct Harrington's mountain goat (*Oreamnos harringtoni*). J. Mamm., 67(2): 284–293.

Mehringer, P.J. Jr., 1967. Pollen analysis of the Tule Springs area, Nevada. Nevada State Museum Archeological Papers, Carson City, 13: 130–200.

Mehringer, P.J. Jr., Haynes, C.V., 1965. The pollen evidence for the environment of early man and extinct mammals at the Lehner Mammoth Site, southeastern Arizona. American Antiquity, 31(1): 17–23.

Mehringer, P.J. Jr., Petersen, K.L., Hassan, F.A., 1979. A pollen record from Birket Qarun and the recent history of the Fayum, Egypt. Quat. Res., 11: 238–256.

Meigs, P., 1953. World distribution of arid and semi-arid homoclimates. In: Reviews of research on arid zone hydrology, UNESCO arid zone research, 1: 203–209.

Meyer, E.R., 1977. A reconnaissance survey of pollen rain in Big Bend National Park, Texas: modern control for a paleoenvironmental study. In: Wauer, R.H., Riskind, D.H. (Editors), Chihuahuan Desert. US Dept. of the Interior, Alpine, Texas, 115–123.

Michel, M., 1955. Techniques utilisées en analyse pollinique pour l'étude des sédiments quaternaires. Unpubl. PhD Dissertation, Fac. des Sciences, University de Paris, 98 pp.

Michelson, H., 1972. The hydrogeology of the southern Golan plateau. Water Planning for Israel, Tel Aviv, 89 pp. (in Hebrew).

Moar, N.T., 1969. Possible long-distance transport of pollen to New Zealand. N.Z. J. Bot., 7: 424–426.

Moe, D., 1983. Palynology of sheep's faeces: relationship between pollen content, diet and local pollen rain. Grana, 22: 105–113.

Moore, A.M.T., 1979. A pre-Neolithic farmers' village on the Euphrates. Scientific American, 241(2): 50–58.

Moore, P.D., Webb, J.A., Collinson, M.E., 1991. Pollen analysis. Blackwell Scientific Publ., Oxford, 224 pp.

Morzadec-Kerfourn, M.T., 1980. Les kystes de dinoflagellés et le climat: *Hemicystodinium zoharyi* (Rossignol), espèce indicatrice d'aridité au Quaternaire. Mém. Mus. Nat. d'Histoire Naturelle, Paris, B(XXVII): 283–290.

Morzadec-Kerfourn, M.T., 1988. Paléoclimats et paléoenvironnements, du tardiglaciaire au récent, en Méditerranée orientale, à l'est du delta du Nil; l'apport des microfossiles à membrane organique. Bull. Centre Rech. Explor.-Prod. Elf Aquitaine, 12(1): 267–275.

Mossiman, J.E., 1965. Statistical methods for the pollen analyst: multinomial and negative multinomial techniques. In: Kummel, B., Raup, D. (Editors), Handbook of paleontological techniques. Freeman, San Francisco, 636–673.

Muller, J., 1959. Palynology of recent Orinoco delta and shelf sediments. Micropaleontology, 5: 1–32.

Naveh, Z., 1990. Fire in the Mediterranean — a landscape ecological perspective. In: Goldammer, J.G., Jenkins, M.J. (Editors), Fire in ecosystem dynamics. SPB Academic Publishers, The Hague, pp. 1–20.

Neev, D., 1960. A pre-Neogene erosion channel in the southern coastal plain of Israel. Geol. Surv. Israel Bull., 25: 1–21.

Nieto, M.A., D'Antoni, H.L., 1984. Pollen analysis of sediments of the Atlantic shore at Mar Chiquita (Buenos Aires Province, Argentina). Zbl. Geol. Paläont. Teil I, H. 11/12: 1731–1738.

Ogden, E.C., Raynor, G.S., Hayes, J.V., Lewis, D.M., Haines, J.H., 1974. Manual for sampling airborne pollen. Hafner Press, New York, 182 pp.

Oller, M.R., D'Antoni, H.L., Nieto, M.A., 1984–1985. Contribuciones a la arqueoecologìa de Pampa Grande, Provincia de Salta. Relaciones Soc. Argentina de Antropologìa, XVI: 153–163.

O'Rourke, M.K., 1982. Pollen content and character of adobe brick from Fort Lowell. Arizona State Museum Archeological Ser., Tucson, 163: 75–85.

O'Rourke, M.K., 1983. Pollen from adobe brick. J. Ethnobiol., 3(1): 39–48.

O'Rourke, M.K., Mead, J.I., 1985. Late Pleistocene and Holocene pollen records from two caves in the Grand Canyon of Arizona, USA. In: Jacobs, B.F., Fall, P.L., Davis, O.K. (Editors), Late Quaternary vegetation and climates of the American Southwest. AASP Contr. Ser., 16: 169–186.

Ouyang, S., 1982. Upper Permian and lower Triassic palynomorphs from eastern Yunnan, China. Canadian J. Earth Sci., 19: 68–80.

Ouyang, S., Utting, J., 1990. Palynology of upper Permian and lower Triassic rocks, Meishan, Changxing County, Zhejiang Province, China. Rev. Palaeobotan. Palynol., 66(1/2): 65–103.

Palynologie et climats, 1980. Mém. Mus. Nat. d'Historie Naturelle, Paris, B(XXVII), 302 pp.

Pearson, R., 1978. Climate and evolution. Academic Press, London, 274 pp.

Peck, R.M., 1973. Pollen budget studies in a small Yorkshire catchment. In: Birks, H.J.B., West, R.G. (Editors), Quaternary plant ecology. Blackwell Scientific Publ., Oxford, 43–60.

Petit-Maire, N., 1990. Natural aridification or man-made desertification? a question for the future. In: Paepe, R., Fairbridge, R.W., Jelgersma, S. (Editors), Greenhouse effect, sea level and drought. Kluwer Academic Publ., Dordrecht, 281–285.

Pianka, E.R., 1979. Diversity and niche structure in desert communities. In: Goodall, D.W., Perry, R.A., Howes, K.M. (Editors), Arid-land ecosystems: structure, functioning and management. Cambridge Univ. Press, Cambridge, 1: 321–341.

Pons, A., Quezel, P., 1958. Premières remarques sur l'étude palynologique d'un guano fossile du Hogger. C. R. Acad. Sci. Paris, 244: 2290–2292.

Pons, A., Thinon, M., 1987. The role of fire from palaeoecological data. Ecologia Mediterranea, XIII(4): 3–11.

Potter, L.D., Krenetsky, J.C., 1967. Plant succession with released grazing on New Mexico range lands. Journal of Range Management, 20(3): 145–151.

Prance, G.T. (Editor), 1982. Biological diversification in the tropical. Columbia Univ. Press, New York, 714 pp.

Prell, W.L., Van Campo, E., 1986. Coherent response of Arabian Sea upwelling and pollen transport to late Quaternary monsoonal winds. Nature, 323: 526–528.

Pye, K., Tsoar, H., 1987. The mechanics and geological implications of dust transport and deposition in deserts with particular reference to loess formation and dune sand diagenesis in the northern Negev, Israel. In: Frostick L., Reid, I. (Editors), Desert sediments: ancient and modern. Geol. Soc. Am. Sp. Publ. 35: 139–156.

Red Sea Charts, 1848. Sailing directions for the Red Sea and the Gulf of Aden. H.O., Publ. 157, London.

Reille, M., 1979. Analyse pollinique du lac de Sidi Bou Rhaba, littoral atlantique (Maroc). Ecologia Mediterranea, 4: 61–65.

Renault-Miskovsky, J., Bui-Thi-Maï, Girard, M., 1985. Palynologie archéologique. Éditions CNRS, Paris, 502 pp.

Ritchie, J.C., 1984. Analyse pollinique de sédiments Holocènes superiéurs des hauts plateaux du Maghreb oriental. Pollen et Spores, XXVI(3–4): 489–496.

Ritchie, J.C., 1987. A Holocene pollen record from Bir Atrun, northwest Sudan. Pollen et Spores, XXIX(4): 391–410.

Ritchie, J.C., Haynes, C.V., 1987. Holocene vegetation zonation in the eastern Sahara. Nature, 330: 645–647.

Robinson, P.L., 1973. Paleoclimatology and continental drift. In: Tarling D.H., Runcorn S.K. (Editors), Implications of continental drift to the Earth sciences. Academic Press, London, 451–476.

Roeland, J.C., Guiot, J., Bonnefille, R., 1988. Pollen et reconstruction quantitative du climat. Validation des données d'Afrique orientale. C. R. Acad. Sci. Paris, 307(II): 1735–1740.

Ronen, A., 1979. Paleolithic industries. In: Horowitz, A., The Quaternary of Israel. Academic Press, New York, 296–307.

Rossignol, M., 1961. Analyse pollinique de sédiments marins quaternaires en Israël. I. Sédiments récents. Pollen et Spores, III(2): 303–324.

Rossignol, M., 1962. Analyse pollinique de sédiments marins quaternaires en Israël. II. Sédiments pléistocènes. Pollen et Spores, IV(1): 121–148.

Rossignol, M., 1964. Analyse pollinique de sédiments quaternaires dans la plaine de Haifa, Israël. Israel J. Earth Sci., 12(4): 207–214.

Rossignol, M., 1969a. Sédimentation palynologique récente dans la Mer Morte. Pollen et Spores, XI(1): 17–38.

Rossignol, M., 1969b. Sédimentation palynologique dans le domain marin quaternaire de Palestine. Notes et Mémoirs sur le Moyen Orient, X: 1–270.

Rossignol-Strick, M., 1983. African monsoons, an immediate climate response to orbital insolation. Nature, 304: 46–49.

Ruddiman, W.R., Kutzbach, J.E., 1991. Plateau uplift and climatic change. Scientific American, 264(3): 42–50.

Savage, R.J.G., Tchernov, E., 1968. Miocene mammals of Israel. Proc. Geol. Soc. London, 98–101.

Schultz, B.C., Frye, J.C. (Editors), 1968. Loess and related eolian deposits of the world. Univ. of Nebraska Press, Lincoln, 369 pp.

Schulz, E., 1984. The recent pollen rain in the eastern central Sahara — a transect between northern Lybia and southern Niger. Palaeoecology of Africa, 16: 245–253.

Schulz, E., 1987a. Holocene palaeoenvironments in the central and southern Sahara. In: Matheis, G., Schandelmeier, H. (Editors), Current research in African Earth sciences. A.A. Balkema, Rotterdam, 439–442.

Schulz, E., 1987b. Die Holozäne vegetation der zentralen Sahara (N-Mali, N-Niger, SW-Lybien). Palaeoecology of Africa, 18: 143–161.

Schulz, E., 1990. Zwischen Syrte und Tschad, der aktuelle pollenniederschlag. Berliner Geographische Studien, 30: 193–220.

Schulz, E., Whitney, J.W., 1986. Upper Pleistocene and Holocene lakes in the An Nafud, Saudi Arabia. Hydrobiologia, 143: 175–190.

Schulz, E., Whitney, J.W., 1987. Upper Pleistocene and Holocene palaeoenvironments in the An Nafud, Saudi Arabia. In: Matheis, G., Schandelmeier, H. (Editors), Current research in African Earth sciences. A.A. Balkema, Rotterdam, 443–446.

Scott, L., 1975. Preliminary palynological results from the Alexandersfontein basin near Kimberley. Ann. S. Afr. Mus., 71: 193–199.

Scott, L., 1982. A late Quaternary pollen record from the Tranasvaal bushveld, South Africa. Quat. Res., 17: 339–370.

Scott, L., 1987. Pollen analysis of hyena coprolites and sediments from Equus Cave, Taung, southern Kalahari (South Africa). Quat. Res., 28: 144–156.

Scott, L., 1988. Holocene environmental change at western Orange Free State pans, South Africa, inferred from pollen analysis. Palaeoecology of Africa, 19: 109–118.

Scott, L., 1989. Climatic conditions in southern Africa since the last glacial maximum, inferred from pollen analysis. Palaeogeogr., Palaeoclimatol., Palaeoecol., 70: 345–353.

Scott, L., 1990. Hyrax (Procaviidae) and dassie rat (Petromuridae) middens in paleoenvironmental studies in Africa. In: Betancourt, J.L., Van Devender, T.R., Martin, P.S. (Editors), Packrat middens: the last 40,000 years of biotic change. Univ. of Arizona Press, Tucson, 398–407.

Scott, L., Bonnefille, R., 1986. Search for pollen from the hominid deposits of Kromdraai, Sterkfontein and Swartkrans: some problems and preliminary results. S. Afr. J. Sci., 82: 380–382.

Scott, L., Bousman, C.B., 1990. Palynological analysis of hyrax middens from Africa. Palaeogeogr., Palaeoclimatol., Palaeoecol., 76: 367–379.

Scott, L., Klein, R.G., 1981. A hyena-accumulated bone assemblage from late Holocene deposits at Deelpan, Orange Free State. Ann. S. Afr. Mus., 86(6): 217–227.

Seely, M., 1987. The Namib. Shell Oil SWA Ltd., Windhoek, 104 pp.

Shackleton, N.J., Hall, M.A., 1984. Oxygen and carbon isotope stratigraphy of Deep Sea Drilling Project Hole 552A: Plio- Pleistocene glacial history. In: Roberts, D.G., Schnitker, D. (Editors), Initial reports of the Deep Sea Drilling Project. US Government Printing Office, Washington, 81: 599–609.

Siedner, G., Horowitz, A., 1974. Radiometric ages of late Cainozoic basalts from northern Israel: chronostratigraphic implications. Nature, 250: 23–26.

Singh, G., 1971. The Indus Valley culture seen in the context of postglacial climatic and ecological studies in northwest India. Archaeology and Physical Anthropology in Oceania, 6: 177–189.

Singh, G., 1980. Savannah and desert vegetation and man: an introduction to the symposium on "Quaternary vegetational history of savannah and desert". Proc. 4th Intl. Palynological Congr., Lucknow, 3: 43–48.

Singh, G. (Convenor), 1988. Synposium 13: Palynology of dry lands. 7th Intl. Palynological Congr., Brisbane.

Singh, G., Chopra, S.K., Singh, A.B., 1973. Pollen rain from the vegetation of northwest India. New Phytologist, 72: 191–206.

Singh, G., Joshi, R.D., Chopra, S.K., Singh, A.B., 1974. Late Quaternary history of vegetation and climate of the Rajasthan Desert, India. Phil. Trans. R. Soc. Lond., B267: 467–501.

Singh, G., Wasson, R.J., Agrawal, D.P., 1990. Vegetational and seasonal climatic changes since the last full glacial in the Thar Desert, northwestern India. Rev. Palaeobotan. Palynol., 64: 351–358.

Sneh, A., Weissbrod, T., Ehrlich, A., Horowitz, A., Moshkovitz, S., Rosenfeld, A., 1986. Holocene evolution of the northeastern corner of the Nile delta. Quat. Res., 26: 194–206.

Sobolik, K.D., 1988. The importance of pollen concentration values from coprolites: an analysis of southwest Texas samples. Palynology, 12: 201–214.

Sonntag, C., Klitsch, E., Shazly, E.M., Calivke, C., Munnich, K.O., 1978. Paläoklimatische information im isotopegehalt ^{14}C datierter Saharawasser, kontinentaleffect in D und ^{18}O. Geol. Rund., 67: 413–424.

Spaulding, W.G., 1977. Late Quaternary vegetational change in the Sheep Range, southern Nevada. J. Arizona Acad. Sci., 22(2): 3–8.

Spaulding, W.G., 1990. Comparison of pollen and macrofossil based reconstructions of late Quaternary vegetation in western North America. Rev. Palaeobotan. Palynol., 64: 359–366.

Spaulding, W.G., Leopold, E.B., Van Devender, T.R., 1983. Late Wisconsin paleoecology of the American Southwest. In: Wright, H.E. Jr., Porter, S.C. (Editors), The late Quaternary of the United States. Univ. of Minnesota Press, Minneapolis, 1: 259–293.

Stanley, E.A., 1966. The problem of reworked pollen and spores in marine sediments. Marine Geology, 4: 397–408.

Stehli, F.G., 1968a. Taxonomic diversity gradients in pole location: the recent model. In: Drake, E.T. (Editor), Evolution and environment. Yale Univ. Press, New Haven, 163–227.

Stehli, F.G., 1968b. A paleoclimatic test of the hypothesis of an axial dipolar magnetic field. In: Phinney, R.A. (Editor), The history of the Earth's crust. Princeton Univ. Press, New York, 195–205.

Stehli, F.G., McAlester, A.L., Helsley, C.E., 1967. Taxonomic diversity of recent bivalves and some implications for geology. Geol. Soc. Am. Bull., 78: 455–466.

Straka, H., Kappen, L., 1988. Analyse palynologique de coussins de mousse dans l'Antarctide. Inst. Fr. Pondichéry, Trav. Sec. Sci. Tech., XXV: 295–300.

Stockmarr, J., 1971. Tablets with spores used in absolute pollen analysis. Pollen et Spores, XIII(4): 615–621.

Suc, J.P., 1978. Analyse pollinique de dépots Plio- Pléistocènes du sud du massif basaltique de l'Escandorgue (Site de Bernasso, Lunas, Hérault, France). Pollen et Spores, XX(4): 497–512.

Suc, J.P., 1989. Distribution latitudinale et étagement des associations végétales au Cénozoïque supérieur dans l'aire ouest-méditerranéenne. Bull. Soc. Géol. France, 8V(3): 541–550.

Swain, A.M., Kutzbach, J.E., Hastenrath, S., 1983. Estimates of Holocene precipitation for Rajasthan, India, based on pollen and lake level data. Quat. Res., 19: 1–17.

Tauber, H., 1967. Investigations of the mode of transfer of pollen in forested areas. Rev. Palaeobotan. Palynol., 3: 277–286.

Tchernov, E., 1987. The age of the Ubeidiya formation, an early Pleistocene hominid site in the Jordan Valley, Israel. Israel J. Earth Sci., 36(1–2): 3–30.

Thompson, R.S., 1985. Palynology and *Neotoma* middens. In: Jacobs, B.F., Fall, P.L., Davis, O.K. (Editors), Late Quaternary vegetation and climates of the American Southwest. AASP Contr. Ser., 16: 89–112.

Thompson, R.S., Van Devender, T.R., Martin, P.S., Foppe, T.M., Long, A., 1980. Shasta ground sloth (*Nothrotheriops shastense* Hoffstetter) at Shelter Cave, New Mexico: environment, diet, and extinction. Quat. Res., 14: 360–376.

Thornthwaite, W.C., 1948. An approach toward a rational classification of climate. Geogr. Rev., 38: 55–94.

Traverse, A., 1988. Paleopalynology. Unwin Hyman, Boston, 600 pp.

Traverse, A., 1990. Studies of pollen and spores in rivers and other bodies of water, in terms of source-vegetation and sedimentation, with special reference to Trinity River and Bay, Texas. Rev. Palaeobotan. Palynol., 64: 297–303.

Traverse, A., Ginsburg, R.N., 1966. Palynology of the surface sediments of Great Bahama Bank, as related to water movement and sedimentation. Mar. Geol., 4: 417–459.

Trevor-Deutsch, B., Bryant, V.M. Jr., 1978. Analysis of suspected human coprolites from Terra Amata, Nice, France. J. Archaeol. Sci., 5: 387–390.

Tschudy, R.H., Scott, R.A. (Editors), 1969. Aspects of palynology. Wiley, New York, 510 pp.

Usinger, H., 1980. Une relation entre la taille du pollen et le climat? Mém. Mus. Nat. d'Histoire Naturelle, Paris, B(XXVII): 51–56.

Van Campo, E., 1986. Monsoon fluctuations in two 20,000 yr. B.P. Oxygen isotope/pollen records off southwest India. Quat. Res., 26: 376–388.

Van Campo, E., Duplessy, J.C., Rossignol-Strick, M., 1982. Climatic conditions deduced from a 150,000 ky oxygen isotope–pollen record from the Arabian Sea. Nature, 296: 56–59.

Van Campo, E., Duplessy, J.C., Prell, W.L., Barrat, N., Sabatier, R., 1990. Comparison of terrestrial and marine temperature estimates for the past 135 ky off southeast Africa: a test for GCM simulations of paleoclimate. Nature, 348: 209–212.

Van Campo, M., 1964. Quelques pollen Pléistocènes nouveaux pour le Hoggar. C. R. Acad. Sci. Paris, 258(4): 1297–1299.

Van Campo, M., 1975. Pollen analyses in the Sahara. In: Wendorf, F., Marks, A.E. (Editors), Problems in Prehistory: North Africa and the Levant. SMU Press, Dallas, 45–64.

Van Campo, M., Quet, L., 1982. Transport par les vents de pollens et de poussières rouges du sud au nord de la Méditerranée. C. R. Acad. Sc. Paris, 295(II): 289–292.

Van Campo, M., Guinet, Ph., Cohen, J., 1968. Fossil pollen from late Tertiary and middle Pleistocene deposits of the Kurkur Oasis. In: Butzer, K.W., Hansen, C.L. (Editors), Desert and river in Nubia. Univ. of Wisconsin Press, Madison, 515–520.

Van Der Hammen, T., 1957. Climatic periodicity and evolution of South American Maestrichtian and Tertiary floras. Boletin Geologico, V(2): 49–91.

Van Devender, T.R., 1988. Pollen in packrat (*Neotoma*) middens: pollen transport and the relationship of pollen to vegetation — discussion. Palynology, 12: 221–229.

Van Gijzel, P., 1971. Review of the UV-fluorescence microphotometry of fresh and fossil exines and exosporia. In: Brooks, J., Grant, P.R., Muir, M., van Gijzel, P., Shaw, G. (Editors), Sporopollenin. Academic Press, London, 659–681.

Van Zeist, W., 1967. Late Quaternary vegetation history of western Iran. Rev. Palaeobotan. Palynol., 2: 301–311.

Van Zeist, W., Bottema, S., 1988. Late Quaternary vegetational and climatic history of southwest Asia. Proc. Indian Nat. Sci. Acad., 54A(3): 461–480.

Van Zinderen Bakker, E.M., 1957. A pollen analytical investigation of the Florisbad deposits (South Africa). In: Clark, J.D. (Editor), Proceedings of the third Pan-African congress on prehistory, 1955. Cambridge Univ. Press, Cambridge, 56–57.

Van Zinderen Bakker, E.M., 1984. A late- and post glacial pollen record from the Namib Desert. Palaeoecology of Africa, 16: 421–428.

Van Zinderen Bakker, E.M., Mercer, J.H., 1986. Major late Cainozoic climatic events and palaeoenvironmental changes in Africa viewed in a world wide context. Palaeogeogr., Palaeoclimatol., Palaeoecol., 56: 217–235.

Van Zinderen Bakker, E.M., Müller, M., 1987. Pollen studies in the Namib Desert. Pollen et Spores, XXIX(2–3): 185–206.

Veblen, T.T., Markgraf, V., 1988. Steppe expansion in Patagonia? Quat. Res., 30: 331–338.

Velitzelos, E., Gregor, H.J., 1987. Preliminary correlation of Oligocene to Pleistocene phytostratigraphic units of the Mediterranean and the Paratethys area. Ann. Inst. Geol. Publ. Hung., LXX: 79–83.

Velitzelos, E., Gregor, H.J., 1990. Some aspects of the Neogene floral history in Greece. Rev. Palaeobotan. Palynol., 62: 291–307.

Vidal, G., 1988. A palynological preparation method. Palynology, 12: 215–220.

Vincens, A. 1986. Diagramme pollinique d'un sondage Pléistocène supérieur–Holocène du Lac Bogoria (Kenya). Rev. Palaeobotan. Palynol., 47: 169–192.

Vishnu-Mittre, 1974. Plant remains and climate from the late Harappan and other Chalcolithic cultures of India — a study in inter-relationship. Geophytology, 4: 46–53.

Vishnu-Mittre, 1976. Palaeoecology of the Rajasthan Desert during the last 10,000 years. The Palaeobotanist, 25: 549–558.

Vishnu-Mittre, Sharma, C., 1979. Pollen analytical studies at the Nal Lake (Nalsarovar), Gujarat. The Palaeobotanist, 26(1): 95–104.

Visscher, H. 1971. The Permian and Triassic of the Kingscourt Outlier, Ireland. Geological Survey of Ireland, Dublin, Sp. Paper 1, 114 pp., 42 plates.

Visscher, H. and Krystyn, L., 1978. Aspects of Late Triassic palynology, 4. Palynological assemblage from ammonoid-controlled late Carnian (Tuvalian) sediments of Sicily. Rev. Palaeobotan. Palynol., 26: 93–112.

Visscher, H., Van Der Zwan, C.J., 1981. Palynology of the circum- Mediterranean Triassic: phytogeographical and palaeoclitological implications. Geol. Rund., 70(2): 625–634.

Von Post, L., 1916. Om skogsträdpollen i sydsvenska torfmosslagerföljder. Geol. Fören. Stockh. Förh. 38: 384–390.

Vrba, E.S., 1974. Chronological and ecological implications of the fossil Bovidae at the Sterkfontein Australopithecine site. Nature, 250: 19–23.

Walch, K.M., Rowley, J.R., Norton, N.J., 1970. Displacement of pollen grains by earthworms. Pollen et Spores, XII(1): 39–44.

Walker, T.R., 1967. Formation of red beds in modern and ancient deserts. Geol. Soc. Am. Bull., 78: 353–368.

Walton, W.R., 1964. Recent foraminiferal ecology and paleoecology. In: Imbrie, J., Newell, N., (Editors), Approaches to paleoecology. John Wiley and Sons, New York, 151–237.

Weinstein, M., 1976. The late Pleistocene vegetation and climate of the northern Golan. Pollen et Spores, XVII(4): 553–562.

Weinstein, M., 1979. Airborne pollen. In: Horowitz, A., The Quaternary of Israel. Academic Press, New York, 180–185.

Weinstein, M., 1981. The influence of slope direction on the pollen spectra. Pollen et Spores, XXIII (3–4): 381–387.

Weinstein-Evron, M., 1983. The paleoecology of the early Würm in the Hula basin, Israel. Paléorient, 9(1): 5–19.

Weinstein-Evron, M., 1986. Pollen spectra from the Acheulean site of Mitzpeh Yiron, Israel: a cautionary tale. Pollen et Spores, XXVIII(2): 157–166.

Weinstein-Evron, M., 1987a. Palynology of Pleistocene travertines from the Arava Valley, Israel. Quat. Res., 27: 82–88.

Weinstein-Evron, M., 1987b. Subsampling of palynological sequences: techniques and implications. Palynology, 11: 67–72.

Weinstein-Evron, M., 1987c. Paleoclimatic reconstructions of the late Pleistocene in the Hula Basin. Israel J. Earth Sci., 36(1–2): 59–64.

Weinstein-Evron, M., 1988. A Middle Paleolithic sequence from the Hula Valley, Israel. L'Homme de Néandertal, 2: 207–222.

Weinstein-Evron, M., 1990. Palynological history of the last pleniglacial in the Levant. In: Kozlowski, J.K. (Editor), Les industries à pointes foliacées en Europe. Eraul 42, Liège: 9–25.

Weissbrod, T., 1981. The Paleozoic of Israel and adjacent countries. Geol. Surv. Israel, Jerusalem, Rep. MP 600/81, 276 pp. (in Hebrew, English summary).

Wellman, H.T., Wilson, A.T., 1965. Salt weathering, a neglected geological erosive agent in coastal and arid environments. Nature, 205: 1097–1098.

West, R.G., 1968. Pleistocene geology and biology. Longmans, London, 377 pp.

Williams-Dean, G., Bryant, V.M. Jr., 1975. Pollen analysis of human coprolites from Antelope House. The Kiva, 41(1): 97–111.

Williams, M.A.J., Faure, H. (Editors), 1980. The Sahara and the Nile. A.A. Balkema, Rotterdam, 451 pp.

Yaalon, D.H., 1967. Factors affecting the lithification of eolianite and interpretation of its environmental significance in the coastal plain of Israel. J. Sediment. Petrol., 37(4): 1189–1199.

Yaalon, D.H., Dan, J., 1967. Factors controlling soil formation and distribution in the Mediterranean coastal plain of Israel during the Quaternary. Proc. VII INQUA Congr., 9: 321–338.

Zohary, M., 1959. Geobotany. Sifriat Poalim, Merhavia, Israel, 589 pp. (in Hebrew).

Zohary, M., 1980. A new analytical flora of Israel. Am Oved Publishers, Tel Aviv, 540 pp. (in Hebrew).

References index

Subject index *

* Page numbers in boldface denote appearance in plates; italics, in text-figures